ANNUAL REVIEW
OF GENETICS

ANNUAL REVIEW OF GENETICS

HERSCHEL L. ROMAN, *Editor*
The University of Washington, Seattle

ALLAN CAMPBELL, *Associate Editor*
Stanford University, Stanford

LAURENCE M. SANDLER, *Associate Editor*
The University of Washington, Seattle

VOLUME 8

1974

ANNUAL REVIEWS INC.   4139 EL CAMINO WAY   PALO ALTO, CALIFORNIA 94306

ANNUAL REVIEWS INC.
Palo Alto, California, USA

International Standard Book Number: 0-8243-1208-2
Library of Congress Catalog Card Number: 67-29891

REPRINTS

The conspicuous number aligned in the margin with the title of each article in this
volume is a key for use in ordering reprints. Available reprints are priced at the
uniform rate of $1 each postpaid. Effective January 1, 1975, the minimum acceptable
reprint order is 10 reprints and/or $10.00, prepaid. A quantity discount is available.

PRINTED AND BOUND IN THE UNITED STATES OF AMERICA

CONTENTS

ANNUAL REVIEWS INC. is a nonprofit corporation established to promote the advancement of the sciences. Beginning in 1932 with the *Annual Review of Biochemistry,* the Company has pursued as its principal function the publication of high quality, reasonably priced Annual Review volumes. The volumes are organized by Editors and Editorial Committees who invite qualified authors to contribute critical articles reviewing significant developments within each major discipline.

Annual Reviews Inc. is administered by a Board of Directors whose members serve without compensation.

Annual Reviews are published in the following sciences: Anthropology, Astronomy and Astrophysics, Biochemistry, Biophysics and Bioengineering, Earth and Planetary Sciences, Ecology and Systematics, Entomology, Fluid Mechanics, Genetics, Materials Science, Medicine, Microbiology, Nuclear Science, Pharmacology, Physical Chemistry, Physiology, Phytopathology, Plant Physiology, Psychology, and Sociology (to begin publication in 1975). In addition, two special volumes have been published by Annual Reviews Inc.: *History of Entomology* (1973) and *The Excitement and Fascination of Science* (1965).

ANALYSIS OF GENETIC REGULATORY MECHANISMS

❖3062

Jon Beckwith and Peter Rossow

Department of Microbiology and Molecular Genetics, Harvard Medical School,
Boston, Massachusetts 02115

Studies on the control of gene expression in bacteria have provided a number of models for how regulation might occur in higher organisms. These models have been applied to explain, among other things, various developmental processes, certain genetic diseases, and mechanisms of viral development and transformation. While many regulatory systems are known, very few have been characterized as to their molecular nature. This may be due, in part, to an approach that still relies on the success of early studies in bacteria which were notable for their simplicity.

In this review we do not give a comprehensive survey of known regulatory mechanisms. Rather, we present an overview of the approaches applicable to the characterization of regulatory genes. We point out, particularly, the limitations of many of these approaches and suggest the necessary criteria for defining such mechanisms. For more detailed reviews, we refer you to other recent articles (1, 2).

DIPLOID ANALYSIS

One of the major approaches to the analysis of regulatory mechanisms in bacteria has been the study of the expression of alleles of regulatory genes in diploid situations. In *E. coli* these studies have been done, for the most part, by examining the behavior of cells that are partially diploid due to the presence of F' factors. The results of such diploid experiments have been considered to distinguish between two models. According to the first of these models, negative control, regulatory genes code for products that *prevent* the synthesis of the gene products in question. According to the second model, positive control, the regulatory gene synthesizes a product *necessary* for the synthesis of the regulated gene products.

In 1959, Pardee, Jacob & Monod analyzed lactose (*lac*) operon regulation by merodiploid experiments [the PaJaMo experiments (3)]. From their results they concluded that the *lac* operon was subject to negative control by a repressor. Both the methodological breakthroughs and the brilliant conceptual intuitions involved in the development of the model of *lac* operon control were major steps in the progress of molecular biology. Following the PaJaMo paper, biologists began to extend these investigations to other systems in bacteria and to extend the operator-repressor control mechanism to examples of gene regulation in higher organisms. There was initially a tendency to assume that one mechanism was responsible for all gene control in all organisms. This attitude may have been a result of the increasing recognition that many biochemical mechanisms are common to all organisms from bacteria to humans. For instance, in the PaJaMo paper itself, the authors state, "according to this scheme, the basic mechanism common to all protein-synthesizing systems would be inhibition by specific repressors formed under the control of particular genes" (3). Later Jacob & Monod developed a general model for developmental processes that invoked sequential activation or inactivation of repressor molecules (4).

However, in the last few years, it has become clear that repressor-operator interactions represent only one out of an ever increasing spectrum of transcriptional control mechanisms. These other mechanisms include specific positive control factors that affect RNA polymerase specificity, and different RNA polymerase molecules serving different functions. These mechanisms relate to the control over initiation of transcription of specific genes. Also, some control mechanisms may act at the level of termination of mRNA synthesis. For all we know, what we have seen so far of regulatory mechanisms may be only the tip of the iceberg.

In addition, as detailed evidence has accumulated on the regulation of more and more systems, including the *lac* operon itself, it has become apparent that the simple diploid analysis which originally broke open the field has many pitfalls. In this review, we wish to cover mainly the genetic criteria used to establish the basis for characterizing mechanisms of regulation in particular systems. It will be interesting to observe how lucky, in many respects, Jacob, Monod, and co-workers were, both in the system they chose and the mutants with which they worked.

PATHWAY-SPECIFIC REGULATION BY SINGLE REGULATORY GENES

In many catabolic pathways, synthesis of the gene products of the pathway is induced by addition to the growth media of the substrate of the pathway. For example, lactose induces the synthesis of β-galactosidase and a galactoside permease. In many biosynthetic pathways, synthesis of the biosynthetic enzymes is repressed upon addition to the medium of the endproduct of the pathway. For example, addition to growth media of tryptophan causes the lowering of the rates of synthesis of the five enzymes specific to the tryptophan (*trp*) biosynthetic pathway (5). In many of these systems, the control can be ascribed to a single genetically defined regulatory gene. Despite the usual semantic distinction between endproduct-repressible and substrate-inducible pathways, the regulatory genes for pathways within one class can operate very differently. Within each group, there are probably both positive and negative control mechanisms operating. Among catabolic pathways, lactose is regulated by negative control and arabinose (*ara*) by positive control (6). Among biosynthetic pathways, tryptophan is negatively controlled and recent evidence suggests that isoleucine-valine is positively controlled. The last mechanism is the least well defined and if established presumably reflects a situation in which the endproduct inhibits an activator protein (7).

The original implicit motivation of the PaJaMo experiments and certainly their subsequent extension to other systems was the concept that simple diploid analysis of regulatory mutations would distinguish between positive and negative control mechanisms. The first step in the analysis was to isolate mutants (R^c) from wild-type strains (R^+) that make products of the regulated genes constitutively. Cells that are diploid for the regulatory gene (R^c/R^+) are then constructed. These diploid cells are examined, usually by enzyme assays for one of the gene products, to see whether the constitutive or the inducible phenotype is dominant. If the regulatory locus is unlinked to the genes it controls, then it can be assumed to code for a diffusible product. If the R gene is linked to an operon, the construction of doubly mutant strains and more careful diploid analyses are necessary to insure that this gene codes for a diffusible product.

It is expected then that wild-type synthesis (R^+) of a repressor protein should be dominant over the loss of the presumed repressor (R^c) in a negative control system. In a positive control system, a mutation in the R gene that eliminates the need for inducer should result in an activator that will turn on the operon in question under all conditions.

Simple Models

negative control	R^+/R^c	inducible
positive control	R^+/R^c	constitutive

However, in practice, life turns out not to be that simple. In the first positive control system to be studied in any detail, the arabinose operon, the R^c mutations are recessive to R^+(8). The current explanation for this surprising result is that the wild-type activator protein has, in addition, a repressive activity. This repression by R^+ allele is thought to counteract the constitutivity caused by the mutant R^c activator. (However, other explanations such as subunit mixing, discussed below, could equally well explain this result.)

Lucky Break Number One: Choice of Operon

If Pardee, Jacob & Monod had initially worked with the arabinose instead of the lactose operon, they would have obtained very similar results and might very well have wrongly concluded that *ara* was negatively controlled by a repressor molecule. The lesson here is that regulatory proteins do not necessarily act in the very simple ways in which they were originally visualized.

A second aspect of regulatory proteins that can confuse diploid analysis is their multimeric nature. Within the *lacI* gene, it has been shown that a sizable fraction of mutations (I^{-d}) that cause constitutivity still permit the synthesis of proteins similar to wild-type repressor. In diploids of the I^+/I^{-d} type, there is evidence that mixed multimers of the repressor are formed. Further, it appears that these mixed I^+-I^{-d} multimers tend to be much less efficient repressors. As a result, these I^+/I^{-d}

diploids exhibit substantial constitutive behavior. Thus, I^{-d} mutants are *trans*-dominant constitutive (9, 10). I^{-d} mutations were at first confused with O^c mutations (11), since they were not recessive to a wild-type I gene. Only careful diploid analysis with various combinations of double mutants can establish their nature (12).

Lucky Break Number Two: Choice of Mutation

If Pardee, Jacob & Monod had chosen an I^- mutation of the I^{-d} type, they might very well have concluded that a positive control system operated on the *lac* operon.

"Negative complementation" may be observed only when the regulatory gene promotes the synthesis of small numbers of protein molecules. For example, in an I^Q/I^{-d} diploid, in which the I^Q allele results in overproduction of *lac* repressor protein, the inducible phenotype is again dominant (10). The dependence on amounts of gene product is seen in an even more confusing way in the regulation of mucopolysaccharide biosynthesis in *E. coli* (13). In this case, the presumed regulatory mutation (*capR*$^-$, constitutive) is recessive to *capR*$^+$ when the latter is included on an episome. The *capR*$^-$, however, is dominant when on the episome. It was suggested that the gene on the episome existed in greater number of copies than the chromosomal gene. Thus, whichever allele was on the episome predominated in the diploid. These data could be explained if the *capR*$^-$ mutants were producing a negatively complementing subunit. In fact, it was in this early work of Markovitz & Rosenbaum that the idea of negatively complementing *lacI* gene mutants was first suggested. The dominance of inducibility over constitutivity in the arabinose operon may be due to this type of subunit mixing of mutant and wild-type activator subunits (see above and 8).

In addition to mutations to constitutivity in regulatory genes, there also exist mutations to uninducibility. In the negatively controlled *lac* system, the uninducible mutations (I^s) are due to altered repressor molecules that have lost their affinity for inducer but can still bind to operator DNA (14, 15). In the positively controlled *ara* system, these mutations are due to inactivation of the activator protein (16). Again, the two simple models predict opposite results for the diploid studies.

Simple Models

negative control	R^+/R^u	uninducible
positive control	R^+/R^u	inducible

Subunit mixing of multimeric proteins can again interfere with interpretation of diploid experiments. For instance, while an I^+/I^s diploid is uninducible and thus Lac$^-$, an I^Q/I^s diploid is Lac$^+$ (10). Apparently the excess of good repressor subunits results in a population of mixed multimers, none of which is unaffected by inducer.

Many of the problems we have raised regarding diploid analysis result from dealing with regulatory gene mutations that make altered regulatory products or altered amounts of normal regulatory products. A simpler approach to the problem of distinguishing between positive and negative control systems is to work only with those mutations presumed to abolish synthesis of the regulatory gene product. Classes of mutations that generally result in absolute loss of gene activity include chain-terminating mutations, frameshift mutations, insertions, and deletions. In a negatively controlled system, the complete inactivation of the regulatory gene product should result in *trans*-recessive constitutivity, while in the positively controlled system it should result in *trans*-recessive uninducibility.

However, such criteria which in the past seemed satisfactory are no longer so. Work with the *lacI* gene has shown that both chain-terminating mutants and deletions of the gene can still code for products that negatively complement. In the case of chain-terminating mutants, it was shown that after certain amber or ochre mutations that caused termination early in the *lacI* gene, reinitiation of protein synthesis was possible (17, 18). As a result, a fragment of the *lac* repressor which is missing an N-terminal sequence is made. Apparently, an internal AUG, GUG, or UUG codon can serve as an initiation codon for such protein synthesis reinitiation. If the reinitiation occurs early in the *I* gene, the "restart" fragment has enough of the subunit structure to become incorporated into multimeric *lac* repressor molecules. However, fragments of repressor missing the N-terminal sequence lack operator-binding activity (19). These hybrid molecules then have no or only partial repressor activity.

Certain deletion mutations that fuse the *lacI* gene to the *trp* operon result in the remaining portion of the *lacI* gene being brought under the control of tryptophan. With several of these fusion strains, it was shown that derepression of the *trp* operon causes production of an *I* gene product that can again negatively complement (20). Preliminary studies suggest that these negative complementation fragments are also restart fragments.

Another example that does not relate to regulatory systems, but that illustrates the insufficiency of using chain-terminating mutations as an indication of complete loss of gene activity has to do with *E. coli* DNA polymerase I. Several nonsense mutations of the gene for DNA polymerase I have been found (21). None of them is lethal, but they retain an exonuclease activity normally associated with the enzyme, despite the absence of polymerizing activity. These mutants were taken as evidence suggesting that polymerase I was not essential. However, recently a temperature-sensitive conditional lethal mutation of this gene was found that retained the polymerizing activity but lacked the exonuclease (22). Again, the phenotype of chain-terminating mutations in a particular gene is not always sufficient evidence to establish the function of that gene.

Therefore, even mutations that supposedly abolish gene function can exhibit behavior in diploid strains that confounds analysis. Ideally, then, the isolation of a complete deletion of a regulatory gene and a study of its properties is the only really safe approach to answer the questions we have been considering. This is not to say that the less ideal approach is not satisfactory. Such studies certainly tend to strengthen the formulations of one hypothesis over another. There are available techniques for selecting for extensive deletions of genes in *E. coli.* Sometimes it is possible to select for deletions of genes close to regulatory genes, and in some cases these deletions include the regulatory gene. For instance, selection of λ^R-mal$^-$ mutations in *E. coli* occasionally generates deletions that include the nearby regulatory gene for glycerol catabolism (23).

In addition, it is possible to transpose genes in *E. coli* from their normal positions on the chromosome to positions where selections for such deletions are possible. For instance, the *lac* operon has been transposed to a site adjacent to the locus (*tonB*) determining sensitivity to phage T1 (24). Selection of *tonB* mutants in the appropriate strains can result in extensive deletions of the *lacI* gene leaving the *lac* operon intact (20).

A method essentially the converse of this last technique is to transpose a "deletion selecting system" to a position on the chromosome close to the gene of interest. Shimada and co-workers (25, 26) have described a technique for inserting the genome of bacteriophage λ at various positions on the *E. coli* chromosome. Using a thermo-inducible derivative of λ in such strains, selection of temperature resistance often yields mutants that have deleted λ and neighboring genes.

One problem, of course, with such an approach is that in certain cases regulatory genes or the genes they control are essential cell components, so that regulatory gene deletions might be lethal.

One final problem in diploid analysis of regulatory genes is the possible existence of genes that code for protein products acting only in *cis,* i.e. proteins that act only on the genomes coding for them. While the mechanisms are not understood, there is evidence that in both phages P2 and λ such *cis*-acting proteins operate (27, 28).

CONFOUNDING MUTATIONS AFFECTING METABOLISM OF EFFECTORS WITH REGULATORY MUTATIONS

In most if not all regulatory systems studied so far, a small molecule effector either activates or inactivates the regulatory protein. In the case of some catabolic, inducible pathways such as arabinose utilization, the substrate of the first enzyme is that effector (6). Moreover, in the tryptophan biosynthetic pathway, tryptophan seems to be the corepressor (29, 30). However, in many pathways the substrate or endproduct is not itself the effector, but must be converted by one or more metabolic steps to the actual effector. Lactose, for example, is not the inducer of the *lac* operon, i.e. it does not inactivate repressor. The enzyme β-galactosidase converts lactose to allolactose, the true inducer (31). Thus, while all three gene products of the *lac* operon are induced by lactose in the wild-type Lac$^+$ cell, there is no induction of permease or transacetylase in *lacZ* mutants missing β-galactosidase. We can imagine that, lacking certain synthetic, gratuitous inducer molecules such as isopropyl-β-D-thiogalactoside, studies on the *lac* operon might have led at one point to the hypothesis that β-galactosidase was itself directly involved in regulation.

An analogous situation occurs in other pathways. The effector molecule for the induction of enzymes determining glycerol catabolism in *E. coli* is glycerol-3-phosphate, not glycerol itself (32). The enzyme glycerol kinase is necessary to permit the induction by glycerol of this pathway. Similarly, histidine is not an inducer of the enzymes determining histidine catabolism in *Salmonella typhimurium* and *Klebsiella aerogenes.* Rather the enzyme histidase must first convert histidine to urocanic acid, which is the true effector (33, 34).

In biosynthetic pathways, similar problems can be seen. Constitutive mutants of the histidine biosynthetic pathway define five separate "regulatory" genes (35). It appears possible that all five of these are involved in the conversion of histidine to what may be the true effector, histidyl-tRNA (36–38). It is not clear that any of these genes code for the actual regulatory protein, whether it be a repressor or an activator.

Transport of effector molecules is also important for regulation. In the *lac* operon, mutations (*lacY*) defective in β-galactoside transport are uninducible by lactose, because they are not able to concentrate the precursor of the true inducer. *LacY⁻* mutations were never confused with regulatory mutations, due to the existence of gratuitous inducers not dependent on the lactose transport system. However, it appears that genes affecting transport have been mistaken for regulatory genes in the case of alkaline phosphatase. Mutations to constitutivity for alkaline phosphatase map in two loci: *phoR* and *phoS* (39). It has recently been shown that mutations in the *phoS* locus, which probably comprises at least two genes (*phoS* and *phoT*), greatly reduce transport of phosphate (40). The constitutivity may then represent failure to achieve high concentrations of phosphate or a phosphate derivative in the cell. With the present evidence, it appears that the *phoR* gene may be the only true regulatory gene for this system.

Thus, without a thorough understanding of the transport and metabolism of supposed effector molecules, mutations that affect these processes can be mistaken for regulatory mutations.

SINGLE OPERONS UNDER MULTIPLE REGULATORY CONTROLS

So far we have been discussing genes and operons controlled by regulatory genes specific for the particular pathway. It is becoming apparent that many such regulated genes also respond to more global controls. The *lac* operon and other inducible genes for catabolic pathways not only have regulatory genes peculiar to themselves but are also further subject to a more general control by CAP protein and 3'5'-cAMP (41, 42). Even in the presence of an inducer of these operons, the genes will not be expressed unless these two catabolite-activating molecules are present.

In the *hut* operon in *Salmonella* and *Klebsiella* it appears that there are at least three distinct transcriptional regulatory mechanisms. First, there is the *hut* repressor that responds to urocanic acid (43); second, since histidine can be utilized as a carbon source, the pathway is subject to CAP-cAMP regulation (44); and finally, since histidine can also be used as a source of nitrogen, it, along with other pathways that serve the same purpose (45), is subject to a third nitrogen-regulating transcriptional control (46). Mutants in glutamine synthetase can eliminate this control, and the unadenylated glutamine synthetase molecule itself can stimulate transcription of the *hut* operon in vitro (47).

Another example of threefold control of a pathway is the regulation of the galactose operon. Here again a *gal*-operon specific repressor and the CAP-cAMP complex are involved (48–50). In addition, as we have mentioned, the operon is regulated by a separate system along with other genes determining mucopolysaccharide synthesis (51). Markovitz and his co-workers have proposed that the *capR* gene determines the synthesis of a repressor that acts at a second operator within the *gal* operon (51). At the present time, there is no direct evidence to support this model. Further, since mutations in three genes, *capR (lon)*, *capS*, and *capT*, control mucopolysaccharide synthesis, it may be that one of these latter two genes actually codes for the regulatory protein. There is no evidence to indicate which of these genes is involved in the metabolism of an effector and which is a true regulatory gene.

Certain pieces of evidence from this system could, in fact, be interpreted to suggest that mucopolysaccharide synthesis is under positive rather than negative control. First, in vivo, the *gal* operon is only partly under CAP-cAMP control (52). In the absence of cAMP, the operon is still expressed at, at least 50% of the maximum. In contrast, in the in vitro Zubay system, the operon is completely dependent on CAP-cAMP (49, 50). These apparently contradictory pieces of information could be reconciled if there were a positive control factor missing in the in vitro system. Based on what is known so far of control of mucopolysaccharide biosynthesis, we suggest that the *capT* gene is a possible candidate for a true (positive) regulatory gene. This discussion is not to suggest that this is the correct model for control, but rather to point out that whenever there are multiple regulatory genes in a single system, a very detailed genetic analysis along with physiological studies are necessary before specific mechanisms can be seriously entertained.

In the *trp* operon, a classical operator and repressor gene have been characterized (53, 54). However, recently, deletion analysis of the *trp* operon has shown that there is a previously undefined site, between the operator and the first structural gene, which appears to play a regulatory role (55). Deletion of this site results in an increased expression (3–8 X) of the operon. One possible explanation for this result is that there exists a hitherto undetected regulatory molecule that interacts with this new site to reduce operon expression.

It is clear from these examples that the existence of several different regulatory proteins, both positive and negative, interacting with the genes of a single pathway can confound both genetic and in vitro analysis.

CONTROLLING SITE MUTANTS

In the early days following the formulation of the repressor-operator mechanism of control, the existence of *cis*-dominant constitutive mutations (O^c) was taken as evidence for the presence of an operator (56). It is now clear that such mutants, by themselves, establish nothing about the control mechanism. For one thing it is possible to find the same class of mutations in a positively controlled operon such as arabinose. In this case I^c mutations allow some degree of expression of the *ara* operon in the absence of the activating regulatory protein, the C gene product (8). Since we do not yet understand the mechanism of C gene protein action, we cannot describe the nature of the I^c mutation. At any rate I^c mutations have altered the controlling site region of the *ara* operon in such a way that transcription can be initiated and will proceed without C gene product present.

In the *lac* operon a class of *cis*-constitutive mutants have been found in addition to O^c mutations. These are mutations (P^r) that alter the *lac* promoter region so that RNA polymerase can initiate transcription in the absence of the CRP-cAMP complex (57–59). These mutations are constitutive in the sense that they no longer require cAMP for induction of the *lac* operon.

Such *cis*-constitutive mutants are conceivable for almost any regulatory system, and therefore their existence cannot define the type of regulatory mechanism operating.

In addition, certain mutations that result in *cis*-constitutive behavior in operons might be due to grosser chromosomal changes than simple point mutations. For instance, it is possible to cause the *lac* operon to function constitutively by fusing it to a *trp* operon being expressed (60). In a *cis-trans* test, such a fusion will behave like an O^c mutation. In other cases, insertion of a new promoter may give the same results. No example of this latter class of mutation has been verified, although it might explain mutations that appear to introduce new high efficiency promoters for the *trp* and *lac* operons (61, 62).

In the case of down-promoter mutations, problems arise in distinguishing them from strong polar mutations. An initial approach was to look for *cis*-acting mutations pleiotropically leaky for all of the genes of an operon (63). In the *lac* operon it was found that the operator mapped between such mutations and the *lac* structural genes (64). This mapping, in itself, suggested that the mutations were promoter mutations rather than polar mutations. However, where that mapping relationship does not exist, the distinction between promoter and polar mutations is more difficult. One approach is to look at revertants of presumed promoter mutants. When polar mutants are reverted by intracodon or linked mutations, the polar effects should be completely abolished and the expression of distal gene products returned to wild-type levels. In contrast, promoter mutants can revert to levels of activity that are intermediate between the mutational level and the wild-type level (65). In the case of the *ara* operon, certain pleiotropically negative mutations that map in or close to the promoter revert to intermediate levels of operon activity (66). Although more complicated classes of revertants of polar mutants might be imagined that would behave in the same way, these results are strong suggestive evidence for the mutations being in the promoter.

Another complexity to the characterization of promoter mutants in certain operons is revealed by the previously described multiplicity of regulatory controls. For instance, in the *lac* operon at least two classes of down-promoter mutants have been found, distinguished by both their map position and their phenotype (67). One class affects the interaction of the CAP-cAMP complex with the promoter, while the other appears to reduce the interaction of RNA polymerase directly. In operons such as *ara, gal, hut,* and others, presumably even more distinct classes of P^- mutants will be found, since there may be three or more factors interacting with these promoters.

In general, great care must be taken in ascribing promoter effects to regulatory mutants. In the *lac* operon, Smith & Sadler (68) described O^c mutations induced with the mutagen 2-aminopurine, which in addition to causing constitutivity also either reduce or increase the maximal level of expression of the *lac* operon. It was suggested, as a result, that these single base-change mutations were affecting both operator and promoter activity, and that, therefore, these two elements overlapped. However, other explanations are possible. For instance, alterations in the structure of the operator may slow down transcription already started or the mRNA in certain O^c mutants might be more or less stable than normal. Carlson & Smith (69) have now presented evidence that the latter explanation may account for the supposed promoter effects of *lac* operator mutants.

While evidence has accumulated that in the *lac* operon the operator and promoter are distinct nonoverlapping elements, the situation with operons of bacteriophage λ appears to be different. The two operons controlled by the λ repressor are regulated by complex operator regions with multiple binding sites for repressor. Both genetic studies and studies with restriction enzyme cleavage of these regions have shown that the promoter regions in both cases probably overlap or are actually part of the operator sequences (70–72).

IN VITRO ANALYSIS

Genetic analyses of regulatory systems have been essential in generating new ideas about control mechanisms. Without this approach, very little would be known about how genes are transcribed in mRNA and how this transcription is affected by various controlling factors. However, it is clear from our discussion that the new hypotheses developed from genetic studies retain ambiguities and uncertainties. It is ultimately the combination of in vitro analysis with genetics and physiology that helps to establish firmly the basis of a particular regulatory mechanism.

In vitro studies have been dependent, for the most part, on the isolation of DNA preparations that are highly enriched for the genes of interest. These preparations are usually obtained from specialized transducing phage lysates. A variety of techniques are now available for obtaining λ and φ80 transducing phages for genes of *E. coli* (25, 26, 73–75). It seems likely that nearly any gene in *E. coli* and related organisms (76, 77) can be obtained on transducing phages by one or another of these techniques.

More detailed studies on the interaction of RNA polymerase and regulatory proteins with promoter and operator regions are making use of purified gene preparations. Techniques with rather limited application for purifying genes of *E. coli* have been described (78, 79). However, the growing collection of restriction enzymes being isolated from many organisms raises the possibility of purifying many different genes. The range of specificities of these enzymes allows one to seek out that enzyme that will generate a pure gene preparation from transducing phage or episomal DNA preparations. For example, one could treat φ80 *lac* DNA with different restriction enzymes, looking for those that cleaved the DNA into a number of fragments, one of which still retained the *lacZ* gene and *lac*-controlling elements. The integrity of such a fragment could be determined by seeing whether the restriction enzyme-treated DNA would still promote properly controlled β-galactosidase synthesis in an in vitro protein synthesizing system. The desired *lac* fragment could then be purified by one of a variety of techniques that separate DNA molecules on the basis of size.

Once the appropriate transducing phages or purified gene preparations are available, they can be used in the in vitro protein synthesizing system developed by Zubay and co-workers to assay regulatory effects (80). [One caution in such studies is that some transducing phage DNA preparations do not respond to appropriate controls in these systems (58).] Finally, a purified transcription system allows direct characterization of the effect of regulatory proteins and controlling element mutations. One flaw in a number of such transcription studies (47, 81) is that, in addition to the mRNA transcribed from the genes of interest, there may also be RNA from neighboring genes. There are a number of ways of determining that the mRNA being assayed corresponds to the genes under study. If the appropriate transducing phages can be constructed, it may be possible to establish a hybridization assay where only the desired mRNA is assayed (58). This would obviously be feasible with a purified gene preparation. Lacking these tools, competition with mRNA made in vivo, which is known to be transcribed from the correct genes, would help identify the appropriate mRNA.

EXTENSIONS TO HIGHER ORGANISMS

The development of techniques for fusing animal cells has allowed the study of gene regulation to extend to higher organisms. These techniques enable one to artificially fuse cells of differing types and then to examine the expression of particular genes in the hybrid. One important area of this work has been the study of the control of differentiation. Basically, the idea is to fuse two cells, one of which is expressing a differentiated function and one of which is not, and to examine the expression of this function in the resulting hybrid. In the majority of cases the hybrid cell is unable to express the differentiated function (82, 83). This negative result has been interpreted as evidence for a negative control of the differentiated function by a diffusible regulatory product. Other interpretations are also possible considering the limitations we have discussed above for prokaryotic systems. One might suggest that the undifferentiated cell contained an inducer-destroying molecule that interfered with a positive control system or perhaps a feedback effector that altered a critical enzyme in the differentiated function. Other possible explanations might be that there are nonspecific interactions between RNA polymerase molecules, altering their specificity for particular genes. All of these examples can perfectly well accommodate these observations and serve to point out that a postulate of negative control says nothing more than that the hybrid is undifferentiated (82). Perhaps even more important, the process of fusing such very highly specialized cell types has been known to cause drastic changes in phenotypes that are almost immediate and thus cannot be the result of genetic interactions (82).

CONCLUSION

The purpose of this paper is to illustrate some limitations of the various approaches now in use for the characterization of regulatory mechanisms. It should be clear that several different approaches used together are necessary to unambiguously establish a mechanism. This is equally true of the study of regulatory mechanisms in higher organisms. The characterization of mechanisms in these systems awaits the development of more sophisticated genetic techniques than are now available.

ACKNOWLEDGMENTS

Jon Beckwith was supported by a Career Development Award and grant from the NIH and grants from NSF and the American Cancer Society. Peter Rossow was supported by a training grant from NIH.

Literature Cited

1. Reznikoff, W. S. 1972. *Ann. Rev. Genet.* 6:133–56
2. Goldberger, R. F. 1974. *Science* 183:810–16
3. Pardee, A. B., Jacob, F., Monod, J. 1959. *J. Mol. Biol.* 1:165–78
4. Jacob, F., Monod, J. 1963. In *Cytodifferentiation and Macromolecular Synthesis,* ed. M. Locke, 30–64. New York: Academic. 274 pp.
5. Cohen, G., Jacob, F. 1959. *C. R. Acad. Sci. Paris* 248:3490–92
6. Englesberg, E. 1971. In *Metabolic Pathways,* Vol. V: *Metabolic Regulation,* ed. D. M. Greenberg, 257–96. New York: Academic. 576 pp.
7. Levinthal, M., Williams, L. S., Levinthal, M., Umbarger, H. E. 1973. *Nature New Biol.* 246:65–68
8. Englesberg, E., Squires, C., Meronk, F. 1969. *Proc. Nat. Acad. Sci. USA* 62:1100–7
9. Davies, J., Jacob, F. 1968. *J. Mol. Biol.* 36:413–17
10. Muller-Hill, B., Crapo, L., Gilbert, W. 1968. *Proc. Nat. Acad. Sci. USA* 59:1259–64
11. Jacob, F., Ullman, A., Monod, J. 1964. *C. R. Acad. Sci. Paris* 258:3125–28
12. Beckwith, J. R. 1970. In *The Lactose Operon,* ed. J. R. Beckwith, D. Zipser. Cold Spring Harbor, New York: Cold Spring Harbor Lab. 437 pp.
13. Markovitz, A., Rosenbaum, N. 1965. *Proc. Nat. Acad. Sci. USA* 54:1084–91
14. Willson, C., Perrin, D., Cohn, M., Jacob, F., Monod, J. 1964. *J. Mol. Biol.* 8:582–94
15. Jobe, A., Riggs, R. D., Bourgeois, J. 1972. *J. Mol. Biol.* 64:181–99
16. Sheppard, D. E., Englesberg, E. 1967. *J. Mol. Biol.* 25:443–54
17. Platt, T., Weber, K., Ganem, D., Miller, J. H. 1972. *Proc. Nat. Acad. Sci. USA* 69:897–901
18. Ganem, G., Miller, J. H., Files, J. G., Platt, T., Weber, K. 1973. *Proc. Nat. Acad. Sci. USA* 70:3165–69
19. Platt, T., Files, J. G., Weber, K. 1973. *J. Biol. Chem.* 248:110–21
20. Miller, J. H., Beckwith, J., Muller-Hill, B. 1968. *Nature* 220:1287–90
21. Gross, J., Gross, M. 1969. *Nature* 224:1166–68
22. Konrad, E. B., Lehman, I. R. 1974. *Proc. Nat. Acad. Sci. USA* 71:2048–57
23. Schwartz, M. 1966. *J. Bacteriol.* 92:1083–89
24. Beckwith, J. R., Signer, E. R., Epstein, W. 1966. *Cold Spring Harbor Symp. Quant. Biol.* 31:393–401
25. Shimada, K., Weisberg, R. A., Gottesman, M. E. *J. Mol. Biol.* 1972. 63:483–503
26. Shimada, K., Weisberg, R. A., Gottesman, M. E. *J. Mol. Biol.* 1973. 80:297–314
27. Lindahl, G. 1970. *Virology* 42:522–33
28. Kleckner, N. 1974. *Plasmid formation by bacteriophage lambda.* PhD thesis. MIT, Cambridge, Mass. 277 pp.
29. Squires, C. L., Rose, J. K., Yanofsky, C., Yang, H. L., Zubay, G. 1973. *Nature New Biol.* 245:131–37
30. McGeoch, D., McGeoch, J., Morse, D. 1973. *Nature New Biol.* 245:137–40
31. Jobe, A., Bourgeois, S. 1972. *J. Mol. Biol.* 69:397–408
32. Cozzarelli, N. R., Freedberg, W. B., Lin, E. C. C. 1968. *J. Mol. Biol.* 31:371–87
33. Brill, W. J., Magasanik, B. 1969. *J. Biol. Chem.* 244:5392–5402
34. Schlesinger, S., Scotto, P., Magasanik, B. 1965. *J. Biol. Chem.* 240:4331–37
35. Goldberger, R. F., Kovach, J. S. 1972. *Curr. Top. Cell. Regul.* 5:285–308
36. Roth, J. R., Ames, B. N. 1966. *J. Mol. Biol.* 22:325–34
37. Silbert, D. F., Fink, G. T., Ames, B. N. 1966. *J. Mol. Biol.* 22:335–47
38. Singer, C. E., Smith, G. R., Cortese, R., Ames, B. N. 1972. *Nature New Biol.* 238:72–74
39. Echols, H., Garen, A., Garen, S., Torriani, A. 1961. *J. Mol. Biol.* 3:435–38

40. Willsky, G. R., Bennett, R. L., Malamy, M. H. 1973. *J. Bacteriol.* 113:529–39
41. Zubay, G., Schwartz, D., Beckwith, J. 1970. *Proc. Nat. Acad. Sci. USA* 66:104–10
42. Emmer, M., deCrombrugghe, B., Pastan, I., Perlman, R. 1970. *Proc. Nat. Acad. Sci. USA* 66:480–87
43. Smith, G. R., Magasanik, B. 1971. *Proc. Nat. Acad. Sci. USA* 68:1493–94
44. Prival, M. J., Magasanik, B. 1971. *J. Biol. Chem.* 246:6288–96
45. Magasanik, B. et al 1974. *Curr. Top. Cell. Regul.* 8:119
46. Prival, M. J., Brenchley, J. E., Magasanik, B. 1973. *J. Biol. Chem.* 248: 4334–44
47. Tyler, B., Deleo, A. B., Magasanik, B. 1974. *Proc. Nat. Acad. Sci. USA* 71:225–29
48. Nakanishi, S., Adhya, S., Gottesman, M. E., Pastan, I. 1973. *Proc. Nat. Acad. Sci. USA* 70:334–38
49. Wetekam, W., Staack, K., Ehring, R. 1971. *Mol. Gen. Genet.* 112:14–27
50. Parks, J. S., Gottesman, M., Perlman, R. L., Pastan, I. 1971. *J. Biol. Chem.* 246:2919–29
51. Hua, S. S., Markovitz, A. 1972. *J. Bacteriol.* 110:1089–99
52. Rothman-Denes, L. B., Hesse, J. E., Epstein, W. 1973. *J. Bacteriol.* 114: 1040–44
53. Morse, D. E., Yanofsky, C. 1969. *J. Mol. Biol.* 44:185–93
54. Hiraga, S. 1969. *J. Mol. Biol.* 39:159–79
55. Jackson, E. N., Yanofsky, C. 1973. *J. Mol. Biol.* 76:89–101
56. Jacob, F., Perrin, D., Sanchez, C., Monod, J. 1960. *C. R. Acad. Sci. Paris* 250:1727–29
57. Silverstone, A. E., Arditti, R. R., Magasanik, B. 1970. *Proc. Nat. Acad. Sci. USA* 66:773–79
58. Eron, L., Block, R. 1971. *Proc. Nat. Acad. Sci. USA* 68:1828–32
59. Arditti, R., Grodzicker, T., Beckwith, J. 1973. *J. Bacteriol.* 114:652–55
60. Reznikoff, W. S., Miller, J. H., Scaife, J. G., Beckwith, J. R. 1969. *J. Mol. Biol.* 43:201–13
61. Morse, D. E., Yanofsky, C. 1969. *J. Mol. Biol.* 41:317–28
62. Bruenn, J., Hollingsworth, H. 1973. *Proc. Nat. Acad. Sci. USA* 70:3693–97
63. Scaife, J. G., Beckwith, J. R. 1966. *Cold Spring Harbor Symp. Quant. Biol.* 31:403–8
64. Miller, J. H., Ippen, K., Scaife, J. G., Beckwith, J. R. 1968. *J. Mol. Biol.* 38:413–20
65. Arditti, R., Scaife, J. G., Beckwith, J. R. 1968. *J. Mol. Biol.* 38:421–26
66. Eleuterio, M., Griffin, B., Sheppard, D. E. 1972. *J. Bacteriol.* 111:383–91
67. Hopkins, J. 1974. *J. Mol. Biol.* In press
68. Smith, T. F., Sadler, J. R. 1971. *J. Mol. Biol.* 59:273–305
69. Carlson, H., Smith, T., Submitted to *J. Mol. Biol.*
70. Maniatis, T., Ptashne, M., Maurer, R. 1973. *Cold Spring Harbor Symp. Quant. Biol.* 38:857–68
71. Maurer, R., Maniatis, R., Ptashne, M. 1974. *Nature* 249:221–23
72. Ordal, G. W., Kaiser, A. D. 1973. *J. Mol. Biol.* 79:709–22
73. Gottesman, S., Beckwith, J. R. 1969. *J. Mol. Biol.* 44:117–27
74. Press, B. et al 1971. *Proc. Nat. Acad. Sci. USA* 68:795–98
75. Konrad, B., Kirschbaum, J., Austin, S. 1973. *J. Bacteriol.* 116:511–16
76. Smith, G. R. 1971. *Virology* 45:208–23
77. Voll, M. J. 1972. *J. Bacteriol.* 109: 741–50
78. Shapiro, J. et al 1969. *Nature* 224: 768–74
79. Gilbert, W., Maxam, A. 1973. *Proc. Nat. Acad. Sci. USA* 70:3581–84
80. Zubay, G., Chambers, D. A., Cheong, L. C. 1970. See Ref. 12, pp. 375–91
81. Blasi, F. et al 1973. *Proc. Nat. Acad. Sci. USA* 70:2692–96
82. Harris, H. 1970. *Cell Fusion.* London: Oxford Univ. Press. 108 pp.
83. Davidson, R. L., de la Cruz, F. F., Eds. 1974. *Somatic Cell Hybridization.* New York: Raven. 295 pp.

CONTROLLING ELEMENTS IN MAIZE

❖3063

J. R. S. Fincham and G. R. K. Sastry
Department of Genetics, University of Leeds, Leeds, England

The subject of this review is often regarded as complicated, difficult, and bizarre. In fact, as we hope to show, the evidence for the main phenomena is clear, but the different parts of the evidence are so mutually dependent that it seems best to start with an outline of the overall picture before considering different aspects in detail. First of all, however, we summarize the essential features of the maize genetic system and clarify some questions of genetic terminology.

Essentials of the Maize Genetic System

Like other Angiosperms, *Zea mays* is a free-living diploid, producing by meiosis haploid male and female spores which, in turn, develop into abbreviated male and female haploid gametophytes (the 3-nucleate germinating pollen grain and the 8-nucleate embryo sac respectively). The diploid embryo within the seed (or kernel) develops from the fusion of the egg nucleus of the embryo sac with a gamete nucleus of the pollen tube. Enveloping the embryo and paralleling it genetically is a *triploid* nutritive tissue, the endosperm, which develops from the fusion of *two* embryo sac nuclei with the second gamete nucleus of the pollen tube. Seed characters, and especially endosperm characters, are greatly used in maize genetics because they are convenient to score. It is important to remember that the phenotype of the triploid endosperm will normally (aside from mutation during gametophytic development) indicate the potential phenotype of the underlying embryo, but that the endosperm differs from the embryo in inheriting *two* sets of chromosomes from the female side. Broadly speaking, the endosperm consists of a white core, which is packed with starch at maturity, and a thin surface layer (the aleurone layer), which is heavily pigmented in strains of appropriate genotype. The seed coat, or pericarp, is diploid tissue of maternal origin. If pigmented, the pericarp can obscure the scoring of endosperm color.

A Note on Genetic Terminology

Definitions of genetic units are of necessity less precise in higher plants than in microorganisms. In this review we adopt the usual maize practice of using the term

15

locus to refer to a short chromosome segment of more or less specific phenotypic effect. It is not always clear whether a maize locus is a single cistron coding for a single protein or polypeptide chain; some loci, such as A_1 and R, are, in fact, known to include at least two cistrons of related function. We use the term *allele* to refer to an inherited and phenotypically distinct state of a locus without necessarily implying that it differs from other alleles at the same locus only in a single cistron, still less in a single base-pair change. By *mutation* we mean an abrupt and inherited change in a locus to give a new allele, again without any implication as to molecular mechanism.

Table 1 lists those maize loci and alleles referred to in this review, together with their map positions and effects on phenotype.

Table 1 Summary of loci and alleles (102)

Chromosome and map position		Alleles	Phenotypes
1	26	$P^{rr}/P^{vv}/P^{wr}$ or $P^{ww}=p$	Red/variegated red-on-white/white pericarp
3	111	A_1/a_1	Anthocyanin/no anthocyanin, in aleurone and plant[a]
	111.2	Sh_2/sh_2	Endosperm full/shrunken (collapses on drying)
4	123	C_2/c_2	Anthocyanin/no anthocyanin, in aleurone; c_2 also dilutes plant color
5	15	A_2/a_2	Anthocyanin/no anthocyanin, in aleurone and plant
	22	Bt_1/bt_1	Endosperm full/collapsed (brittle) on drying
	46	Pr/pr	Purple/red anthocyanin, in aleurone
6	17	Y/y	Yellow/white aleurone (seen in absence of anthocyanin)
	37	Pg/pg	Full green/pale green seedling
9	0	py/Py	Pale yellow/normal green seedling (*py* is a terminal chromosome deficiency)
	26	$C_1^I=I/C_1/c_1$	Dominant inhibition of anthocyanin/anthocyanin/no anthocyanin, in aleurone
	29	Sh_1/sh_1	Endosperm full/shrunken (collapses on drying)
	31	Bz/bz	Anthocyanin/pale brown pigment, in aleurone
	59	Wx/wx	Amylose (black with iodine)/amylopectin (red-brown with iodine), in endosperm and pollen grain
10	57	$R^r/R^{nj}/R^{st}/$ r^r or r^g	Anthocyanin uniform/in crown only/stippled/absent, in aleurone

[a]A note on gene interactions in anthocyanin pigment formation: Anthocyanin in the aleurone depends on the simultaneous presence of A_1, A_2, R^r, C_1, C_2, or equivalent alleles at the same loci. Alleles at other loci (*B*, *Pl*) intensify plant color. Red pericarp requires both A_1 and P^{rr} or equivalent alleles and tends to obscure scoring of aleurone pigment.

OUTLINE OF THE PROPERTIES
OF CONTROLLING ELEMENTS

The term "controlling element" was coined by McClintock (1) to describe transposable elements, of apparently sporadic occurrence, which make themselves visible through their abnormal control of the activities of standard genes. Most simply, a controlling element may inhibit activity of a gene through becoming integrated in, or close to, that gene. From time to time, either in germinal or somatic tissue, it may be excised from this site and, as a result, the activity of the gene is often more or less restored, while the element may become reintegrated elsewhere in the genome where it may affect the activity of another gene.

In the simplest examples (the autonomous or one-element systems), the only element that needs to be considered is the one that resides in or close to the affected gene, apparently acting autonomously with regard to its inhibition of gene action and its occasional transposition. Frequently, however, the element inhibiting gene action is itself controlled by another element, in some way complementary to it, located elsewhere in the genome. In such cases the excision of the first element, with consequent release of gene activity, does not occur except in the presence of the second element, which appears to supply some missing excision function. A situation of this kind is called a *two-element* system, and McClintock (2) has termed the element at the affected locus the *operator* and that acting on it from a distance the *regulator* element. The analogy with bacterial regulator-operator systems should not be pushed too far, and we propose to use the term *receptor,* rather than operator which has a precise meaning in molecular biology that may not be appropriate. The term regulator is convenient and does not seem misleading, and we will continue to use it in this review.

In the autonomous systems, the regulator and receptor components are integral parts of the same element, both residing at the locus under control. An extremely important conclusion, which will be justified later, is that a nonautonomous (two-element) system can originate from an originally autonomous one through loss of the regulator function; following such an event the receptor component may continue to respond to a regulator elsewhere in the genome. Regulators are, in fact, more often than not identified through their effects on nonautonomous receptors. An active regulator promotes not only excision of a responsive receptor but (at least in many cases) its own excision and transposition as well.

The receptor-regulator relationship is a highly specific one and three classes of elements have been recognized on the basis of this specificity. These have been given the names *Dotted (Dt)* (3, 4), *Activator (Ac)* (6, 7), and *Suppressor-mutator (Spm).* Peterson's *Enhancer (En)* appears to be the same as *Spm,* while Brink's *Modulator (Mp)* is homologous with, if not identical to, *Ac.* Receptors of each class respond only to their own regulator and are quite unaffected by the other two. While its status is not altogether clear, the determinant of the instability of *R-stippled* (see p. 24) may represent a fourth class.

The *Spm* element differs from *Dt* and *Ac* in being commonly capable of regulating the activity of a gene even in the absence of mutation-like (excision) events. The receptor component of *Spm* may, in the absence of the regulator, bring about only a partial or even quite slight inhibition of the activity of the associated gene. In such cases the regulator may act at a distance on the receptor to suppress the gene activity in most cells as well as to release full gene activity, presumably by receptor excision, in some of them. The *suppressor* and activity-releasing (*mutator*) functions of the *Spm* regulator are separable, in as much as derivatives of *Spm* are known that have lost the second while retaining the first.

The three classes of element differ strikingly in their dosage effects on the frequency of excision. In the case of *Ac* this frequency is maximal with one copy of the element, is strikingly reduced by a second copy, and is much lower again with a third (note that the endosperm is triploid). *Dt* shows the opposite dosage effect; release of activity of the susceptible allele a_1 increases in frequency at least in proportion to the number of *Dt* copies present (4). *Spm* shows no dosage effect, at least for mutation frequency of the susceptible allele, and behaves in this respect as a simple dominant.

Finally, it must be mentioned that both receptor and regulator components of controlling elements frequently change their properties. These changes, though they may be transmitted through many cell divisions like genetic mutations, are often referred to as changes in *state* or *phase,* in token of their high frequency, their strong tendency to revert, and (in some cases) their high degree of predictability at certain stages of plant development.

ORIGINS OF CONTROLLING ELEMENTS

There is no doubt that active controlling elements are widespread in cultivated maize stocks of diverse geographical origins. The original *Dt,* and the a_1 allele responding to it, turned up spontaneously in Black Mexican sweet corn (4) while the variegated pericarp allele P^{vv}, first studied by Emerson (8) and later shown by Barclay & Brink (9) to be due to the association of P^{rr} with an element resembling *Ac,* was also found in a Central American stock.

On the other hand, the numerous new unstable alleles in McClintock's stocks, which led to her very extensive later work, appeared in plants subjected to a breakage-fusion-bridge cycle (Figure 1) during the early part of their development. Several of the genes in which instability appeared (e.g. C_1, Yg_2, and Wx) were on the short arm of chromosome 9 which was directly involved in the cycle, though some others (A_2, Y) were not (7). Even more important for later work was the discovery, again in the short arm of chromosome 9, of a locus at which breaks, anaphase bridge formation, and translocations tended to occur. This behavior was attributed by McClintock to an element called *Dissociation (Ds),* and it was shown that *Ds* caused chromosome instability only in the presence of *Ac.* Subsequently the appearance of an *Ac*-controlled mutable allele of C_1 (c_1^{m-1}) in one pollen grain of a plant homozygous for C_1 and also carrying both *Ac* and *Ds* was shown to be accompanied by the loss of the propensity for breakage from the original locus of *Ds* and the new appearance of this property at a point at or close to the locus of C_1 (Figure 2). McClintock concluded that *Ds* had been transposed from its original

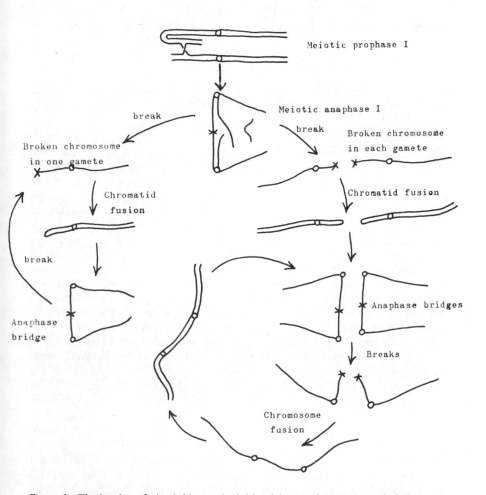

Figure 1 The breakage-fusion-bridge cycle, initiated by crossing-over at meiosis between chromosomes 9, one of which has an inverted terminal duplication in its short arm. Fusion between broken sister chromatids leads to the "chromatid" type of cycle (*left*) in the gameto-phyte and endosperm but not, usually, in the embryo. Fusion between whole chromosomes, when both fusing gametes have a broken chromosome, leads to the "chromosome" type of cycle (*right*) which may continue during the early part of development of the sporophyte, after which the broken ends tend to "heal." Diagonal crosses indicate breaks or recently broken ends. Adapted from McClintock (7).

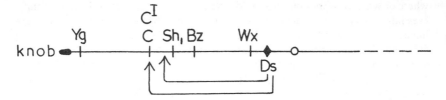

Figure 2 Transpositions of *Ds* within the short arm of chromosome 9. Only those transpositions discussed in the text are included; McClintock (7) observed several others.

location near the tip of the short arm of chromosome 9 to a new position some 20 map units away on the same chromosome (7, 10). It looks as if *Ds* was generated during the breakage-fusion-bridge cycle in the chromosome region undergoing breakage and that it was subsequently able to move away from its locus of first appearance to cause both structural instability and reversible blockage of gene action at the C_1 locus. The origin of *Ac,* which was only identified because of its induction of breakage at the locus of *Ds,* was less clear. It may have been in the stocks before the inception of *Ds,* or it may have been generated at the same time as *Ds,* an alternative that fits well with the idea, elaborated below, that *Ds* is essentially an incomplete *Ac.* McClintock (7) states that early attempts to map *Ac* were abandoned because of its frequent transposition from one locus to another, and so it seems entirely possible that *Ac* arose, like *Ds,* in the short arm of chromosome 9 but had already moved to another chromosome by the time *Ds* was discovered.

The apparent effect of the breakage-fusion-bridge cycle in inducing, or uncovering, determinants of chromosome and gene instability encouraged McClintock (7, 11) to search for new origins of *Dt*-like elements. She used plants, homozygous for a_1 (susceptible to the action of *Dt*) and with the chromosome constitution needed to generate the breakage-fusion-bridge cycle in the short arm of chromosome 9, to pollinate plants that were also homozygous for a_1 but of normal chromosome constitution; neither parent carried *Dt.* Among over 90,000 kernels formed, 117, or a little over 0.1%, showed larger or smaller areas containing spots of A_1 (deeply pigmented) phenotype. This is the appearance expected if sectors carrying *Dt* were being formed during endosperm development, but since these sectors were confined to the endosperm they could not be tested genetically. More recently Doerschug (12) has repeated the experiment in a similar way and, in addition to finding spotted regions in 0.16% of the kernels, he also observed two kernels that were entirely *Dt*-like in phenotype and transmitted a *Dt*-like element to subsequent sexual generations. These new *Dt* elements must have been due to events occurring during meiosis. In one case the new *Dt* was very close to the locus of Rhoades' original *Dt* (close to the tip of the short arm of chromosome 9), but the other was on a different chromosome.

As McClintock's work progressed it became clear that several mutable states of genes that originally seemed to be in the same category as c_1^{m-1} and other manifestations of the *Ds-Ac* system were really due to a quite different class of controlling element, later called *Suppressor-mutator (Spm)* (13, 14). There is nothing to show where or when this *Spm* originated; like *Ac* it may have been silently present in the breeding stocks for an indefinite time, detected only when its receptor component happened to become associated with a gene of suitable phenotypic effect. Peterson (15) had already identified an element which he called *Enhancer (En)* and later showed (16) to be similar to if not identical with *Spm*. This element was discovered through its effect on the pale-green (*Pg*) locus. An unstable *pg* allele was found in the progeny of plants grown from seeds exposed to radiation from an atomic test explosion, but whether it owed its origin to the radiation is not known.

Neuffer (17) designed experiments to detect induced reversion of nonautonomous mutable alleles in the absence of their respective regulators. The results were negative in this regard [as also were those of an experiment of similar intent by Mouli & Notani, (18)], but indications were obtained of new occurrences of *Dt* and *Spm* following radiation treatments of pollen from plants apparently devoid of these elements. In one experiment at least one convincing sector of $a_1^{m-1}Spm$ phenotype appeared after X-ray treatment in a stock carrying a_1^{m-1}, but initially no overt *Spm*, while in another experiment in which ultraviolet irradiation was used, a few sectors apparently showing the effect of *Dt* on the susceptible allele a_1^m(Neuffer) were obtained though no *Dt* was originally present. These apparently new initiations of controlling elements were not obtained in germ cells and so could not be further investigated.

Granting that new occurrences of overt controlling elements may sometimes result from chromosome damage, one still finds it difficult to believe that they arise de novo from undifferentiated DNA or from DNA of quite different function. It seems more likely that they were hidden in the genome, perhaps buried in heterochromatin where they could neither transmit nor respond to regulatory signals. McClintock's emphasis (5, 6) on the frequent involvement of heterochromatin, especially the terminal knob on the short arm of chromosome 9, in the aberrations generated in the breakage-fusion-bridge cycle is particularly interesting in this connection.

TRANSFER OF INSTABILITY TO NEW SITES

The first appearance of instability at a particular locus in a maize stock has often been followed by further inceptions of instability in other parts of the genome. Thus, following Peterson's (15, 19, 20) first detection of instability at the *pg* locus, due to the *En* (= *Spm*) element, he isolated a new mutable a_1 from his mutable *pg* stock (21) and, later, a new mutable a_2 from a plant carrying the mutable a_1 (22). Again, Greenblatt (quoted in Brink & Williams, 23) obtained a mutable derivative of R^{nj} from plants carrying variegated pericarp (*Pᵛᵛ*), which had been shown to consist of P^{rr} combined with Brink's element *Mp* (= *Ac*); a mutable *wx* has been isolated from the same material (24). In each of these cases the new instability was shown to be due to the same controlling element as was the preexisting one. The largest collection of mutable alleles has been made by McClintock from stocks carrying either *Ac* or *Spm* or both. The earliest example, already referred to, involved the element *Ds*. A new mutable allele, c_1^{m-1}, was associated with an apparent transposition of

Ds, identified as the site of frequent chromosome breaks, to a site close to the C_1 locus. Later, a large number of other mutable alleles turned up in McClintock's stocks; some proved to be controlled by *Ac* and some by *Spm,* and they are listed in Table 2. What all these occurrences have in common is that the transposable element afterwards shown to be controlling the instability—*Ac* or *Spm*—was known to be already present somewhere in the genome of the plant in which the new mutable allele arose.

In one experiment McClintock (25) deliberately set out to look for new mutable a_1 and a_2 alleles by crossing homozygous $A_1A_1A_2A_2$ plants, carrying *Ac* at a third locus, to homozygous stable a_1a_1 or a_2a_2 testers. Such alleles were in fact found; a_1^{m-4} and a_2^{m-4} had this origin (Table 2). The frequency was of the order of 10^{-4} for new occurrences of instability at each locus.

A later attempt by Peterson (26) to demonstrate transfer of instability due to *En* (= *Spm*) from A_1 to either A_2 or C_1 had a much lower frequency of success. One new *En*-controlled mutable allele at each locus was detected in the experiment, but over five million gametes had been screened. Peterson suggested that McClintock's higher success rate may have been due, at least so far as the origin of a_2^{m-4} was concerned, to the fact that what was being looked for was transposition of *Ac* to another site on the same chromosome. Orton (27) has provided very good evidence, to be discussed below (see Figure 5), that *Mp* (= *Ac*) transposes most frequently relatively short distances along the same chromosome, less frequently to more distantly linked sites, and least frequently to other chromosomes.

A generalization that can be made with some assurance is that each new mutable allele arose from a functional dominant allele and thus represented an unstable *inhibition* of what was previously a functional gene. This is clearly demonstrated in many cases (indicated in Table 2) and is consistent with the genotype of the progenitor plant in all the others. The inhibition of gene activity is attributed, on the basis of the evidence discussed below, to the insertion of the controlling element, or a derivative of it, in or adjacent to the gene.

Almost all known loci with clear effects on the pigmentation, shape, or texture of the seed are represented among the mutable alleles listed in Table 2. In other words, they include practically all the genes amenable to study. There is no reason to doubt that, in a plant carrying one of the transposable elements, any gene can become "infected" by the element.

RELATION BETWEEN THE CONTROLLING ELEMENT AND THE STRUCTURAL GENE

It is clear that controlling elements can block, partially or totally, the expression of genes at whose loci they reside. A very good example of such blocking at the level of protein synthesis was obtained by Schwartz (28) in the case where *Ds* was associated with Sh_1. We do not know, however, whether the blocking elements are usually inserted within the DNA sequence that codes for the gene

Table 2 Mutable alleles and their origins

Locus	Allele	Derived from	Control by	Whether regulator at locus	Phenotype in absence of regulator	References
A_1	a_1	A_1	Dt	no	no pigment	3, 4
	a_1^{m-1}	?	Spm	no	pale pigment	7, 11, 13, 14, 35, 49, 66
	a_1^{m-2}	?	Spm	yes, no[a]	pale, irregular pigment[a]	7, 11, 13, 34, 43
	a_1^{m-3}	?	Ac	no	no pigment	11, 35
	a_1^{m-4}	A_1	Ac	no	no pigment	25, 40
	a_1^{m-5}	A_1	Spm	yes, no[a]	pale or medium pigment[a]	43
	a_1^m	$A_1 (+ pg^m)$	En (= Spm)	yes, no[a]	no pigment	20, 21
	$a_1^{m(dense)}$	a_1^m	En	yes	—	20
	$a_1^{m(pa-pu)}$	$a_1^{m(dense)}$	En	yes	—	64
	$a_1^{m(N)}$	A_1	Dt	no	no pigment	17, 65
A_2	a_2^{m-1}	?	Spm	yes, no[a]	medium-dark pigment[a]	49
	a_2^{m-4}	A_2	Ac	no	no pigment	25, 40
	a_2^{m-5}	?	Spm	no	?	35
	a_2^m	$A_2 (+ a_1^m)$	En (= Spm)	yes, no[a]		22
Bt_1	bt_1^m	?	Spm	no	?	R. L. Phillips, (pers. comm.)
Bz_1	bz_1^{m-1}	?	Ac	no (Ds at locus)	bronze	7, 11
	bz_1^{m-2}	?	Ac	yes, no[a]	bronze[a]	7, 11, 41, 43
	bz_1^{m-4}	?	Ac	no	bronze	35
C_1	c_1^{m-1}	C_1	Ac	no (Ds at locus)	no pigment	7, 10, 55
	c_1^{m-2}	C_1	Ac	no	no pigment	10, 55
	c_1^{m-4}	?	Ac	no	no pigment	35
	c_1^{m-5}	?	Spm	yes	—	59
C_2	c_2^{m-1}	?	Spm	yes	—	36
	c_2^{m-2}	?	Spm	no	?	36
P	p^{vv}	?	Mp (= Ac)	yes	—	8, 9, 37, 44
Pg	pg^m	Pg	En (= Spm)	yes, no[a]	?	15
Pr	pr^{m-2}	?	Spm	no	?	35
	pr^{m-3}	?	Spm	no	?	35
R	mutable-R^{nj}	$R^{nj} (+ P^{vv})$	Mp (= Ac)	yes	—	23
	spotted-dilute-R	?	Spm	no	pale pigment	Sastry (unpublished)
Wx	wx^{m-1}	Wx	Ac	no	no amylose starch	56
	wx^{m-5}	?	Ac	no	low level of amylose, higher near periphery of endosperm	35
	wx^{m-6}	?	Ac	no	?	35
	wx^{m-7}	Wx	Ac	yes	medium level of amylose	57
	wx^{m-8}	?	Spm	no	low level of amylose	34
	wx^{m-9}	?	Ac	yes, no[a]	little or no amylose	59
	wx^{m-1} (Ashman)	? $(+ P^{vv})$	Mp (= Ac)	yes	—	24

[a]Following transposition of the controlling element.

product (the structural gene) or in a nearby segment of chromatin with a regulatory function.

Neuffer (29) was not able to demonstrate separability by crossing-over of blocking receptor elements from the gene (A_1) whose action they were inhibiting; he obtained some indication that the presence of the element at the gene locus reduced the frequency of crossing-over in the immediate vicinity. Kermicle, on the other hand (30, 31), did appear to resolve the mutable allele R-stippled (R^{st}) by rare crossing-over into a stable "self-colored" component R^{sc} and an adjacent inhibiting component I^R. From heterozygotes of constitution R^{st}/R^{nj} it was possible to recover reciprocal crossover products, namely R^{sc} and stippled-R^{nj} (i.e. $R^{nj}I^R$). These results, however, do not disprove the possibility that I^R was initially inserted within R and that the crossovers concerned were intragenic, falling between the site of insertion of I^R and the site(s) differentiating R^{sc} from R^{nj}. Incidentally, I^R is not known to be at all related to Dt, Ac, or Spm, and it may represent a fourth class of excisable element. Its instability is enhanced by a dominant linked modifier M^{st} (30), rather as Spm is complemented by Mod (see p. 36).

Reasonably precise mapping of inserted controlling elements has been carried out only in one system. The waxy character of maize affords a unique opportunity for fine structure analysis because it can be scored in the pollen grains. The normal (Wx) amylose starch, which stains black with iodine-potassium iodide reagent, is replaced in homozygous or haploid wx by amylopectin which stains a light red-brown. Hence one can look for and count Wx pollen grains formed by recombination in heteroallelic wx^a/wx^b plants and, since the black-stained grains stand out strikingly among large numbers of light brown ones, it is possible to detect recombination down to frequencies of 10^{-5} to 10^{-6}. In the absence of the appropriate regulator element a nonautonomous mutable wx allele is completely stable and can be mapped just like any stable allele.

The results of Nelson's (32) recombinational analysis of several wx alleles are shown in Figure 3. The striking conclusion is that many of the wx sites, including three due to inserted elements (Ac in two cases and Spm in the other), mapped as points, in the sense that they recombined with one another in all pairwise crosses. Furthermore, the mutable wx alleles are intermingled with the others without any suggestion that they occupy a special region of the locus. There was, of course, no way of checking the apparently recombinant pollen grains to confirm that they really were associated with crossing-over, but such a check was made on a more limited scale on recombinant Wx kernels (it is possible to stain a small portion of the endosperm and grow the embryo), and here it was shown that the appearance of Wx did indeed correlate with crossing-over of flanking markers.

Accepting Nelson's map, it is still not necessary to conclude that the receptor component of the controlling element is actually within the structural gene in each case. It could be that all the wx alleles studied have mutant sites within a segment with control function that is outside the DNA sequence coding for the Wx-dependent enzyme [ADP glucose-starch glucosyltransferase (33)]. It would be most interesting to know whether the various wx mutations determine qualitative or merely quantitative changes in the enzyme.

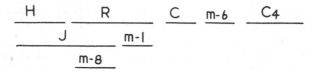

Figure 3   Part of Nelson's (32) map of the Wx locus, showing the order of the sites of several allelic markers including three mutable alleles: *m-1* and *m-6* controlled by *Ac* and *m-8* controlled by *Spm*. Overlaps of segments indicate no recombination detected between the mutations concerned; the sequence of nonoverlapping segments was deduced from recombination frequencies. Nelson's full map includes many more sites than those shown here.

EVIDENCE FOR TRANSPOSITION OF CONTROLLING ELEMENTS

Relation between Autonomous and Two-Element Systems

The status of a controlling element as an entity in its own right depends on the demonstration of its independent transposability. This in turn depends on the element having a readily scorable effect on the phenotype independent of its position in the genome.

We have already introduced the idea of regulator and receptor components of controlling elements, the former transmitting signals and the latter responding to them. Regulator and receptor components can often be shown to reside together to form an autonomous system at the locus of the affected gene (instances of this situation are specified in Table 1), but loss of the regulator component (or at least of its activity), leaving a receptor element showing instability only in response to an independently located regulator, is a rather common event, and some examples are reviewed below. It seems reasonable to suggest that all two-element systems originated from what were originally autonomous systems, though this cannot be proved. In two-element systems, both the receptor and the regulator are regarded by McClintock as being transposable, though this is more difficult to demonstrate for the receptor since its visible effect depends on its being located in or adjacent to a convenient gene. It is, in fact, particularly difficult to obtain evidence for the receptor as an element independent of the gene that shows high mutability; generally speaking it is just something that one postulates to explain the aberrant behavior of the gene. The main reason for regarding the receptor as an entity additional to the gene is that it may be present at a gene locus as a residual effect of a previously resident autonomous element. The latter, as we shall see, is much more readily shown to be an independently transferable entity, and it is tempting to regard the receptor in a derived two-element system as a fragment of what was originally a more complete element, though this cannot be claimed to be a rigorous argument.

Transposition of Ds

The only case in which a receptor component can be identified by a test independent of its effect on the activity of an immediately adjacent gene is that of *Ds* which, at least in its originally identified state, causes breaks at the locus at which it resides. In its original location, just proximal to Wx in the short arm of chromosome 9, Ds, in the presence of Ac, caused frequent loss of a distal chromosome fragment including the dominant markers C_1 (or C_1^I), Sh_1, Bz, and Wx. In endosperm of suitably heterozygous genotypes, this led to the formation of sectors showing the effects of any of the corresponding recessives (5, 7). Transposition of Ds to a new location in the same chromosome arm, between C_1 and Sh_1, was detected (5) by a different pattern of variegation. Chromosome breaks during endosperm development at this new Ds site caused immediate loss of the C_1^I marker only. Subsequent anaphase bridge formation, due to fusion of broken chromatids following replication, set up a breakage-fusion-bridge cycle which resulted in variegation with respect to Sh_1, Bz, and Wx within the C_1^I-deficient sector. For example, breakage of the bridge between Bz and Wx could give rise to twin subsectors, one of which will be uniformly bz and sh_1 in phenotype while continuing to show sectoring with respect to Wx/wx. Such a complex pattern could not have arisen had Ds remained in its original position (see Figure 2).

Not long after the discovery of Ds, the mutable allele c_1^{m-1} originated in a single germ cell of a plant carrying C_1 together with both Ds (in its original location) and Ac. This new instability was found by McClintock to be associated with frequent chromosome breaks, not at the original Ds site, but at another site which, within the limits of resolution of the pachytene chromosome analysis, corresponded to the locus of C_1. The disappearance of the breaking tendency from one locus and its simultaneous appearance at another, with the concurrent onset of genetic instability at or close to the latter locus, is at least strongly suggestive of the transposition of some destabilizing physical element.

Transposition of Ac or Spm from a Gene Locus

The most favorable situation for detecting the transposition of a controlling element exists when the element, including an active regulator component, is initially present at the locus of the gene that shows instability. Several such cases are listed in Table 1. The loss of the element from the gene locus is signaled by the restoration of stable gene activity and one can then test whether, in kernels showing such restoration, the controlling element is now present at a different locus, either on the same chromosome or on another.

In order to identify a controlling element at the locus of a mutable allele it is necessary to have an independent tester system that responds to the element in *trans*. In the case of Ac, the original state of Ds, which responds to Ac by becoming the site of breaks, can be used. Again, both Ac and Spm can be detected and mapped through the use of suitable nonautonomous mutable alleles. It is most convenient if the mutation of the tester allele has a phenotypic effect distinct from that of the mutable allele at the locus at which the element in question is residing. Sometimes, as in the case of a_1^{m-1} and a_1^{m-2} reviewed below, the patterns of mutation are sufficiently distinct to allow independent scoring of the two in the same kernel, even though they both affect the same feature of the phenotype (aleurone pigment in this case). The nonautonomous alleles wx^{m-5} and wx^{m-8}, which respond to Ac and Spm

respectively (34, 35), have been particularly useful as testers since the nature of the starch, easily determined by sectioning the endosperm and staining with I_2-KI reagent, can be scored entirely independently of the aleurone pigment involved in so many of the mutable systems. For example, the association of *Spm* with the C_2 locus in c_2^{m-2} was demonstrated with the aid of wx^{m-8} (35, 36).

The Use of Twinned Sectors and the Mechanism of Transposition

Of all the examples of transposition of a controlling element away from a gene locus, that involving P^{vv} has been the most informative. This unstable allele, described by Emerson (8), determines medium-variegated (red sectors on white) pericarp and cob. It mutates at a frequency of a few percent in heterozygotes to stable full-red pericarp and cob (P^{rr}) and, at a similar or somewhat lower frequency, to a lower grade of variegation (fewer red sectors) called light-variegated.

Brink & Nilan (37) observed that medium-variegated ears frequently showed groups of full-red kernels twinned with symmetrically placed and similar sized groups of light-variegated kernels, as if some nondisjunctional process was responsible for the simultaneous generation of the two phenotypes. They were able to show that, unlike the full-red mutants, the light-variegated type did not breed true but segregated up to 50% medium-variegated on crossing to a stable recessive homozygous tester such as P^{ww}. It seemed that the decrease in the grade of variegation was due to the presence of a dominant factor, unlinked or only loosely linked to P^{vv}, which appeared in one product of a cell division early in ear development coincident with the mutation of P^{vv} to P^{rr} in the sister product. Brink & Nilan postulated an element called *Modulator-of-P (Mp)* which was supposed to be present at the *P* locus in P^{vv} causing the instability and which, when present as an additional copy together with P^{VV}, reduced the frequency of variegation (cf Wood & Brink, 38). They interpreted P^{vv} as a complex $\overline{P^{rr}Mp}$ and the twinning event as the transposition of *Mp* away from P^{rr} to generate the sister products P^{rr} (full-red) and $\overline{P^{rr}Mp}$ + *Mp* (light-variegated). Later work fully substantiated this interpretation and brought to light some further important features of the system.

First, Barclay & Brink (9) showed that *Mp* had properties indistinguishable from those of *Ac*. Their demonstration made use of *Ds* as an independent test system. Tester plants, homozygous for C_1 and *Ds* (the latter in its "standard" position just proximal to *Wx*) were crossed to plants homozygous for c_1 and, in some cases, heterozygous for P^{vv}. The inheritance of *Mp* (= *Ac*) along with the P^{vv} allele was shown by frequent colorless aleurone sectors due to *Ds*-breaks and loss of C_1. There was a perfect correlation between segregation of P^{vv} and segregation of *Ac* by this test. Six other alleles of *P* showing different patterns of pericarp pigmentation were tested in the same way and none showed the presence of *Ac*.

Secondly, Brink (39) showed that one type of change in P^{vv} did not involve removal of the *Mp* element but rather its stabilization. Analysis of a stable white-pericarp-and-cob (P^{ww}) mutant, which had arisen from P^{vv}, showed that it was still closely associated with *Mp*. Brink supposed that the mode of insertion of *Mp* into the chromosome was changed in some way so that, while it continued to block gene action, it could no longer be transposed. A similar example of stabilization of a blocking element, in this case Ds at the Sh_1 locus, was reported by McClintock (13, 40, 41) (cf Schwartz, 28).

A further most important conclusion, demonstrated by van Schaik & Brink (42), was that when Mp was transposed from the P locus it showed a very strong tendency to go to another site on the same chromosome. Analysis of 87 different transpositions obtained from light-variegated sectors showed that 56 of them were to another site on chromosome 1. Furthermore, there was a marked preference for sites close to P. In 25 cases the additional Mp was not separated by crossing-over from the $\overline{P^{rr}Mp}$ complex, in 13 cases it was separated from within 5 map units of P, and in another 7 cases it was within 10 map units. This tendency of Mp to *short-range* transposition was later amply confirmed by Orton's (27) study of transpositions *back* to P (Figure 4). A similar, though somewhat weaker, preference was shown by McClintock in several cases, such as that of transposition of Ac from the Bz locus (41, 43).

Finally, Greenblatt & Brink (44) showed that, as a result of transposition, Mp very often underwent an extra replication. They found that in 65% of cases of twinned full-red and pale-variegated ear sectors the full-red seeds (the embryonic genotype of which generally accorded with the pericarp phenotype) carried one copy of Mp, though not at the P locus. In 13 such cases the Mp in the red sector and also the extra Mp in the twinned pale-variegated sector were both mapped by three-point test crosses. The striking result was that, in each of 12 of the 13 twins analyzed, the Mp in the red sector and the additional Mp in light-variegated sector were at the same locus within the error of the mapping. This locus was, however, different in different pairs—unlinked to P in five cases and linked, at various distances from 2 to 30 map units, in the other seven. The thirteenth case, where Mp was linked to P in the red sector and unlinked in the twin, was attributed to a secondary transposition.

In a continuation of the analysis Greenblatt (45) concluded that virtually all red sectors were, at least potentially, twinned with light-variegated sectors. The basis for this conclusion was that red sectors, whether twinned or apparently untwinned, had about the same probability (about 65%) of carrying the Mp; hence he felt that all red sectors probably had the same kind of origin. Greenblatt's model to account for the data was ingenious but now seems implausible, since it depends on the assumption of conservative DNA replication. We offer an alternative interpretation in Figure 4. The essence of this hypothesis is that transposition of Mp occurs from a recently replicated P^{vv} (i.e. $\overline{P^{vv}Mp}$) either to an unreplicated site, most likely on the same chromosome, or, about equally frequently, to an already replicated site; in a proportion of cases of transposition (about 10%) the Mp element is supposed to be lost or inactivated. Occasional transposition from unreplicated P^{vv} is not ruled out.

Transposition of a Transposed Element Back to its Former Locus

Following the demonstration that red-pericarp ear sectors arising from transposition of Mp away from P^{vv} frequently carried a transposed Mp, studies were made to see whether red pericarp stocks of this origin could regenerate P^{vv} from the

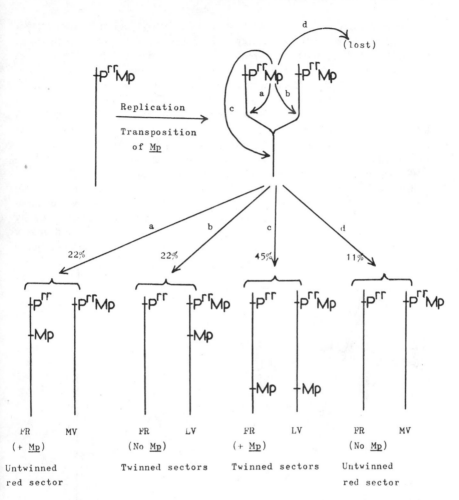

Figure 4 Interpretation of Greenblatt's (45) data on twinned and untwinned sectors on P^{vv}/P^{ww} ears. The same consequences will follow if *Mp* is transposed to the replicated or unreplicated region of another chromosome rather than to another site on the same chromosome, as shown here. FR = full red, MV = medium variegated, LV = light variegated.

separated P^{rr} and *Mp* components. Orton & Brink (46) showed that such reconstitution did indeed occur, giving occasional medium-variegated sectors on otherwise red kernels or, with lower frequency, whole medium-variegated kernels. Some of the wholly variegated kernels proved to carry a new P^{vv} allele in the embryo as well as in the pericarp, and hence a large number of secondarily derived P^{vv} lines were established. Some of these were indistinguishable from the original P^{vv} but most were different, with various grades of variegation. A few gave red sectors on a lighter orange background, an interesting case of an inserted element of the *Ac* type permitting a reduced, rather than a zero level of gene activity. In an accompanying paper, Orton (27) showed that the frequency of reconstitution of the $\overline{P^{rr}Mp}$ complex was critically dependent on the distance that the *Mp* element needed to be transposed. In $P^{rr} + Mp$ cultures in which *Mp* was closely linked to *P,* the frequency was far higher than when it was more loosely linked. Figure 5 summarizes the data, which confirm the conclusion drawn from studies of transpositions *away* from a locus. Short-range transpositions are so much more frequent than longer range ones that, extrapolating to very short distances, we may speculate that intralocus, or intracistron, transpositions may be very frequent indeed; such may be the explanation of the frequent "changes of state" of mutable alleles (see pp. 34ff).

Brink & Williams (23) selected a range of mutable R^{nj} alleles formed by transposition to R^{nj} of a linked *Mp.* These secondarily derived alleles constituted a continuous spectrum with regard to mutation frequency and timing. By the test of activity in the *Ds* system all of them were associated with an *Mp* of the same degree of effectiveness, and so the authors concluded that the varying effects at the *R* locus

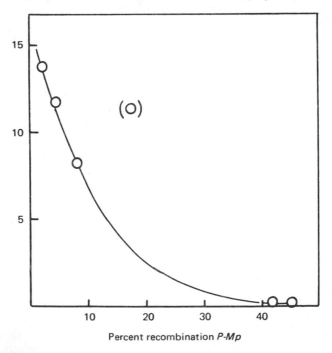

Figure 5   Relation between frequency of transposition of *Mp* (= *Ac*) back to the *P* locus from sites at different distances from *P.* Plotted from the data of Orton (27). There was reason to suspect that the recombination frequency was overestimated in the case of the point shown in parenthesis.

were due to different modes of insertion of the *Mp* element within the locus rather than to differences in the element itself.

Other Cases of Transposition of Spm or Ac from a Gene Locus: the Origin of Two-Element from Autonomous Systems

MUTATION OF a_1^{m-2} Among other examples of transposition of a controlling element away from a gene locus, that involving the mutable allele a_1^{m-2} is one of the best documented. McClintock (43) mapped *Spm* at or very close to a_1^{m-2} using another allele a_1^{m-1}, which is stably active in the absence of *Spm*, as a tester. In the presence of *Spm*, a_1^{m-1}, which is a nonautonomous allele, gives dark spots on a white background; in the isolate used in this study the spots tended to be small. The characteristic phenotype of the original state of a_1^{m-2} was a coarsely variegated aleurone pigmentation due to both early and late mutations to either higher or lower A_1 gene expression. In many plants carrying a_1^{m-2} there is a high frequency of mutation in the sporogenous cell layer; mutant germ cells transmit either a stable allele similar in effect to standard A_1 (full color) or, more frequently, one of a continuous spectrum of alleles giving different grades of rather mottled pigmentation. The heterozygote a_1^{m-1}/a_1^{m-2}, assuming that *Spm* is present, shows independent effects of the two alleles, with the small dark a_1^{m-1} spots superimposed on the a_1^{m-2} variegation. The crosses that showed that a_1^{m-1} segregated along with *Spm* were of the form

$$\frac{a_1^{m-2}Sh_2}{a_1^{m-1}sh_2} \times \frac{a_1^{m-1}sh_2}{a_1^{m-1}sh_2} \quad \text{(No } Spm\text{)},$$

sh_2 and a_1 being separated by only about 0.1 map units. In a few crosses of this type there were, among nearly a thousand progeny, no exceptions to the rule that all the Sh_2 kernels showed both the a_1^{m-2} variegation and the *Spm*-dependent a_1^{m-1} late mutations, while all the sh_2 kernels showed the uniform (rather pale) pigmentation characteristic of a_1^{m-1} in the absence of *Spm* (top three lines of Table 2 in McClintock, 43). Hence, at least in some plants heterozygous for a_1^{m-2}, *Spm*, defined as the factor causing a_1^{m-1} mutation, segregated with a_1^{m-2} without exception. In several other crosses, however, from 5 to 70% of the Sh_2 progeny kernels were either fully colored or mottled in phenotype. In the case of the fully colored kernels it could not be seen whether mutations of a_1^{m-1} were still occurring, but in about half of the mottled ones it appeared that they were not, indicating that *Spm* had been lost or inactivated. In addition to the mutant Sh_2 progeny there were, in the same ears, a minority of sh_2 kernels that did show the small dark spots characteristic of a_1^{m-1} but on a colorless background. Since Sh_2 can be regarded as a marker for the very closely linked A_1 locus, it was reasonable to interpret the Sh_2 mutants as due to loss of *Spm* from the A_1 locus, and the spotted sh_2 kernels as due to transposed *Spm* segregating into the same spore as a_1^{m-1}. The frequencies of the a_1^{m-2} mutants and the a_1^{m-1} *Spm* recombinants were correlated, being both very low in some plants and both high in others; overall, the ratio of the two classes was a little more than 2:1, in good accord with a random transposition model.

The case of $a_1{}^{m-2}$ just discussed provided a good example of the derivation of a two-element system from an originally autonomous one. One pale-mottled mutant allele derived from $a_1{}^{m-2}$, and detected in a segregant lacking *Spm,* gave the variegated pattern characteristic of the original $a_1{}^{m-2}$ when *Spm* was reintroduced into the same genome by an appropriate cross. The segregation of *Spm* was followed by virtue of its close linkage, in this instance, to *Pr* on chromosome 5. It was as if the receptor component of *Spm* had been left behind at the A_1 locus after the regulator, originally also at the locus, had been lost or inactivated.

Another example of an initially autonomously mutable allele capable of giving a two-element system is bz^{m-2}, with which *Ac* was originally closely associated (41, 43). Crosses of plants carrying bz^{m-2} to plants homozygous for the stable recessive *bz* yielded a range of mutant derivatives of bz^{m-2} ranging in phenotype from dark to very pale. All the stable mutants tested, of whatever level of expression, proved to have had *Ac* transposed away from the *Bz* locus either to another site on the same chromosome or, with roughly equal frequency, to another chromosome. Thirty-five independently isolated mutant alleles of null expression were analyzed. In 33 cases the new allele was stable; in two, *Ac* had been transposed to a closely linked site, in 12 to a relatively distant site on the same chromosome, and in the remainder to another chromosome. The other two alleles of null expression were originally found in segregants lacking *Ac* and showed restored mutability when an unlinked *Ac* was reintroduced into the same genome. Other examples of the origin of nonautonomous *Ac*-controlled alleles from various states of bz^{m-2} were also described by McClintock (43).

Several other well-documented examples of loss of autonomy by unstable alleles have been provided by Peterson. As has been mentioned, his element *En* seems to have the same specificity as *Spm.* However, it does seem to be especially prone to loss or transposition of its regulator component. Originally pg^m (19), $a_1{}^m$ (20), and $a_2{}^m$ (22) all had *En* at or very close to the affected gene locus, and all gave rise to two classes of mutant alleles of null or very low expression. These two classes were distinguished by the superscripts *m(nr)* (nonresponsive) and *m(r)* (responsive) respectively, depending on whether or not they mutated in response to *En* elsewhere in the nucleus; both classes were stable in its absence. These examples are evidently closely parallel to those studied by McClintock.

It seems, then, that *Ac* and *Spm* (or *En*) are capable of losing the regulator function from the locus of the gene under control (though, interestingly enough, such behavior has not been reported for Brink's *Mp,* in other respects apparently identical to *Ac*). Whether such loss is a consequence of transposition of part of an originally "complete" element, or whether it results from some sort of mutation without transposition, cannot be decided. Although there are several descriptions, some of them reviewed above, of the transposition of *Ac* or *Spm* to new loci at the same time as, or correlated with, the origin of *stable* mutants from an autonomously mutable allele, there seem to be no clear examples of similar transpositions accompanying the origin of a *responsive* mutant. If a case of the latter kind were found it would be interesting to see whether the element transposed was a complete one, equivalent to standard *Ac* or *Spm,* or whether it consisted of the regulator component only, in which case it might be defective in its ability to undergo further transpositions. The detection of a new responsive allele in one of a pair of twinned sectors, of which the other might contain the transposed regulator as an extra copy, could lead to a better understanding of this question.

Transposition of the Regulator Function Independent of the Locus Controlled

In two-element systems, with nonautonomous alleles responding to the regulator function of a distant controlling element, transpositions of the latter may be rather common. McClintock (10) reported that two or three percent of the gametes formed in a plant initially homozygous for Ac at a particular locus may have a transposed Ac, either on the same chromosome or a different one, or may have no Ac. These different situations were easily distinguished by test-crosses to a Ds tester stock.

One can specify the initial and the new locus of a transposing element only if there happen to be segregating markers linked to these loci, and this is not usually the case. However, McClintock was able to learn a good deal about transpositions of Spm. The first identified locus of this element was on chromosome 6, about 40 map units from Y, though transpositions to unknown sites were frequent (13, 14). Later, in another culture, Spm was mapped by linkage to Wx on chromosome 9, a_1^{m-1} being used as a tester. In this culture further transpositions continued to occur in most plants during vegetative development to various new loci, including one linked to Pr on chromosome 5 (47). In contrast to this very labile behavior, the Spm present closely linked to Y in another culture showed a relatively high stability, with evidence of transposition in only one ear out of a total of 44 on 17 plants (49). We have already noted Brink's report (39) of the fixation of Mp at the P locus. It seems that the tendency to transpose can vary considerably, but whether this is due to variation intrinsic to the controlling element or is a function of the site at which it is inserted is not known.

CONTROL OF TRANSPOSITION FREQUENCY BY EXTRINSIC FACTORS

There is considerable though rather scattered evidence for control of the frequency of mutation of unstable alleles by factors other than the controlling elements themselves.

Temperature

Temperature is the only environmental variable shown to have a strong effect on the mutable systems. Rhoades (4) reported a negative relationship between temperature and the frequency of Dt-induced mutations in a_1—the higher the temperature the lower the frequency. This result is in line with the case of mutability in *Antirrhinum majus* described by Harrison & Fincham (50) and with a number of other instances in flowering plants briefly reviewed by the same authors (e.g. Sand, 51). Peterson (52) compared the frequencies of mutation of pg^m to Pg, giving sectors of dark green on pale green, in seedlings raised at 28°C and at 16°C and found a 5-

to 13-fold *greater* frequency at the higher temperature. Thus, although temperature may have a strong effect, there is no general rule about its direction. One may guess that this depends on the specific response to temperature of the particular segment of chromatin within which the controlling element is inserted. In *Antirrhinum,* too, different mutable systems respond differently to temperature (53, 54).

Effects of Heterochromatin

Williams & Brink (55) investigated the effect on mutation frequency of mutable R^{nj} (i.e. $\overline{R^{nj}\ Mp}$) to stable R^{nj} of the heterochromatic chromosome knob K10; the knob is linked to R at a distance of about 35 map units. They showed a significant increase in the mutations frequency when the mutable allele was in coupling with K10 but not when the knob was on the homologous chromosome in *trans* relationship to the mutable locus. The effect was not a spectacular one (twofold or less) but it appeared to be consistent, being stronger for earlier than for later-occurring mutations. Williams (24) found no strong effect of different numbers of heterochromatic B chromosomes on the mutation frequency of Ashman's wx^{m-1}.

CHANGES OF STATE OF CONTROLLING ELEMENTS

Perhaps the strangest of the properties of controlling elements is their tendency to undergo frequent changes in state. The changed state may be transmitted stably through many cell or sexual generations but may then change again. In extreme cases there may be a frequent and almost predictable oscillation between two more or less well-defined states, a phenomenon described by McClintock as a cyclical change of phase. The states may be distinguished by the frequency of mutational events in the responsive alleles or by the timing of the events or by both. In the case of *Spm,* which has the dual role of controlling both mutability and stable gene activity, the two functions may be altered differently by the changes in state. Some of the clearest examples are reviewed in the following paragraphs.

Different States of Ac

In a relatively early report McClintock (56) described sectoring behavior in kernels carrying *Ac* in single dose, together with *Ds* in the short arm of chromosome 9. These kernels mostly showed a fairly coarse pattern of variegation due to *Ds*-breaks occurring at different times during development, some of them early. However, in some crosses most of the kernels showed sectors in which the *Ds*-breaks were more or less delayed and reduced in frequency, giving various grades of fine variegation. The effect was as if the potency of *Ac* was increased in these sectors, since it was well established that the incidence of *Ds*-breaks and other *Ac*-induced events was reduced and delayed by extra copies of *Ac.* Support was lent to the interpretation of the sectors in terms of *Ac* variants of increased "strength" by the recovery of such variants in whole plants, so that their unit nature could be established and their dosage effects investigated. Thus a "weak" isolate of *Ac* was shown in some instances to give about the same effect in two doses as a "strong" isolate in a single dose.

These early investigations of change in state of Ac seemed to imply an almost continuous quantitative variation in the potency of the regulator function of Ac, but they were only reported briefly and have not been much emphasized in later work. The more recent reports of instability in Ac have concerned reversible ("cyclic") changes between fully active and almost or quite inactive states of the element (35, 57, 58). The study was facilitated by the availability of the allele wx^{m-7} which, as originally identified, had Ac closely associated with it. As long as the regulator function of the Ac element remained active, kernels carrying wx^{m-7} (either homozygous or heterozygous with stable wx) showed striking variegation with respect to amylose starch; this chould readily be shown by sectioning the kernel and staining with I_2-KI reagent. Some kernels, which from their parentage should have been of this type, showed, instead, a low level of amylose increasing regularly in amount from the center of the endosperm to the periphery. Within such kernels occasional sectors were observed that had regained the typical wx^{m-7} variegation. That this reversible loss of mutability was due to a reversible inactivation of Ac was shown by the simultaneous action of Ac on a responsive allele at the unlinked A_1 locus, a_1^{m-3}. Mutations of a_1^{m-3} resulted in clones of cells with anthocyanin pigmentation, which could be scored independently of the variegation with respect to starch type. In the kernels in which wx^{m-7} had become stable, there was also an absence of a_1^{m-3} mutations, while the endosperm sectors showing restored wx^{m-7} variegation were precisely overlaid by sectors of a_1^{m-3} spotting in the aleurone layer. The concerted behavior of the two mutable alleles showed that each was responding to changes in activity of the common controlling element.

Different States of Spm

The greater complexity of function of Spm as compared with Ac is reflected in the greater diversity of states of the former element. As its name implies, Spm has two functions, one the suppression or (in the one case of a_1^{m-2}) the promotion of the stable activity of the gene brought under its control, and the other the induction of mutational changes, most commonly to full gene activity. One particularly significant change in state of Spm involves the loss of the mutator function only, called by McClintock the loss of component 2 with retention of component 1 of Spm. McClintock (58) has a brief report of a change of this type in the Spm present at the A_1 locus in a_1^{m-2}. Losses of the mutator function may be reversible, and in her review of the same year McClintock (35) published a striking photograph of a kernel in which both c_2^{m-2} and wx^{m-8} showed null activity throughout the greater part of the endosperm, due to the presence of suppressor without mutator activity, but had both recovered mutability simultaneously in a well-defined sector in which component 2 of Spm had apparently become reactivated. A similar demonstration of concurrent control of two mutable alleles (in this case a_2^{m-1} and wx^{m-8}) by the same cyclically changing Spm element had been mentioned in an earlier report (34).

In some changes of state of Spm, the mutator activity was only partly lost, and a few mutations were still induced very late in endosperm development. McClintock called this state Spm^w (w for weak), and she regarded it as being altered in component 2 (the mutator component) only (58, 59). However, in earlier references to

Spm^w, McClintock (49) describes this state as also weakened with respect to suppression of gene action, implying that component 1 is affected in some degree as well. At all events it seems clear that some of the changes of state of *Spm* can virtually eliminate the mutator activity while leaving the suppressor function more or less intact.

MOD, A TRANSPOSABLE MODIFIER OF *SPM* ACTION   Following her description of the "weak" states of *Spm,* McClintock (47) reported the discovery of a dominant factor, called *Modifier (Mod),* which was able to restore full mutator activity when in combination with a weak *Spm.* A later review (35), illustrated by photographs, reported that *Mod* was fully complementary to *Spm* derivatives which, by themselves, had no mutator activity at all (inactive component 2). *Mod* had no effect in the absence of *Spm,* nor did it enhance the effect of strong *Spm*'s. It is perhaps best regarded as itself a derivative of *Spm* with component 2 activity only; this interpretation is consistent with McClintock's view (35) that component 2 activity depends on the presence of an active component 1. No dosage effect of *Mod* has been described, and it seems that it is completely dominant.

When first discovered, in a single densely spotted kernel on a plant carrying a_1^{m-1} and a weak state of *Spm, Mod* was linked to *Wx* and unlinked to the *Spm.* In the course of further experiments designed to confirm the monofactorial basis of the enhanced mutability, a few plants, or parts of plants, were found in which *Mod* had been transposed to new positions; either a second independently located *Mod* element was present in addition to the one linked to *Wx,* or a single *Mod* was unlinked to *Wx.* Thus *Mod* shares the transposability of *Spm,* its putative progenitor. It may be noted here that M^{st}, a modifier of the mutator element in R^{st} (48), complements this element in much the same way as *Mod* complements *Spm,* and is also transposable.

CYCLIC PHASE CHANGES OF *SPM*   In the absence of active *Spm,* a_2^{m-1} produces a moderately deep grade of aleurone pigmentation, while in the presence of the active controlling element it produces no pigment except in scattered dark mutant spots. When phase changes in *Spm* cause both suppressor and mutator functions to vary simultaneously, variegation patterns due to changes in *Spm* itself may be confused with those due to changes in the susceptible allele. McClintock was able to study the frequency and timing of inactivations of initially active *Spm* in the absence of change in the responsive allele by making use of a derived state of a_2^{m-1} (49, discussed again below) which was stable to the mutator action of *Spm* but still susceptible to its suppressor activity. The particular state of *Spm* present with this special derivative of a_2^{m-1} was one that showed frequent inactivations during endosperm development, releasing the activity of a_2^{m-1} in clones of aleurone cells that appeared as pigmented sectors. That these sectors were due to change in *Spm* and not to mutation of a_2^{m-1} was shown by the simultaneous release of activity of wx^{m-8} in plants also homozygous for this second *Spm*-controlled allele (60). In her 1971 report McClintock (60) documents much more fully an effect, mentioned several years previously (47), of increasing dosage of *Spm* in delaying its own loss of activity. Thus, a plant homozygous for the special state of a_2^{m-1}, and with one

Spm, showed many pigmented aleurone sectors, some of them large; corresponding plants with two *Spm*'s had only rather few small endosperm spots, while the presence of a third *Spm* almost entirely eliminated the spotting. The straightforward explanation of this dosage effect would be that, where there are two or three *Spm*'s, they both, or all three, need to be inactivated independently before the activity of the suppressed allele can be released. There seems to be more to the phenomenon than this, however, since McClintock (60, 61) observed that an inactive *Spm* together with an active one gave the same dosage effect as two active *Spm*'s, even though the inactive element still appeared to be inactive when segregated from its active partner following meiosis. McClintock explains this contribution of an inactive *Spm* as due to its being temporarily activated by its association with an active *Spm,* and this seems to imply a third regulator function, supplying positive signals for the maintenance of component-1 (suppressor) activity.

Inactive *Spm* elements may reveal their presence not only by contributing to the dosage effect just discussed but also by occasionally returning to activity [see Fig. 4A in McClintock (60)]. There is evidence, which unfortunately has never been fully presented, that the frequency with which an *Spm* undergoes the cycle of changes from active to inactive and back again is governed by the element itself. McClintock (34) described a case in which a plant carried two *Spm* elements, one closely linked to *Wx* and the other close to the recessive allele *wx.* The former was in a prolonged state of inactivity; the latter was active but in immediately preceding generations had been observed to undergo rather frequent inactivation, both in plant and kernel. The two differing states of *Spm,* marked by the *Wx* alleles, segregated from each other at meiosis. The implication seems to be that the *Spm* in the prolonged phase of inactivity was similarly stable in the active state following rare reactivation, in other words, that this state was characterized by a slow cycle of activation/inactivation and the other by a relatively rapid cycle, but one would like to see more evidence on this point. It should be noted that these two states of *Spm* arose from a single ancestral element.

PHASE CHANGES CORRELATED WITH DEVELOPMENT   While the evidence that changes of phase of *Spm* are controlled by some kind of biological clock inherent in the element itself is more tantalizing than satisfying, it is very clear that they can be closely tied to plant development. The most striking and fully described example is due to Peterson (62, 63). He observed two different derivatives of his mutable allele a_1^m, in which *En* (= *Spm*) is closely associated with the A_1 locus. In one, called $a_1^{m(crown)}$, the pigmented spots due to clones of A_1 mutant aleurone cells were concentrated in the crown of the kernel (i.e. the area diametrically opposite the insertion of the kernel on the cob). In the other, $a_1^{m(flow)}$, exactly the opposite situation prevailed, there being no spots on the crown but a considerable number round the waist of the kernel. The best evidence that these two strikingly different patterns were due to controlled variation in the activity of *Spm* was obtained by combining $a_1^{m(crown)}$ and $a_1^{m(flow)}$ respectively in heterozygotes with McClintock's a_1^{m-1} which, as already mentioned, gives an intermediate grade of uniform pigmentation in the absence of *Spm* and dark mutant spots on a colorless background in its presence. In the *crown* heterozygote, the crown area of the kernel showed the mutational spots on a colorless background due, presumably, to mutations both of a_1^m and a_1^{m-1}, while the aleurone away from the crown had the uniform pigmentation characteristic of a_1^{m-1} in the absence of *Spm*. In the *flow* heterozygote, the crown area was uniformly pigmented and the rest of the aleurone showed the mutational spots on the colorless background. These effects, exercised in the *trans* configuration on a_1^{m-1}, a well-characterized test allele for *Spm* activity, make it certain that the *crown* and *flow* phenotypes are due to controlled changes in the activity of *Spm* according to the position of the cells or the time of their differentiation. The endosperm of the crown of the kernel is thought to be differentiated at a different time from the remainder, and so the *crown* and *flow* states of *Spm* (*En*) could be regarded as having been programmed to become active at these different times during development. It should be noted that in this example, as in the cyclical variation in *Spm* activity investigated by McClintock, both components of action of *Spm* vary together.

VARIED RESPONSES OF THE MUTABLE ALLELES

Mutations induced by controlling elements are seen as sectors in the endosperm or other parts of the kernel or plant. Mutation is seen also, at least in most cases, in occasional whole kernels which, when germinated and grown, transmit a mutant allele to a whole plant and hence to further generations. Mutation frequencies in germinal cells vary greatly from one case to another. At one extreme, Peterson's (64) a_1^m derivative $a_1^{m(pu-pa)}$, referred to again below, gave rise to stable mutants in the germ cells with a frequency of 50% or more, most of the mutants being colorless or pale rather than fully colored. Most other mutable alleles seem to give much lower frequencies, of the order of 0.1 to 1% for wx^{m-9} (59), a_1^{m-1} (1, 48, 49), and Neuffer's Dt-controlled a_1^m (65). Such estimates can only be very approximate and subject to sporadic high values because of the "jackpot" effect when a mutation in the sporogenous cell layer happens to occur early in plant development.

Many, perhaps most, mutable alleles under *Ac* or *Spm* control mutate not only to full gene expression but also to a graded series of stable alleles of lower expression. For example, Peterson's $a_1^{m(pa-pu)}$ (64) mutates constantly to stable colorless and to one grade of pale purple. In one case, the *Ac*-controlled c_1^{m-1}, McClintock (56) reported that only mutations to full C_1 expression were observed, and it is probably true that for most mutable alleles full release of gene activity is the commonest consequence of mutation.

Mutability of Mutability

Mutable alleles often show abrupt changes in the nature, frequency, and timing of the mutations they undergo. Peterson's mutable a_1^m, formed through the association of *En* (= *Spm*) with A_1, started its career with a simple pattern of late mutation back to A_1, giving small densely pigmented aleurone spots in a colorless background. Subsequently a derivative $a_1^{m(dense)}$ that gave a denser pattern of spots was isolated, and from this in turn there arose $a_1^{m(pa-pu)}$, which as mentioned above gave extremely frequent mutation, sometimes quite early in development, to stable color-less and to pale purple as well as to full pigmentation [see Figure 1 in Peterson (64)].

Since, in Peterson's example, the controlling element was at the locus controlled, it was not clear whether the changes in mutational pattern should be attributed to the receptor or to the regulator component of the system, though it would be possible to check the regulator for altered activity on other responsive alleles. Analogy with McClintock's $a_1{}^{m-1}$, another Spm-controlled allele without the regulator at its own locus, suggests that it was the receptor component that was varying. Since its first isolation, $a_1{}^{m-1}$ has given rise to numerous derived states differing from one another not only in the frequency and timing of mutation but also in the level of stable aleurone pigmentation that each gives in the absence of Spm. The original state gave a fairly dark grade of pigment in the absence of Spm and many fully colored spots and sectors on an unpigmented background in its presence. The various derived states showed a wide range of autonomous stable activities, from practically no pigment to an almost maximal level, and, independent of this varia-tion, a wide range of mutation frequencies in the presence of Spm, from a few spots per kernel to almost confluent spotting (35, 66, 67).

McClintock's second Spm-controlled mutable allele at the A_1 locus, $a_1{}^{m-2}$, varied even more extravagantly. Of its derivatives, McClintock states, "The patterns are so varied that they defy a meaningful classification." Even in its original state $a_1{}^{m-2}$ poses a special problem since, uniquely among known Spm-controlled alleles, its basal level of activity before mutation is *enhanced* rather than suppressed by the presence of Spm (photographic evidence in reference 57, 67). Most studies have been made on states of $a_1{}^{m-2}$ in which Spm is closely associated with the A_1 locus, so again we have the difficulty of distinguishing between variation in the receptor and regula-tor elements of the system. It appears that this necessary distinction was, in fact, made using the nonautonomous allele $a_1{}^{m-1}$ to detect variation in the regulator. McClintock's analysis (67) indicated that both receptor and regulator were undergo-ing mutation; in some $a_1{}^{m-2}$ derivatives, regulator function had been reduced or lost (sometimes component 2 was inactive while component 1 remained active), while in others it was the response to the regulator that was changed. Among the latter cases, some had lost their responsiveness to component 2 (the mutator function) while continuing to respond to component 1, while in others the reverse was the case (67). These observations, if correctly interpreted, point strongly to the existence of two functionally independent receptors of signals in the responsive allele. We have already referred (p. 36) to the special state of $a_2{}^{m-1}$ which had become stable to mutation induced by Spm component 1 while continuing to be suppressed by component 2.

Perhaps the most interesting of the derived states of $a_1{}^{m-2}$ was one (state 8004) in which mutation appeared to occur from one nonpigmented phenotype to another, also nonpigmented but capable of complementing the first to produce pigment. Each variant sector formed by this state of the allele was delimited by a pigmented rim of cells in which, it appeared, some pigment precursor was diffusing into the sector and there being converted to anthocyanin. It was as if there were two adjacent or at least closely linked genes of complementary action and that one was being "switched on" as the other was "switched off." The rimmed pattern has been well illustrated by published photographs (35, 67). McClintock attributes the pattern to a change in phase of activity of the regulator, which could be monitored by means of the tester allele wx^{m-8}. The complementation effect could be explained if state 8004 represents a short-range transposition of Spm to a closely linked complementary cistron with its retention, in one segregating product, at the original site. If the receptor at the new site now responded to the Spm regulator signal in the more usual negative way, while that remaining at the original site continued to show its unique positive response to the regulator, then inactivation of component 1 of the regulator would result in repression of the second cistron accompanied by release of activity of the first. Further rimmed areas seen within the primary ones could well be due to cyclic reactivations of Spm. It is interesting to note that a somewhat similar situation, with a mutable allele giving functionally different mutant derivatives capable of cross-feeding, was noted by McClintock (10) many years earlier in the case of the Ac-controlled c_1^{m-2}. This earlier case does not seem to have been followed up.

So far as the published reports allow one to say, mutations affecting mutability occur only in the presence of the active regulator component of the appropriate controlling element; in the absence of the regulator, the potentially mutable allele neither mutates nor changes in its potential for mutation.

"Presetting" Effects

The phenomenon described by McClintock as "presetting" was first observed by her during studies on a_1^{m-2}. In its original state, closely associated with Spm, this allele gave a medium-high level of pigmentation with many mutant sectors of darker or lighter pigment. The derivatives of a_1^{m-2} in which the regulator of Spm had been either inactivated or removed by transposition gave the same phenotype as the original state if Spm was present elsewhere in the nucleus, but normally produced no color in its absence (43). However, one derivative of a_1^{m-2} continued to give a more or less pronounced pattern of pigment in the aleurone even after Spm had been removed by meiotic segregation (57, 59). These patterns, some of which are illustrated in McClintock (57), consisted of patches of color ranging, in different kernels, from pale and indistinct to fairly dark, on either a colorless or a pale background. The patches tended to be rather uniform in size, as if they were due to switching on of pigment production at rather precise times; different kernels on the same ear differed from one another in the number and intensity of the colored patches, as if there were many different ways in which the system could be preset to show A_1 activity after Spm had been segregated away. Plants grown from the kernels showing these patterns gave no evidence for the presence of Spm when crossed to Spm tester stocks. The "preset" pattern was, in general, not transmitted to a further generation (57). In one line, however, in which another derived state of a_1^{m-2} was involved, the preset variegation was transmitted with low frequency through two generations of crossing to plants homozygous for a_1 and lacking Spm (41). Even here, a_1^{m-2} reverted to its normal near-colorless expression in the absence of Spm

in the great majority of the cells in each generation, but the fact that the preset pattern was transmitted at all raises difficult questions.

One might wonder whether a maternal effect could account for these observations, with a carry-over of the *Spm* regulator product(s) from maternal plant to embryo sac. If this were the explanation one might expect a stronger presetting when *Spm* was present in the maternal rather than in the pollen parent. Whether there was any such difference is not clear, but at all events some transmission of presetting through the pollen is stated to occur (35, 57). It is difficult to avoid the conclusion that the "preset" state of the allele maintains itself through some kind of positive feedback even if not by self-replication based on DNA.

A feature of the presetting phenomenon that strongly supports McClintock's own view that it is due to special states of $a_1{}^{m-2}$ receptor, rather than to a slow diluting out of the *Spm* regulator signal, is the variety of patterns shown by different "preset" kernels. These are very reminiscent of the different "states" of mutable alleles reviewed in the preceding section, and they certainly strongly suggest that a specific pattern of activity of $a_1{}^{m-2}$ can be transmitted through mitotic divisions throughout the development of the endosperm. Rather than attributing these metastable states to presetting of the receptor, however, it seems to us equally reasonable to suppose that they represent transient reactivations of an almost silenced *Spm* regulator element still residing at the A_1 locus in the derived state of $a_1{}^{m-2}$. There is, in fact, nothing to show that the regulator was physically removed from $a_1{}^{m-2}$ when it changed from an autonomous to a two-element system. It may merely have lapsed into a prolonged phase of inactivity from which it may be stimulated to a state of precariously self-perpetuating activity as a result of the presence in the same nucleus of an active *Spm*. If, as suggested above (p. 36), *Spm* has a function concerned with maintenance of its own activity, this view of the presetting effect seems quite consistent.

In a later report, McClintock (36) described what she regarded as another case of presetting, this time involving the *Spm*-controlled $c_2{}^{m-1}$ and $c_2{}^{m-2}$ alleles. Most of the results reported concern $c_2{}^{m-2}$. In the presence of other appropriate dominant alleles for pigmentation (A_1, A_2, *Pl*, *B*), C_2 permits the development of strong anthocyanin in the vegetative and floral parts of the plant, including the cob and kernel, while c_2 largely blocks this pigmentation. In the presence of *Spm*, $c_2{}^{m-2}$, a nonautonomous allele, mutates to C_2 to give deeply pigmented endosperm spots, pigmented streaks in the pericarp (if P^{rr} is present), and red patches and sectors on the cob. Thus $c_2{}^{m-2}$ affords a good opportunity of observing changes in gene action in many parts of the plant. Two types of observation were interpreted by McClintock in terms of presetting. First, in some plants, active *Spm* appeared, as judged by test-crosses, to have been lost either from a whole ear or a large sector of it; nevertheless, the sectors sometimes contained patches of pigmented tissue in cob and pericarp as if $c_2{}^{m-2}$ was still undergoing sporadic release of pigmenting activity. This effect seems quite comparable to the $a_1{}^{m-2}$ presetting discussed above. Second, some ears borne on plants carrying both $c_2{}^{m-2}$ and *Spm* showed large pigmented cob sectors with all the kernels on these sectors exhibiting the original $c_2{}^{m-2}$ endosperm phenotype of small dark spots on a colorless background. Thus an early mutational event having the effect, as far as cob color is concerned, of a mutation to C_2, did not appear in the endosperm of the kernels that would have been expected to have descended from the same mutant clone. Possible explanations of this observation do not come readily to mind; much depends on how certain one can be of the ontogenetic relationship between the germinal cells and the adjacent floral tissue.

RELATION OF CONTROLLING ELEMENTS TO PARAMUTATION

Paramutation was reviewed comprehensively in the last volume of this series (68). The connection between this subject and that of the present review is twofold: Brink has made seminal contributions to both, and certain paramutagenic alleles show patterns of instability similar to those associated with controlling elements. However, it has been clearly shown (31, 69, 70) that the paramutagenicity of *R-stippled* (R^{st}) is due to an element separable by crossing-over from that responsible for the stippling pattern. It seems best not to assume that the two phenomena have any direct connection.

COMPARABLE SYSTEMS IN OTHER EUKARYOTES

Mutable systems that have features in common with maize controlling elements and that may have a similar basis have been reported in a number of other flowering plants. Since these examples hardly extend the range of phenomena that have been studied, usually much more thoroughly, in maize, we do not consider them individually here. It must suffice to mention the studies on *Antirrhinum majus* (50, 54, 71, 72), on *Impatiens balsamina* (Sastry, in preparation), on *Glycine max* (73), on interspecific hybrids in *Nicotiana* (51, 74–77), and on *Nicotiana tabacum* (78). The earlier literature was reviewed by Demerec (79).

The fullest parallel to maize controlling elements is provided, in fact, not by another plant but by *Drosophila melanogaster.* Green (80) has recently published a good brief review of *Drosophila* transposing elements that cause high mutability; all the cases so far studied involve the *white* locus. It seems that something very like the maize phenomenon is at work here. For further information the reader is referred to Green (80–83), Gethmann (84), Kalisch & Becker (85), and Judd (86). As compared with maize, *Drosophila* has the advantage that the polytene chromosomes permit the identification of some of the small chromosomal deletions and rearrangements associated with the mutational events.

POSSIBLE MECHANISMS—PARALLELS IN PROKARYOTES

Until a few years ago maize controlling elements appeared completely outlandish and without parallel in the rest of biology. Now, however, thanks to bacterial and bacteriophage geneticists, an abundance of analogies can be found in prokaryote systems, and there is hardly a property of a controlling element that does not find a parallel in either a real or a plausibly extrapolated provirus, plasmid, or transposable inserted sequence of DNA. In this section we briefly review some of the relevant prokaryote systems, not because any real mechanistic similarity has in any case been actually demonstrated, but rather because the analogies may serve as an aid to thought and a stimulus to experiment.

The possible analogy between maize controlling elements and bacterial episomes has been realized for some years (87) and has more recently been developed in some detail by Peterson (88) with particular emphasis on bacterial prophages. Many, if not all of the features of controlling elements can be described by a model making use only of known properties of bacteriophages. The most interesting phage from this point of view is *Mu-1* of *E. coli*, which has recently been the object of intensive study (89).

The recently described transposable inserted sequences (IS) in *E. coli* DNA (90), with their very small size (barely sufficient for one gene), seem likely to be simpler than maize controlling elements, but they are not associated with virus production and, in this respect provide an even closer analogy.

Integration and Inhibition of Gene Action

E. coli Mu phage and the IS elements can integrate anywhere in the chromosome and probably, indeed, at any site within a gene in either orientation (91, 92). When inserted within a gene, they greatly reduce or entirely eliminate mRNA transcription from the operator-distal side of the insertion (93, 94) and so exert a strong polar effect within an operon. Similar properties on the part of the maize controlling elements could account for their ability to block the activity of any gene, though the fact that the receptor element of *Spm* often, in the absence of the regulator, permits some level of gene activity would suggest that the insertion of this element is more often in an adjacent sequence exercising quantitative control of gene activity rather than within the structural gene itself.

Removal of the Blocking Element and the Role of the Regulator

Both *Mu* phage and the IS elements can be excised from the *E. coli* chromosome either cleanly or inexactly; in the latter case they leave behind deletions of various extents. Analogous behavior on the part of controlling elements would account for mutation either to full gene activity or to one of a series of quantitatively graded alleles. The latter consequence would be particularly likely if the deletions generated by inexact excision were in a segment with a regulatory function or in a sequence of repeated gene copies. It seems likely that *Mu*, like λ, integrates into and is excised from the chromosome by virtue of its own specific recombination function(s). If we suppose that controlling elements also control their own excision we can understand both autonomous mutability, in the case where a complete element resides at the locus under control, and two-element systems, where the element at the locus is deficient in the excision function and needs to be complemented by another element of the same class elsewhere in the nucleus. The dosage effects on excision frequency of *Ac* and *Dt* could be due to regulation, either positive or negative, of the element's own excision function by a protein coded for by another gene of the element. Parallels for almost any kind of interaction could be found in the regulatory circuits of λ bacteriophage (95).

Regulation of Gene Activity

Whereas *Ac* seems usually, though not always, to block gene activity completely, *Spm* often brings the affected gene under the quantitative control of the *Spm* regulator, capable of acting in *trans*. Perhaps this aspect of *Spm* can be best understood as the replacement of the promoter of transcription normal to the maize gene by a promoter in the inserted element. The latter could well be controlled by a regulator protein or proteins specific for the particular receptor element and, just as for control of excision, one would expect a complete element to act autonomously and a regulator-deficient one to need complementing by another *Spm*. It may well be, of course, that regulation of gene activity by *cis*-acting elements (such as the *Spm* receptor) can occur in a number of different ways in eukaryotes. One can imagine effects on the stability of DNA-histone complexes or on chromatin packing, and signals acting in *trans* on these levels of chromosome organization as well as others acting in the known manner of bacterial repressor proteins on exposed DNA.

Changes in State of Controlling Elements

It seems likely that the changes in state of the receptor of controlling element signals, classically exemplified by McClintock's a_1^{m-1}, are a rather different phenomenon from the changes of state or phase of the regulator transmitting the signals. The former changes may be due to rather frequent very short-range transpositions resulting in shifts in the position of the inserted element within a single locus. Depending on its exact position within the locus the effects of an inserted element both on gene activity and on its own frequency of excision could well vary. Furthermore, in the course of such short-range transpositions, small deletions or rearrangements in the normal structure of the locus may occur, making reversion to the pristine state impossible even if the element is excised.

Changes in phase of the regulator function(s), often of a long-term and inherited character but ultimately reversible, are an intriguing problem. Even here there may be an analogy in bacteriophage *Mu*. One of the more novel features of *Mu* is the existence of a sequence near the end of its DNA that is subject to controlled inversion (96). This inversion seems to depend on a function of the integrated prophage inasmuch as it does not occur when the phage is propagated lytically. It apparently occurs through crossing-over within a reverse duplication (97). Whatever the significance of this curious feature for the bacteriophage, a controlled rearrangement of this type, especially if controlled by a prophage gene itself subject to regulation, can provide a model for heritable states of activity or inactivity that could apply to controlling elements.

Two-Element Versus Autonomous Systems

As we have seen, *Spm* and *Ac* both tend to lose their regulator functions leaving a receptor element at the locus controlled, capable of responding to a distant regulator. It is interesting to note that Brink's *Mp*, in other respects apparently identical to *Ac*, has never been reported to behave in this way; it is not clear whether this apparent difference is inherent in the elements themselves or rather reflects a difference in genetic background. The proneness to dissociate into functionally distinct moieties recalls the behavior of certain plasmids (resistance transfer factors) of enteric bacteria that constantly dissociate in one host while remaining in one piece in another (98).

The Complexity of Controlling Elements

In the course of our speculations, based, we believe, on plausible analogies with prokaryote systems, we have postulated a considerable number of distinct and specific functions of controlling elements. In the case of *Spm* we need an excision function, a promoter of the activity of the adjacent gene (possibly but not necessarily acting directly on transcription), very possibly a function bringing about controlled structural changes within *Spm* itself, and *trans*-regulation of all these functions probably mediated by several distinct signals acting both positively and negatively. A detailed complementation analysis of a sufficiently large collection of *Spm* variants might reveal how many regulator functions this element possesses. Altogether, *Spm* could well contain as many genes as a small virus. *Ac* may be somewhat less complex, but this is not certain. When the sensitivity of the techniques available for the analysis of eukaryote DNA is sufficient to justify the search it will be worthwhile to look for excisable, perhaps circular, DNA elements of at least 5000 base pairs and perhaps ten times that size, the presence of which might correlate with the presence of controlling elements. Unfortunately, even with several copies per nucleus, such elements need only constitute a minute fraction of the DNA present.

NORMAL COMPONENTS OR INVADERS?

The comparisons that we have been concerned to make in the last section between maize controlling elements and mutagenic phage need not be taken as implying that the former are of viral origin or without function for the maize plant. It does indeed seem likely that the maize elements and mutagenic phage may have something in common as regards *mechanism* of transposition and of interference with gene action, but the questions of origin and present function are logically independent of mechanism. For many years McClintock has (7, 99) argued persuasively for an important role for controlling elements in the regulation of normal gene activity. In her view, the events that release the activity of mutable alleles in a relatively uncontrolled and sporadic way are of a similar kind to normal events which, at precisely controlled times and places during development, switch on normal gene activity. It is certainly possible that, with their known high degree of versatility, controlling elements could become involved in the evolution of patterns of genuinely adaptive kinds, and it might even be considered rather surprising if natural selection were never to take advantage of such opportunities. This argument is independent of any particular view about the origins of controlling elements. There is no reason why something that started as a virus could not first be reduced to a relatively harmless provirus, lose its capacity for making infective particles, and ultimately become enrolled in the apparatus of normal control. However, according to this scenario, controlling elements ought only to play a rather peripheral role in regulation. Whether they are, in fact, involved in essential developmental processes is a question that cannot be answered at the present time, but the balance of argument seems rather against the idea.

There is nothing inherently implausible in a hypothesis involving controlled transpositions of DNA segments as a basis for at least some kinds of cellular differentiation. Indeed, this notion has achieved a good measure of respectability as a result of recent developments in immunogenetics of mammals (100). In the genesis of specific immunoglobulin chains, the fusion of initially separate nucleic acid (probably DNA) sequences coding respectively for the "constant" and "variable" regions of both heavy and light polypeptide chains seems inescapable. The mechanism involved could be essentially the same as for short-range transpositions of maize controlling elements, but with a specificity as regards site of reintegration that the maize elements lack.

It is, indeed, their lack of specificity for any particular chromosome site that casts the most doubt on controlling elements as normal regulatory components. As we have seen, controlling elements have a strong tendency to move short distances, but longer range transpositions can also occur and, even within a short range, one finds reintegration at many different sites. It is hard to reconcile this nearly random movement with a normal regulatory role unless one supposes that the elements have in some way lost their sense of direction through being removed from their normal milieu. If they are representative of normal components they must be quite aberrant either in their inherent properties or in their chromosome position and exposure to normal controls.

Again, if controlling elements were normal components they should be present in all plants. The apparently sporadic occurrence of Ac, Spm, and Dt may perhaps be misleading since it is possible that they are present universally but often "buried" in heterochromatin or in sequences that do not achieve expression in the tissues one is normally looking at. That this may be so is suggested by the success of McClintock (6, 7) and Doerschug (12) in evoking Dt activity in plants, in which it had not been known to be present, by assaults on the chromosomes through the breakage-fusion-bridge cycle. It is conceivable that chromosome breakage alters patterns of chromatin condensation so as to make available for transcription DNA sequences which are normally masked. Of course, if controlling elements were permanently buried throughout development they could play no part in regulation, but it is possible to postulate, without much risk of disproof, that they normally are transcribed, but only in some tissue or at some stage of development in which the activity of the available tester alleles cannot in any case be expressed. The argument can be maintained, but it shows signs of strain.

The hypothesis that controlling elements are proviruses, defective in the sense of being unable to form infective particles, would carry more weight if plant virus genomes (or their DNA transcripts) were known to be capable of integration into the genome. We know of only one study pointing in this direction, but that is a very suggestive one. Sprague & McKinney (101) have provided evidence of an element, mappable at a chromosome locus but transposable to other loci, that causes aberrant Mendelian ratios and that apparently turns up repeatedly following infection of maize plants with barley stripe mosaic virus. This work, if extended and confirmed, could provide a strong hint as to the origin of controlling elements. On the other hand, one should not lose sight of the possibility that a chromosomally integrated provirus-like element might evolve from the plant's own DNA. In order for such an element to evolve on its own account, rather than for the benefit of the host plant, it would be necessary for it to be able to multiply more rapidly than the rest of the genome. We have seen, from studies of Mp ($= Ac$) by Brink and his colleagues, how a controlling element can undergo extra replications through being transposed from an already replicated to an unreplicated locus. Thus a plant receiving the element from only one parent could, in the extreme case, transmit it to all of its progeny. This differential multiplication could provide sufficient basis for independent evolution of the element. In a domesticated and artificially protected species like maize, a cryptic chromosomal parasite doing only minor damage and occasionally producing attractive patterns would, moreover, bring little disadvantage to the host plant.

Even if the questions of origin and normal function (if any) must for the present remain unanswered, there is no doubt that maize controlling elements have great potential importance as a means of probing genetic structure. The selection, mapping, and phenotypic (including biochemical) characterization of short-range transpositions within a genetic region may well provide the best information on the organization of maize loci including (and perhaps most important) the relation of structure-determining and regulatory sequences.

Literature Cited

1. McClintock, B. 1956. Controlling elements and the gene. *Cold Spring Harbor Symp. Quant. Biol.* 21:197–216
2. McClintock, B. 1961. Some parallels between gene control systems in maize and in bacteria. *Am. Natur.* 95:265–77
3. Rhoades, M. M. 1938. Effect of the *Dt* gene in the mutability of the a_1 allele in maize. *Genetics* 23:377–97
4. Rhoades, M. M. 1941. The genetic control of mutability in maize. *Cold Spring Harbor Symp. Quant. Biol.* 9:138–44
5. McClintock, B. 1950. Mutable loci in maize. *Carnegie Inst. Washington Yearbook* 49:157–67
6. McClintock, B. 1950. The origin and behavior of mutable loci in maize. *Proc. Nat. Acad. Sci. USA* 36:344–55
7. McClintock, B. 1951. Chromosome organization and gene expression. *Cold Spring Harbor Symp. Quant. Biol.* 16:13–47
8. Emerson, R. A. 1917. Genetical studies on variegated pericarp in maize. *Genetics* 2:1–35
9. Barclay, P. C., Brink, R. A. 1954. The relation between modulator and activator in maize. *Proc. Nat. Acad. Sci. USA* 40:1118–26
10. McClintock, B. 1949. Mutable loci in maize. *Carnegie Inst. Washington Yearbook* 48:157–67
11. McClintock, B. 1951. Mutable loci in maize. *Carnegie Inst. Washington Yearbook* 50:174–81
12. Doerschug, E. B. 1973. Studies of *Dotted*, a regulatory element in maize. I. Inductions of *Dotted* by chromosome breaks. II. Phase variation of *Dotted*. *Theor. Appl. Genet.* 43:182–89
13. McClintock, B. 1953. Mutation in maize. *Carnegie Inst. Washington Yearbook* 52:227–37
14. McClintock, B. 1954. Mutations in maize and chromosomal aberrations in *Neurospora*. *Carnegie Inst. Washington Yearbook* 53:254–60
15. Peterson, P. A. 1953. A mutable *pale-green* locus in maize. *Genetics* 38:682–83
16. Peterson, P. A. 1965. A relationship between the *Spm* and *En* control systems in maize. *Am. Natur.* 99:391–98
17. Neuffer, M. G. 1966. Stability of the suppressor element in two mutator systems at the A_1 locus in maize. *Genetics* 53:541–49
18. Mouli, C., Notani, N. K. 1970. Absence of a detectable change in *Ds* at the locus

in maize following mutagenic treatments. *Can. J. Genet. Cytol.* 12:436–42

19. Peterson, P. A. 1960. A *pale green* mutable system in maize. *Genetics* 45:115–33
20. Peterson, P. A. 1961. Mutable a_1 of the *En* system in maize. *Genetics* 46:759–71
21. Peterson, P. A. 1956. An a_1 mutable arising in pg^m stocks. *Maize Co-op. News L.* 30:82
22. Peterson, P. A. 1968. The origin of a new unstable locus in maize. *Genetics* 59:391–98
23. Brink, R. A., Williams, E. 1973. Mutable *R-Navajo* alleles of cyclic origin in maize. *Genetics* 73:273–96
24. Williams, E. 1972. The effects of sexual differentiation and B-chromosomes on the rate of transposition of modulator from the *Wx* locus in maize. *Maize Co-op. News L.* 46:189–93
25. McClintock, B. 1953. Induction of instability of selected loci in maize. *Genetics* 38:579–99
26. Peterson, P. A. 1963. Influence of mutable genes on induction of instability in maize. *Proc. Iowa Acad. Sci.* 70:129–34
27. Orton, E. R. 1966. Frequency of reconstitution of the variegated pericarp allele in maize. *Genetics* 53:17–25
28. Schwartz, D. 1960. Electrophoretic and immunochemical studies with endosperm proteins of maize mutants. *Genetics* 45:1419–27
29. Neuffer, M. G. 1965. Crossing over in heterozygotes carrying different mutable alleles at the A_1 locus in maize. *Genetics* 52:521–28
30. Kermicle, J. L. 1970. Somatic and meiotic instability of *R-stippled*, an aleurone spotting factor in maize. *Genetics* 64:247–58
31. Kermicle, J. L. 1973. Organization of paramutational components of the *R* locus in maize. In *Basic Mechanisms in Plant Morphogenesis. Brookhaven Symp. Biol.* 25:In press
32. Nelson, O. E. 1968. The *waxy* locus in maize. II. The location of the controlling element alleles. *Genetics* 60:475–91
33. Akatsuka, T., Nelson, O. E. 1966. Starch granule-bound adenosine diphosphate glucose-starch glucosyltransferase of maize seeds. *J. Biol. Chem.* 241:2280–86
34. McClintock, B. 1961. Further studies of the suppressor-mutator system of control of gene action in maize. *Carnegie Inst. Washington Yearbook* 60:469–76
35. McClintock, B. 1965. The control of gene action in maize. *Brookhaven Symp. Biol.* 18:162–84
36. McClintock, B. 1967. Regulation of gene expression by controlling elements in maize. *Carnegie Inst. Washington Yearbook* 65:568–78
37. Brink, R. A., Nilan, R. A. 1952. The relation between light variegated and medium variegated pericarp in maize. *Genetics* 37:519–44
38. Wood, D. R., Brink, R. A. 1956. Frequency of somatic mutation to self color in maize plants homozygous and heterozygous for variegated pericarp. *Proc. Nat. Acad. Sci. USA* 42:514–19
39. Brink, R. A. 1958. A stable somatic mutation to colorless from variegated pericarp in maize. *Genetics* 43:435–47
40. McClintock, B. 1952. Mutable loci in maize. *Carnegie Inst. Washington Yearbook* 51:212–19
41. McClintock, B. 1956. Mutations in maize. *Carnegie Inst. Washington Yearbook* 55:323–32
42. van Schaik, N. W., Brink, R. A. 1959. Transpositions of modulator, a component of the variegated pericarp allele in maize. *Genetics* 44:725–38
43. McClintock, B. 1962. Topographical relations between elements of control systems in maize. *Carnegie Inst. Washington Yearbook* 61:448–61
44. Greenblatt, I. M., Brink, R. A. 1962. Twin mutations in medium variegated pericarp in maize. *Genetics* 47:489–501
45. Greenblatt, I. M. 1968. The mechanism of modulator transposition in maize. *Genetics* 58:585–97
46. Orton, E. R., Brink, R. A. 1966. Reconstitution of the variegated pericarp allele in maize by transposition of modulator back to the *P* locus. *Genetics* 53:7–16
47. McClintock, B. 1958. The suppressor-mutator system of control of gene action in maize. *Carnegie Inst. Washington Yearbook* 57:415–29
48. Ashman, R. B. 1960. Stippled aleurone in maize. *Genetics* 45:19–34
49. McClintock, B. 1957. Genetic and cytological studies of maize. *Carnegie Inst. Washington Yearbook* 56:393–401
50. Harrison, B. J., Fincham, J. R. S. 1964. Instability at the *pal* locus in *Antirrhinum majus*. 1. Effects of environment on frequencies of somatic and germinal mutation. *Heredity* 19:237–58
51. Sand, S. A. 1971. A mutable allele at the *E* locus in *Nicotiana*. *Genetics* 67:61–73
52. Peterson, P. A. 1958. The effect of temperature on the mutation rate of a mutable locus in maize. *J. Hered.* 49:121–24

53. Harrison, B. J. 1965. Mutability in *Antirrhinum majus*. *John Innes Inst. Ann. Rept.* 56:15–18
54. Harrison, B. J., Carpenter, R. 1973. A comparison of the instabilities at the *nivea* and *pallida* loci in *Antirrhinum majus*. *Heredity* 31:309–23
55. Williams, E., Brink, R. A. 1972. The effect of abnormal chromosome 10 on transposition of Modulator from the *R* locus in maize. *Genetics* 71:97–110
56. McClintock, B. 1948. Mutable loci in maize. *Carnegie Inst. Washington Yearbook* 47:155–69
57. McClintock, B. 1964. Aspects of gene regulation in maize. *Carnegie Inst. Washington Yearbook* 63:592–602
58. McClintock, B. 1965. Components of action of the regulators *Spm* and *Ac*. *Carnegie Inst. Washington Yearbook* 64:527–36
59. McClintock, B. 1963. Further studies of gene-control systems in maize. *Carnegie Inst. Washington Yearbook* 62:486–93
60. McClintock, B. 1971. The contribution of one component of a control system to versatility of gene expression. *Carnegie Inst. Washington Yearbook* 70:5–17
61. McClintock, B. 1959. Genetic and cytological studies of maize. *Carnegie Inst. Washington Yearbook* 58:452–56
62. Peterson, P. A. 1965. Phase variation of controlling elements in maize. *Genetics* 52:466
63. Peterson, P. A. 1966. Phase variation of controlling elements in maize. *Genetics* 54:249–66
64. Peterson, P. A. 1970. The *En* mutable system in maize. III. Transposition associated with mutational events. *Theor. Appl. Genet.* 40:367–77
65. Neuffer, M. G. 1961. Mutation studies at the A_1 locus in maize. A mutable allele controlled by *Dt*. *Genetics* 46:625–40
66. McClintock, B. 1955. Controlled mutation in maize. *Carnegie Inst. Washington Yearbook* 54:245–55
67. McClintock, B. 1968. The states of a gene locus in maize. *Carnegie Inst. Washington Yearbook* 66:20–28
68. Brink, R. A. 1973. Paramutation. *Ann. Rev. Genet.* 7:129–52
69. Ashman, R. B. 1970. The compound structure of the R^{st} allele in maize. *Genetics* 64:239–45
70. Gavazzi, G. 1967. Control of gene action in the synthesis of anthocyanin in maize. *Mol. Gen. Genet.* 99:151–64
71. Harrison, B. J., Fincham, J. R. S. 1968. Instability at the *Pal* locus in *Antirrhinum majus*. 3. A gene controlling mutation frequency. *Heredity* 23:67–72
72. Fincham, J. R. S., Harrison, B. J. 1967. Instability at the *Pal* locus in *Antirrhinum majus*. 2. Multiple alleles produced by mutation of one original unstable allele. *Heredity* 22:211–24
73. Peterson, P. A., Weber, C. R. 1969. An unstable locus in soybeans. *Theor. Appl. Genet.* 39:156–62
74. Smith, H. H., Sand, S. A. 1957. Genetic studies on somatic instability in cultures derived from crosses between *Nicotiana landsgorffii* and *N. sanderae*. *Genetics* 42:560–82
75. Sand, S. A. 1957. Phenotypic variability and the effect of temperature on somatic instability in cultures derived from hybrids between *Nicotiana langsdorffii* and *N. sanderae*. *Genetics* 42:685–703
76. Sand, S. A. 1969. Origin of the *v* variegated allele in *Nicotiana*: basic genetics and frequency. *Genetics* 61:443–52
77. Gerstel, D. U., Burns, J. A. 1967. Phenotypic and chromosomal abnormalities associated with the introduction of heterochromatin from *Nicotiana otophora* into *N. tabacum*. *Genetics* 56:483–502
78. Deshayes, A. 1973. Mise en évidence d'une corrélation entre la fréquence de variations somatique sur feuilles et l'état physiologique d'un mutant chlorophyllien monogénique chez *Nicotiana tabacum*. *Mutation Res.* 17:323–34
79. Demerec, M. M. 1935. Unstable genes. *Bot. Rev.* 1:233–48
80. Green, M. M. 1973. Some observations and comments on mutable and mutator genes in *Drosophila*. *Genetics* 73: (Suppl.) 187–94
81. Green, M. M. 1967. The genetics of a mutable gene at the *white* locus of *Drosophila melanogaster*. *Genetics* 56:467–82
82. Green, M. M. 1969. Mapping a *Drosophila melanogaster* "controlling element" by interallelic crossing over. *Genetics* 61:423–28
83. Green, M. M. Controlling element mediated transpositions of the *white* gene in *Drosophila melanogaster*. *Genetics* 61:429–41
84. Gethmann, R. C. 1971. The genetics of a new mutable allele at the *white* locus in *Drosophila melanogaster*. *Mol. Gen. Genet.* 114:144–55
85. Kalisch, W. -E., Becker, H. J. 1970. Über eine Reihe mutabler Allele des *white*-locus bei *Drosophila melanogaster*. *Mol. Gen. Genet.* 107:321–35

86. Judd, B. H. 1969. Evidence for a transposable element which causes reversible gene inactivation in *Drosophila melanogaster*. *Genetics* 62:s29
87. Dawson, G. W. P., Smith-Keary, P. F. 1963. Episomic control of mutation in *Salmonella typhimurium*. *Heredity* 18: 1–20
88. Peterson, P. A. 1970. Controlling elements and mutable loci in maize: their relationship to bacterial episomes. *Genetics* 41:33–56
89. Abelson, J. et al 1973. Summary of the genetic mapping of prophage *Mu*. *Virology* 54:90–92 (followed by five other papers)
90. Hirsch, H. -J., Starlinger, P., Brachet, P. 1972. Two kinds of insertions in bacterial genes. *Mol. Gen. Genet.* 119: 191–206
91. Daniell, E., Roberts, R., Abelson, J. 1972. Mutations of the lactose operon caused by bacteriophage *mu*. *J. Mol. Biol.* 69:1–8
92. Fiandt, M., Szybalski, W., Malamy, M. H. 1972. Polar insertions in *lac, gal* and phage λ consist of a few IS-DNA sequences inserted in either orientation. *Mol. Gen. Genet.* 119:223–31
93. Daniell, E., Abelson, J. 1973. *lac* messenger RNA in *lacZ* gene mutants of *Escherichia coli* caused by insertion of bacteriophage *mu*. *J. Mol. Biol.* 76: 319–21
94. Starlinger, P. et al 1973. mRNA distal to polar nonsense and insertion mutations in the *gal* operon of *E. coli*. *Mol. Gen. Genet.* 122:279–86
95. Echols, H. 1972. Developmental pathways for the temperate phage: lysis vs. lysogeny. *Ann. Rev. Genet.* 6:157–90
96. Daniell, E., Boram, W., Abelson, J. 1973. Genetic mapping of the inversion loop in bacteriophage *mu* DNA. *Proc. Nat. Acad. Sci. USA* 70:2153–56
97. Hsu, M. T., Davidson, N. 1972. Structure of inserted bacteriophage *Mu-1* DNA and physical mapping of bacterial genes by *Mu-1* insertion. *Proc. Nat. Acad. Sci. USA* 69:2823–27
98. Cohen, S. N., Miller, C. A. 1970. Nonchromosomal antibiotic resistance in bacteria II. Molecular nature of R-factors isolated from *Proteus mirabilis* and *Escherichia coli*. *J. Mol. Biol.* 50:671–87
99. McClintock, B. 1967. Genetic systems regulating gene expression during development. *Develop. Biol.* 1:(Suppl.)84–112
100. Hood, L. 1973. The genetics, evolution and expression of antibody molecules. *Stadler Symp.* 5:74–142
101. Sprague, G. F., McKinney, H. H. 1971. Further evidence on the genetic behavior of AR in maize. *Genetics* 67:533–42
102. Neuffer, M. G., Jones, L., Zuber, M. S. 1968. *The mutants of maize*. Madison, Wis.: Crop Sci. Soc. Am. 74 pp.

THE RELATIONSHIP BETWEEN GENES AND POLYTENE CHROMOSOME BANDS

♦3064

George Lefevre Jr.
Biology Department, California State University, Northridge, California 91324

INTRODUCTION

From the time forty years ago when the genetic significance of the giant banded chromosomes found in many Dipteran tissues was first recognized (1–6), most Drosophila cytogeneticists have accepted almost as an article of faith that each individual polytene chromosome band (or chromomere, if you will) is associated with a single gene locus. Painter (4) said, "it was clear that we had within our grasp the material of which everyone had been dreaming . . . that the highway led to the lair of the gene." Bridges (6) more specifically inferred that "each of the faint cross-bands . . . corresponds to one locus." Further, that "certain of the heavy-walled capsules . . . correspond to three loci, since they enclose between them a line of dots or dashes."

These views quickly triggered attempts to validate or discredit this "one band – one gene" hypothesis. This review describes subsequent reports that bear on the central question: Can genes and polytene chromosome bands be equated, one for one?

THE NUMBER OF GENES

Accepting for the moment that the number of bands can be counted accurately, the problem reduces to determining whether the number of genes corresponds well or poorly with the band count.

For the X chromosome of *Drosophila melanogaster,* the number of loci capable of mutating to lethality under the influence of X rays can be theoretically estimated by comparing the total incidence of sex-linked lethal mutations elicited by a given dose and dividing by the average frequency of various specific, single gene mutations elicited by the same dose. The same sort of calculation can be done with spontaneous mutation frequencies. Computations of this sort (7, 8) yielded values of 500 and 1280, and a later, more sophisticated calculation yielded a value of 838 (9). Muller

& Altenberg (10, 11) pioneered a conceptually more appropriate procedure, whereby the frequency of allelism among a sample of mutants could be used to estimate the total gene number. Alikhanian (12), correcting Muller's (11) formula, experimentally attacked the problem by determining the proportions of loci in a defined segment of the X chromosome "hit" once, twice, or more. These data were used to estimate the saturation level when all loci in the target segment would have mutated. The ratio of the size of the measured segment to that of the whole chromosome formed the basis for his estimate that the total number of sex-linked lethal loci was 968. These various early estimates are of interest in light of the band counts by Bridges. In his original map, Bridges (6) depicted 725 bands in the X chromosomes; in the revised, definitive map (13), 1012 bands. Thus, confidence grew that the one band – one gene hypothesis was reasonable.

MODERN "SATURATION" STUDIES

The work of Alikhanian (12) was fraught with imperfections, and his estimate has a large inherent error. More recently, Judd and his associates (14, 15) have reported an ambitious experiment wherein a restricted interval of the X chromosome of D. melanogaster delimited by the giant (gt) and white (w) loci was selected for study. Well over 100 single-site lethal and semilethal mutants were recovered following X ray and chemical mutagenesis. These distributed themselves into just 13 complementation units, a number that, with the addition of the zeste (z) locus (which has no lethal alleles), matches exactly the number of bands that Bridges (14) drew between band 3A1, the locus of gt, and 3C2, the locus of w. This apparently perfect correlation must be viewed with some reservation, however, because the accompanying cytological analysis of the chromosome locations of the complementation groups required that two bands that Bridges drew in the 3A region (3A5 and 3A10) were either "empty" or nonexistent, but two bands had to be added to the 3B region, where Bridges drew only 4 bands, in order to accommodate the 6 complementation groups located there. Each of the bands in question, however, is of the most delicate variety, so Bridges' map might be in error.

Beermann (16), in a recent review on chromomeres and genes, published electron microscope (EM) photographs of the 3A-3B region taken by V. Sorsa. These pictures fail to resolve band 3A5, but 3A10 is visible. Furthermore, bands 3A3 and 3A4 are both shown to be double bands, but only one "extra" band is visible in 3B. Berendes (17), however, also using EM, reported two extra delicate bands in 3B. On the basis of this new EM information, the "correct" band count between 3A1 and 3C2 should be 17, rather than 14, but Sorsa, Green & Beermann (18) have shown that band 3C1 has no lethal content. Since publishing his results, B. H. Judd (personal communication) has identified one more complementation group that would correspond to the 3A10 band. However, the region's bands cannot yet be said to have been fully saturated with lethal or semilethal loci.

A related question tackled by Shannon et al (15) was whether each different complementation group actually corresponds to a separate and distinct genetic function. That is to say, could any two adjacent groups represent allelic complementation of mutants at the same locus? This question was studied by carrying out a developmental analysis of the abnormalities associated with each group. No two adjacent groups were similar; indeed, each group exhibited a distinctive syndrome of developmental defects. Thus the total number of complementation groups appears to correspond to an equivalent number of distinct genetic functions, no more numerous, in fact, than the number of bands available to accommodate the functions, one for one.

Other saturation studies, less prodigious in scale, support the basic thesis that there are no more identifiable gene functions than bands. Hochman, Gloor & Green (19–21), over a number of years, accumulated a large number of mutants located in the small chromosome 4 of D. melanogaster. These included spontaneous as well as X ray and EMS-induced visible and lethal mutants. Taking his latest reports together, Hochman (20, 21) has identified a total of 37 different lethal and 7 visible loci in chromosome 4. Bridges' (6) drawing shows a total of 45–50 bands. In passing, I should note that Slizynski (22) mapped 137 bands in chromosome 4, but this value simply is not realistic.

In a smaller study, reported only in abstract, Rayle (23) claimed to have completely saturated a short interval of 12 bands at the tip of the X chromosome included in the stub-arista deficiency, Df(1)sta. Here, too, the number of visible and lethal loci corresponded perfectly with the band count, if double bands are counted as two, not one.

An even shorter, but more interesting, region was investigated by Lefevre & Green (24). In the five-band interval between w and split (spl), from 3C2 to 3C7 on Bridges' (13) map, only two genetic functions, roughest (rst) and verticals (vt), were found to be associated with the pair of double bands, 3C2-3 and 3C5-6. The w locus itself might be associated with a faint previously undescribed band between 3C1 and 3C2-3 or with no more than a tenth of band 3C2 (see also reference 18), and 3C4 could not be visualized. However, bands 3C2-3 and 3C5-6 acted as though they were genetic duplicates of one another in that deficiency for either one was compensated for by the presence of the other. Thus, two functions were related to a (duplicated) double band.

No published report, to my knowledge, has clearly placed two or more distinct genetic functions in one and the same single band, or more than two in any double band. However, Lefevre (unpublished) has identified in region 1B at the tip of the X chromosome 15 or 16 lethal sites, plus 4 visible, where Bridges showed 14 bands. Here "supersaturation" may have occurred, or else allelic complementation is frequent. On the other hand, in the 8-band interval between N in 3C7 and diminutive (dm) in 3D4, no mutant of any sort has been found.

THE ACCURACY OF THE BAND COUNT

In equating genes to bands, we cannot ignore the possibility that Bridges' (13) map, showing 1012 bands on the X chromosome, is seriously in error. Beermann (25) pointed out that the "doubleness" of bands is an artifact of methodology. Incubation of the salivary glands in saline before fixation favors the double appearance, whereas dissection directly into fixative yields "solid" bands. Beermann (16), Berendes (17), and Sorsa (26) have reported on EM studies of Drosophila polytene chromosomes in which many of Bridges' double bands could not be resolved. In fact, Berendes would reduce the total number of X chromosome bands by about 25%, or to a number equivalent to that in the original Bridges' (6) map. However, there is little indication that EM resolves any significant number of submicroscopic bands not represented on Bridges' (13) revised map. Indeed, Beermann (16) noted that Bridges drew bands 3A7 and 3A10, for example, with approximately correct interband separations and in correct relative thickness, even though each of these bands, as seen in EM, ranges below 0.05 μ in thickness, theoretically incapable of being resolved with the light microscope. Nonetheless, Bridges apparently missed one or two very delicate bands in 3B, even though he did draw two there of equivalent size, and he drew 3A5, which seems to be nonexistent.

Altogether, then, the problem of the "correct" band number in Drosophila remains to be resolved. Bridges' determination of approximately 1000 for the X chromosome and something over 5000 for the genome represents the best available estimate, if his double bands are real, EM to the contrary. If his double bands are mostly, in fact, single, then the total number of bands (and genes?) would be significantly reduced. However, the reported saturation studies that appear to associate genes with bands, one for one, do so only on the basis that bands depicted by Bridges as double contain two gene loci, and single bands, one locus.

THE CYTOLOGICAL LOCALIZATION OF GENES

Saturation studies determine the number of independent loci in a delimited interval containing a known number of bands. By contrast, cytological studies try to pinpoint the location of specific genes on the polytene chromosome. Painter (4) and Mackensen (27) early used chromosome rearrangements to determine the cytological locations of genes. If a chromosome is broken at a particular gene locus, its mutant phenotype will be expressed. The problem reduces, then, to determining the spread of breakpoint positions producing the same mutant expression. Demerec (28) reported the exact position of 37 breakpoints associated with Notch (N) expression. Of these, 15 were involved in euchromatic rearrangements, and all but one were reported to occur just to the left or to the right of band 3C7 (see reference 16). In view of the dimensions of 3C7, which is closely apposed on its left to the thicker 3C5-6 double band, it seems reasonable to conclude that N effects are elicited by interruptions of band 3C7 or perhaps of the interbands to its right and left. More distant breaks probably were produced independently of the N effect, which is especially evident when the breakpoints of heterochromatic rearrangements associated with N effects are plotted. Only 6 of 25 such breakpoints were immediately adjacent to band 3C7, the rest removed from it by as many as 16 bands. A spread of variegated position effect explains these cases.

In addition to the N locus, rearrangement mutants associated with the cut (ct) locus have been produced on a scale large enough to be meaningful. Many of these were analyzed by Hannah (29) and, together with others, are described by Lindsley

& Grell (30). (Also see reference 31.) Here inversion and translocation breakpoints associated with *ct* effects are invariably found just to the right of the prominent band 7B1-2. Thus, the *ct* locus has been assigned to the delicate band 7B3 and/or 4. At both the *N* and *ct* loci, male-viable and lethal alleles occur in nonrearranged chromosomes, but all of the *N* and *ct* mutants associated with rearrangements that affect 3C7 or 7B3/4 are male lethal, as a review of the mutants listed in Lindsley & Grell (30) attests.

Similar studies have been made on various other loci, but none as extensive as those above. At the yellow (*y*) locus, rearrangement mutants appear to be male viable, perhaps because the *y* locus can be deleted without causing lethality to males (32). A recent study of induced *y* mutants by Roberts (33) described several cases where a rearrangement breakpoint was not at the presumed position of the *y* locus (1A8) but removed from it by some distance. He emphasized the possibility of drawing erroneous conclusions regarding gene positions from breakpoint data, unsupported by accompanying genetic analysis of the mutants.

"CRYPTIC" MUTANTS

Since saturation studies have used mutants identified by their readily detectable effects on viability or phenotype, many other kinds of mutants producing no easily identified alteration obviously escape the usual mutant screen. O'Brien (34) argued cogently that numerous enzyme-controlling loci produce no discernible consequence when homozygous for completely "null" alleles. Even double null mutants, in two specific cases, proved to be viable, fertile, and phenotypically normal. If many loci are selectively neutral, as O'Brien suggests, studies of lethal mutants alone cannot lead to conclusive statements about saturation and gene-band relationships.

Further, there are behavioral, fertility, neurological, and "clock" mutants that would not likely be included in a mutant collection. For example, Konopka & Benzer (35) have reported a clock mutant known to be located in the interval studied by Judd et al (14). If such mutant loci have no lethal alleles, if the most extreme allele—the null allele—remains viable and ostensibly normal, then the reported saturation studies cannot be considered to have demonstrated the equality between the number of bands and the number of genetic functions.

Cytogenetic studies by Lefevre (36) addressed themselves to this problem. He analyzed the cytological location of about 500 sex-linked euchromatic rearrangement breakpoints found after 2000 and 3000 r irradiations. These were divided into two groups: those associated with lethal or mutant effects (350) and those with no ostensible genetic consequence (150). Many of the latter were discovered by analyzing X chromosomes of progeny that had passed through the mutant screen and were judged phenotypically normal (nonmutant). No case was identified in which breakpoints from the two different categories were clearly identical. That is to say, the genetic consequence of breakage was a function of exactly *where*, not *how*, the chromosome was broken. Estimates of the probability that a rearrangement breakpoint would be nonmutant ranged from 35–65%, depending on different assumptions. A study of X-Y translocations by Nicoletti & Lindsley (37) showed that about 75% of random, euchromatic X-chromosome breakpoints were nonmutant. Without question, half or more of all euchromatic rearrangement breakpoints fail to elicit a readily detectable mutant expression.

The breakpoint analysis makes it unlikely that a band associated with a lethal function also has one or more neutral functions associated with it. If such were the case, then examples of lethal and ostensibly normal breakpoints affecting the same band, particularly if it were large, should be commonplace; they are not. On the other hand, the alternative that half or more of the bands are associated exclusively with neutral, cryptic functions also seems ruled out by the results of saturation studies where far more than half of the bands have been shown to have lethal functions. Existing evidence in fact suggests that most genes are essential, that rearrangements affecting a given gene characteristically elicit its maximum mutant expression, and that breakpoints at a particular point on the chromosome produce the same unique genetic response, that is, they do not elicit double or multiple mutants.

CROSSING OVER AND BANDS

In an attempt to correlate band structure with recombination frequency, Lefevre (38) compared the amount of crossing over between genes separated by thin, delicate bands with that exhibited by genes separated by large, dark-staining bands. In particular, two lethal mutants, *1(l)Q54* and *1(l)L12*, immediately the left and right of the large, thick band 10A1-2 [with which the vermilion (*v*) locus is associated] exhibited about 0.7% recombination. Two lethals, *1(l)L12* and *1(l)L8*, confined to the faintly staining region just to the right of *10A1-2*, but, on a band basis, separated from each other by about as many bands as between *1(l)Q54* and *1(l)L12*, exhibited only 0.1% recombination. This prompted Lefevre to postulate that crossover frequencies could be correlated with band quality and that linkage relationships between genes of known cytological position could be predicted on the basis of the kinds of bands between them.

As an alternative to the interpretation that more crossovers occur in regions of the meiotic chromosome represented by dark bands than in regions represented by equal numbers of thin bands (implying that crossovers occur in bands in proportion to their size), Beermann (16) noted that crossing over might be stimulated in the immediate vicinity of bands to a degree proportional to their size, but that crossovers need not be frequent within the band region. Such a view would perhaps follow from the conclusions of Schultz & Redfield (39), who suggested that where pairing is interrupted, "more frequent overlap will occur and crossing over should be increased." Perhaps bands produce regions of pairing "interruption" at their edges to a degree according to their size, stimulating crossing over nearby, but not participating significantly in crossing over themselves. Bands, then, might contain "synaptic DNA."

A partial test of this possibility comes from an as yet unpublished continuation of Lefevre's (38) studies. Originally, he had been unable to find a locus sufficiently near v to be contained in band 10A1-2, despite the large size of this double band. Now, however, a new lethal mutant, *1(l)L68*, not allelic with v, appears by all cytological tests to be associated with 10A2, possibly at its right edge. The v locus must be in 10A1. The new lethal shows a very low level of recombination with v (0.04%) but slightly more than 0.5% with lethal *1(l)L12*, which itself showed about 0.5%, recombination with v (38). As it stands, either v and *1(l)L68* are literally close together within the substance of band 10A1-2 (unlikely on cytological grounds), or they are separated by most of the substance of 10A1-2 and yet show little recombination. The possibility that large, nonpuffing bands represent areas of strong synaptic attraction promoting recombination at their edges, but not internally, has to be given serious consideration.

BAND-DNA RELATIONSHIPS: THE DILEMMA

If the number of different loci in *D. melanogaster* is the same as the number of bands, then the amount of DNA per gene could be readily determined if the total DNA content of the haploid genome were known. Such values have been reported by Rudkin (40) and Laird (41). They agree quite well, providing an estimate of about 30,000 base pairs (bp), or 30 kbp, per average band. Berendes' (17) band count, being lower than Bridges', would give a value of about 40 kbp per band.

Such large amounts of DNA seem difficult to reconcile with single gene functions, for each band would have enough DNA to code for 30–40 average cistrons. To be sure, bands vary in size, the most delicate having fewer than 5 kbp and the largest perhaps 100 kbp (16).

Finding two functions associated with a large double band such as 10A1-2, however, does not solve the problem of the seemingly excessive amount of DNA per function. Speculations on this score are abundant. Most attractive at face value is the possibility that band material represents repetitive copies of structural cistrons, some of which are more highly duplicated than others. A "master-slave" concept of cistronic reiteration was advanced by Callan & Lloyd (42, 43) following their studies on lampbrush chromosomes of salamanders. Here a master copy forms a model for producing a series of slave copies and additionally for rectifying any mutational changes that might arise in the slaves. Difficulties in visualizing an accurate and persistent mechanism for rectification and for maintaining constancy in the number of copies in the face of mutation and recombination have led several authors to propose sophisticated derivative models (44–48). One effect of the Thomas model (46) would be to allow any copy of the series to serve as the master for the next round of rectification. The reiterative models require that from one fourth to one half of the polytene chromosome DNA be in repetitive form, with the number of reiterated cistronic families presumably being of the same order as the number of genes.

Direct measurements fail to identify high levels of repetitive DNA in Drosophila and show even less in DNA extracts from salivary glands than from adult tissues (41, 49–54). Highly repetitive sequences are largely confined to the centric heterochromatic regions which are not polytenized in salivary gland chromosomes and comprise about 5–10% of nuclear DNA from diploid (nonpolytene) cells of Drosophila. Moderately repetitive sequences, found mostly in euchromatic regions, account for another 5–15%, depending on the species (41, 55–58). For *D. melanogaster,* 80–85% of the total genome must be in the form of unique sequences, as measured by reannealing procedures, so that by no stretch of the imagination can polytene bands be formed entirely of repeated sequences. Since only 5% or less of polytene chromosomes can be composed of interband DNA (16), the substance of the bands must consist, for the most part, of unique DNA sequences. Repetitive DNA can form only a relatively small proportion of bands, if evenly distributed (see, for example, reference 57, 58). The alternative that some bands might be entirely formed of repetitive sequences, but the majority entirely of unique sequences, seems unlikely (59). The question of the dispersion of repeated sequences among the euchromatic bands of polytene chromosomes is yet to be firmly resolved. The relationship, if any, of repetitive DNA to regions of genetic duplication, as in 3C (24), remains to be explored.

THEORETICAL MODELS OF BAND ORGANIZATION

In addition to the reiterative models mentioned above, several other theoretical models of DNA organization have recently been advanced (60–63). Three in particular might be described as "superoperon" models and propose, in brief, that the large amount of DNA often associated with individual bands represents mainly recognition, regulatory, and control nucleotide sequences governing the activity of the relatively short structural sequences (see discussion, reference 16).

Britten & Davidson (60) assume a number of "producer" genes (structural cistrons), perhaps no more than exist in prokaryotes, that are integrated in a complex system of regulatory sequences involving "receptor" genes (promoters) sensitive to activation by specific "activator" RNA sequences. The latter are produced by "integrator" genes, which in turn are under the control of "sensor" genes responsive to various developmental signals. Receptors and integrator genes are thought to be redundant so that one sensor gene could control "batteries" of producer genes. Here two separate types of functional groups, producer-receptor and sensor-integrator, must exist; but, as Beermann (16) notes, cytogenetic evidence does not identify two different kinds of bands. However, Britten & Davidson call attention to the fact that the pleiotropic effects of Notch mutants, for example, are consistent with mutations in integrator gene sets.

Georgiev (61) proposed a somewhat similar model. Each "unit" would contain a "promoter-proximal" acceptor zone and a "promoter-distal" structural (informational) zone. Acceptor loci would specifically interact with regulatory proteins. The informational zone would contain not only structural cistrons but also regulatory genes. The entire unit would be transcribed as one giant heterogeneous RNA molecule, but the noninformational portions would be degraded in the nucleus so that only the informational component would be transported to the cytoplasm. Further, according to Georgiev, "different operons may contain identical or similar acceptor loci, the most multiple acceptor being localized in the proximal part. On the other hand, one operon contains a number of different acceptor loci, reacting with different regulatory proteins." In this model there is no separation, as in the Britten and Davidson model, of structural (producer-receptor) and regulatory (sensor-integrator) cistrons. Both models incorporate repeated sequences and fit the facts derived from DNA studies.

Another model, advanced by Crick (62), incorporates the cytogenetic data derived from studies on polytene chromosomes. Crick assumes that the complexity of gene control requires a relatively large amount of DNA to be associated with each gene. This DNA he considers to constitute the band material, and the controlled gene occupies the adjacent interband region. Thus, mutations of either a structural cistron or its associated (band) control sequences would give the same effect and could form a single complementation group. He further suggests that the control regions might have evolved from originally tandemly repeated sequences.

This theme has been amplified by Paul (63) in a "chromomere" theory that provides a function for nonhistone proteins and an explanation for the large size of eukaryotic genomes, repetitive sequences in DNA, heterogeneous RNA, and "processing" of nuclear RNA. Paul assumes that highly compacted nucleoprotein (band regions) alternates with loosely compacted nucleoprotein (interbands) according to a fixed pattern and that the two kinds of regions have different chemical properties. In particular, interband regions contain associated polyanionic macromolecules which prevent supercoiling and which bind at specific sites. The transcribable unit would contain a promoter locus closely linked to an "address" site to which destabilizing polyanionic macromolecules, probably nonhistone protein, would temporarily bind and produce a localized reduction in supercoiling. Adjacent regulator sites as well as an initiator site would follow before the structural cistron. He suggests that such complex units have been tandemly duplicated in general, followed by evolutionary degradation. Thus, the eventual unit of transcription need retain only address, promoter, regulator, and initiator sites at the beginning, vestiges of altered structural sequences, but at least one "sensible" gene, probably several to many remaining effective address loci, and a terminator locus at the end. The resulting transcript would be large, needing further processing before release into the cytoplasm. A selective advantage must be visualized for an initially large transcript to prevent the system from degrading back to one essential unit.

In most respects this model fits the facts of both DNA and cytogenetic studies on eukaryotic chromosome organization. In particular, it agrees well with the chromosome breakage studies where all breaks affecting the same place (band?) yield the same genetic result and fail to complement with one another or with chromosomally normal mutants of the same locus. If a transcribable unit were disrupted by breakage anywhere between initiator and terminator, it is reasonable to assume that the resulting garbled transcripts would not be properly "processed." Thus the breakage mutant would likely exhibit maximum mutant expression and would appear to be allelic with all other mutants, of whatever kind, of the same locus, i.e. show *cis*-dominant relationships, in Judd's (14) terminology.

CONCLUDING THOUGHTS

The theories of eukaryotic chromosome organization are in accord with the facts regarding DNA repetitiveness, and they can explain why some bands are bigger than others and why there should appear to be a one gene – one band relationship. None, however, seem specifically disposed to explain band-crossover relationships or nonmutant rearrangement breakpoints, that is, why the same band could not easily be involved in both mutant and nonmutant rearrangements. Additional properties of band structure are required, properties that must reflect underlying molecular structure. I suggest that a mechanism must exist that results in a tendency for both crossover and induced chromosome breakpoints to move before reunion from an initial position within a band to its edge. Such movement must involve a considerable resection of nucleotides, which is more likely to be successful in a small band than in a large one. At or near the edge, as in Paul's theory, the sensible locus is most likely to be found. Thus, mutant breakpoints, such as those that give rise to *ct* effects, will not only be allelic but will seem to be at the edge of the band, not in it. Many other loci besides cut, including yellow, scute, white, echinus, singed, and forked, seem, on the basis of breakage data, to be at the edge of a band. Nonmutant breakpoints are most often found in faintly banded territories (36). If the effective reunion position following a break in a thin band is transferred to the interband region, a nonmutant rearrangement could result. I do not consider it likely that interbands or thin bands would be preferentially affected by X rays; rather X rays should act at random along the length of the DNA molecule. But the piling up of crossovers and breakpoints alongside of easily analyzed (large) bands and a deficit of breakpoints and crossovers within large bands call for a mechanism for shifting the apparent point of effect from an internal to a peripheral position when the band substance itself is initially affected, but not if an interband area were broken in the first place.

Whatever the true situation, the bulk of current evidence generally supports the concept of a single genetic function being associated with each polytene chromosome band. Nonetheless, the possibility still exists that the distribution of gene functions and the distribution of chromomeres (bands) are independent, though approximately the same in number. Supersaturated regions (1B) and "geneless" regions (3C8-3D3) almost surely exist. Existing theoretical models of eukaryotic chromosome organization, though incorporating the evidence from DNA studies, fall short of explaining all of the interrelated facts concerning genes and bands, including mutation, breakage, and crossover frequencies and distribution.

Literature Cited

1. Heitz, E., Bauer, H. 1933. Beweise für die Chromosomennatur der Kernschleifen in den Knäuelkernen von *Bibio hortulanus* L. *Z. Zellforsch.* 17:67–82
2. Painter, T. S. 1933. A new method for the study of chromosome rearrangements and the plotting of chromosome maps. *Science* 78:585–86
3. Painter, T. S. 1934a. A new method for the study of chromosome aberrations and the plotting of chromosome maps in *Drosophila melanogaster. Genetics* 19:175–88

4. Painter, T. S. 1934b. Salivary chromosomes and the attack on the gene. *J. Hered.* 25:465–76

5. King, R. L., Beams, N. W. 1934. Somatic synapsis in *Chironomus,* with special reference to the individuality of the chromosomes. *J. Morphol.* 56: 577–91

6. Bridges, C. B. 1935. Salivary chromosome maps. *J. Hered.* 26:60–64

7. Demerec, M. 1934. The gene and its role in ontogeny. *Cold Spring Harbor Symp. Quant. Biol.* 2:110–15

8. Gowen, J. W., Gay, E. H. 1933. Gene number, kind, and size in Drosophila. *Genetics* 18:1–31

9. Lea, D. E. 1947. *Actions of Radiations on Living Cells.* New York: MacMillan

10. Muller, H. J., Altenberg, E. 1919. The rate of change of heredity factors in Drosophila. *Proc. Soc. Exp. Biol.* 17: 10–14

11. Muller, H. J. 1929. The gene as the basis of life. *Proc. Int. Congr. Plant Sci., 4th, 1926* 1:897–921

12. Alikhanian, S. I. 1937. A study of the lethal mutations in the left end of the sex-chromosome in *Drosophila melanogaster. Zool. Zh.* 16:247–79 (In Russian with English summary)

13. Bridges, C. B. 1938. A revised map of the salivary gland X chromosome of *Drosophila melanogaster. J. Hered.* 29: 11–13

14. Judd, B. H., Shen, M. W., Kaufman, T. C. 1972. The anatomy and function of a segment of the X chromosome of *Drosophila melanogaster. Genetics* 71: 139–56

15. Shannon, M. P., Kaufman, T. C., Shen, M. W., Judd, B. H. 1972. Lethality patterns and morphology of selected lethal and semi-lethal mutations in the zeste-white region of *Drosophila melanogaster. Genetics* 72:615–38

16. Beermann, W. 1972. Chromomeres and genes. In *Results and Problems in Cell Differentiation,* ed. W. Beermann, 4: 1–33. Berlin: Springer

17. Berendes, H. D. 1970. Polytene chromosome structure at the submicroscopic level. *Chromosoma* 29:118–30

18. Sorsa, V., Green, M. M., Beermann, W. 1973. Cytogenetic fine structure and chromosomal localization of the white gene in *Drosophila melanogaster. Nature New Biol.* 245:34–37

19. Hochman, B., Gloor, H., Green, M. M. 1964. Analysis of chromosome *4* in *Drosophila melanogaster.* I. Spontaneous and X-ray induced lethals. *Genetics* 35:109–26

20. Hochman, B. 1971. Analysis of chromosome *4* in *Drosophila melanogaster.* II. Ethylmethanesulfonate induced lethals. *Genetics* 67:235–52

21. Hochman, B. 1972. The detection of four more vital loci on chromosome *4* in *Drosophila melanogaster. Genetics* 71:s71

22. Slizynski, B. M. 1944. A revised map of salivary gland chromosome *4. J. Hered.* 35:322–25

23. Rayle, R. E. 1972. Genetic analysis of a short X-chromosomal region in *Drosophila melanogaster. Genetics* 71:s50

24. Lefevre, G., Green, M. 1972. Genetic duplication in the white-split interval of the X chromosome in *Drosophila melanogaster. Chromosoma* 36:391–412

25. Beermann, W. 1962. *Riesenchromosomen. Protoplasmatologia* VI/C:1–161. Wien:Springer

26. Sorsa, M. 1969. Ultrastructure of the polytene chromosome in *Drosophila melanogaster. Ann. Acad. Sci. Fenn. A, IV Biol.* 151:1–18

27. Mackensen, O. 1935. Locating genes on the salivary chromosomes. Cytogenetic methods demonstrated in determining position of genes on the X chromosome of *Drosophila melanogaster. J. Hered.* 26:163–74

28. Demerec, M. 1941. The nature of changes in the white-Notch region of the X chromosome of *Drosophila melanogaster. Proc. Int. Genet. Congr., 7th, 1939. J. Genet.,* Suppl., 99–105

29. Hannah, A. 1949. Radiation mutations involving the cut locus in Drosophila. *Proc. Int. Congr. Genet,, 8th, 1948. Hereditas,* Suppl., 588–89

30. Lindsley, D. L., Grell, E. H. 1968. Genetic variations of Drosophila melanogaster. *Carnegie Inst. Wash. Publ.* 627

31. Lefevre, G., Johnson, T. K. 1973. Evidence for a sex-linked haplo-inviable locus in the cut-singed region of *Drosophila melanogaster. Genetics* 74:633–45

32. Muller, H. J. 1935. A viable two-gene deficiency phenotypically resembling the corresponding hypomorphic mutations. *J. Hered.* 26:193–96

33. Roberts, P. A. 1974. A cytogenetic analysis of X-ray induced "visible" mutations at the *yellow* locus of *Drosophila melanogaster. Mutat. Res.* 22:139–44

34. O'Brien, S. J. 1973. On estimating functional gene number in eukaryotes. *Nature New Biol.* 242:52–54

35. Konopka, R., Benzer, S. 1971. Clock mutants of *Drosophila melanogaster. Proc. Nat. Acad. Sci. USA* 68:2112–16

36. Lefevre, G. Jr. 1974. The one band – one gene hypothesis: Evidence from a cytogenetic analysis of mutant and non-mutant rearrangement breakpoints in *Drosophila melanogaster. Cold Spring Harbor Symp. Quant. Biol.* 38:591–99

37. Nicoletti, B., Lindsley, D. 1960. Translocations between the X and the Y chromosomes of *Drosophila melanogaster. Genetics* 45:1705–22

38. Lefevre, G. Jr. 1971. Salivary chromosome bands and the frequency of crossing over in *Drosophila melanogaster. Genetics* 67:497–513

39. Schultz, J., Redfield, H. 1951. Interchromosomal effects on crossing over in *Drosophila. Cold Spring Harbor Symp. Quant. Biol.* 16:175–97

40. Rudkin, G. T. 1965. The relative mutabilities of DNA in regions of the X chromosome of *Drosophila melanogaster. Genetics* 52:665–81

41. Laird, C. D. 1971. Chromatid structure: relationships between DNA content and nucleotide sequence diversity. *Chromosoma* 32:378–406

42. Callan, H. G., Lloyd, L. 1960. Lampbrush chromosomes of crested newts *Triturus cristatus* (Laurenti). *Phil. Trans. Roy. Soc. London Ser. B* 243:135–219

43. Callan, H. G. 1967. The organization of genetic units in chromosomes. *J. Cell Sci.* 2:1–7

44. Thomas, C. A. Jr. 1970. The theory of the master gene. In *The Neurosciences: Second Study Program,* ed. F. O. Schmitt. New York: Rockefeller Univ. Press

45. Smith, G. P. 1974. Unequal crossing over and the evolution of multigene families. *Cold Spring Harbor Symp. Quant. Biol.* 38:507–13

46. Thomas, C. A. Jr. 1974. The rolling helix. *Cold Spring Harbor Symp. Quant. Biol.* 38:347–52

47. Thomas, C. A. Jr., Zimm, B. H., Dancis, B. M. 1973. Ring theory. *J. Mol. Biol.* 77:85–100

48. Whitehouse, H. L. K. 1973. Hypothesis of post-recombination resynthesis of gene copies. *Nature* 245:295–98

49. Laird, C. D., McCarthy, B. J. 1969. Molecular characterization of the *Drosophila* genome. *Genetics* 63:865–82

50. Rae, P. M. M. 1970. Chromosomal distribution of rapidly reannealing DNA in *Drosophila melanogaster. Proc. Nat. Acad. Sci. USA* 67:1018–25

51. Rae, P. M. M. 1972. The distribution of repetitive DNA sequences in chromosomes. In *Advan. Cell Mol. Biol.* 2:109–49

52. Gall, J., Cohen, E., Polan, M. 1971. Repetitive DNA sequences in *Drosophila. Chromosoma* 33:319–44

53. Entingh, T. D. 1970. DNA hybridization in the genus *Drosophila. Genetics* 66:55–68

54. Laird, C. D. 1973. DNA of *Drosophila* chromosomes. *Ann. Rev. Genet.* 7:177–204

55. Dickson, E., Boyd, J., Laird, C. D. 1971. Sequence diversity of the polytene chromosome DNA from *Drosophila hydei. J. Mol. Biol.* 61:615–27

56. Hennig, W. 1972. Highly repetitive DNA sequences in the genome of *Drosophila hydei.* I. Preferential localization in the X chromosome heterochromatin. *J. Mol. Biol.* 71:407–17

57. Wu, J. R., Hurn, J., Bonner, J. 1972. Size and distribution of the repetitive segments of the *Drosophila* genome. *J. Mol. Biol.* 64:211–19

58. Bonner, J., Wu, J. R. 1973. A proposal for the structure of the Drosophila genome. *Proc. Nat. Acad. Sci. USA* 70:535–37

59. Thomas, C. A. Jr., Lee, C. S., Pyeritz, R. E., Bick, M. D. 1972. Closing the ring. In *Molecular Genetics and Developmental Biology,* ed. M. Sussman. Englewood Cliffs, NJ:Prentice-Hall

60. Britten, R. J., Davidson, E. H. 1969. Gene regulation for higher cells: A theory. *Science* 165:349–57

61. Georgiev, G. P. 1969. On the structural organization of operon and the regulation of RNA synthesis in animal cells. *J. Theor. Biol.* 25:473–90

62. Crick, F. 1971. General model for the chromosomes of higher organisms. *Nature* 234:25–27

63. Paul, J. 1972. General theory of chromosome structure and gene activation in eukaryotes. *Nature* 238:444–46

GENETIC POLYMORPHISM OF ❖3065
THE *HISTOCOMPATIBILITY-2* LOCI
OF THE MOUSE

Jan Klein
Department of Microbiology, The University of Texas Southwestern Medical School,
Dallas, Texas 75235

PHENOTYPIC MANIFESTATION OF *H-2* LOCI

Each mammalian species that has been thoroughly tested has been shown to carry a cluster of loci known as the major histocompatibility complex (MHC). In the mouse the MHC is called *histocompatibility-2* (*H-2*). The *H-2* loci are phenotypically manifested in several ways, which are briefly enumerated below.

First, inoculation of cells (usually of lymphoid origin) from a donor into an *H-2–*disparate recipient elicits the production of humoral antibodies that agglutinate the donor's erythrocytes (1) or, in the presence of complement, kill the donor's lymphocytes (2). The antibodies can be absorbed from the serum by cells of practically any mouse tissue (3), indicating that the serologically detectable H-2 antigens have a wide tissue distribution.

Second, transplantation of tissue (most commonly skin) between mice differing in the *H-2* complex stimulates the proliferation of the host's lymphocytes and the production of effector cells that destroy the graft, usually within the first three weeks after the operation (4). Graft rejection is often accompanied by the production of humoral antibodies, which are detectable by serological methods. The antigens responsible for the tissue incompatibility (histocompatibility antigens) are present in nearly all mouse tissues, and the *H-2* loci coding for them are believed to be identical (with one exception) to loci coding for the serologically detectable antigens. The antigens detected by serological and transplantation methods are summarily designated H-2 antigens.

Third, immunization of recipients with *H-2–*disparate lymphocytes may lead to the production of humoral antibodies that differ from classical H-2 antibodies in that they react only with a subpopulation of lymphocytes (5, 6). The antigens defined by such antibodies, summarily designated *I* region-associated or Ia antigens (7), are serologically (8, 9), genetically (5, 6), and biochemically (10, 11) distinct from classical H-2 antigens.

63

Fourth, inoculation of immunocompetent cells (mature lymphocytes) into an *H-2–* disparate immunoincompetent recipient (for example, a newborn, immunologically immature mouse) results in the donor cells' attacking the recipient's tissues (graft-vs-host reaction, or GVHR, see reference 12). The symptoms of the GVHR may range from mere enlargement of lymphoid organs to the death of the host.

Fifth, mixing of lymphocytes from two mice differing at their *H-2* loci in tissue culture results in mixed lymphocyte reaction (MLR), characterized by enlargement of the lymphocytes (blast transformation) and their proliferation (13). The determinants responsible for the GVHR and MLR appear to be identical and are summarily designated *lymphocyte-activating determinants (Lad)*.

Sixth, mixing of sensitized lymphocytes (i.e. lymphocytes from preimmunized donors or cells obtained from mixed lymphocyte culture) with appropriate target cells in tissue culture leads to cell-mediated lymphocytotoxicity (CML), manifested by killing of the target cells. The CML seems to be directed against the serologically detectable H-2 antigens on target cells (15).

Seventh, immunization of inbred mouse strains with a variety of apparently unrelated antigens (synthetic polypeptides, serum proteins, or cell surface antigens) induces high levels of antibodies in some strains and low levels (or no response) in other strains. The quantity of antibodies produced is controlled by *Immune response-1 (Ir-1)* loci, which are part of the *H-2* complex (16, 17).

Eighth, inoculation of mice with leukemia viruses induces cancer in some strains more easily than in others (18). These differences in leukemogenesis are controlled by genes which, like the *Ir-1* genes, are a part of the *H-2* complex and may actually be identical to the *Ir-1* loci.

Ninth, the serum of normal mice contains proteins that display quantitative and qualitative variation; some of these are controlled by loci in the *H-2* complex. One such protein, called serologically detectable serum substance (*Ss*), is present in a high level in some strains (allele *Ss^h*) and in a low level (allele *Ss^l*) in others (19). Another protein, called sex-limited protein (Slp), is present in males of some inbred strains (allele *Slp^a*) and absent in males of other strains (allele *Slp^o*) and in all females (20). Both the *Ss* and *Slp* loci map within the *H-2* complex (21, 22).

Tenth, loci in or near the *H-2* complex also determine the level of complement in normal serum (23), the level of androgens in mouse blood (24), the physiological cooperation between thymus-derived lymphocytes (T cells) and bone-marrow–derived lymphocytes (B cells) in the immune response (25), and resistance of irradiated F_1 hybrids to parental bone marrow grafts (26). The genetic determinants of these traits have not been mapped with respect to markers in the *H-2* complex, and their identity or nonidentity to the various *H-2* loci is an open question.

GENETICS

The genetic distance between the extreme ends of the *H-2* complex is about 0.4 map units, which corresponds to about 2000 cistrons, each cistron coding for a polypeptide chain about 100 amino acid residues long (27). Whether the complex indeed contains so many cistrons remains to be resolved, but it is probably safe to predict that the number of *H-2* loci is quite large. The *H-2* complex is currently divided into four regions, designated *K, I, S,* and *D* (Figure 1), and the *I* region is further divided into two subregions, *Ir-1A* and *Ir-1B* (14). The regions and subregions are separable by crossing over and are associated with distinct functions. Each region and subregion is believed to consist of at least one locus. The individual *H-2* loci are distributed among the regions as follows (14). The *H-2K* and *H-2D* loci are located in the *K* and *D* regions, respectively, and are concerned with the control of antigens detectable by serological and transplantation methods (classical H-2 antigens, cf references 28, 29). The *Ir-1A* locus is in the subregion of the same name and is concerned with the control of the immune response to synthetic polypeptides (T, G)-A--L, (H, G)-A--L (16), ovalbumin (OA), ovomucoid (OM), and bovine gammaglobulin (BGG); the *Ir-1B* locus is in the *Ir-1B* region and controls the immune response to IgG myeloma protein MOPC173 (17) and to porcine LDH_B (30). The *I* region also contains the *Ia* (5, 6) and *Lad* (31) loci, as well as at least one strong histocompatibility locus (*H-2I*) responsible for rapid rejection of skin grafts (32), but the mutual relationship of these loci and their relationship to the *Ir-1A* locus is not clear. A separate *Lad* locus also seems to be located between the *S* and *D* regions (33). The *Ss* and *Slp* loci are located in the *S* region in the middle of the *H-2* complex (21, 22). Whether they represent one common genetic determinant or two distinct loci has not yet been determined. A specific combination of alleles at the individual loci of the *H-2* complex is called the *H-2* haplotype.

H-2 POLYMORPHISM IN INBRED STRAINS

This review is primarily concerned with the peripheral *H-2* regions *K* and *D*. The serologically detectable antigens controlled by these two regions can be divided somewhat artificially into two classes: public and private (34). The public *H-2*

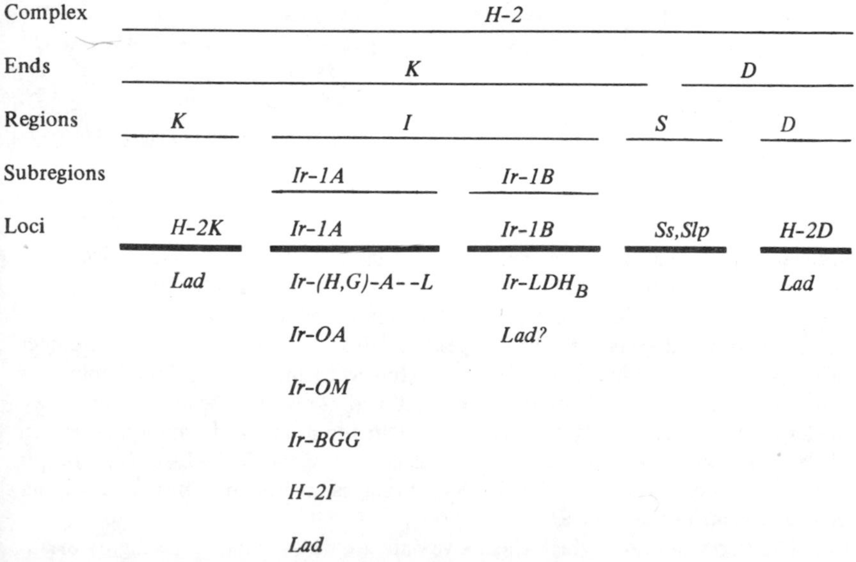

Complex				*H-2*			
Ends			*K*			*D*	
Regions	*K*		*I*			*S*	*D*
Subregions		*Ir-1A*		*Ir-1B*			
Loci	*H-2K*	*Ir-1A*		*Ir-1B*		*Ss,Slp*	*H-2D*
	Lad	*Ir-(H,G)-A--L*		*Ir-LDH$_B$*			*Lad*
		Ir-OA		*Lad?*			
		Ir-OM					
		Ir-BGG					
		H-2I					
		Lad					

Figure 1   Genetic map of the *H-2* complex. (The genetic relationship of the individual loci in each region and subregion is not known.)

Table 1 List of inbred-strain derived *H-2* haplotypes and private antigens characterizing them[a]

H-2 Haplotype Symbol	Private Antigens Determined by Regions	
	K	D
b(ba,bb,bc,bd,be)	33	2
d(da)	31	4
f(fa)	9	9[b]
k	23	32
p(pa,n)	16	?
q	17	30
r	18[a]	
s	19	12
a(a1)	23	4
an1	19	9[b]
ap1(ap2,ap3,ap4)	9	4
aq1	23	9[b]
by1	17	2
g(g1,g2,g3,g4,g5)	31	2
h(h1,h2,h3,h4,h15,h18)	23	2
i(i3,i5,i7,ia)	33	4
j(ja)	15	2
m(m1)	23	30
o1(o2)	31	32
qp1	17	12
sq1(sq2)	19	30
t1(t2,t3)	19	4
u	20	4
v	21	?
y1(y2)	17	4

[a] Antigens not yet assigned to either the *K* or *D* region are placed in between.
[b] Modified antigen.

antigens are shared by several different *H-2* haplotypes; the private antigens are restricted to a single haplotype and its genetic derivatives. One *H-2* haplotype usually codes for two private antigens (one controlled by the *K* and the other by the *D* region) and several public antigens. In the inbred strains the total number of known *H-2* haplotypes is 60 (Table 1). However, since many of the haplotypes were derived by recombination between the *K* and *D* regions, the number of known alleles at the *H-2K* and *H-2D* loci is less than the number of haplotypes: eleven alleles are known to exist at the *H-2K* locus and ten at the *H-2D* locus (27). In this respect, the polymorphism of the *H-2* loci surpasses the polymorphism of any other genetic system in the mouse.

The inbred strains, on which the above data are based, represent a highly biased sample of the mouse population. Not only were the strains derived from relatively few sources, but the sources could not be considered independent. Furthermore, interbreeding, deliberate or accidental, also contributed to gene exchange among the lines in the early stages of inbred mice development. For a realistic estimate of *H-2* polymorphism, the inbred mice are quite insufficient and the analysis must turn to natural mouse populations.

WILD FORMS OF THE HOUSE MOUSE

The house mouse, *Mus musculus* L., is believed to have originated in the steppes of central Asia (35), where aboriginal mice still live in their original habitats, digging burrows in the ground and feeding on grass, seeds, and grain (36). In prehistoric times, mice began to associate with man and to accompany him all over the world. The last continent to be settled by mice was North America. Mice arrived in North America in the post-Columbian era, probably with settlers from central and northern Europe (for mice in Canada and the northern United States) or with settlers from the Mediterranean areas (for mice in Mexico and the southern United States). Commensal mice depend on man for food and shelter. Occasionally commensal mice revert to a more feral existence and once again become more or less independent of man. All studies described in this review are concerned with commensal mice.

NUMBER OF *H* LOCI SEGREGATING AMONG COMMENSAL WILD MICE

In addition to the *H-2* complex, the mouse possesses a number of other histocompatibility (*H*) loci. More than 30 of these loci have been identified in inbred mice so far (27), and the total number has been estimated at several hundred (37). Two inbred strains usually differ in at least 15 *H* loci (38). The number of *H* loci segregating among wild mice can be estimated by crossing individual mice to an inbred strain and exchanging skin grafts among the hybrids (40). If the wild mouse differs from the inbred strain at both alleles of any given *H* locus, the cross is *aa* X *bc,* and the probability (*S*) of graft survival among the siblings is $S = (½)^L$ where *L* is the number of segregating *H* loci. If the wild mouse shares one allele at a particular *H* locus with the inbred strain, the cross is *aa* X *ab*, and the probability of graft survival is $S = (¾)^L$. In the former case, $L = \log S / \log 0.5$; in the latter case, $L = \log S / \log 0.75$ (43).

Equations for calculating the effective number of alleles per locus and the effective number of segregating *H* loci were derived by Klein & Bailey (41). This approach has demonstrated (41, 44) that skin grafts exchanged between mice sired by males from the same locality have relatively long mean survival times and broad rejection ranges; some survive permanently. In contrast, grafts exchanged between mice sired by males from different localities have short mean survival times and narrow rejection ranges; none survive permanently. Apparently, mice from the same locality have a more similar histocompatibility endowment than do mice from different localities. The percentage of permanently surviving grafts in the former group has been used to calculate the effective number of alleles per locus and the effective number of segregating histocompatibility loci. The estimates range from 2 to 9 heterozygous H loci per mouse (39–41, 43, 44), depending on the particular locality, the particular inbred strain used to produce the hybrids, the sex of the recipient, and other factors. The H heterozygosity of wild mice is thus considerably lower than that of an F_1 hybrid between two inbred strains of the laboratory mouse.

H-2 POLYMORPHISM IN COMMENSAL WILD MICE

H-2 typing of non-inbred mice was first attempted by Rubinstein & Ferebee (42), who used the hemagglutination assay to test red cells of some 50 randomly bred Swiss-Webster animals against 14 H-2 antisera. The typing suggested that there are probably many more H-2 haplotypes and H-2 antigens than those known to exist in inbred strains. Serological analysis of mice captured in the wild has recently begun in the laboratories of P. Iványi (44–45) and J. Klein (34, 46–48). In the latter laboratory, a long-term multiphasic program to obtain information regarding H-2 population behavior is in progress.

In the first phase of this program (34, 46), operationally monospecific antisera prepared by cross-immunization of inbred strains were applied to panels of wild mice sampled from different localities of a single geographic area (Ann Arbor, Michigan). The typing, done by the direct PVP hemagglutination test and complemented in a few instances by absorption analysis, demonstrated that the public antigens of inbred strains are widely distributed among wild mice, whereas inbred private H-2 antigens occur rarely, if at all, in wild populations. For example, public antigen H-2.5 was found in 95% of wild mice trapped in the Ann Arbor area, while private antigens H-2.4, H-2.23, and others were not present in a sample of over 2000 mice. Some wild mice did not react with any inbred-derived H-2 antisera, a result interpreted as being caused by the absence of all known H-2 antigens ("null H-2 haplotypes").

In general, the typing of wild mice with inbred-derived antisera provides only limited information about the population behavior of the H-2 system. The inbred private antigens are too rare and the public antigens too common among wild mice to be useful for differentiation of wild H-2 haplotypes. For this reason an attempt to produce antisera identifying wild-specific H-2 antigens is being made in the second phase of population studies. The simplest way to obtain anti-H-2 wild reagents might seem to be by immunization of inbred strains with tissues from wild mice. Such immunization, however, is complicated by several factors: antisera must be tested against mice other than the donor (the donor is sacrificed), the immunizing tissue is in short supply, and the same antiserum cannot be reproduced at will. Furthermore, antisera produced by immunization with wild mice as donors are usually complex, containing not only multiple H-2, but also numerous non-H-2 antibodies. In view of these difficulties, an alternative method of producing anti-H-2 wild antisera has been designed. The H-2 haplotypes of wild mice are first transferred onto a defined genetic background of strain C57BL/10, and the ensuing B10.W congenic or semicongenic lines are used as tissue donors for immunization.

Currently, more than 80 B10.W lines, many nearing their completion, are being developed in my laboratory. Upon completion, the B10.W lines are cross-immunized with each other as well as with inbred strains, and the antisera produced are serologically analyzed, using panels of inbred strains and B10.W lines.

The feasibility of the multiphasic approach has been demonstrated on a small scale, and some information pertinent to the extent of H-2 polymorphism has already been obtained (47, 48). So far, antisera against some 30 semicongenic B10.W lines (i.e. lines that have undergone a limited number of backcrossing to the B10 background strain) have been produced and tested against a panel of inbred and B10.W strains (Table 2). In 12 of these antisera, a prominent component is an antibody directed against private antigens of the B10.W donor. These 12 antisera react either with only the B10.W donor and no other cells in the panel, or with cells of a few other B10.W lines, usually those derived from the same locality as that of the donor (Table 3). The remaining 18 antisera contain a mixture of antibodies directed against public H-2 antigens and, in a few instances, against non-H-2 antigens as well. These antisera react not only with a number of B10.W lines, but also with several inbred strains. The reaction patterns of the anti-H-2 public antisera are sufficiently diverse to permit the tentative conclusion that the H-2 haplotypes of their respective B10.W donors are probably all different. In five instances, the H-2^{wild} haplotypes have been analyzed with all available antisera and their antigenic configuration has been determined (Table 4). The new H-2 haplotypes were originally designated H-2^{wa} through H-2^{we}, but these symbols were later changed to H-2^{w1} through H-2^{w5}.

Each H-2^{wild} haplotype is characterized by the presence of a particular private antigen and an array of public H-2 antigens, though the assignments of public antigens to the individual H-2^{wild} haplotypes are questionable because of the problems encountered during serological analysis. For example, it is quite common to obtain a strong reaction from a particular H-2^{wild} haplotype with one anti-H-2.1 antiserum and no reaction with another anti-H-2.1 antiserum, so that the typing of H-2^{wild} haplotypes for H-2 public antigens is of dubious value. Since private antigens sufficiently characterize a given H-2 haplotype, it may become necessary to limit H-2 typing of wild mice to the determination of private antigens (or, more generally, to less cross-reactive antigens). Each of the 12 B10.W lines possesses only one private H-2 antigen, but the absence of a second private antigen is almost certainly caused by the unavailability of H-2 recombinants, permitting separation of private K and D antibodies.

Altogether, some 40 different H-2 haplotypes are now known: 30 derived from wild and 10 from laboratory mice (excluding the recombinant H-2 haplotypes). The indications are that many more new H-2 haplotypes exist among the 80 B10.W lines maintained in my laboratory, and when haplotypes being identified in Dr. Iványi's laboratory are taken into account, the number of known H-2 haplotypes can safely be estimated at at least 100. The actual number of existing H-2 haplotypes is difficult to predict, the minimal estimate being several hundred. H-2 polymorphism already far exceeds any other known polymorphism in the mouse; if the number of different

Table 2 Reactivity in hemagglutination test of antisera produced in inbred strains against tissues from various B10.W semicongenic lines

Erythrocyte Donor	Reciprocal of Hemagglutinin Titer of Antiserum Directed Against Indicated Donor Cells																		
	B10.BUB19	B10.CHA2	B10.CHB57	B10.GAA20	B10.GAB3	B10.KEB5	B10.KPA42	B10.KPB44	B10.KPB68	B10.SAA48	B10.STA13	B10.TOB1	B10.KPA28	B10.WOPI	B10.WOA25	B10.STA15	B10.KEA28	B10.KPA32	B10.BUAI
B10.BUB19	≥640	0	0	0	0	0	0	0	0	0	0	0	0	0	0	0	0	0	0
B10.CHA2	0[a]	≥640	0	0	0	0	0	0	0	0	0	0	0	0	0	0	≥640	≥640	0
B10.CHB57	0	0	≥640	0	0	0	0	0	0	0	0	0	0	0	0	0	0	0	0
B10.GAA20	0	0	0	≥640	0	0	0	0	0	0	0	0	0	0	0	0	80	≥640	0
B10.GAB3	0	0	0	0	320	0	0	0	0	0	0	0	0	0	160	0	0	0	0
B10.KEB5	0	0	0	0	0	≥640	0	0	0	0	0	0	0	0	0	0	0	0	0
B10.KPA42	0	0	0	0	0	0	≥640	0	0	0	0	0	0	0	0	0	0	0	0
B10.KPB44	0	0	0	0	0	0	0	≥640	0	0	0	0	0	0	160	0	≥640	≥640	0
B10.KPB68	0	0	0	0	0	0	0	0	≥640	0	0	0	0	0	0	0	160	0	80
B10.SAA48	0	0	0	0	0	0	0	0	0	320	0	0	0	0	0	0	0	0	0
B10.STA13	0	0	0	0	0	0	0	0	0	0	160	0	0	0	0	0	0	0	0
B10.TOB1	0	0	0	0	0	0	0	0	0	0	0	160	0	0	0	0	0	0	0

Strain	1	2	3	4	5	6	7	8	9	10	11	12	13	14	15	16	17	18	19	20
B10.BUA1	0	0	0	0	0	0	0	0	0	0	0	0	0	0	0	0	0	160	>640	160
B10.BUA16	0	0	0	0	0	0	0	0	0	0	0	0	0	0	320	0	0	0	0	0
B10.CHR51	0	0	0	0	0	0	0	0	0	0	0	0	0	0	0	0	0	160	320	320
B10.DIA68	0	0	0	0	0	0	0	0	0	0	0	0	0	0	0	0	0	0	>640	0
B10.DRA60	0	0	0	0	0	0	0	0	0	0	0	0	0	0	0	0	0	0	>640	160
B10.DRB62	0	0	0	0	0	0	0	0	0	0	0	0	0	0	0	320	320	320	>640	0
B10.GAA19	0	0	0	0	0	0	0	0	0	0	0	0	0	0	80	0	0	0	0	0
B10.KEA2	0	0	0	0	0	0	0	0	0	0	0	0	0	>640	0	0	0	160	0	0
B10.KEA4	0	0	0	0	0	0	0	0	0	0	0	0	0	0	0	0	0	320	0	0
B10.KEA28	0	0	0	0	0	0	0	0	0	0	0	0	>640	0	0	0	>640	>640	>640	0
B10.KPA32	0	0	0	0	0	0	0	0	0	0	0	0	0	0	0	0	0	40	>640	0
B10.KPB78	0	0	0	0	0	0	0	0	0	0	0	0	0	>640	0	0	0	0	0	0
B10.PTA52	0	0	0	0	0	0	0	0	0	0	0	0	0	0	0	0	0	0	0	40
B10.STA15	0	0	0	0	0	0	0	0	0	0	0	0	>640	>640	0	320	0	0	>640	0
B10.WOA25	0	0	0	0	0	0	0	0	0	0	0	0	0	0	0	320	320	320	>640	0
B10.WOP1	0	0	0	0	0	0	0	0	0	0	0	0	0	0	80	0	>640	0	0	0
B10.WRA66	0	0	0	0	0	0	0	0	0	0	0	0	0	0	0	0	0	0	0	40
B10.SAB41	0	0	0	0	0	0	0	0	0	0	0	0	0	0	0	0	0	0	0	0

[a] Titer of < 20.

Table 3 Reactions of H-2^w antisera with panel of B10.W strains

Wild-Derived Congenic Line	Reciprocal of Hemagglutinin Titer with Antiserum				
	K-74[a]	K-78[b]	K-53[c]	K-57[d]	K-72[e]
B10.KPA42	1280	0	0	0	0
B10.KPA74	1280	0	0	0	0
B10.KPA88	640	0	0	0	0
B10.KPA100	1280	0	0	0	0
B10.KPA132	640	0	0	0	0
B10.KPB68	0	1280	0	0	0
B10.KPB44	0	640	0	0	0
B10.KPB28	0	640	0	0	0
B10.SAA48	0	0	640	0	0
B10.SAA39	0	0	640	0	0
B10.GAA20	0	0	0	2560	0
B10.GAA22	0	0	0	1280	0
B10.KEA5	0	0	0	0	1280
B10.KEA7	0	0	0	0	1280

[a] K-74:B10.A   anti-B10.KPA42.
[b] K-78:(C3HxB10.D2)F$_1$ anti-B10.KPB68.
[c] K-53:C3H   anti-B10.SAA48.
[d] K-57:C3H   anti-B10.GAA20.
[e] K-72 B10.A   anti-B10.KEA5.

H-2 haplotypes does indeed run into the hundreds, the H-2 complex will become the most polymorphic system known.

MECHANISMS MAINTAINING H-2 POLYMORPHISM

Although the mechanisms maintaining H-2 polymorphism are not known, four factors are worth considering: high tolerance of mutational changes, peculiarities of population structure, selection at closely linked loci, and function of the H-2 products.

Tolerance of Mutational Changes

The H-2 glycoprotein may differ from other proteins, particularly enzymes, in that it has a high degree of tolerance to mutational changes. In an enzyme, almost any mutation affecting binding with a substrate and thus causing partial or complete loss of enzymatic activity is harmful and is therefore selected against. On the other hand, in H-2 molecules, which are perhaps relatively inert biologically in that they display no enzymatic or similar activities, there is greater tolerance of genetic variability. Many mutations that would be deleterious to most enzymes may have no effect on the function of the H-2 glycoprotein. Supporting this hypothesis is the observation that the mutation rate of H-2 loci is higher than that of other loci in the mouse (see 27 for references). It may be that this seemingly increased rate is caused not by higher mutability of H-2 loci, but rather by the fact that more H-2 mutants survive.

Table 4 The H-2 chart of wild mice (wild-derived H-2 haplotypes, antigens determined by the haplotypes, and strains that carry them)

H-2 Haplotype	H-2 Antigens[a]																									Strain
	1	2	3	4	5	6	7	8	9	13	15	16	17	18	19	23	30	31	32	33	101	102	103	104	105	
w1	-	-	3	-	5	6?	-	NT	-	-	-	-	-	-	-	-	-	-	?	-	101	-	-	-	-	B10.KPA42
w2	-	-	-	-	-	?	?	NT	-	-	-	-	-	-	-	-	-	-	-	-	-	102	-	-	-	B10.KPB68
w3	-	-	-	-	5?	6?	7	8?	-	-	-	-	-	-	-	-	-	-	-	-	-	-	103	-	-	B10.SAA48
w4	-	-	3	-	5?	6?	7	?	-	-	-	-	-	-	-	-	-	-	-	-	-	-	-	104	-	B10.GAA20
w5	-	-	-	-	NT	?	?	NT	-	-	-	-	-	-	-	-	-	31?	?	-	-	-	-	-	105	B10.KEA5

[a] - = absence of an antigen; ? = presence or absence of an antigen is uncertain; NT = not tested.

Peculiarities of Population Structure

The populations of commensal wild mice are divided into small subpopulations or demes (breeding units, family units, tribes, colonies), each with a rigid social structure (see 27 for reference). Each deme occupies a small territory probably not larger than a few square meters. Demarcation of the territory is behavioral rather than physical, since the territories are established even in a single room with no physical barriers. A territory is vigorously defended from intrusion, primarily by one male. The rate of immigration into established demes is thus negligible, with the demes behaving as isolates open in one direction (out) and virtually closed in the opposite direction (in).

An average deme consists of seven to twelve adult mice (three to five males and four to seven females) and is ruled by a dominant male who fights to achieve and maintain his superior position. All other males present in the deme are subordinate and probably do not sire any progeny. On the other hand, the females in the deme seem to be equal socially. Each pregnant female builds a separate nest in which she bears and rears her young. The average litter size is six to eight, with each female bearing three to five litters per year. Reproductive activity of wild mice is highest in prevernal and autumnal seasons, the population density being maximal around April and November and minimal around January and June. Most of the young leave the deme after reaching maturity, settle in unoccupied territories, and establish new demes; if no new territories are available in the buildings, they move into fields surrounding the buildings and establish transitory feral populations. The pressure to leave the deme is higher for males than for females, and consequently more males are found among the feral mice in the vicinity of human dwellings. As the weather becomes colder and the food supply more limited, the feral mice attempt to move back into buildings. Dwindling of the food supply leads to decimation of the indoor populations, with the subordinate males dying first and dominant males last. During extremely cold winters, a large proportion of indoor mice vanishes (the population is said to go through a bottleneck phase). When the weather improves, the few survivors start new demes and a new population cycle begins.

In populations of this size and of this structure, random genetic drift becomes an extremely important factor leading to rapid fixation of some *H-2* haplotypes and extinction of others, and so to genetic diversification of the demes. The effect of random drift is demonstrated by studies in which mice collected at the same and at geographically separated localities were typed with both anti-H-2 inbred and anti-H-wild sera. The typing revealed a great similarity of *H-2* haplotypes among mice from the same locality as compared to mice from different localities (46–48).

Selection at Closely Linked Loci

Linked to the *H-2* complex is the *T-t* system, which consists of a series of dominant (*T*) and recessive (*t*) factors whose common denominator is an effect on the development of the axial system of the mouse embryo, manifested in its mildest form as a shortening of the tail (for review and references see 49 and 50). The *T-t* factors possess a series of peculiar characteristics, of which the following are relevant to the subject discussed here. First, the recessive *t* factors are present in practically all populations of commensal mice that have been thoroughly sampled, though the frequency with which they occur may vary from locality to locality and from year to year. The overall frequency calculated from samples collected at widely different geographic areas is 0.11, indicating that approximately 22% of mice in any given population are *t*/+ heterozygotes. Second, all *t* factors found in wild populations are homozygous lethals or homozygous semiviables. The *t*/*t* homozygotes usually die in utero, at a particular stage of development characteristic for each *t* factor. Third, all *t* factors extracted from wild populations distort segregation ratios when present in males. A mating of a *t*/+ male with a *T*/+ female may produce progeny of which up to 99% carry the *t* factor. Fourth, the lethal factors suppress crossing over on a relatively long segment of chromosome 17. In the presence of *t*, the distance between *T* and *H-2* is reduced from its normal 15 map unit value to about 0.5 (51).

As first pointed out by Snell (52), the crossing-over suppressing effect and the high male transmission ratios provide favorable conditions for fixation of *H-2* mutations in natural populations. The conditions become even more favorable when the peculiar structure of the mouse population is taken into account. It is easy to imagine how an *H-2* mutation carried by a dominant male in association with a *t* factor spreads rapidly through a deme, and how this leads to *H-2* diversification of the demes.

Function

The *H-2* product function is perhaps the strongest factor maintaining *H-2* polymorphism at a high level. It is possible that *H-2* variability is actually required for the *H-2* molecules to perform their function. What the function might be is not known, but the possibility that the *H-2* complex may represent a cell-to-cell recognition system (53, 54) or even a primitive immune system (27) has been raised. For further discussion of the possible relationships between H-2 function and H-2 polymorphism see reference 27.

In conclusion, combination of the four factors described above could provide conditions favorable to the development of a polymorphism which, perhaps with the exception of immunoglobulins, is not possible in any other genetic system.

ACKNOWLEDGMENTS

The experimental work cited in this review was supported by US Public Health Service research grants GM 15419 and DE 02731. I thank Ms. Pamela Erbe Wiener for critical reading of the manuscript.

Literature Cited

1. Gorer, P. A. 1936. The detection of antigenic differences in mouse erythrocytes by the employment of immune sera. *Brit. J. Exp. Pathol.* 17:42–50
2. Gorer, P. A., O'Gorman, P. 1956. The cytotoxic activity of isoantibodies in mice. *Transplant. Bull.* 3:142–43
3. Basch, R. S., Stetson, C. A. 1962. The relationship between hemagglutinogens and histocompatibility antigens in the mouse. *Ann. NY Acad. Sci.* 97:83–94
4. Gorer, P. A. 1937. The genetic and antigenic basis of tumour transplantation. *J. Pathol. Bacteriol.* 44:691–97
5. Hauptfeld, V., Klein, D., Klein, J. 1973. Serological identification of an Ir-region product. *Science* 181:167–69
6. David, C. S., Shreffler, D. C., Frelinger, J. A. 1973. New lymphocyte antigen system (Lna) controlled by the *Ir* region of the mouse H-2 complex. *Proc. Nat. Acad. Sci. USA* 70:2509–14
7. Shreffler, D. et al 1974. Genetic nomenclature for new lymphocyte antigens controlled by the *I* region of the *H-2* complex. *Immunogenetics.* In press
8. Hämmerling, G. J., Deak, B. D., Mauve, G., Hämmerling, U., McDevitt, H. O. 1974. B lymphocyte alloantigens controlled by the *I* region of the major histocompatibility complex in mice. *Immunogenetics* 1:68–81
9. Hauptfeld, V., Hauptfeld, M., Klein, J. 1974. Tissue distribution of *I* region associated antigens in the mouse. *J. Immunol.* 113:181–88
10. Vitetta, E. S., Klein, J., Uhr, J. W. 1974. Partial characterization of Ia antigens from murine lymphoid cells. *Immunogenetics* 1:82–90
11. Cullen, S. E., David, C. S., Shreffler, D. C., Nathenson, S. G. 1974. Membrane molecules determined by the H-2 associated immune response region: Isolation and some properties. *Proc. Nat. Acad. Sci. USA* 71:648–52
12. Simonsen, M. 1962. Graft-versus-host-reactions. Their natural history and applicability as tools of research. *Progr. Allergy* 6:349–467
13. Dutton, R. W. 1965. Further studies of the stimulation of DNA synthesis in cultures of spleen cell suspensions by homologous cells in inbred strains of mice and rats. *J. Exp. Med.* 122:759–70
14. Klein, J. et al 1974. Genetic nomenclature for the *H-2* complex of the mouse. *Immunogenetics.* In press
15. Mauel, J., Rudolf, H., Chapius, B., Brunner, K. T. 1970. Studies of allograft immunity in mice. II. Mechanism of target cell inactivation *In Vitro* by sensitized lymphocytes. *Immunology* 18:517–35
16. McDevitt, H. O. et al 1972. Genetic control of the immune response, mapping of the *Ir-1* locus. *J. Exp. Med.* 135:1259–78
17. Lieberman, R., Humphrey, W. Jr. 1972. Association of H-2 types with genetic control of immune responsiveness to IgG (γ2a) allotypes in the mouse. *J. Exp. Med.* 136:1222–30
18. Lilly, F. 1966. The inheritance of susceptibility to the Gross leukemia virus in mice. *Genetics* 53:529–39
19. Shreffler, D. C., Owen, R. D. 1963. A serologically detected variant in mouse serum: Inheritance and association with the histocompatibility-2 locus. *Genetics* 48:9–25
20. Passmore, H. C., Shreffler, D. C. 1971. A sex-limited serum protein variant in the mouse: Hormonal control of phenotypic expression. *Biochem. Genet.* 5:201–9
21. Shreffler, D. C. 1964. A serologically detected variant in mouse serum: Further evidence for genetic control by the histocompatibility-2 locus. *Genetics* 49:973–78
22. Passmore, H. C., Shreffler, D. C. 1970. A sex-limited serum protein variant in the mouse: Inheritance and association with the H-2 region. *Biochem. Genetics* 4:351–65
23. Démant, P., Čapková, J., Hinzová, E., Voráčová, B. 1973. The role of the histocompatibility-2-linked Ss-Slp region in the control of mouse complement. *Proc. Nat. Acad. Sci. USA* 70:863–64
24. Iványi, P., Gregorová, S., Micková, M., Hampl, R., Stárka, L. 1973. Genetic association between a histocompatibility gene (*H-2*) and androgen metabolism in mice. *Transplant. Proc.* 5:189–91
25. Katz, D. H., Hamaoka, T., Dorf, M. E., Benacerraf, B. 1973. Cell interactions between histoincompatible T and B lymphocytes. The *H-2* gene complex determines successful physiologic lymphocyte interactions. *Proc. Nat. Acad. Sci. USA* 70:2624–28
26. Cudkowicz, G., Stimpfling, J. H. 1964. Hybrid resistance to parental marrow grafts: Association with the *K* region of *H-2*. *Science* 144:1339–40

27. Klein, J. 1974. *Biology of the Mouse Histocompatibility-2 Complex.* New York: Springer-Verlag. In press
28. Klein, J., Shreffler, D. C. 1971. The H-2 model for major histocompatibility systems. *Transplant. Rev.* 6:3–29
29. Klein, J., Shreffler, D. C. 1972. Evidence supporting a two-gene model for the H-2 histocompatibility system of the mouse. *J. Exp. Med.* 135:924–37
30. Melchers, I., Rajevsky, K., Shreffler, D. C. 1973. Ir-LDH$_B$: Map position and functional analysis. *Eur. J. Immunol.* 3:754–61
31. Bach, F., Widmer, M. B., Segal, M., Klein, J. 1972. Genetic and immunological complexity of major histocompatibility regions. *Science* 176:1024–27
32. Klein, J., Hauptfeld, M., Hauptfeld, V. 1974. Evidence for a third, Ir associated, histocompatibility region in the *H-2* complex of the mouse. *Immunogenetics* 1:45–56
33. Widmer, M. B., Peck, A. B., Bach, F. H. 1973. Genetic mapping of H-2 LD loci. *Transplant. Proc.* 5:1501–5
34. Klein, J. 1971. Private and public antigens of the mouse H-2 system. *Nature* 229:635–37
35. Schwarz, E., Schwarz, H. K. 1943. The wild and commensal stocks of the house mouse, *Mus musculus Linnaeus. J. Mammal.* 24:59–72
36. Tupikova, N. V. 1947. Ecology of the house mouse in the central part of the USSR. *Mater. Rodents* 2:7–67
37. Bailey, D. W. 1968. The vastness and organization of the murine histocompatibility-gene system as inferred from mutational data. In *Advances in Transplantation,* ed. J. Dausett, J. Hamburger, G. Mathé, 317–23. Copenhagen: Munksgaard
38. Barnes, A. D., Krohn, P. L. 1957. The estimation of the number of histocompatibility genes controlling the successful transplantation of normal skin in mice. *Proc. Roy. Soc. London* 146: 505–26
39. Iványi, P., Démant, P., Vojtíšková, M., Iványi, D. 1969. Histocompatibility antigens in wild mice (*Mus musculus*). *Transplant. Proc.* 1:365–67
40. Micková, M., Iványi, P. 1972. An estimate of the degree of heterozygosity at histocompatibility loci in wild popula-

tions of house mouse (*Mus musculus*). *Folia Biol. Praha* 18:350–59
41. Klein, J., Bailey, D. W. 1971. Histocompatibility differences in wild mice; further evidence for the existence of deme structure in natural populations of the house mouse. *Genetics* 68:287–97
42. Rubinstein, P., Ferrebee, J. W. 1964. The H-2 phenotypes of random-bred Swiss-Webster mice. *Transplantation* 6:715–21
43. Iványi, P., Démant, P. 1970. Preliminary studies on histocompatibility antigens in wild mice (*Mus musculus*). In *Proc. 11th Eur. Conf. Anim. Blood Groups Biochem. Polymorphism,* ed. W. Junk, 547–50. The Hague: N. V. Publ.
44. Micková, M., Iványi, P. 1971. Histocompatibility antigens in the wild house mouse (*Mus musculus*) In *Immunogenetics of the H-2 System,* ed. A. Lengerová, M. Vojtísková, 20–34. Basel: Karger
45. Iványi, P., Micková, M. 1972. Testing of wild mice for "private" H-2 antigens. Cross reactions or nonspecific reactions? *Transplantation* 14:802
46. Klein, J. 1970. Histocompatibility-2 (*H-2*) polymorphism in wild mice. *Science* 168:1362–64
47. Klein, J. 1972. Histocompatibility-2 system in wild mice. I. Identification of five new H-2 chromosomes. *Transplantation* 13:291–99
48. Klein, J. 1973. Polymorphism of the *H-2* loci in wild mice. In *Int. Symp. HL-A Reagents,* 251–56. Basel: Karger
49. Dunn, L. C. 1964. Abnormalities associated with a chromosome region in the mouse. *Science* 144:260–63
50. Bennett, D. 1964. Embryological effects of lethal alleles in the *t*-region. *Science* 144:260–67
51. Green, M. C., Snell, G. D. 1969. Private communication. *Mouse News Lett.* 40:29
52. Snell, G. D. 1968. The *H-2* locus of the mouse observations and speculations concerning its comparative genetics and its polymorphism. *Folia Biol. Praha* 14:335–58
53. Bodmer, W. F. 1972. Evolutionary significance of the *HL-A* system. *Nature* 237:139–45
54. Benacerraf, B., McDevitt, H. O. 1972. Histocompatibility-linked immune response genes. *Science* 175:273–79

BIOCHEMICAL GENETICS OF BACTERIA

❖3066

Joseph S. Gots and Charles E. Benson

Department of Microbiology, School of Medicine, University of Pennsylvania, Philadelphia, Pennsylvania 19174

INTRODUCTION

The general topic of biochemical genetics of bacteria encompasses a wide variety and assortment of gene-enzyme interactions. In this review we focus on the biochemical genetics of metabolic pathways, particularly the biosynthesis and, in a few cases, the catabolism of nitrogenous metabolites. The systems considered are those in which the metabolic steps in the pathway are essentially complete and gene assignments for the reactions are known and well documented (1, 2). These and other earlier developments that were involved in identifying the nature of the mutations, the modifications they created, and their consequences on biochemical activities are not considered in detail here. The necessary background information and details of the systems considered may be obtained from the comprehensive reviews cited in each case. Other general reviews have recently been published that cover special features of biochemical genetics of bacteria that are not dealt with in detail here. They include the biochemical genetics of carbohydrate metabolism (3), ribosomes (4), drug resistance (5), and suppression (6).

In the past few years, a major impetus in extension of the already well-defined systems has been directed toward elucidation and evaluation of mechanisms of genetic control and regulation of gene expression. These developments and our own research interests have prompted us to focus this review on the present status of the genetic regulation of the metabolic pathways to be considered. We examine these in terms of organization and elements of genetic regulatory units, unusual and unique features that have been revealed, divergence from accepted models, and unification of concepts. Our considerations deal primarily with the enteric bacteria, *Escherichia coli* and *Salmonella typhimurium,* (hereafter referred to only as *Salmonella*) because both the biochemical and genetic information relevant to regulation is more complete in these organisms than in others. The reader is also directed to a number of recent reviews on various aspects of genetic regulation (7–11).

79

ORGANIZATION AND ELEMENTS OF GENETIC REGULATORY UNITS

The Operon

This is the functional unit of the original Jacob-Monod model. It is usually considered to be a multigenic cluster comprised of adjacent structural genes that control related functions, are transcribed as a unit, and are regulated together. The controlling elements consist of a regulatory signal (repressor) produced by a separate gene and a site (operator) in the operon complex that responds to the signal. In the systems considered below, the initial revelation of a gene cluster invited superficial comparison with an operon. Too often the comparison dissipated when further analyses did not meet certain basic criteria such as gene contiguity, coordinate expression, or constitutive mutations that affect all of the genes of the cluster. In practice, a functional operon may be assumed if these criteria are met and, for completion, the existence of polarity and pleiotropic effects are usually added. Other criteria include the existence of a polycistronic mRNA and sequential appearance and equimolar synthesis of the operon enzymes. Among the biosynthetic systems, the histidine operon is one of the few that meets all of these criteria (12).

In the process of accumulating criteria for assignment of a gene cluster to a functional operon, a number of variations and complexities have become evident. In a few cases the assumed single operon has been found to be a complex of adjacent tandem operons each functioning independently with its own operator site. In other cases (e.g. the *argECBH* cluster, see below) an internal operator site has been found that suggests a complex of two overlapping operons each sharing a common operator and transcribing in opposite directions.

The Regulon

A constant feature of genes controlling related functions is that they may respond to a common effector with similar, although usually not coordinated expression even though they may be unlinked and widely scattered. While studying the common effect of a regulatory mutation, *argR,* on the scattered genes of the arginine pathway, Maas & Clark (13) introduced the term "regulon" to define a physiological unit of regulation. This is described as a group of genes controlling physiologically related functions that are regulated together by the same repressor substance, regardless of whether or not they are clustered together. This implies a common recognition site that is repeated on all of the operons that comprise the regulon.

A survey of the metabolic pathways controlled in this manner indicates that the regulon as a regulatory unit is the rule rather than the exception. Other complications and modifications become evident in the systems described below. Some involve more than one regulon for control of the overall pathway; some consist of a super-regulon in which two physiologically related regulons interact, either unidirectionally or reciprocally, in terms of response to repressors.

A case of interaction of physiologically unrelated regulons has been described in the cross regulation of the histidine and aromatic amino acid pathways in *Bacillus subtilis* (14). Regulatory mutants are derepressed for the enzymes in both of these pathways, but the nature of the common regulatory element is not known. An explanation for cross regulation, and a possible lead to explain the above, has been found in one case where a regulatory mutation leads to derepression of both the histidine and branched amino acid pathways (15). This mutation causes the loss of a nonspecific enzyme required for maturation of several tRNAs so that they can no longer function as effectors in their respective systems.

Regulatory Elements

The known regulatory elements that participate in control of operon expression are those that either generate or recognize the repressor signal. The use of mutants that are unable to carry out regulatory functions in bacteria has proved invaluable in recognizing these elements and how they function in the normal regulatory pattern. Isolation of such mutants usually relies on their ability to overcome metabolic blocks. The blocks are commonly imposed by metabolite analogs, and the use of such agents for this purpose has been reviewed by Umbarger (16). Other methods, including the use of bradytrophs, have been applicable for special cases. The regulatory mutants isolated may be defective in control by feedback inhibition or in regulation of operon expression. Among the latter are those with altered repressor or operator functions. These two types are usually distinguished by linkage studies and by cis-trans dominance tests. Operator mutations have altered recognition sites, are genetically linked to the operon, and are cis dominant. Mutations that affect the repressor substance are usually unlinked, are trans recessive to the wild-type allele, and thus specify a diffusible cytoplasmic factor. The existence of nonsense and temperature-sensitive mutant types gives presumptive evidence that this factor is a protein. However, these criteria alone are not sufficient to implicate the factor as a repressor protein. In several systems, mutants of this phenotype have been identified as being defective in the generation of an active corepressor (see below for histidine, isoleucine-valine, and methionine systems). Ultimate proof that a true aporepressor is involved requires its isolation and in vitro demonstration that it can prevent transcription and interact with both the endproduct and operator site on the operon. The Zubay system of in vitro enzyme synthesis (17) has been useful in identifying repressor proteins. To date, the products of argR and trpR, the repressors of the arginine and tryptophan pathways, respectively, are the only ones among the biosynthetic systems that meet most criteria.

Repressor proteins appear to be allosteric in nature so that the configuration that allows for interaction with the operator is influenced by ligands. There may be more than one ligand that may act alternately, cumulatively, or in a multivalent fashion (8) on specific operators. Operator sites may be nondiscriminating in their response to different repressor complexes modified by alternative ligands. These and other interactions are discussed in the systems described below.

Structural Genes as Regulatory Elements and Autoregulation

The original Jacob-Monod model of regulation of gene expression had an implicit requirement for a regulatory gene whose product had no other function except to control the expression of structural genes. A major divergence from this aspect of the original model has become increasingly evident with the repeated recognition that products of structural genes may have regulatory functions. Goldberger (18) has recently reviewed the surprisingly wide variety and number of systems in which regulatory functions for structural genes have been found. The product of the structural gene often is the first enzyme in the pathway that participates in control by feedback inhibition. In some cases, the structural gene is a component of the regulation unit controlled by its bifunctional product, so that it controls its own expression. This concept has led to the development of such terms as autoregulation (19), self-regulation (20), and autogenous regulation (18). Several of these systems are described below.

BIOSYNTHESIS OF ISOLEUCINE, VALINE, AND LEUCINE

The *ilv* regulon consists of five genes which are divided into three closely linked operons: *ilvADE, ilvC,* and *ilvB* (21). Operator genes have been identified for each of the operons (22–24). The operons are subject to multivalent repression by leucine, isoleucine, and valine when supplied in excess, and noncoordinate derepression when any of the amino acids are limiting. The *ilvC* operon is inducible, responding to either acetohydroxybutyric acid or acetolactate (25), the alternate products of the *ilvB* enzyme and substrates of *ilvC* enzyme. Leucine is formed in a three-step process from ketomethylbutyrate, an intermediate in the valine pathway. Four contiguous genes are involved and they are organized as the leucine operon (*leuABCD*) under unit control by leucine (26).

The control of the *ilv* operon was suggested to be autoregulatory with the structural gene product of *ilvA,* threonine deaminase, as the central protein moiety in the regulatory mechanism (27). The concept has now developed to become a multifunctional system to include repression and induction (19). The *ilvA* gene product is translated as a monomeric unit which develops through several stages of sulfhydryl bondings to a tetrameric stage, consisting of four *ilvA* monomeric subunits. At this point two molecules of pyridoxal phosphate are bound to each tetramer (28, 29) to produce the immature holotetramer moiety of threonine deaminase. Maturation to an active catalytic enzyme requires isoleucine or valine and is prevented when both amino acids are present (27). The in vivo importance of pyridoxal phosphate for maturation of the enzyme as a repressor and as a catalytic unit was demonstrated by Guirard et al (30) with a mutant that had an alternative requirement for pyridoxine, isoleucine, α-ketobutyrate, or α-ketomethylvalerate. The genetic lesion in this mutant was in *ilvA,* resulting in a higher K_m for pyridoxine for enzyme maturation. In addition, Wasmuth et al (31) provided evidence that immature threonine deaminase accumulates in the absence of pyridoxine even when the cells were supplied ample quantities of the branched-chain amino acids. A sharp increase in threonine deaminase levels occurred when pyridoxine was added to the culture. This "maturation" was chloramphenicol resistant, indicating the *ilvA* gene product was present in an inactive state. Calhoun & Hatfield (19) confirmed that pyridoxine was necessary for the formation of a functional repressor for *ilvA* and *ilvD.* They also noted that while induction of *ilvC* by either substrate occurred in the absence of the vitamin, a significantly greater rate of enzyme synthesis was achieved when the vitamin was included in the induction system.

Several authors have reviewed the evidence that tRNAs and their corresponding charging enzymes play important roles in the repression of the *ilv* regulon and *leu* operon (8, 9, 11, 21). The aminoacylated tRNAs of all three branched-chain amino acids apparently act as corepressor effectors in the *ilv* system, and evidence that all three bind to the immature threonine deaminase (aporepressor) was cited by Calhoun & Hatfield (19). Accordingly, the prediction could be made that any mutation that affects the charging reaction or the structure of the tRNA would alter the repression levels of the *ilv* operons. This was realized by Rizzino et al (15), who described the derepression of *ilv* and *leu* biosynthesis in a *hisT* mutant of *S. typhimurium*. The *hisT* gene is responsible for a tRNA-modifying enzyme which produces pseudouridine in the anticodon region of the tRNA. In this instance, the alteration was in the *leu*-tRNA. McGinnis & Williams (32) provided evidence that the aminoacyl-tRNA synthetases are themselves under control of their respective amino acids.

Wasmuth & Umbarger (33) showed that the isoleucine analog, thiaisoleucine, which is ordinarily able to replace isoleucine in the multivalent repression of the *ilv* regulon, was unable to do so in a strain bearing *ilvA466* mutation which has an altered threonine deaminase and is hyperderepressible for *ilvADE*. The dipeptide, glycyl leucine, which ordinarily cannot replace isoleucine in the repression system can nevertheless prevent accumulation of threonine deaminase-forming potential in normal cells but not in strain *ilvA466*. The authors interpreted this to mean that glycyl leucine in combination with *ilvA* gene product prevents initiation of new threonine deaminase polypeptides in nascent mRNA.

Levinthal et al (34) gave additional support to the role of threonine deaminase as a regulatory element and suggested a few modifications of the Calhoun-Hatfield-Burns autoregulation model. They isolated a leucine-sensitive strain (growth inhibited by 1.0 μM leucine) which mapped in the *ilvA* region and had considerably lower levels of threonine deaminase and hydrolase than the wild-type strain when grown under repressing conditions. The *ilvC* gene was inducible at two thirds the wild-type levels. Another anomaly of this mutant was the lower levels of all three branched-chain aminoacyl-tRNA synthetases and their differential levels during derepression in limiting isoleucine. A new concept of positive control was introduced by these findings which suggests that the *ilvA* gene product may augment mRNA synthesis and that this is prevented by the charged tRNA corepressors.

HISTIDINE BIOSYNTHESIS

Histidine biosynthesis, primarily studied in *Salmonella*, is accomplished by ten enzymes coded for by nine genes. The genes are contiguous and comprise a regulatory element that meets all the criteria for a classical operon (12). Regulatory deficient mutants have been isolated by the use of histidine analogs, and all of the cases that were not linked to the *his* operon have been identified as participants in maintaining the concentration, formation, or modification of histidyl-tRNA, the known corepressor. For example, *hisR* is the gene that codes for histidine tRNA, *hisS* controls the synthesis of histidyl-tRNA synthetase, and *hisU, W,* and *T* are involved in modifying the structure of histidine tRNA (12, 35). *HisT* codes for an enzyme required for the conversion of uridine to pseudouridine residues in the anticodon region of several tRNAs (36).

The absence of a regulatory gene coding for a true aporepressor has stimulated Goldberger and his associates to look for an alternative. Compelling evidence has been accumulated by this group of investigators that implicates the product of the *hisG* gene, phosphoribosyl-ATP synthetase, as the regulatory protein that participates with the corepressor in regulating the entire *his* operon, including itself. The following evidence meets the predictions for this alternative model: 1. Genetic events in the *hisG* gene create modifications that render the enzyme resistant to feedback inhibition, with or without loss of catabolic activity, and at the same time cause the rest of the enzymes to become derepressed (37, 38). 2. The mutations that cause this alteration in regulation are *trans* recessive as shown by *cis-trans* tests in merodiploids where the wild-type *hisG* gene acts in a dominant manner on the expression of a *his* operon carrying the defective *hisG* gene (39). 3. The recognized corepressor, histidyl-tRNA, specifically binds to the normal *hisG* enzyme but not to the genetically altered forms (40–42). 4. The *hisG* enzyme inhibits transcription of a *his* operon DNA template carried on a specialized transducing phage, *φ80dhis* (43). 5. The *hisG* enzyme binds to the *his* phage DNA and this binding is inhibited by neither the normal phage DNA nor *his* DNA carrying an operator constitutive mutation (18). Another prediction is obligatory participation of the corepressor in the inhibition of transcription by the *hisG* enzyme. This has not been realized in that inhibition is obtained with the *hisG* enzyme in the absence of added histidyl-tRNA (43). Another detail that must be considered in the eventual understanding of the histidine repression mechanism is the recent observation by Wyche et al (44) that histidyl-tRNA synthetase may act as a positive effector in the expression of the *his* operon. In strains that showed an increased synthetase activity, there was a concomitant increase in histidine operon expression measured by activity of the *hisB* and *hisD* enzymes.

TRYPTOPHAN BIOSYNTHESIS

A recent excellent and comprehensive review by Margolin (45) dealing with tryptophan synthesis and regulation is highly recommended for anyone interested in studying the various interactions of the aromatic amino acids and their regulation. Our emphasis is on recent advances dealing with tryptophan regulation in the *E. coli* system.

In *E. coli* the structural genes for biosynthesis of tryptophan from chorismate constitute a functional operon, map at 27 minutes (1) in the order *trpOEDCBA,* in the sequence of their role in the biosynthetic pathway. The first enzyme of the pathway is sensitive to feedback inhibition by tryptophan (45). An operator region preceding *trpE* had been defined, as well as two promoter regions: one preceding the operator and the second an internal promoter throught to be within *trpD* (45a).

Thus the architecture of the *trp* operon is more properly noted as *pOEDpD'CBA*. Regulation of this operon is controlled by a regulatory gene, *trpR,* which is completely separate from the *trp* operon and regulates the semicoordinate repression and derepression of the operon. In addition, *trpR* controls one of three isoenzymes for the first step in the general chorismic acid pathway, deoxy-D-arabinoheptulosonic acid-7-phosphate synthetase [DAHP synthetase (*trp*)] (46, 47). This enzyme is coded by the *aroH* gene, specifically repressed by tryptophan, and derepressed in the *trpR* mutant. Thus, a *trp* regulon exists comprising the five genes of the *trp* operon plus *aroH.*

The protein nature of the *trpR* product was implied by the existence of temperature-sensitive mutants (48) and amber-suppressible *trpR* mutants (49). Zubay et al (50) isolated the *trp* repressor using an in vitro transcription and translation system containing a DNA template, λ*dtrp-lac,* which carried a *lac-trp* fusion so that β-galactosidase expression was controlled by the *trp* operator. The synthesis of β-galactosidase from the λ*dtrp-lac* was decreased approximately 90% when partially purified *trpR*+ but not *trpR* proteins were added to the system. β-galactosidase synthesis directed by λ*dlac* was unaffected by the addition of either extract, indicating the specificity of the inhibition. The repressor interacted strongly with phosphocellulose, as do many proteins which bind to nucleic acid. The concentration of the repressor was calculated to be approximately 10 molecules per cell. Shimizu et al (51) demonstrated inhibition of mRNA synthesis from λ*pt60-30* DNA carrying the entire *trp* operon with a *trp* repressor isolated from a phosphocellulose column. Acrylamide gel electrophoresis analysis and Bio-gel separation indicated a molecular weight of approximately 60,000. Rose et al (52) have developed an in vitro system for measuring transcription of the *trp* operon in which *trp* mRNA synthesis is initiated with purified RNA polymerase. This transcription was blocked by a combination of tryptophan plus the *trpR* product from *TrpO*+ but not from *trpO*c templates. Thus, an intact operator site is necessary for the repressor protein to inhibit.

There have been several suggestions that the *trpS* gene product, tryptophanyl tRNA synthetase, may play a role in repression of the *trp* operon (53). The in vitro studies of Squires et al (54) substantially proved that the *trpS* gene product is not involved in regulation. They found repression of in vitro transcription by partially purified (55-fold) preparations from *trpR*+ cells identical with that of a preparation from wild-type or *trpS* mutants. Further evidence that tryptophanyl-tRNA synthetase does not play a role in repression is found in the fact that 5-methyltryptophan, which does not substitute for tryptophan in the charging reaction by the synthetase, does stimulate repression. In contrast, 7-azatryptophan, which does charge to tRNA, is completely inactive in repression. The addition of tryptophanyl-tRNA and repressor to the transcription system did not stimulate repression; only tryptophan (or its methylated analogs) were active. Further purification of the repressor away from the synthetase did not reduce the effectiveness of the repression. McGeoch et al (55) have reported the same conclusions concerning the absence of the *trpS* influence and the requirement of only repressor plus tryptophan. However, their preparation had a half-maximal repression value approximately 10 times lower than the preparation of Squires. McGeoch suggests the repressor in Squires' preparation may be altered during the purification or that some active component is missing. Whatever the resolution of this point, it is relatively certain charged tryptophanyl-tRNA or its synthetase are not components of the repressor complex.

Somerville & Yanofsky (56) found that mutations mapping within the *trpE* region alter the regulation of the operon, and it was suggested that the *trpE* gene product, anthranilate synthetase, may have a role in the regulation of the operon in addition to its catalytic function. Lack of support for this concept came with the demonstration by Hiraga & Yanofsky (57) that a mutant that had almost the entire *trpE* region deleted still repressed or derepressed tryptophan synthetase (TS) in a normal fashion. In addition, when tryptophan was supplied to tryptophan-starved cells, the synthesis of *trp*-mRNA ceased sequentially. They concluded that neither the *trpE* gene product nor the *trpE* gene message was required for regulation of the *trp* operon. Also relevant is the finding by Jackson & Yanofsky (58) that the region between the operator and the first structural gene has regulatory functions. Two deletion mutants of the *trpED* region had increased rates of synthesis of tryptophan synthetase. They cited unpublished evidence indicating that both deletion mutants retained the oligonucleotide sequence preceding the *trpE* initiator codon but not the first part of the *trpE* message. Therefore, the authors propose that the deletion removed *trpE* and a region to the left of *trpE* initiator codon responsible for regulation of transcription initiations.

In addition to the regulator-mediated control, there is an important control concept independent of an operator-repressor interaction. Several lines of evidence indicate the importance of metabolic regulation such as differences in the frequency of initiation of transcription in *trpR* cells compared to repressed and derepressed *trpR*$^+$ (59), or the observation that transcription and translation of *trpR* cultures can be modulated by various nutritional sources (60). In the same category are the observations that addition of tryptophan to derepressed cultures resulted in immediate cessation of transcription, at a rate faster than expected to occur by repression (61). Hiraga & Yanofsky (62) observed the same rapid reduction when rifampicin was added to block transcription in regulatory constitutive and derepressed cultures.

ARGININE BIOSYNTHESIS

The arginine regulon is composed of four unlinked structural genes and four (*argECBH*) that are clustered together, suggesting an operon. The cluster is apparently a complex of overlapping operons as suggested by lack of coordinate control of *argE* and *argH* (63–66). Polar and constitutive mutations mapping within this cluster have been used to suggest that (*a*) an internal operator exists, (*b*) *argE* has a counterclockwise and *argCBH* a clockwise direction of transcription, and (*c*) a region between *argB* and *argH* may function as a secondary promotor (67, 68). Although separate operator mutations for *argE* or *argCBH* have not been described, recognition of the interrelationship between promoter and operator genes have led the authors to propose the gene order: $argE \cdot p_{CBH} \cdot O_{CBH}/O_E \cdot p_E \cdot argCBH$.

Krzyzek & Rogers (69) have found that the mRNA of this cluster is synthesized in short pieces averaging 1800 to 2400 nucleotides long. If divergent transcription occurs and the secondary promotor between *argB* and *H* is of physiological significance, then the mRNA data support the gene order. The bidirectional transcription proposed here is not new and has been shown to be the mechanism of transcription of the biotin locus (70, 71). The recent experiments of Panchel et al (72) indicate that the *argECBH* cluster is indeed divergently transcribed. They used a system of the type used for the biotin investigation (73) to demonstrate that RNA isolated from *E. coli* (*arg*$^+$), grown in the absence of arginine, hybridized with the leftward and rightward transcribing DNA strands of arginine-transducing ϕ80 phage with a ratio of 30:70. Arginine, supplied in excess to growing cells, completely repressed both directions of transcription, while ornithine and citrulline completely repressed the leftward transcription of *argE*. The rightward transcription was reduced 80–95% by ornithine and over 98% in the presence of citrulline. It was proposed that a gradient of repression of *argECBH* can be achieved by intermediates of arginine relative to their position in the biochemical pathway.

It is known that arginine does act as a feedback inhibitor of the first enzyme of the biosynthetic pathway and also affects repression in cooperation with the product of the regulatory gene *argR* (74). Jacoby & Gorini (64), following the original observations of Maas & Clark (13), have demonstrated that the product of the *argR*$^+$ gene was a protein that functions as a *trans*-dominant entity and participates in negative control of the *arg* regulon. In addition, Kadner & Maas (75) have isolated a variety of regulatory mutants and classified them into three categories. The first class constitutes altered gene products while the other two classes are thought to have altered stabilities or reduced affinities for arginine and operator regions. These latter two classes complement in vivo to effect intermediate levels of regulation. This observation could imply that the regulatory gene product is composed of at least two subunits or that the operator regions bind more than one repressor molecule.

A method to study the nature of the repressor is found in work of Urm et al (77). They used a ϕ80 bacteriophage carrying *argECBH* (76) as template for in vitro transcription and translation of the *argE* gene product in the presence and absence of partially purified extracts of *argR*$^+$ and *argR* cells. The amount of the *argE* enzyme produced was quantitated by direct assay for enzyme activity. The *argR*$^+$ extract was 6 times more active than the *argR* extract in inhibiting transcription. Neither extract caused a decrease in β-galactosidase synthesis from mRNA transcribed from $\lambda plac$. The author calculated that approximately 206 repressor molecules are present per cell, which accounts for about 0.03% of the total cell protein. The cofactor of the repressor could not be defined since both arginine and arginyl-tRNA were present in the synthesis mixture.

The possibility that arginyl-tRNA might be a corepressor was examined by Celis & Maas (78). They found no difference in the levels of the five arginyl-tRNA fractions of wild-type cells grown under conditions of repression and derepression. This was only indirect evidence that charged tRNA did not play a role in repression, and further work is required to prove this at the transcriptional level. A role for arginyl-tRNA synthetase as a direct participant in the repression mechanism has been suggested by Williams (79, 80).

There is some evidence that regulation is also exerted at the translational level (81–83) as well as the transcription site (66, 83). For instance, some investigators have reported evidence that *argS* (arginyl-tRNA synthetase) must be functional for repression by canavanine, an arginine analog (79, 80), while others (84) indicate that the state of the *argS* gene is inconsequential for inhibition by canavanine. In fact, Faanes & Rogers (84) felt that canavanine probably exerts its effect at the level of translation. Lavelle (81) noted that the sharp decrease in enzyme activity after the addition of arginine to a culture was too fast for a repression effect and suggested a translational control system. All this must be reinterpreted in view of the recent evidence regarding metabolic regulation and the influence of the metabolic state of the cell on the synthesis of inducible and repressible systems.

METHIONINE AND CYSTEINE BIOSYNTHESIS

A comprehensive review dealing with the biochemistry and genetics of the sulfur amino acids has been recently published by Smith (85). Methionine is one of the four amino acids derived from aspartic acid. It arises from the branch point intermediate, homoserine, which also serves as a precursor to threonine. For convenience of later discussions, the conversion of homoserine to methionine is considered in two parts. The first part deals with the straightforward formation of homocysteine from homoserine in three steps with cystathionine as a key intermediate. These three sequential enzymes are derived from the unlinked *metA, metB,* and *metC* genes respectively. The second part involves the transmethylase complex required for conversion of homocysteine to methionine (86). This can be divided into three elements, controlled by three separate genes. There are two transmethylase enzymes both of which use the methylation cofactor, N_5-methyltetrahydrofolate. This cofactor is formed by the *metF* enzyme, N_5,N_{10}-methylene tetrahydrofolate reductase. One of the transmethylases (*metH* enzyme) requires exogenous Vitamin B_{12} (cobalamine) for its activation; the other is the product of the *metE* gene and functions without Vitamin B_{12}.

None of the six genes are adjacent, and methionine represses all of the enzymes in a noncoordinate manner (85). In addition, methionine also represses the aspartokinase-homoserine dehydrogenase complex II, specified by *metLM* (87). The isolation of regulatory mutants with altered repressibility of the enzymes of the methionine pathway now allows the assignment of a *met* regulon to at least those genes involved in the branch pathway from homoserine. These regulatory mutants were first isolated in *Salmonella* by Lawrence & Smith as mutants resistant to the methionine analogs, α-methyl methionine and/or ethionine (88). Three distinct genetic types were obtained: one was a feedback-resistant alteration of the *metA* enzyme, and the others (*metK* and *metJ*) classify as controlling regulatory cytoplasmic factors on the basis of *cis-trans* analyses (88, 89). Equivalent mutants have since been isolated in *E. coli* (90–93). The existence of two such genes suggested that one may represent an aporepressor and the other may be needed for the conversion of methionine to the actual corepressor. Chater (89) suggested S-adenosylmethionine (SAM) as the possible corepressor, and Greene et al (90, 93) independently realized this prediction by isolating ethionine-resistant mutants of *E. coli* that were depressed for the methionine pathway, and had high intracellular methionine pools and low SAM synthetase activity. Additional evidence that *metK* is the gene for SAM synthetase in *Salmonella* has been reported by Hobson & Smith (94). They found that those *metK* that excreted methionine were deficient in SAM synthetase and those that did not excrete methionine had defective synthetases with low substrate binding capacity (K_m mutants). One *metK* (excretor) mutant had normal SAM synthetase activity, and this was found to be in a different genetic complementation group than the other *metK* mutants. Complete agreement that *metK* mutants are defective in SAM synthetase activity has not been obtained in all cases (95, 96). Since bacterial cells are impermeable to SAM, it is not possible to obtain direct evidence that this is the true corepressor, and ultimate clarification must await more direct tests involving in vitro transcription systems.

With the assignment of *metK* to corepressor function, *metJ* remains as a likely candidate for the aporepressor though other possibilities remain viable (94). It is interesting, and a bit perplexing, that *metJ* also regulates *metK* since SAM synthetase levels are unusually elevated in *metJ* mutants (91, 92, 94). Since *metJ* is also involved in the repression of methylene tetrahydrofolate reductase (*metF*), Meedel & Pizer (97) looked into the possible role of S-adenosylmethionine as a repressor of other enzymes of one-carbon metabolism. They found that *metK* mutants showed elevated levels of serine transhydroxymethylase, but no change in the activities of methylene tetrahydrofolate dehydrogenase or the generation of one-carbon units from glycine.

An additional complexity in regulation of the methionine pathway was introduced with the discovery of enzyme repression by Vitamin B_{12} (98). Whereas methionine can repress all of the enzymes of the pathway, Vitamin B_{12} selectively represses only two of the elements of the homocysteine methylating system: the *metE* and *metF* enzymes (98–100). In *metK* and *metJ* mutants, repression of these enzymes by methionine is altered, but repression by Vitamin B_{12} still occurs (99, 100). In *metH* mutants deficient in the B_{12}-dependent transmethylase, repression by methionine is normal but repression by Vitamin B_{12} is abolished (100). This finding strongly suggests that repression by B_{12} requires the product of *metH*, i.e. the B_{12}-transmethylase controls the expression of the other transmethylase. The model includes the activation of the *metH* apoenzyme by Vitamin B_{12} to give a holoenzyme that is now catalytically active and also functions as a repressor for the non-B_{12}-transmethylase. Thus, the *metE* and *metF* genes differ from the other *met* genes in that they are repressed by either the S-adenosylmethionine-*metJ* complex or the B_{12}-*metH* repressor. This difference may reside in a repressor recognition element that responds alternately to either of the two repressor complexes.

Cysteine enters the methionine pathway at the second enzyme (*metB*) as a substrate for the formation of cystathionine. The biosynthesis of cysteine from serine and exogenous sulfate requires a sulfate permease (*cysA*) and six enzymes in a branched convergent pathway (85). Four of the enzymes are involved in the reduction of sulfate to sulfide, and six genes, including a three-gene complex (*cysIJG*), have been identified for these reactions in *S. typhimurium*. The final step is a sulfhydrylation of O-acetylserine to yield cysteine. No auxotrophic mutants have been found for this reaction. An apparent gene that controls the sulfhydrylase was recently identified by mutation to triazole resistance (101). The resistant mutants have defective sulfhydrylase activity but are still prototrophic, thus indicating an alternate pathway for the final step in the conversion of serine to cysteine. The unusual feature of the cysteine pathway that makes it different from other biosynthetic pathways is that all of the genes, including the *cysA* permease, are under positive cascade control requiring O-acetylserine, the product of *cysE*, as inducer. Thus *cysE* mutants unable to generate O-acetylserine are pleiotropic for the other reactions (102, 103). *CysB* mutants are also pleiotropic, and this gene has been identified as playing a regulatory role in control by coding for the positive protein effector (102, 103).

BIOSYNTHESIS OF PURINE NUCLEOTIDES

The biochemical and genetic aspects of the purine pathway in *Salmonella* has recently been reviewed by Gots (104). Ten steps are required for the synthesis of the first complete nucleotide, inosine-5-monophosphate (IMP), and two each for its conversion to adenosine-5-monophosphate (AMP) and guanosine-5-monophosphate (GMP) to give a total of 14 enzymatic reactions. Corresponding genes have been identified for all of these reactions except for the third enzyme, phosphoribosyl-glycinamide formyltransferase (105). The genes are scattered throughout the chromosome, but some are linked and represent members of functioning operons. One of these consists of three contiguous genes in the order *purJ–purH–purD*, controlling enzymes 10, 9, and 2, respectively (106, 107). These enzymes are coordinately controlled, and polarity and pleiotropic effects have been observed. Similar effects have been observed for the other functional operon which contains the two genes, *guaA* and *guaB*, that control the two enzymes required for the conversion of IMP to GMP (108–110).

Though the formation of all of the enzymes is repressible (and derepressible) to varying degrees, neither the nature of the repressor substance nor the metabolic corepressor has yet been determined for any of the enzymes. Attempts to determine the true corepressor have been frustrated by interconversion reactions. In mutants where some interconversion reactions are blocked, a guanine derivative has been implicated as the corepressor for the two enzymes of the *gua* operon, an adenine derivative for adenylosuccinate synthetase, and either derivative for the early enzymes (104). Whether or not this implies three or more separate regulatory units controlled by different repressors with varying affinities for a common corepressor must await the isolation and characterization of true regulatory mutants.

Regulatory mutants of the purine pathway have been recently isolated in our laboratory (111, 112; Gots and Benson, unpublished observations). These mutants, designated *purR*, are derepressed for at least five of the enzymes that participate in the early synthesis of IMP. At least one member of the *gua* operon, IMP dehydrogenase, is unaffected, as is also adenylosuccinate lyase, a bifunctional enzyme required for both synthesis of IMP and its conversion to AMP.

Isolation of the *purR* mutants was accomplished by several methods. Earlier attempts to isolate such mutants by use of analogs were unsuccessful when purine bases were used, but they finally were found among mutants resistant to purine nucleoside analogs such as mercaptoguanosine, 6-chloropurine riboside, and 6-methylmercaptopurine riboside. They were also found spontaneously as a secondary mutation in some *purA* mutants (see below). Still another and most productive method for isolation of *purR* mutants took advantage of the discovery by Jacob, Ullmann & Monod (113) of an episome carrying a deletion that fuses the *lac* operon with the *purE* operon. The deletion cuts through the *lacZ* gene of the *lac* operon and one of the two genes of the *purE* operon so that the expression of the intact *lacY* (permease) is now under control of the *purE* operator and hence repressible by purines. Melibiose is an alternative substrate for the *lac* permease so that in a *Salmonella* mutant (*melB*) lacking the melibiose permease (114), the ability to use melibiose as a growth factor can be restored by introducing the *E. coli* episome carrying the *lac-purE* fusion. However, since *lacY* is now under control of the *purE* operator, purines, such as adenine, strongly inhibit growth on melibiose in such a strain. Mutants resistant to this inhibition are readily isolated and have been shown by *cis-trans* tests to be due either to a chromosomal mutation of the *purR* type or to an episomal mutation involving the *purE* operator site. The *purR* mutants isolated by all of the above methods are phenotypically identical, but genetic identity in terms of allelism has not been determined as yet. It is also not yet known whether the *purR* mutants are defective in an aporepressor or in a common corepressor.

The nature of the *purR* mutants and earlier observations on patterns of repression now suggest that the purine pathway is composed of at least three separate regulatory units: 1. the *gua* operon controlling the two enzymes in the conversion of IMP to GMP, 2. an AMP regulon containing the two unlinked genes (*purA* and *purB*) for conversion of IMP to AMP, and 3. the IMP regulon consisting of the *purJHD* operon and other scattered genes required for the synthesis of IMP and under control of the *purR* product.

A role for one of the purine nucleotide biosynthetic enzymes, adenylosuccinate synthetase, as a possible regulatory protein for control of the early enzymes in the pathway has been revealed in studies with yeast mutants (115–118). The regulatory defect was implied by showing that in mutants defective in the synthetase enzyme, synthesis of an early intermediate was no longer prevented by growth in adenine. Similar mutants that were unaltered in enzyme activity mapped in the gene that controlled the synthetase. Direct measurement of the effect of these mutations on the activities of the early enzymes was not done, so it remained uncertain whether the regulatory modification was related to repression control or feedback inhibition.

Using these observations in the yeast system as a lead, Benson & Gots (111) examined equivalent mutants of *Salmonella*, namely *purA*, for their ability to repress the first two enzymes of the pathway. Suggested confirmation for the potential regulatory function of the *purA* enzyme was obtained by showing that some of the *purA* mutants were abnormally derepressed for the early enzymes. However, further genetic analyses showed that this regulatory defect was not linked to *purA* and represented a secondary unrelated mutation that appeared spontaneously in the strains. This mutation, designated *purR,* is phenotypically indistinguishable from the *purR* mutants obtained in other ways (see above). The reason for the propensity of such mutations among *purA* strains is still not clear.

PYRIMIDINE BIOSYNTHESIS

Six enzymes controlled by six (*pyrA* to *pyrF*) separate unlinked genes (119) are involved in the de novo biosynthesis of uridine monophosphate (UMP). Although the last four enzymes appear to be coordinately controlled, the existence of a functional operon is precluded because the six genes are unlinked (1, 2). No clearcut evidence for unit control by a regulon has yet been obtained. In fact, a multiplicity of controlling elements is suggested by experiments designed to assess corepressor function.

A variety of interconversion reactions ordinarily makes it difficult to determine whether a cytosine or uracil derivative participates as a corepressor. In order to allow independent assessment of the two pyrimidines as corepressors, Williams & O'Donovan (120) used a pyrimidine (*pyrA*) auxotroph of *Salmonella* that was also deficient in interconversion reactions such as cytidine deaminase, CTP synthetase, and uridine phosphorylase. This mutant thus provided a unique system where two separate and noninterconvertible channels were maintained for the independent flow of cytosine and uracil metabolites. The results indicated that aspartyl transcarbamylase (*pyrB*) was uniquely controlled by a uridine derivative and that the others were primarily affected by a cytidine derivative with some minor participation of uridine in two cases. The control of the first enzyme of the pathway, carbamoyl phosphate synthetase (*pyrA*), is further complicated because of its role in providing an intermediate for arginine biosynthetase. Cumulative repression by both arginine and pyrimidines has been shown and the pyrimidine effector appears to be a cytidine derivative (121, 122). A role for the *argR* repressor product has also been implicated since *argR* mutation reduces repressibility of carbamoyl phosphate synthetase by both arginine and pyrimidines (123).

O'Donovan & Gerhart (124) have reported the isolation of regulatory mutants of the pyrimidine pathway, designated *pyrR,* in *Salmonella.* These mutants excrete pyrimidines and are derepressed for five of the enzymes. A similar type of mutant, designated *pyrH,* was isolated independently by Ingraham & Neuhard (125) as a cold-sensitive conditional mutant defective in UMP kinase. *PyrH* mutants have low UTP pool levels, excrete pyrimidines, and are derepressed for several biosynthetic enzymes as well as for CMP kinase. Since *pyrH* and *pyrR* map in the same place, these may well represent two independent isolations of the same type of mutant with regulatory defects due to corepressor deficiencies. The patterns of repression so far observed suggest that several repressor proteins with different ligand binding specificities act together for overall regulation of the six biosynthetic enzymes. A mutation affecting a single repressor protein would thus be expected to be constitutive for some but not all enzymes of the pathway. As yet, no such mutant has been found.

NUCLEOSIDE CATABOLISM

A genetic unit for the control of nucleoside catabolism was first revealed by the biochemical and genetic analyses of the catabolic utilization of thymidine and purine deoxyribonucleosides as carbon sources for the growth of *Salmonella* and *E. coli* (see reviews 104, 119). Thymidine phosphorylase (*tpp*) and purine nucleoside phosphorylase (*pup*) in concert with two other enzymes, deoxyribose-1-phosphate mutase (*drm*) and deoxyribose-5-phosphate aldolase (*dra*), participate in the ultimate catabolic degradation. All four enzymes are induced by growth in thymidine, and the induction patterns in *drm* and *dra* mutants imply that the product of the mutase and substrate of the aldolase, namely deoxyribose-5-phosphate, is the true inducer (126, 128). Mapping and ordering of the genes show contiguity of the four genes in the order, *deoC(dra)–deoA(tpp)–deoB(drm)–deoD(pup)* (127, 128). The *deo* designations have been used in *Salmonella,* and the enzyme abbreviations have been used to name the genes in *E. coli,* though it has been suggested that the latter be called *nucA, nucB, nucC, nucD* (129). The existence of a common inducer and the gene clustering implied that the unit of control was that of a classical type operon. However, further studies with coordinate induction, polarity effects, and operator constitutive mutations suggest that the *deo* cluster is a regulon composed of two tandem operons, *deoCA* and *deoBD* (127–131). Both are regulated by deoxyribose-5-phosphate derived from deoxyribonucleosides, but *deoBD* is also induced by purine ribonucleosides (128, 130, 132). A more accurate survey of coordinate effects and the existence of a constitutive mutant specific for *deoA* (P. Hoffee, personal communication) suggest that the *deoCA* region contains separate operator sites for the two genes.

An unlinked regulatory gene that controls the entire *deo* regulon has been found both in *Salmonella,* where it is designated *deoR* (133), and in *E. coli,* designated *nucR* (129). Mutations in this gene create constitutive expression of all four of the enzymes.

Another regulon consisting of two unlinked genes has been implicated in the catabolic utilization of cytidine. The two enzymes, cytidine deaminase (*cdd*) and uridine phosphorylase (*udp*), are induced by cytidine and under control of a single regulatory gene, *cytR* (134–137). An unusual relationship between the *cyt* regulon and the *deo* regulon was revealed by showing that the four *deo* enzymes are partially induced by cytidine and become partially constitutive in the *cytR* mutant (135–137). The *deoR* mutation has little or no effect on expression of the two *cyt* enzymes. Thus, a complicated interaction exists involving a super-regulon composed of at least three, and possibly four, separate regulatory units. The *deoR* product controls two, and possibly three, recognized operons of the *deo* regulon, with deoxyribose-5-phosphate as effector for all units and purine ribonucleosides as effector for the *deoBD* pair. This control is apparently insensitive to catabolite repression (135, 136). The *cytR* product, in addition to regulating the two enzymes of the *cyt* regulon with cytidine as effector, can also participate in the regulation of all elements of the *deo* cluster. This regulation is sensitive to catabolite repression, and adenosine may also participate as an effector (134, 136). Though cytidine, and not uridine, is the apparent inducer in *E. coli* (137), uridine may also act as an inducer in *Salmonella* (136).

HISTIDINE UTILIZATION

The biochemistry and genetics of the histidine-utilizing enzymes (*hut*) for *S. typhimurium, Klebsiella aerogenes,* and *B. subtilis* have been studied by Magasanik and co-workers. Histidine can be utilized as a carbon and/or nitrogen source by degradation to glutamic acid, ammonia, and formamide. Four enzymes, coded by four genes (*hutH, U, I, G*), are involved. In *Salmonella* the genes are clustered together and function as two closely linked tandem operons (138, 139). The left-hand operon contains the two structural genes (*I–G*) for the last two enzymes and its promoter *M* (140). The right-hand operon consists of promoter *P, R* (the site of catabolite resistance), and *Q*, the operator for structural genes *U-H* for the first two enzymes. The *hutC* regulatory gene is part of the *I-G* operon, specifies the repressor protein controlling both operons (141), and does not appear to serve any other function in the cell. Mutations in *hutC* result in constitutive expression of both operons. Repressibility is dominant in merodiploids (*hutC⁻/F'hutC⁺*) and thus depends on a diffusible entity (20, 141). The *hutC* gene product is unique in that it is a repressor that regulates its own synthesis.

Induction of the *hut* operon, i.e. counteracting negative repressor control, occurs when excess histidine is added to the culture medium. The true inducer of the operons has been shown to be urocanate (138) as evidenced by facts that (*a*) a *hutH⁻* mutant (lacks histidase) was induced by imidazole propionate, a nonmetabolizable analog of urocanate, and (*b*) a *hutU⁻* mutant (lacks urocanase) had increased levels of histidase and enzyme I. The mechanism appears to be that excess histidine is metabolized by low residual levels of histidase to form urocanic acid, which interacts with the repressor to remove inhibition, thus permitting transcription and translation of the *hut* operon. Consequently, more *hutC* gene product is produced and, in the absence of additional urocanate, repression occurs. Residual histidase in noninduced cells is thought to be adequate to produce quantities of urocanate when the histidine levels are elevated. Exogenously supplied urocanate is a poor inducer of the *hut* operon because of poor permeability into the cell.

The *hut* repressor has been isolated and partially purified (142). Activity was determined by DNA–protein binding to nitrocellulose filter experiments with $\lambda phut-^{32}$ P labeled DNA. Pronase removed activity from the filter, confirming the protein nature of the repressor. Extracts prepared from *hut* deletion mutants and *hutC* strains did not have specific DNA binding activity. Thus the *hutC* gene product is the only factor involved in the binding mechanism and seems to be solely responsible for repression of the *hut* operons, including its own expression.

The preceding physiological and biochemical data also apply to *Klebsiella* and *Bacillus.* Although a repressor protein has not been isolated from these strains, this appears to be an academic problem at this time. There is a regulatory difference which should be considered. While *Salmonella* and *Klebsiella* have similar genetic maps, the catabolite repression is different in *Klebsiella* (143). *S. typhimurium* is sensitive to catabolite repression when histidine is the sole nitrogen source, while *Klebsiella* is not. The histidase of *K. aerogenes* is sensitive to catabolite repression when the bacteria are grown with a good nitrogen source, and this repression is relieved by cAMP. When a *Salmonella* episome bearing the *hut* operon was introduced into *Klebsiella* (*hut* operon deleted) the histidase was controlled like the normal histidase of *Klebsiella.* Thus the factor involved in the resistance of *Klebsiella* to catabolite repression was a cytoplasmic entity which could operate in *trans.*

Prival & Magasanik (143) concluded that certain enzymes which help provide the cell with carbohydrate and nitrogen escape catabolite repression under conditions of limiting nitrogen. Presumably some system involved in nitrogen metabolism was responsible for this escape mechanism. It was not surprising that Prival et al (144) extended the original observations by studying histidase regulation in *Klebsiella* with elevated levels of glutamine synthetase. They found that escape of histidase from catabolite repression occurred only in cells with elevated levels of glutamine synthetase. In addition, they isolated and partially characterized three different types of glutamine-requiring strains. The *glnA* mutants, thought to bear mutations in the structural gene for glutamine synthetase, and *glnB* mutants, suggested to bear mutations in the structural gene for an activator of synthesis or a regulator of glutamine synthetase, were unable to produce histidase when grown with glucose and a limited nitrogen source. The *glnC*, linked by transduction to *glnA*, probably has an operator function since this type of mutation results in constitutive but catabolite-resistant synthesis of glutamine synthetase, histidase, and proline oxidase. The important role of glutamine synthetase in this phenomenon was illustrated by the requirement for activation by either cAMP and CAP or nonadenylylated (active) glutamine synthetase for in vitro transcription of *hut* operon DNA of *S. typhimurium* (145). Glutamine synthetase is proposed to function as a positive control element in the regulation of the *hut* operon and probably any catabolic system that contributes to the nitrogen pool during nitrogen limitation.

ACKNOWLEDGMENTS

The work in the authors' laboratory was supported by Public Health Science grant CA-02790 from the National Cancer Institute and by a research grant GB-25357 from the National Science Foundation. C. E. Benson is a Pennsylvania Plan Scholar.

Literature Cited

1. Taylor, A. L., Trotter, C. D. 1972. Linkage map of *Escherichia coli* K-12. *Bacteriol. Rev.* 36:504–24
2. Sanderson, K. E. 1972. Linkage map of *Salmonella typhimurium,* Edition IV. *Bacteriol. Rev.* 36:558–86
3. Fraenkel, D. G., Vinopal, R. T. 1973. Carbohydrate metabolism in bacteria. *Ann. Rev. Microbiol.* 27:69–100
4. Davies, J., Nomura, M. 1972. The genetics of bacterial ribosomes. *Ann Rev. Genet.* 6:203–34
5. Benveniste, R., Davies, J. 1973. Mechanisms of antibiotic resistance in bacteria. *Ann. Rev. Biochem.* 42:471–506
6. Hartman, P. E., Roth, J. R. 1973. Mechanisms of suppression. *Advan. Genet.* 17:1–105
7. Epstein, E., Beckwith, J. R. 1968. Regulation of gene expression. *Ann. Rev. Biochem.* 37:411–31
8. Umbarger, H. E. 1969. Regulation of amino acid metabolism. *Ann. Rev. Biochem.* 38:323–70

9. Calvo, J. M., Fink, G. R. 1971. Regulation of biosynthetic pathways in bacteria and fungi. *Ann. Rev. Biochem.* 40:943–68
10. Reznikoff, W. S. 1972. The operon revisited. *Ann. Rev. Genet.* 6:133–56
11. Smith, J. D. 1972. Genetics of transfer RNA. *Ann. Rev. Genet.* 6:235–56
12. Brenner, M., Ames, B. N. 1971. The histidine operon and its regulation. In *Metabolic Pathways,* Vol. 5, *Metabolic Regulation,* ed. H. Vogel, Chap. 11. New York: Academic
13. Maas, W. K., Clark, A. J. 1964. Studies on the mechanism of repression of arginine biosynthesis in *Escherichia coli:* II. Dominance of repressibility in diploids. *J. Mol. Biol.* 8:385–70
14. Chapman, L. F., Nester, E. W. 1968. Common element in the repression control of enzymes of histidine and aromatic amino acid biosynthesis in *Bacillus subtilis. J. Bacteriol.* 96:1658–63
15. Rizzino, A. A., Bresalier, R. S., Freundlich, M. 1974. Derepressed levels of the isoleucine-valine and leucine enzymes in *hisT1504,* a strain of *Salmonella typhimurium* with altered leucine transfer ribonucleic acid. *J. Bacteriol.* 117:449–55
16. Umbarger, H. E. 1971. Metabolic analogs as genetic and biochemical probes. *Advan. Genet.* 17:119–40
17. Zubay, G. 1973. *In vitro* synthesis of protein in microbial systems. *Ann. Rev. Genet.* 7:267–87
18. Goldberger, R. F. 1974. Autogenous regulation of gene expression. *Science* 183:810–16
19. Calhoun, D. H., Hatfield, G. W. 1973. Autoregulation: A role for a biosynthetic enzyme in the control of gene expression. *Proc. Nat. Acad. Sci. USA* 70:2757–61
20. Smith, G. R., Magasanik, B. 1971. Nature and self-regulated synthesis of the repressor of the *hut* operons in *Salmonella typhimurium. Proc. Nat. Acad. Sci. USA* 68:1493–97
21. Umbarger, H. E. 1971. The regulation of enzyme levels in the pathways of branched-chain amino acids. See Ref. 12, Chap. 13
22. Pledger, W. J., Umbarger, H. E. 1973. Isoleucine and valine metabolism in *Escherichia coli.* XXII. A pleiotropic mutation affecting induction of isomeroreductase activity. *J. Bacteriol.* 114:195–207
23. Ramakrishnan, T., Adelberg, E. A. 1965. Regulatory mechanisms in the biosynthesis of isoleucine and valine II. Identification of two operator genes. *J. Bacteriol.* 89:654–60
24. Ramakrishnan, T., Adelberg, E. A. 1965. Regulatory mechanisms in the biosynthesis of isoleucine and valine. III. Map order of the structural genes and operator genes. *J. Bacteriol.* 89:661–64
25. Arfin, S. M., Ratzkin, B., Umbarger, H. E. 1969. The metabolism of valine and isoleucine in *Escherichia coli.* XVII. The role of induction in the derepression of acetohydroxyl acid isomeroreductase. *Biochem. Biophys. Res. Commun.* 37:902–8
26. Freundlich, M., Burns, R. O., Umbarger, H. E. 1962. Control of isoleucine, valine and leucine biosynthesis. I. Multivalent repression. *Proc. Nat. Acad. Sci. USA* 48:1804–8
27. Hatfield, G. W., Burns, R. O. 1970. Specific binding of leucyl-transfer RNA to an immature form of L-threonine deaminase: Its implications in repression. *Proc. Nat. Acad. Sci. USA* 66:1027–35
28. Hatfield, G. W., Burns, R. O. 1970. Ligand-induced maturation of threonine deaminase. *Science* 167:75–76
29. Hatfield, G. W., Burns, R. O. 1970. Threonine deaminase from *Salmonella typhimurium.* III. The intermediate structure. *J. Biol. Chem.* 245:787–91
30. Guirard, B. M., Ames, B. N., Snell, E. E. 1971. *Salmonella typhimurium* mutants with alternate requirements for vitamin B_6 or isoleucine. *J. Bacteriol.* 108:359–63
31. Wasmuth, J., Umbarger, H. E., Dempsey, W. B. 1973. A role for a pyridoxine derivative in the multivalent repression of the isoleucine and valine biosynthetic enzymes. *Biochem. Biophys. Res. Commun.* 51:158–64
32. McGinnis, E., Williams, L. S. 1971. Regulation of synthesis of the aminoacyl-transfer ribonucleic acid synthetases for the branched-chain amino acids of *Escherichia coli. J. Bacteriol.* 108:254–62
33. Wasmuth, J. J., Umbarger, H. E. 1974. Role for free isoleucine or glycyl-leucine in the repression of threonine deaminase in *Escherichia coli. J. Bacteriol.* 117:29–39
34. Levinthal, M., Williams, L. S., Levinthal, M., Umbarger, H. E. 1973. Role of threonine deaminase in the regulation of isoleucine and valine biosynthesis. *Nature New Biol.* 246:65–68
35. Brenner, M., Ames, B. N. 1972. Histidine regulation in *Salmonella ty-*

phimurium: IX. Histidine transfer ribonucleic acid of the regulatory mutants. *J. Biol. Chem.* 247:1080–88

36. Singer, C. E., Smith, G. R., Cortise, R., Ames, B. N. 1972. Mutant tRNA[his] ineffective in repression and lacking two pseudouridine modifications. *Nature New Biol.* 238:72–74

37. Kovach, J. S., Berberich, M. A., Venetianer, P., Goldberger, R. F. 1969. Repression of the histidine operon: Effect of the first enzyme on the kinetics of repression. *J. Bacteriol.* 97:1283–90

38. Kovach, J. S., Phang, J. M., Ference, M., Goldberger, R. F. 1969. Studies on the repression of the histidine operon: The role of the first enzyme in the control of the histidine system. *Proc. Nat. Acad. Sci. USA* 63:481–88

39. Kovach, J. S., Ballesteros, A. O., Meyers, M., Soria, M., Goldberger, R. F. 1973. A cis/trans test of the effect of the first enzyme for histidine biosynthesis in regulation of the histidine operon. *J. Bacteriol.* 114:351–56

40. Kovach, J. S. et al 1970. Interaction between histidyl transfer ribonucleic acid and the first enzyme for histidine biosynthesis in *Salmonella typhimurium. J. Bacteriol.* 104:787–92

41. Blasi, F., Barton, R. W., Kovach, J. S., Goldberger, R. F. 1971. Interaction between the first enzyme for histidine biosynthesis and histidyl transfer ribonucleic acid. *J. Bacteriol.* 106:508–13

42. Vogel, T., Meyers, M., Kovach, J. S., Goldberger, R. F. 1972. Specificity of interaction between the first enzyme for histidine biosynthesis and aminoacylated histidine transfer ribonucleic acid. *J. Bacteriol.* 112:126–30

43. Blasi, F. et al 1973. Inhibition of transcription of the histidine operon by the first enzyme of the histidine pathway. *Proc. Nat. Acad. Sci. USA* 70:2692–96

44. Wyche, J. H., Ely, B., Cebula, T. A., Snead, M. C., Hartman, P. R. 1974. Histidyl-transfer ribonucleic acid synthetase in positive control of the histidine operon in *Salmonella typhimurium. J. Bacteriol.* 117:708–16

45. Margolin, P. 1971. Regulation of tryptophan synthesis. See Ref. 12, Chap. 12

45a. Jackson, E. N., Yanofsky, C. 1972. Internal promoter of the tryptophan operon of *Escherichia coli* is located in a structural gene. *J. Mol. Biol.* 69:307–13

46. Brown, K. D., Somerville, R. L. 1971. Repression of aromatic amino acid biosynthesis in *Escherichia coli* K-12. *J. Bacteriol.* 108:386–99

47. Camakaris, J., Pittard, J. 1971. Repression of 3,deoxy-D-arabinoheptulosonic acid-7-phosphate synthetase (*trp*) and enzymes of the tryptophan pathway in *Escherichia coli K-12. J. Bacteriol.* 107:406–14

48. Ito, J., Imamoto, F. 1969. Sequential derepression and repression of the tryptophan operon. *Nature* 220:441–44

49. Morse, D. E., Yanofsky, C. 1969. A transcription-initiating mutation within a structural gene of the tryptophan operon. *J. Mol. Biol.* 41:317–28

50. Zubay, G., Morse, D. E., Schrank, W. J., Miller, J. H. M. 1972. Detection and isolation of the repressor protein for the tryptophan operon of *Escherichia coli. Proc. Nat. Acad. Sci. USA* 69:1100–3

51. Shimizu, Y., Shimizu, N., Hayashi, M. 1973. *In vitro* repression of transcription of the tryptophan operon by *trp* operator. *Proc. Nat. Acad. Sci. USA* 70:1990–94

52. Rose, J. K., Squires, C. L., Yanofsky, C., Yang, H.-L., Zubay, G. 1973. Regulation of *in vitro* transcription of the tryptophan operon by purified RNA polymerase in the presence of partially purified repressor and tryptophan. *Nature New Biol.* 245:133–37

53. Ito, K. 1972. Regulatory mechanism of the tryptophan operon in *Escherichia coli:* Possible interaction between *trpR* and *trpS* gene products. *Mol. Gen. Genet.* 115:349–63

54. Squires, C. L., Rose, J. K., Yanofsky, C., Yang, H.-L., Zubay, G. 1973. Tryptophanyl-tRNA and tryptophanyl-tRNA synthetase are not required for in vitro repression of the tryptophan operon. *Nature New Biol.* 245:131–33

55. McGeoch, D., McGeoch, J., Morse, D. 1973. Synthesis of tryptophan operon RNA in a cell-free system. *Nature New Biol.* 245:137–40

56. Somerville, R. L., Yanofsky, C. 1965. Studies on the regulation of tryptophan biosynthesis in *Escherichia coli. J. Mol. Biol.* 11:747–59

57. Hiraga, S., Yanofsky, C. 1972. Normal repression in a deletion mutant lacking almost the entire operator-proximal gene of the tryptophan operon of *E. coli. Nature New Biol.* 237:47–49

58. Jackson, E. N., Yanofsky, C. 1973. The region between the operator and first structural gene of the tryptophan operon of *Escherichia coli* may have a regulatory function. *J. Mol. Biol.* 76:89–101

59. Baker, R., Yanofsky, C. 1972. Transcription initiation frequency and translational yield for the tryptophan operon of *Escherichia coli. J. Mol. Biol.* 69:89–102

60. Rose, J. K., Yanofsky, C. 1972. Metabolic regulation of the tryptophan operon of *Escherichia coli:* Repressor-independent regulation of transcription initiation frequency. *J. Mol. Biol.* 69:103–18

61. Imamoto, F. 1968. Immediate cessation of transcription of the operator-proximal region of the tryptophan operon in *E. coli* after repression of the operon. *Nature* 220:31–34

62. Hiraga, S., Yanofsky, C. 1973. Inhibition of the progress of transcription on the tryptophan operon of *Escherichia coli. J. Mol. Biol.* 79:339–49

63. Cunin, R., Elseviers, D., Sand, G., Freundlich, G., Glansdorff, N. 1969. On the functional organization of the *argECBH* cluster of genes in *Escherichia coli K-12. Mol. Gen. Genet.* 106:32–47

64. Jacoby, G. A., Gorini, L. 1969. A unitary account of the repression mechanism of arginine biosynthesis in *Escherichia coli.* I. The genetic evidence. *J. Mol. Biol.* 39:73–87

65. Baumberg, S., Ashcroft, E. 1971. Absence of polar effect of frameshift mutations in the *E* genes of the *Escherichia coli argECBH* cluster. *J. Gen. Microbiol.* 69:365–73

66. Bollon, A. P., Vogel, H. J. 1973. Regulation of *argE-argH* expression with arginine derivatives in *Escherichia coli.* Extreme non-uniformity of repression and conditional repression action. *J. Bacteriol.* 114:632–40

67. Jacoby, G. A. 1972. Control of the *argECBH* cluster in *Escherichia coli. Mol. Gen. Genet.* 117:337–48

68. Elseviers, D., Cunin, R., Glansdorff, N., Baumberg, S., Ashcroft, E. 1972. Control regions within the *argECBH* gene cluster of *Escherichia coli K-12. Mol. Gen. Genet.* 117:349–66

69. Krzyzek, R., Rogers, P. 1972. Arginine control of transcription of *argECBH* messenger ribonucleic acid in *Escherichia coli. J. Bacteriol.* 110:945–54

70. Guha, A., Saturen, Y., Szybalski, W. 1971. Divergent orientation of transcription from the biotin locus of *Escherichia coli. J. Mol. Biol.* 56:53–62

71. Cleary, P. P., Campbell, A., Chang, R. 1972. Location of promoter and operator sites in the biotin gene cluster of *Escherichia. Proc. Nat. Acad. Sci. USA* 69:2219–23

72. Panchal, C. J., Bagchee, S. N., Guha, A. 1974. Divergent orientation of transcription from the arginine gene *ECBH* cluster of *Escherichia coli. J. Bacteriol.* 117:675–80

73. Vrancic, A., Guha, A. 1973. Evidence of two operators in the biotin locus of *Escherichia coli. Nature New Biol.* 245:106–8

74. Vogel, R. H., McLellan, W. L., Hirvonen, A. P., Vogel, H. J. 1971. The arginine biosynthetic system and its regulation. See Ref. 12, pp. 463–88

75. Kadner, R. J., Maas, W. K. 1971. Regulatory gene mutations affecting arginine biosynthesis in *Escherichia coli. Mol. Gen. Genet.* 111:1–14

76. Press, R. et al 1971. Isolation of transducing particle of φ80 bacteriophage that carry different regions of the *Escherichia coli* genome. *Proc. Nat. Acad. Sci. USA* 68:795–98

77. Urm, E., Yang, H., Zubay, G., Kalker, N., Maas, W. 1973. *In vitro* repression of N-acetyl-L-ornithinase synthesis in *Escherichia coli. Mol. Gen. Genet.* 121:1–7

78. Celis, T. F. R., Maas, W. K. 1971. Studies on the mechanism of repression of arginine biosynthesis in *Escherichia coli.* IV. Further studies on the role of arginine transfer RNA repression of the enzymes of arginine biosynthesis. *J. Mol. Biol.* 62:179–88

79. Williams, A. L., Williams, L. S. 1973. Control of arginine biosynthesis in *Escherichia coli:* Characterization of arginyl-transfer ribonucleic acid synthetase mutants. *J. Bacteriol.* 113:1433–41

80. Williams, L. S. 1973. Control of arginine biosynthesis in *Escherichia coli:* Role of arginyl-transfer ribonucleic acid synthetase in repression. *J. Bacteriol.* 113:1419–32

81. Lavelle, R. 1970. Regulation at the level of translation in the arginine pathway of *Escherichia coli K-12. J. Mol. Biol.* 51:449–51

82. McLellan, W. I., Vogel, H. J. 1970. Translational repression in the arginine system of *Escherichia coli. Proc. Nat. Acad. Sci. USA* 67:1703–9

83. Rogers, P., Krzyzek, R., Kaden, T. M., Arfman, E. 1971. Effect of arginine and canavanine on arginine messenger RNA synthesis. *Biochem. Biophys. Res. Commun.* 44:1220–26

84. Faanes, R., Rogers, P. 1972. Repression of enzymes of arginine biosynthesis by L-canavanine in arginyl-transfer

ribonucleic acid synthetase mutants of *Escherichia coli. J. Bacteriol.* 112: 102–13

85. Smith, D. A. 1971. S-amino acid metabolism and its regulation in *Escherichia coli* and *Salmonella typhimurium. Advan. Genet.* 16:141–65

86. Taylor, R. T., Weissbach, H. 1973. N^5-methyltetrahydrofolate-homocysteine methyltransferases. *The Enzymes,* ed. P. D. Boyer, 9: Chap. 4, 121–65. New York: Academic

87. Theze, J., Margarita, D., Cohen, G. N., Borne, F., Patte, J. C. 1974. Mapping structural genes of the three aspartokinases and of the two homoserine dehydrogenases of *Escherichia coli K-12. J. Bacteriol.* 117:133–43

88. Lawrence, D. A., Smith, D. A., Rowbury, R. J. 1968. Regulation of methionine synthesis in *Salmonella typhimurium:* mutants resistant to inhibition by analogs of methionine. *Genetics* 58:473–92

89. Chater, K. F. 1970. Dominance of the wild-type alleles of methionine regulatory genes in *Salmonella typhimurium. J. Gen. Microbiol.* 63:95–105

90. Greene, R. C., Su, C.-H., Holloway, C. T. 1970. S-adenosylmethionine synthetase deficient mutants of *Escherichia coli K-12* with impaired control of methionine biosynthesis. *Biochem. Biophys. Res. Commun.* 38:1120–26

91. Holloway, C. T., Greene, R. C., Su, C.-H. 1970. Regulation of S-adenosylmethionine synthetase in *Escherichia coli. J. Bacteriol.* 104:734–47

92. Su, C.-H., Greene, R. C. 1971. Regulation of methionine biosynthesis in *Escherichia coli:* Mapping of the *metJ* locus and properties of a *metJ⁺/metJ⁻* diploid. *Proc. Nat. Acad. Sci. USA* 68:367–71

93. Greene, R. C., Hunter, J. S. V., Coch, E. H. 1973. Properties of *metK* mutants of *Escherichia coli K-12. J. Bacteriol.* 115:57–67

94. Hobson, A. C., Smith, D. A. 1973. S-adenosylmethionine synthetase in methionine regulatory mutants of *Salmonella typhimurium. Mol. Gen. Genet.* 126:7–18

95. Savin, M. A., Flavin, M., Slaughter, C. 1972. Regulation of homocysteine biosynthesis in *Salmonella typhimurium. J. Bacteriol.* 111:547–56

96. Ahmed, A. 1973. Mechanism of repression of methionine biosynthesis in *Escherichia coli.* I. Role of methionine, S-adenosylmethionine and methionyl-tRNA in repression. *Mol. Gen. Genet.* 123:299–324

97. Meedel, T. H., Pizer, L. I. 1974. Regulation of "one-carbon" biosynthesis and utilization in *Escherichia coli. J. Bacteriol.* 118:905–10

98. Dawes, J., Foster, M. A. 1971. Vitamin B_{12} and methionine biosynthesis in *Escherichia coli. Biochim. Biophys. Acta* 237:455–64

99. Kung, H.-F., Spears, C., Greene, R. C., Weissbach, H. 1972. Regulation of the terminal reactions in methionine biosynthesis by vitamin B_{12} and methionine. *Arch. Biochem. Biophys.* 150:23–31

100. Greene, R. C., Williams, R. D., Kung, H.-F., Spears, C., Weissbach, H. 1973. Effects of methionine and vitamin B_{12} on the activity of methionine biosynthetic enzymes in *metJ* mutants of *Escherichia coli. Arch. Biochem. Biophys.* 158:249–56

101. Hulanicka, M. D., Kredich, N. M., Treiman, D. M. 1974. The structural gene for O-acetylserine sulfhydrylase A in *Salmonella typhimurium. J. Biol. Chem.* 249:867–72

102. Jones-Mortimer, M. C. 1968. Positive control of sulphate reduction in *Escherichia coli.* The nature of the pleiotropic cysteineless mutants of *E. coli K-12. Biochem. J.* 110:597–602

103. Kredich, N. M. 1971. Regulation of L-cysteine biosynthesis in *Salmonella typhimurium. J. Biol. Chem.* 246: 3474–84

104. Gots, J. S. 1971. Regulation of purine and pyrimidine metabolism. See Ref. 12, pp. 225–54

105. Westby, C. A., Gots, J. S. 1969. Genetic blocks and unique features in the biosynthesis of 5'-phosphoribosyl-N-formylglycinamide in *Salmonella typhimurium. J. Biol. Chem.* 244:2095–2102

106. Gots, J. S., Dalal, F. R., Shumas, S. R. 1969. Genetic separation of the inosinic acid cyclohydrolase-transformylase complex of *Salmonella typhimurium. J. Bacteriol.* 99:441–49

107. Gots, J. S., Dalal, F. R., Westby, C. A. 1969. Operon controlling three enzymes in purine biosynthesis in *Salmonella typhimurium. Bacteriol. Proc.* p. 131

108. Gots, J. S. 1965. A guanine operon in *Salmonella typhimurium. Fed. Proc.* 24:416

109. Nijkamp, H. J. J., de Haan, P. G. 1967. Genetic and biochemical studies of the guanosine 5'-monophosphate pathway

in *Escherichia coli. Biochim. Biophys. Acta* 145:31–40

110. Nijkamp, H. J. J., Oskamp, A. A. G. 1968. Regulation of the biosynthesis of guanosine 5'-monophosphate: evidence for one operon. *J. Mol. Biol.* 35:103–9

111. Benson, C. E., Gots, J. S. 1971. Regulatory role of the *purA* product in purine biosynthesis in *Salmonella typhimurium. Bacteriol. Proc.* p. 157

112. Thomulka, K. W., Gots, J. S. 1972. Study of regulation of *purE* operon in *Salmonella typhimurium* using a *purE-lac* fusion. *Abstr. Ann. Meet. Am. Soc. Microbiol.*, p. 176

113. Jacob, F., Ullmann, A., Monod, J. 1965. Délétions fusionnant l'operon lactose et un operon purine chez *Escherichia coli. J. Mol. Biol.* 13:704–19

114. Levinthal, M. 1971. Biochemical studies of melibiose metabolism in wild type and *mel* mutant strains of *Salmonella typhimurium. J. Bacteriol.* 105:1047–52

115. Armitt, S., Woods, R. A. 1970. Purine excreting mutants of *Saccharomyces cerevisiae.* I. Isolation and genetic analysis. *Genet. Res.* 15:7–17

116. Dorfman, B. 1969. The isolation of adenylosuccinate synthetase mutants in yeast by selection for constitutive behavior in pigmented strains. *Genetics* 61:377–89

117. Dorfman, B., Goldfinger, B. A., Berger, M., Goldstein, S. 1970. Partial reversion in yeast: genetic evidence for a new type of bifunctional protein. *Science* 168:1482–84

118. Dorfman, B. 1971. Allelic variability in comparative complementation confirming that the *ade-12* specific protein of yeast is bifunctional. *J. Bacteriol.* 107:646–54

119. O'Donovan, G. A., Neuhard, J. 1970. Pyrimidine metabolism in microorganisms. *Bacteriol. Rev.* 34:278–343

120. Williams, J. C., O'Donovan, G. A. 1973. Repression of enzyme synthesis of the pyrimidine pathway in *Salmonella typhimurium. J. Bacteriol.* 115:1071–76

121. Neuhard, J. 1965. Pyrimidine nucleotide metabolism and pathways of thymidine triphosphate biosynthesis in *Salmonella typhimurium. J. Bacteriol.* 96:1519–27

122. Abd-El-Al, A., Ingraham, J. L. 1969. Control of carbamyl phosphate synthesis in *Salmonella typhimurium. J. Biol. Chem.* 244:4033–38

123. Piérard, A., Glansdorff, N., Yashphe, J. 1973. Mutations affecting uridine monophosphate pyrophosphorylase or the *argR* gene in *Escherichia coli.* Effect

on carbamoyl phosphate and pyrimidine synthesis and on uracil uptake. *Mol. Gen. Genet.* 118:235–45

124. O'Donovan, G. A., Gerhard, J. C. 1972. Isolation and partial characterization of regulatory mutants of the pyrimidine pathway in *Salmonella typhimurium. J. Bacteriol.* 109:1085–96

125. Ingraham, J. L., Neuhard, J. 1972. Cold-sensitive mutants of *Salmonella typhimurium* defective in UMP kinase (*pyrH*). *J. Biol. Chem.* 247:6259–65

126. Barth, P. T., Beachem, I. R, Ahmad, S. I., Pritchard, R. H. 1968. The inducer of the deoxynucleoside phosphorylases and deoxyriboaldolase in *E. coli. Biochim. Biophys. Acta* 161:554–57

127. Ahmad, S. I., Pritchard, R. H. 1969. A map of four genes specifying enzymes involved in catabolism of nucleosides and deoxynucleosides in *E. coli. Mol. Gen. Genet.* 104:351–59

128. Robertson, B. C., Jargiello, P., Blank, J., Hoffee, P. A. 1970. Genetic regulation of ribonucleoside and deoxyribonucleoside catabolism in *Salmonella typhimurium. J. Bacteriol.* 102:628–35

129. Ahmad, S. I., Pritchard, R. H. 1971. A regulatory mutant affecting synthesis of enzymes involved in the catabolism of nucleosides in *Escherichia coli. Mol. Gen. Genet.* 111:77–83

130. Bonney, R. J., Weinfeld, H. 1971. Regulation of thymidine metabolism in *Escherichia coli K-12:* evidence that at least two operons control the degradation of thymidine. *J. Bacteriol.* 105:940–46

131. Ahmad, S. I., Pritchard, R. H. 1973. An operator constitutive mutant affecting the synthesis of two enzymes involved in catabolism of nucleosides in *Escherichia coli. Mol. Gen. Genet.* 124:321–28

132. Munch-Petersen, A. 1968. On the catabolism of deoxyribonucleosides in cells and cell extracts of *Escherichia coli. Eur. J. Biochem.* 6:432–42

133. Blank, J., Hoffee, P. A. 1972. Regulatory mutants of the *deo* regulon in *Salmonella typhimurium. Mol. Gen. Genet.* 116:291–98

134. Hammer-Jespersen, K., Munch-Petersen, A., Nygaard, P., Schwarz, M. 1971. Induction of enzymes involved in catabolism of deoxyribonucleosides and ribonucleosides in *Escherichia coli. Eur. J. Biochem.* 19:533–38

135. Munch-Petersen, A., Nygaard, P., Hammer-Jespersen, K., Fiil, N. 1972. Mutants constitutive for nucleoside

catabolizing enzymes in *Escherichia coli K-12. Eur. J. Biochem.* 27:208–15

136. Nygaard, P. 1973. Nucleoside catabolizing enzymes in *Salmonella typhimurium. Eur. J. Biochem.* 36:267–72

137. Hammer-Jespersen, K., Munch-Petersen, A. 1973. Mutants of *Escherichia coli* unable to metabolize cytidine. *Mol. Gen. Genet.* 126:177–86

138. Brill, W. J., Magasanik, B. 1969. Genetic and metabolic control of histidase and urocanase in *Salmonella typhimurium,* strain 15–59. *J. Biol. Chem.* 244:5392–5402

139. Meiss, H. K., Brill, W. J., Magasanik, B. 1969. Genetic control of histidase degradation in *Salmonella typhimurium,* strain LT-2. *J. Biol. Chem.* 244:5382–91

140. Smith, G. R., Halpern, Y. S., Magasanik, B. 1971. Genetic and metabolic control of enzymes responsible for histidine degradation in *Salmonella typhimurium. J. Biol. Chem.* 246:3320–29

141. Smith, G. R., Magasanik, B. 1971. The two operons of the histidine utilization system in *Salmonella typhimurium. J. Biol. Chem.* 246:3330–41

142. Hagen, D. C., Magasanik, B. 1973. Isolation of the self-regulated repressor of the *hut* operon of *Salmonella typhimurium. Proc. Nat. Acad. Sci. USA* 70:808–12

143. Prival, M. J., Magasanik, B. 1971. Resistance to catabolite repression of histidase and proline oxidase during nitrogen-limited growth of *Klebsiella aerogenes. J. Biol. Chem.* 246:6288–96

144. Prival, M. J., Brenchley, J. E., Magasanik, B. 1973. Glutamine synthetase and the regulation of histidase formation in *Klebsiella aerogenes. J. Biol. Chem.* 248:4334–44

145. Tyler, B., Deleo, A. B., Magasanik, B. 1974. Activation of transcription of *hut* DNA by glutamine synthetase. *Proc. Nat. Acad. Sci. USA* 71:225–29

GENETICS OF AMINO ACID TRANSPORT IN BACTERIA

❖3067

Yeheskel S. Halpern
Department of Molecular Biology, Institute of Microbiology,
Hebrew University - Hadassah Medical School, Jerusalem, Israel

INTRODUCTION

The earliest reports on temperature- and energy-dependent concentrative amino acid uptake by bacteria are probably those of Gale and co-workers on *Staphylococcus aureus* and *Streptococcus faecalis* published in 1947 (1, 2). Studies on the nature, specificity, and kinetic behavior of bacterial amino acid transport systems were initiated soon after by the Carnegie Institution group in Washington and at the Pasteur Institute in Paris. This pioneering work was followed by numerous investigations in many laboratories on the biochemical, physiological, and genetic aspects of bacterial transport. These early studies had been summarized and discussed in several reviews and monographs (3–7).

Considerable progress has been made in recent years in our understanding of the molecular mechanisms of transport. Most significant contributions in this area have been due to the studies of Roseman and others on group translocation reactions and their role in carbohydrate transport (8–10). Another major development in the study of bacterial transport was the elaboration by Kaback and co-workers of the isolated membrane vesicles system (10, 11). Studies with these subcellular systems have already improved our picture of the molecular basis of transport, notably in regard to energy coupling (see below). A very active and fruitful aspect of recent work has been the study of the periplasmic binding proteins and their relation to transport. This field has been reviewed by Pardee (12) and more recently by Heppel (13), Kaback (8), Lin (14), and Oxender (15, 16). A review article on the genetics of bacterial transport systems appeared in 1970 (17). A comprehensive chapter on amino acid transport in microorganisms has been written recently by Oxender (16). Peptide transport systems and their relation to the transport of the respective amino acids have also been reviewed recently (18, 19).

In the present review I try to draw attention to some basic but still outstanding problems in bacterial transport. Some of these problems, such as the coupling of energy to amino acid transport, have been the subject of a number of recent investigations and discussion. Other questions that I hope to raise the reader's interest in,

103

for example the identity and interrelation of carriers mediating influx and efflux, have not been posed often and clearly enough. Finally, emphasis is placed on studies dealing with the genetic control of the synthesis and activity of amino acid transport systems and on the important contributions of the genetic approach to the elucidation of amino acid transport in bacteria.

SPECIFICITY OF AMINO ACID TRANSPORT SYSTEMS

One characteristic feature of active transport of amino acids in bacteria is its high specificity (3). This is somewhat different from the situation in eukaryotic microorganisms which on the whole exhibit much broader specificity (20–34), similar to that encountered in mammalian systems (35–45).

The following is a description of amino acid transport in bacteria, stressing specificity and interactions between different systems, with special emphasis on recent contributions.

Glycine, Alanine, Serine, and Threonine

Early studies on the antagonistic effects of L-alanine and glycine on the inhibition of growth of *Escherichia coli* by D-serine had suggested that these amino acids might interfere with the uptake of D-serine (46). Mutants resistant to D-serine were isolated, and the ability of one such mutant to take up radioactive amino acids was examined by Schwartz, Maas & Simon (47). The uptake of glycine, D-serine, and L-alanine was greatly impaired in the mutant (7, 45, and 35% of wild-type activity, respectively), while the uptake of L-arginine, L-lysine, and DL-threonine was not affected. The authors concluded that the D-serine-resistant mutant was defective in a specific concentrative mechanism for glycine, L-alanine, and D-serine. Lubin and co-workers (48) and Kessel & Lubin (49) also succeeded in isolating *E. coli* W mutants defective in glycine transport by selection on plates containing D-serine or D-cycloserine (D-4-amino-3-isoxazolidone). The authors also isolated a glycine-transport-defective mutant, Tr_{gly-}, from a glycine auxotroph by selecting colonies requiring very high concentrations of glycine (500–1000 μg/ml) for growth. The uptake rates of glycine and D-alanine in the mutant were about 5% of the respective rates in the parent, while uptake of L-alanine was reduced to only half of the normal rate. These data, as well as competition studies, led Kessel & Lubin to suggest that glycine, D-alanine, and D-cycloserine are transported by the same system, which also seems to transport L-alanine and D-serine, but the latter two may have an additional route of rapid entry into the cell (49). Levine & Simmonds, working with serine-glycine auxotrophs of *E. coli* K12, isolated mutants that grew well when supplemented with L-serine, but grew slowly and only after an initial lag period on glycine-supplemented media. Growth on glycine peptides was normal. These mutants lack a normal transport system for glycine (50–52).

The existence of a separate transport system common for glycine, serine, and alanine in *E. coli* K12 was also supported by kinetic studies of Piperno & Oxender (53). Their conclusions were based mainly on differences in competitive interactions among amino acids for transport and on the acceleration of loss from the cells of accumulated labeled amino acids by countertransport. Kinetic and genetic studies by Wargel, Shadur & Neuhaus (54, 55) also indicate that the transport system(s) for D-alanine and glycine in *E. coli* are related and separate from that involved in the accumulation of L-alanine. The accumulation of D-alanine, glycine, and D-cycloserine was characterized by a biphasic Lineweaver-Burk plot, whereas the accumulation of L-alanine could be described by a single line segment. D-alanine inhibited glycine uptake more efficiently than the uptake of L-alanine (respective K_i values, 1.4 and 6.5 $\times$ 10^{-4} M). Similarly, D-cycloserine was a better inhibitor of D-alanine and glycine uptake than of the uptake of L-alanine (respective K_i values, 1.1, 1.5, and 8.7 $\times$ 10^{-4} M), whereas L-cycloserine, which was a very good inhibitor of L-alanine uptake ($K_i = 2.2 \times 10^{-4}$ M), hardly affected the uptake of D-alanine or glycine. As could be expected from these results, glycine and D-alanine at 10^{-5}M antagonized the effect of D-cycloserine, whereas this concentration of L-alanine had no effect. Multistep mutants of *E. coli* K12 resistant to increasing concentrations of D-cycloserine were isolated by repeated cycles of growth in the presence of increasing concentrations of the drug. The first-step mutant was characterized by loss of the high-affinity D-alanine-glycine transport; the mutation mapped between 77 and 90 min to the right of the *metB* locus. The second mutation, located 0.5 min from the former, resulted in the loss of the low-affinity component of D-alanine-glycine transport. The transport of L-alanine was decreased only 20–30% in each of these mutants. A multistep mutant of *E. coli* W that was eightyfold resistant to D-cycloserine lost more than 90% of the D-alanine and glycine transport activities, while retaining 75% of the transport activity for L-alanine. This mutant also lost the ability to utilize D-alanine as a carbon source. The authors conclude that a functioning transport system for D-alanine and glycine is required for both D-cycloserine action and growth on D-alanine (55).

The first-step mutation to D-cycloserine resistance was later accurately mapped by transduction with phage P1 and found to be located between *purA* and *pyrB* at min 83 of Taylor's *E. coli* K12 chromosome map; its designation is *cycA* (56).

Mutants affected in D-serine transport have been isolated recently by Cosloy who used a D-serine deaminaseless strain and selection for D-serine resistance (57). Strain EM1302, which carried the *dagA* mutation (*da = d*-alanine, *g* = glycine), showed greatly reduced D-serine uptake. The accumulation of D-alanine and glycine was similarly inhibited. The rate of uptake of L-alanine was partially inhibited in strain EM1302; L-serine uptake was unaffected. Genetic analysis by conjugation between an *ilv* derivative of the F⁻ strain EM1302 and a *metB* Hfr strain and by determining the D-cycloserine resistance levels of the recombinants indicated that there are indeed a minimum of three cistrons involved in the transport of D-serine, D-alanine, glycine, and D-cycloserine. Results obtained upon introduction of various F factors harboring genes from the *ilv* region into strain EM1302 led the authors to conclude that unless *dagA* is dominant in *dagA/dagA⁺* merodiploids, its location is just to the right of *malB*. Precise mapping of the *dagA* locus has been recently accomplished by Robbins & Oxender (58). These authors found the inability of *dagA* strains to utilize D-alanine as a carbon source, rather than antibiotic resistance, to be a useful phenotype in genetic experiments because of the high background of spontaneous resistance. By the use of episomes containing genes of the relevant region, the authors established the dominance of the wild-type allele *dagA*⁺. Precise mapping by transduction with phage P1 revealed a 7–12% linkage between *dagA* and *pyrB,* similar to the location of *cycA* (56) and consistent with the approximate position of *dagA,* as reported by Cosloy (57). Employing the kinetic approach and inhibition studies and comparing the behavior of the *dagA* mutant EM1302 and its parent, Robbins & Oxender (58) showed that at least two different routes serve for the entry of alanine, glycine, and serine into *E. coli* K12. One takes up glycine, D-alanine, and D-serine, and to some extent L-alanine, while the second is used for L-alanine, L-threonine, and perhaps L-serine. Very interestingly, the remaining L-alanine uptake in *dagA* mutants is inhibited by L-serine, L-threonine, and L-leucine. It is also sensitive to osmotic shock and repressed by growth in the presence of L-leucine. These data seem to indicate that L-alanine, L-threonine, and perhaps L-serine are transported by the leucine, isoleucine, and valine (LIV-I) system (see below).

Data supporting the existence of a transport system for L- and D-alanine and a different one for glycine, were reported by Leach & Snell for *Lactobacillus casei* (59). In *S. faecalis* R8043, however, a single transport system seems to be responsible for the uptake of L- and D-alanine and glycine (60). Reitz, Slade & Neuhaus (61) isolated mutants of *Streptococcus* strain Challis resistant to D-cycloserine and O-carbamyl-D-serine by repeated cycles of growth in the presence of increasing concentrations of the analogs. Among the different mutant types obtained, one, D-CS$_b$, manifested greatly impaired transport of L- and D-alanine and failed to accumulate D-cycloserine. Recent work from the laboratory of Harold with cycloserine-resistant mutants of *S. faecalis* strongly suggests a single transport system for glycine, L-alanine, L-serine, and L-threonine (61a). Marquis & Gerhardt (62) studied the transport of the nonmetabolizable α-methyl analog of alanine, α-aminoisobutyric acid (AIB) in *Bacillus megaterium* KM. On the basis of two criteria, competitive effects of amino acids on AIB uptake and exchange of intracellular AIB with extracellular amino acids, it would seem that AIB was taken up by a transport system shared by L- and D-alanine and glycine.

Isoleucine, Valine, and Leucine

The existence of a common transport system for isoleucine, valine, and leucine in *E. coli* had already been indicated by the competition data of Cohen & Rickenberg (63, 64). Further support was later provided by the work of Piperno & Oxender, mentioned above (53). Studies on the inhibition of growth of *E. coli* 15 by O-methylthreonine strongly indicated that this isoleucine and threonine analog is also transported via the branched-chain amino acid permease system (65). Rahmanian & Oxender (66) pointed out the heterogeneity of leucine entry by showing that isoleucine did not completely inhibit the uptake of leucine and that the entry of all three branched-chain amino acids exhibited biphasic kinetics. The multiplicity of transport systems for branched-chain amino acids was also supported by the kinetic and competition analysis of Tager & Christensen (67). They used the four isomers of the synthetic amino acid 2-aminonorbornane-2-carboxylic acid. Their results indicate four separate entry routes for leucine and three for isoleucine. The existence of at least three transport systems responsible for the entry of branched-chain amino acids into *E. coli* K12 was demonstrated in a recent study by Rahmanian, Claus & Oxender (68). By careful kinetic analysis, using the method of Neal (69) for resolving biphasic reciprocal plots into their component systems and determining the kinetic constants for each mechanism, the authors identified two kinetically discernible components of L-leucine transport: a low K_m (8×10^{-8} M) system, LIV-I, and a high K_m ($\sim 2 \times 10^{-6}$ M) system, LIV-II. In wild-type strains more than 90% of the radioactive L-leucine is brought in by the LIV-I system. The LIV-I component of leucine entry is sensitive to osmotic shock and to repression by growth in the presence of leucine. Leucine also represses the synthesis of a LIV-binding protein, which can be isolated from the shock fluid of nonrepressed cultures (70). Transport via the LIV-II system is not affected by osmotic shock or by growth in the presence of leucine. Both the LIV-I and LIV-II systems transport all three branched-chain amino acids. However, whereas the LIV-II system is specific for leucine, isoleucine, and valine, the low K_m LIV-I system, as well as the LIV-binding protein associated with it, have considerable affinity towards L-threonine, L-alanine, and L-serine. An additional saturable component of L-leucine entry with a K_m of about 2×10^{-7} M has been demonstrated in the presence of an excess of unlabeled L-isoleucine. The relative activity of this leucine-specific system rose from less than 10% in wild-type *E. coli* to about 50% of total leucine uptake activity in strain EO 0319, a D-leucine-utilizing (*Dlu*) mutant. The leucine transport activity of strain EO 0319 is still repressible by leucine, while that of another *Dlu* mutant, EO 0312, is not. Strain EO 0319 formed increased amounts of leucine-specific binding protein, which can be distinguished from LIV-binding protein by sensitivity to trifluoroleucine (71). Two leucine-transport-defective mutants, EO 0321 and EO 0323, were isolated from strains EO 0312 and EO 0319 by penicillin selection. Kinetic analysis showed that both mutants lost the LIV-I system, while the LIV-II system remained relatively unaffected. No leucine-specific binding activity was found in strain EO 0321, although the presence of the protein was demonstrated immunologically. In strain EO 0323 the total binding activity was the same as that in the parent strain EO 0312. The possibility that there might be still another system participating in the transport of branched-chain amino acids in *E. coli* K12 is indicated by the detection of an additional isoleucine-leucine-valine binding protein with predominant affinity for isoleucine (68). All three branched-chain amino acid binding proteins of *E. coli* K12 have common antigenic properties (68), perhaps indicating a diverging evolution by gene duplication.

Similar results were obtained in a very recent study by Guardiola et al (72). These authors describe a "very high affinity" system in *E. coli* K12 with K_m values of 1, 2, and 10×10^{-8} M for isoleucine, leucine, and valine, respectively. Methionine, threonine, and alanine inhibit this transport system, probably because they are also substrates. Growth in the presence of methionine represses the very high affinity transport system. Another transport system is the "high affinity" one with an apparent K_m of 2×10^{-6} M for isoleucine, leucine, and valine. The two systems exhibit different sensitivities to inhibition by various structural analogs of the branched-chain amino acids. Thus, D,L-4-azaisoleucine and D,L-methallylglycine at a concentration of 8×10^{-5} M and D-serine and D-leucine at a concentration of 4×10^{-5} M abolished most of the very high affinity transport of valine, while the high affinity transport system was hardly affected by these inhibitors even at concentrations fivefold higher, respectively. Three different "low affinity" transport systems, each specific for one of the three branched-chain amino acids, with apparent K_m values of 0.5×10^{-4} M were also found. These systems disclosed very high specificity: no amino acid or precursor or analog of any branched-chain amino acid inhibited any of the three systems. The results of these kinetic and competition analyses were substantiated by mutant studies (73, 74). As the basis for the isolation of transport mutants served the well-known sensitivity of *E. coli* K12 to inhibition by valine, due to its interference with the biosynthesis of isoleucine. Mutants resistant both to valine (Val^r) and glycylvaline inhibition are defective in the regulation of branched-chain amino acid biosynthesis. Val^r mutants sensitive to glycylvaline inhibition are impaired in the transport of valine. Repression of the very high affinity transport system by growth in the presence of methionine increases the proportion of transport mutants among the Val^r isolates. The frequency of transport mutants recovered is even higher when the uptake of valine by the high affinity system is partially inhibited by leucine. Using this approach Guardiola et al (74) isolated transport mutants mapping (by transduction with phage P1) at three different loci: *brnR*, 93% linked to *lac*, at 9 min; *brnQ*, close to *phoA*, at 9.5 min; and *brnS*, close to *pdxA*, at 1 min. Another mutation, *brnT8*, probably different from the above ones, has not been mapped. The very high affinity transport is missing in strains carrying the *brnT8* mutation or an amber mutation at the *brnR* locus (*brnR6*[am]), but not in a *brnR* missense mutant (*brnR3*). These seem to be pleiotropic effects, since other transport systems are also affected (see below). The high affinity system was resolved into two components: "high affinity-1," which requires the *brnQ* gene product for the transport of isoleucine, leucine, and valine and is unaffected by threonine; and "high affinity-2," requiring a protein encoded in the *brnS* gene, transports isoleucine, leucine, and valine and is inhibited by threonine, which is probably a substrate. Both high affinity transport systems also require a protein produced by gene *brnR*. A mutant lacking the low affinity transport system for isoleucine was isolated from an isoleucine auxotroph carrying a *brnR* mutation and therefore missing the high affinity transport. The mutant lacking the low affinity system could not grow on isoleucine but grew on glycylisoleucine. The low affinity transport of leucine and valine in the mutant was only partially affected. This mutation was designated *brnT8*. A mutation in another gene, *brnP*, located at min 1, also resulted in a Val^r phenotype and caused changes in K_m and a 40% reduction in valine uptake. In spite of the relatively small decrease in uptake, incorporation of exogenous valine into protein was practically abolished by the *brnP* mutation (73). It is therefore possible that the observed transport alterations might be due to secondary effects of a primary lesion at another level. In view of the pleiotropic effects of mutations *brnR6*[am] and *brnT8* the effects of these mutations on the very high affinity transport system might also be secondary ones.

A transport system similar to the LIV-I system of *E. coli* K12, but of broader specificity, has been recently studied in *E. coli* B (74a). In addition to the branched-chain amino acids this system also takes up cysteine, alanine, phenylalanine, tyrosine, homoserine, and threonine. Whereas homoserine is transported only by this system, as borne out by kinetic analysis, mutant studies, and repression upon growth in the presence of leucine, threonine in addition shares a transport system with serine (74a).

Thorne & Corwin (75), using a series of deletions in the *trp* region of *Salmonella typhimurium*, showed that a gene governing leucine transport maps on the side of the *chr* (chromium sensitivity) locus distal to the *trp* operon.

Phenylalanine, Tyrosine, and Tryptophan

An inducible tryptophan transport system in *E. coli* K12 has been described by DeMoss and co-workers (76, 77). Transport was highly specific: tryptophan uptake was not inhibited in the presence of a twentyfold molar excess of DL-valine, DL-phenylalanine, DL-methionine, and glycine. Fifty percent inhibition was obtained in the presence of DL-serine. However, because tryptophan uptake was also inhibited in the presence of pyruvate, the authors suggest that inhibition by serine might be due to its conversion through pyruvate to an inhibitor, rather than to competition for the tryptophan transport system. Tryptophan did not inhibit serine uptake. The D-isomer of tryptophan and a number of tryptophan derivatives (including indole) were potent inhibitors of L-tryptophan uptake. Tryptamine and anthranilate were without effect. Optimum induction of tryptophan transport activity was obtained at a concentration of 5 μg/ml of L-tryptophan in the growth medium. Chloramphenicol prevented induction of tryptophan transport activity.

Brown studied aromatic amino acid uptake and pool formation in *E. coli* K12 (78, 79). Phenylalanine, tyrosine, and tryptophan were taken up by a general aromatic transport system. The apparent K_m's were 4.7, 5.7, and 4.0 $\times$ 10^{-7} M, respectively. High concentrations ($>$0.1 mM) of histidine, leucine, methionine, alanine, cysteine, and aspartic acid were inhibitory. Each aromatic amino acid inhibited entry of the other two into the cell. However, even with a large excess of inhibitor, inhibition was never complete, not exceeding 80%. In fact, three additional uptake systems, each specific for a different aromatic amino acid, could be measured in the presence of a large excess of one of the other two amino acids. The specific permeases showed lower affinities for their respective substrates than did the general permease, the K_m's being 2.2, 2.0, and 3.0 $\times$ 10^{-6} M for tyrosine, phenylalanine, and tryptophan, respectively.

Mutants lacking the general permease were obtained from *E. coli* K12 (W1485) by selection for resistance to β-thienylalanine (10^{-4} M). Two such strains, KB2800 and KB3100, were further examined and found to be also resistant to DL-5-methyltryptophan (2 $\times$ 10^{-4} M) and DL-p-fluorophenylalanine (2 $\times$ 10^{-4} M). These analog-resistant mutants were not excretors of aromatic amino acids, as demonstrated by their inability to support the growth of the aromatic auxotroph AT1359. Thus, the reduced uptake seen in KB2800 and KB3100 was not an artifact of dilution of the label by material excreted by the cells into the medium. The fraction of the transport of each of the three aromatic amino acids sensitive to inhibition by the other two in wild-type cells was almost entirely missing in the mutants. The mutants had reduced pools of aromatic amino acids, but incorporation of external aromatic amino acids into protein was unaffected. The gene specifying the general aromatic permease *aroP*, was mapped by transduction with phage Plkc and found to be 30% cotransducible with *leu*, 43% cotransducible with *pan*, 0% with *thr*, and 21% with *pyrA*. This places *aroP* between *leu* and *pan*.

These results are very similar to those obtained by Ames (80, 81) with *S. typhimurium*. This organism also has four transport systems: a general aromatic permease with K_m values of about 10^{-7} M for all three aromatic amino acids, and three specific permeases, each for one of them, with K_m's of about 10^{-6} M. L-histidine is also taken up via the general aromatic permease, with a K_m of 10^{-4} M (see also below). A mutant with a defective aromatic permease, AZA-3, was isolated by selection for resistance to azaserine. AZA-3 was also resistant to low concentrations of 5-methyltryptophan and *p*-fluorophenylalanine. The aromatic permease gene (*aroP*) was located in the proximity of *proA* by mating; no gene has been found to contransduce with *aroP*.

Two high affinity transport systems were found by Kay & Gronlund (82) to serve for the entry of aromatic amino acids in *Pseudomonas aeruginosa*. By competition and exchange experiments they found one system to function with phenylalanine, tyrosine, and tryptophan (in order of decreasing activity), whereas the second system was active with tryptophan, phenylalanine, and tyrosine. Eighteen nonaromatic amino acids, when present at 100 times the concentration of ^{14}C-phenylalanine or ^{14}C-tryptophan, did not inhibit the uptake of the radioactive compound. A number of aromatic amino acid analogs, including the D-isomers of phenylalanine and tyrosine, competitively inhibited the uptake of L-phenylalanine, L-tyrosine, and L-tryptophan. Three transport-defective mutants were isolated and their transport activities examined. Mutant TC10, isolated as a slow tyrosine utilizer, showed a 50% decrease in the uptake of tyrosine and was also defective in the transport of phenylalanine; tryptophan uptake was only slightly reduced. Mutant 5FT3, selected for 5-fluorotryptophan resistance, lost about 85% of wild-type tryptophan transport activity but retained tyrosine uptake at 66% of wild-type rate. Mutant TA3 was highly defective in both tryptophan and tyrosine uptake. The authors classified mutant TC10 as being defective in transport system I (phenylalanine, tyrosine, tryptophan) and 5FT3 as tentatively defective in system II (tryptophan, phenylalanine, tyrosine). TA3 may be a double mutant, or more likely may be impaired in a step common to the two transport systems. In contrast to *P. aeruginosa, P. acidovorans* has a highly specific tryptophan transport system (83).

Inhibition of growth of *Pseudomonas* sp. (11299a) by *p*-chlorophenylalanine in media with phenylalanine or tyrosine but not asparagine as the carbon source, was reported by Guroff, Bromwell & Abramowitz (84). The analog inhibited competitively phenylalanine and tyrosine uptake. Tyrosine and tryptophan were also competitive inhibitors of phenylalanine uptake. The K_m for phenylalanine was 2 X 10^{-5} M. Cells grown in the presence of phenylalanine showed 5–15-fold higher uptake activity than did asparagine-grown cells. Upon osmotic shock, induced cells released into the medium a protein which bound phenylalanine with a K_m of about 2×10^{-7} M. The amount of binding protein released by osmotic shock increased twofold upon induction. However, the induction of binding protein was slower than that of transport activity (85). The binding protein has a molecular weight of 24,000–27,000; it contains no cysteine and no histidine (86).

An aromatic amino acid transport system highly specific toward tyrosine and phenylalanine has been found recently in *Bacillus subtilis* (86a). Although uptake follows biphasic kinetics, suggesting that more than one transport system may be involved, the authors prefer to interpret their data in terms of a single system with negative cooperativity (86a, 86b). The strongest argument in favor of this interpretation is the finding that two different mutations resulted in pleiotropic effects on both of the kinetically defined phases of transport.

Proline and Hydroxyproline

Mutants isolated from proline auxotrophs of *E. coli* W, requiring very high concentrations of proline for optimal growth, were specifically defective in the active transport of ^{14}C-proline at 37°C. Glycine, phenylalanine, histidine, and lysine were taken up normally by these mutants (48, 87). The mutant cells also lacked the ability to carry out a rapid exchange between an intracellular pool and a low concentration of extracellular ^{14}C-proline. The rates of loss of proline from cells at 37°C and at 0°C were similar in the mutant and parent cells (87). A proline permease with similar properties was later described in strain C4, a derivative of *E. coli* K10 (88). The uptake of proline was competitively inhibited by a number of proline analogs: 3,4-dehydroproline, 4-methyleneproline, *cis* and *trans*-4-chloroprolines, L-thiazolidine-4-carboxylic acid, and L-azetidine-2-carboxylic acid. These analogs also exchanged with preaccumulated intracellular proline. The K_m for proline uptake was 6.4×10^{-7} M; the K_i values were 2.6×10^{-6} and 2.4×10^{-5} M for 3,4-hydroxyproline and L-azetidine-2-carboxylic acid, respectively. Numerous mutants resistant to either dehydroproline or azetidine were isolated, and many of them showed impaired proline uptake. One of the mutants selected for azetidine resistance (AZ642) was of particular interest. The affinity of this strain's permease for proline and for dehydroproline was practically the same as in the parent strain, but uptake of proline by the mutant was much less sensitive to inhibition by azetidine. The K_i for azetidine in strain AZ642 was 3.0×10^{-4} M, approximately tenfold higher than in the parent.

A proline permease inducible by proline, together with the proline catabolic pathway, was described by Kay & Gronlund in *P. aeruginosa* (89). This transport system was highly specific for L-proline; the presence of 18 amino acids at 0.1 mM had no effect on ^{14}C-proline uptake. Thiazolidine, dehydroproline, and azetidine inhibited competitively proline uptake and also exchanged with a preestablished ^{14}C-proline pool. The ability to exchange correlated with the relative degree of inhibition of proline uptake by the analog, suggesting that exchange was a function of proline permease. Hydroxyproline was not detectably transported by induced or uninduced cells. Cells grown on proline as the only source of carbon transported proline more effectively than did cells grown in a proline-supplemented glucose medium, suggesting that proline permease is sensitive to catabolite repression. The kinetics of proline uptake show a break at high proline concentrations (20 μM), giving two K_m values: 10^{-6} and 10^{-5} M; the V_{max} also increases more than fivefold at high proline concentrations, enabling the cell to take up enough proline to serve as a carbon and energy source for growth. A transport-negative mutant (P5) was isolated from a culture induced to utilize proline as a sole source of carbon. The mutant appeared as a very small colony after 48 hr (90). It was unable to transport ^{14}C-proline from a low concentration in the medium. The uptake of glutamate, arginine, tyrosine, and isoleucine was unaffected. Spontaneous revertants were selected on minimal medium with proline as a carbon source. Of 5 such revertants, all had regained the ability to transport proline at wild-type rates.

An inducible permease has also been described in *Pseudomonas putida* for hydroxy-L-proline ($K_m = 3 \times 10^{-5}$ M) and allohydroxy-D-proline ($K_m = 10^{-3}$ M) (91). A number of amino acids interfered with hydroxy-L-proline uptake; the greatest effect was produced by L-alanine and L-proline.

A nonspecific proline transport system was found in *Agrobacterium tumefaciens*. A great number of unrelated amino acids inhibited proline uptake (92).

Methionine

The existence of a distinct transport system specific for L-methionine in *E. coli* K12 was inferred from kinetic data and particularly from inhibition and countertransport experiments (53). The kinetic constants for L-methionine uptake were: $K_m = 2.3 \times 10^{-6}$ M, $V_{max} = 0.8$ nmol/g wet wt/min. The rate of L-methionine uptake was similar to that of D-alanine and about one fourth of the rate of uptake of L-leucine. L-ethionine was a very effective inhibitor ($K_i = 2.3 \times 10^{-5}$ M), while D-methionine had much lower affinity ($K_i = 6 \times 10^{-4}$ M). A similar highly specific L-methionine transport system has been found in *S. typhimurium* (93). The K_m for L-methionine was $1-2 \times 10^{-7}$ M, and there was little or no affinity for other naturally occurring amino acids. Methionine uptake was competitively inhibited by DL-ethionine, α-methyl-DL-methionine, and DL-methionine sulfoximine. Mutants resistant to α-methylmethionine and methionine sulfoximine were isolated and found to possess a severely defective specific methionine permease. Two such mutants, *metP760* and *metP761*, were mapped and neither was linked to any structural or regulatory gene of the methionine biosynthetic pathway; *metP760* mapped at about 7 min, between *leu* and *proB*, while *metP761* mapped near the *gal* locus.

A recent study involving kinetic analysis and the use of mutants of *E. coli* defective in methionine uptake demonstrated the existence of distinct high affinity ($K_T = 7.5 \times 10^{-8} M$) and low affinity ($K_T = 4.0 \times 10^{-5} M$) methionine transport systems (93a).

Cystine and Diaminopimelic Acid (DPA)

Strain 17-325 and other DPA-auxotrophs of *E. coli* W grew slowly on DPA unless also supplemented with lysine. Mutants selected for fast growth on DPA in the absence of lysine (strains D and D_2) showed a fivefold increase in the rate of uptake of ^{14}C-DPA. That the enhanced uptake of DPA did not result from the block in DPA synthesis was shown by the fact that the rate of DPA uptake in strain D_2W, a spontaneous prototroph revertant of strain D_2, remained as high as in strain D_2. However, because most of the intracellularly accumulated radioactivity was in lysine and other metabolites and very little remained as DPA, a lysine-requiring mutant, *DW16*, was isolated from strain D_2W and examined for DPA uptake. Again, the lysine-requiring D-line strain took up DPA at a rate fivefold higher than did a lysine auxotroph, strain 26–26, derived from wild-type *E. coli* W. Most of the accumulated radioactivity (1000-fold concentration) was found in DPA.

The D mutation not only resulted in faster uptake of DPA but also reduced the affinity of the uptake system for DPA, from a K_m well below 10^{-7} to one of 3×10^{-7} M. Hence the mutation altered the nature of the carrier and not simply the amount of carrier available for transport. The affinity for meso-DPA was higher than for L-DPA, although the V_{max} values were about the same. L-cystine was the only compound effectively inhibiting DPA uptake. Comparison of L-cystine uptake by W and D strains showed that L-cystine and DPA were taken up by the same system. The K_m for L-cystine also increased from 3×10^{-8} M in the parent strain, to 3×10^{-7} M in the D mutant. This transport system was neither repressed nor induced by either DPA or lysine. In the original W strain (but not in prototroph revertants) there was an additional transport system strictly specific for L-cystine (94, 95).

All these findings have been corroborated by a recent study of Berger & Heppel (96). The general system (cystine and DPA) in *E. coli* W grown in minimal medium was completely eliminated by osmotic shock, whereas the cystine-specific system was not affected. The shock fluid exhibited cystine-binding activity, which could be totally inhibited by DPA (general binding protein). The binding activity in shockates of mutants of D_2W (which lack the general cystine transport system) grown in minimal medium, was only 10% of that found in strain D_2W. In strain W grown in a very rich medium (3% yeast extract and 4% tryptone) the general system is missing altogether, but the activity of the specific system is threefold higher than in cultures grown in minimal medium. This activity can now be reduced by osmotic shock. The shockate of such cells has no general cystine-binding protein, but instead exhibits cystine-binding activity with a specificity similar to that of the specific cystine transport system. The general cystine-binding protein has been purified to homogeneity (mol wt = 27,000); its specificity closely resembles that of the general cystine transport system. In spite of the fivefold increase in the activity of the general cystine transport system in strain D_2W, the specific binding activities are similar to those found in the parent strain. This may indicate that the mutation affected a component of the system other than the binding protein.

Histidine

A very thorough study of histidine transport in *S. typhimurium* has been carried out by Ames and co-workers (80, 81, 97, 99, 100, 102, 103). Lineweaver-Burk plots of substrate saturation data for L-histidine uptake showed distinct biphasic kinetics with two K_m's for histidine: 10^{-8} M and 10^{-4} M. In the low concentration range histidine uptake was not inhibited by any other amino acid or analog tested. However, at 3×10^{-4} M of L-histidine, its uptake was 70% inhibited by tryptophan, phenylalanine, and tyrosine and by L-1-methylhistidine, 3-pyrazolealanine, and 2-thiazolealanine. The aromatic amino acids were very potent inhibitors: 50% inhibitions were obtained at inhibitor-histidine ratios of 1:100 (80).

These data suggested the operation of two permeation systems for L-histidine: a high-affinity specific histidine permease and a low-affinity histidine transport via the general aromatic permease. Indeed, the histidine analogs 1-methylhistidine and 3-pyrazolealanine also inhibited tyrosine uptake. Resistance to the α-hydrazino analog of histidine, 2-hydrazino-3-(4-imidazolyl) propionic acid (HIPA), which is a potent growth inhibitor for *S. typhimurium*, *E. coli* W, and *E. coli* K12, was successfully used for the isolation of specific histidine permease, *hisP*, mutants (97). Transduction with phage P22 revealed 49% linkage between *hisP* and *purF* and about 0.3% linkage with *aroD*. Because *aroD* and *purF* are 10% cotransducible, the order is *aroD*, *purF*, *hisP*. Three-point crosses by transduction confirmed this order. The *hisP*+ allele was dominant to *hisP* in merozygotes constructed by transfer of the *E. coli* episome F'32. Merozygotes diploid for the *hisP*+ allele were supersensitive to HIPA and yielded no resistant mutants.

Histidine auxotrophs also carrying a mutated *hisP* allele and therefore missing the high affinity component of L-histidine transport were inhibited by tryptophan which blocks histidine entry via the general aromatic permease. This was exploited for the isolation of *hisP* revertants by selection for resistance to tryptophan inhibition. Some of the *hisP* reversions were found to be due to the presence of amber suppressors, thus demonstrating that a protein product of *hisP* is essential for the activity of the specific histidine transport system (81).

Two classes with elevated specific histidine permease activity have been isolated by selection for the ability of histidine auxotrophs to utilize D-histidine (98). Mutations of one class (*dhuA*), which resulted in growth on 0.03 mM D-histidine at similar rates as on 0.03 mM L-histidine, were closely linked to the *hisP* locus. The *dhuB* class of mutants utilized histidine rather poorly; they have not been mapped. *dhu* mutants, in contrast to the parent strain, showed prominent D-histidine uptake with K_m's 1000-fold higher than for L-histidine. The two isomers competed for entry into the cell. The D-histidine-utilizing mutants became supersensitive to HIPA (99).

A number of tertiary mutants that lost the ability to utilize D-histidine were isolated by ampicillin counterselection from *dhuA his−* strains; all of them mapped in the *dhuA hisP* region, and most of them behaved like *hisP* mutants. Upon transduction with a *dhuA+ his−* lysate, all of the 13 mutants tested gave some *dhuA his−* recombinants, indicating that none of them was a true *dhuA+* revertant. When the revertants were used as donors in transduction with a *purF his−* recipient on L-histidine plates, only 2% of the *purF+ his−* recombinants could grow on D-histidine. On this basis all 13 apparent *dhuA* revertants were classified as double mutants *dhuA hisP*.

dhu mutations have also been isolated in the absence of D-histidine. The selection method used (low histidine, high arginine, and high tryptophan) was based on the observation of Ames (cited in reference 98) that the specific histidine permease was sensitive to L-arginine. Only mutants derepressed or altered in either the histidine or the aromatic permease would grow. All his^- strains able to grow on these plates were either $dhuA$ or $dhuB$ mutants. These findings support the notion that dhu mutants have derepressed activity of histidine permease (98).

The regulatory nature of the $dhuA$ locus was further substantiated after the isolation and analysis of $hisJ$ mutants. These mutants were selected in a $dhuA$ his^- strain for the loss of ability to utilize D-histidine, while still retaining normal (wild-type) sensitivity to HIPA. The phenotype of these $dhuA$ $hisJ$ double mutants differed from that of the $dhuA$ $hisP$ double mutants discussed earlier, which lost both the ability to grow on D-histidine and the sensitivity to HIPA. In transductions of a $purF his^-$ strain with phage grown on $dhuA hisJ$ mutants, only 2% of the pur^+ his^- recombinants inherited the $dhuA$ mutation without also inheriting the $hisJ$ mutant gene and could therefore grow on D-histidine. Thus the $hisP$, $hisJ$, and $dhuA$ loci are very closely linked.

Fractionation of the supernatant from osmotically shocked wild-type cells on DEAE-Sephadex revealed two peaks of histidine-binding activity: 95% of the binding activity (at 10^{-8} M L-histidine) peaked at 0.04 M NaCl (J protein), and about 5% was eluted as a peak (K protein) at 0.15 M NaCl. The level of J protein was elevated fivefold in $dhuA$ mutants, and there was no J protein binding activity in strains carrying a $hisJ$ mutation. Mutations at the $hisP$ locus did not affect the amount of J protein. The amount of K protein appeared to be the same in all these strains. The binding protein from dhu strains chromatographed similarly and had the same binding affinity ($K_d \sim 2 \times 10^{-7}$ M at 4°C) for histidine as the protein from the parent strain. Both $hisJ$ and $hisP$ mutants are defective in L-histidine uptake but to a different extent. The respective affinities of the residual transport activities are 2×10^{-7} M and 10^{-6} M even in completely defective mutants (frameshift and amber). The authors conclude that the loss of J binding protein results in a loss of a very high affinity transport component ($K_m \sim 10^{-8}$ M), the remaining, still quite efficient transport ($K_m = 2 \times 10^{-7}$ M), presumably occurring through additional components, possibly the K protein. Mutations in $hisP$ eliminate transport through both the J and the K components. The $dhuA1$ mutant showed increased transport and higher affinity ($K_m = 6.6 \times 10^{-9}$ M) than the wild type ($K_m = 2.6 \times 10^{-8}$ M). A double mutant, $dhuA1 hisP5503$ showed the same rate of transport and low affinity for histidine uptake as did the single mutant $hisP1661$.

Convincing evidence that the J protein is an essential component of the histidine transport system and that $hisJ$ is its structural gene has been provided in an elegant study by Ames & Lever (100). They studied a $hisJ$ revertant that had a temperature-sensitive J component. The mutation that caused the $hisJ$ phenotype to revert also mapped within the $hisJ$ locus but at a site distinct from that of the primary $hisJ$ mutation. Histidine-binding activity in shock fluids of the revertant, growth of the revertant on D-histidine, and L-histidine transport in whole cells decayed as a function of increasing temperature. The J^{ts} protein exhibited markedly different chromatographic properties from those of wild-type J protein. It probably differs from wild-type J protein in a sequence of several amino acids, because the revertant was obtained by ICR-191 treatment of a frameshift mutant. Both proteins have a similar molecular weight ($\sim$26,000) and the mutant protein cross-reacts with antiserum to wild-type J protein. The purification and properties of the histidine-binding protein (J component) have been described in detail by Rosen & Vasington (101) and Lever (102).

In a recent paper by Kustu & Ames (103) *hisP* has been shown to participate as an essential component of an arginine low-affinity transport system. *hisP* mutants are unable to utilize arginine as the sole source of nitrogen, but they grow normally on arginine di- and tripeptides. These strains also have normal high-affinity arginine transport as witnessed by unimpaired growth of double mutants *hisP arg⁻* in media with ammonia as the nitrogen source, supplemented with low concentrations of arginine. Thus, not all the arginine transport systems in *S. typhimurium* require the *hisP* protein. Although the histidine-binding J protein binds arginine in vitro (100–102), experiments with frameshift mutants TA1650 and TA1789 have shown that the J protein is not required for utilization of arginine as a nitrogen source.

Lysine, Arginine, and Ornithine

That lysine, arginine, and ornithine share a common transport system in *E. coli* W was suggested by Schwartz et al in 1959 (47). This was indicated by the behavior of an L-canavanine-resistant mutant that lost the repressibility of ornithine transcarbamylase by exogenous arginine although the enzyme could be further derepressed, as in the wild type, by conditions that reduce the endogenous synthesis of arginine. Measurement of the uptake of radioactive amino acids showed that the mutant's capacity for transporting L-lysine, L-arginine, and DL-ornithine has been reduced to 20% of that in the parent, whereas the uptake of glycine, L-alanine, and D-serine remained the same as in the wild type. However, later studies by Wilson & Holden (104, 105) indicated that lysine and arginine are not transported by identical systems. This was inferred from inhibition studies with intact cells and from experiments with osmotically shocked cells in which the reduced arginine transport could be partly restored by purified protein fractions isolated from the shock fluid, while that of lysine could not.

Basic amino acid transport in *E. coli* K12 has been recently studied by Rosen (106–108). Kinetic analysis showed that L-arginine and L-ornithine uptake gave linear reciprocal plots indicating transport by a single system. The kinetic parameters obtained were: $K_m = 2.6 \times 10^{-8}\ M$ and $V_{max} = 3$ nmol/min/mg cell protein for arginine, and $K_m = 1.4 \times 10^{-6}\ M$ and $V_{max} = 11$ nmole/min/mg cell protein for ornithine. However, the data on L-lysine uptake gave curvilinear reciprocal plots and could be best fitted with the following constants, assuming only two active transport systems and no contribution by diffusion: $K_1 = 5 \times 10^{-7}\ M$, $V_1 = 1.8$ nmol/min/mg cell protein, and $K_2 = 10^{-5}\ M$, $V_2 = 2.8$ nmol/min/mg cell protein. In competition studies with lysine, arginine, and ornithine as substrates none of the other amino acids normally occurring in proteins inhibited significantly at a twentyfold molar excess. However, lysine and arginine each inhibited ornithine uptake, and arginine and ornithine inhibited the uptake of lysine. Arginine uptake was not inhibited by lysine or ornithine even at a 500-fold molar excess of inhibitor. Inhibition of lysine uptake by arginine or ornithine was not complete even at a 500-fold excess of inhibitor, indicating that only one of the lysine transport systems was inhibited. The percentage of inhibition by either inhibitor decreased with increasing concentrations of lysine. This indicates that the high affinity system is common to the three basic amino acids (LAO system) and the low affinity system is specific for lysine. A second system for the transport of ornithine may be indicated by the fact that arginine did not completely inhibit ornithine transport.

Osmotic shock treatment reduced lysine transport to that in control cells in which the LAO system had been inhibited by arginine. Furthermore, the residual lysine transport in shocked cells was totally resistant to arginine inhibition. Thus, only the activity of the LAO system is reduced by osmotic shock; the lysine-specific system is unaffected. Transport of both arginine and ornithine is severely reduced in shocked cells. Fractionation of crude shock fluid on DEAE-cellulose gave two protein peaks with arginine-specific binding activity, as in the work of Wilson & Holden with *E. coli* W (105). A third peak showed binding activity toward all three basic amino acids. This LAO-binding protein was purified by isoelectric focusing. The dissociation constants determined by equilibrium dialysis were: $K_d(\text{Lys}) = 3.0$ μM, $K_d(\text{Arg}) = 1.5$ μM, and $K_d(\text{Orn}) = 5.0$ μM. Each of the three amino acids could inhibit the binding of the other two; no other naturally occurring amino acid was inhibitory. The molecular weight of the LAO-binding protein was between 26,200 and 30,200, depending on the method used (106). The arginine-specific binding protein has also been purified (105). It has a K_d of 3×10^{-8} M and binds 1 mole of arginine per 1 mole of protein (mol wt = 27,700). It does not cross-react immunologically with the LAO-binding protein. Citrulline and canavanine were not bound by the arginine-specific protein, but did bind to the LAO-binding protein. Canavanine and citrulline were also transported by the LAO system, whereas arginine, which binds to the LAO protein and inhibits the activity of the LAO transport system, appears not to be a substrate for it. First, growth of cells in an enriched medium (3% yeast extract and 4% tryptone) abolished ornithine uptake and reduced arginine uptake by 70%, while lysine uptake was even enhanced. Second, the growth of an arginine-ornithine auxotroph was inhibited by lysine when supplemented with ornithine, but not when supplemented with arginine (107). In the light of these observations it seems difficult to explain the severe reduction in arginine transport observed in a canavanine-resistant mutant (47). This problem has been very recently investigated in two laboratories (108, 109). Two canavanine-resistant mutants, *CanR22* and JC182-5, were studied by Rosen (108); Celis et al (109) used strain JC182-5. Both studies reconfirmed the early findings of Schwartz et al (47) that canavanine resistance was accompanied by a decrease in transport of all three basic amino acids, arginine transport being affected more drastically than that of ornithine or lysine.

A primary effect on metabolism rather than on transport was ruled out by the fact that the differences in the rates of ^{14}C-arginine uptake between wild type and mutant persisted in the presence of aminooxyacetic acid, which prevents the decarboxylation of arginine (109). That the apparent decrease in ^{14}C-arginine uptake did not result from excretion of endogenous arginine due to derepression of the arginine biosynthetic enzymes in the mutant was demonstrated by the fact that introduction of a block in arginine biosynthesis in the canavanine-resistant mutant did not improve its poor ability for arginine uptake.

That more than one transport system was affected by the mutation was shown by the different regulation of the uptake of arginine and ornithine from that of lysine, and also by kinetic and inhibition studies of uptake. Growth in the presence of arginine or ornithine repressed the specific transport systems for both these amino acids, but did not affect the transport of lysine. Growth in the presence of lysine repressed the lysine-specific transport system and the common LAO system (108). Substrate saturation kinetics and studies on the inhibition of uptake of each of the basic amino acids by the other two clearly showed that the activities of both the LAO system and the arginine-specific transport system were reduced in the mutant (108). In fact, all four systems, the high affinity LAO system and the three low affinity specific systems, may be affected (109). Rosen compared the properties of the binding proteins from mutant and wild-type strains and found no differences in terms of fractionation profiles, amount, binding constants, or immunological reactivity (108). Genetic analysis of the *CanR22* mutation by transduction showed that the mutation maps at min 56, in or very close to the *argP* gene found by Maas for canavanine resistance (110).

To explain the pleiotropic effect of this mutation on the arginine- and LAO-specific transport systems, Rosen proposes that the lesion occurred in a common step past the level of recognition or binding site of the two systems, possibly in one of the genes of the energy coupling step which is shared by the two systems.

Basic amino acid transport in *Pseudomonas putida* has been studied by Rodwell and co-workers (111, 112). At least three different transport systems have been identified. A basic amino acid transport system takes up L-lysine ($K_m = 7.3 \times 10^{-6}M$), L-arginine ($K_m = 4.8 \times 10^{-6} M$), and L-ornithine. This transport system is induced by L-lysine or DL-pipecolic acid, together with the enzymes of the lysine catabolic pathway. A second system transports L-lysine ($K_m = 4.1 \times 10^{-7} M$) and L-ornithine ($K_m = 1.3 \times 10^{-7} M$). Cells grown in the presence of L-arginine possess an additional high affinity L-arginine-specific system ($K_m = 5.2 \times 10^{-8} M$).

Friede et al (113) described two transport systems for L-lysine in *S. faecalis.* One with higher affinity is specific for L-lysine and L-hydroxylysine. The other system transports both L-lysine and L-arginine and also accepts L-hydroxylysine. The affinities of both systems for L-hydroxylysine are much smaller than for L-lysine. A mutant resistant to L-hydroxylysine appears to have a defective L-lysine-specific transport system.

Glutamic and Aspartic Acids

Some strains of *E. coli* are unable to utilize glutamic acid as the sole source of carbon. Halpern & Umbarger suggested in 1961 (114) that this was due to the low activity of the glutamate transport system in these organisms. Mutants of *E. coli* W which specifically acquired the ability to utilize L-glutamic acid as a carbon source were readily isolated; they were still unable to utilize α-ketoglutarate. Conversely, mutants isolated for their ability to grow on α-ketoglutarate were unable to grow ·on glutamate. The growth of strains capable of utilizing glutamate as a source of carbon was inhibited by α-methyl-DL-glutamic acid, when added to the glucose-ammonia minimal medium. Strains unable to grow on glutamate were resistant to high concentrations (1.5 mg/ml) of the analog. Revertants of glutamate-utilizing mutants lost their sensitivity to α-methylglutamate, and conversely, mutants isolated for the acquisition of resistance to the analog concomitantly lost their ability to grow on glutamate (114). Later detailed studies have shown that the glutamate-utilizing mutants did not differ from the parent strains in the activities of enzymes involved in glutamate metabolism (115, 116). However, the mutants did exhibit severalfold higher rates and greater capacities for the active transport of ^{14}C-glutamate than did the respective E. coli W, H, and K12 parent strains (115, 117, 118).

The uptake of ^{14}C-glutamate was strongly temperature- and energy-dependent and resulted in considerable accumulation against a concentration gradient (up to 1700-fold under optimal conditions). The uptake system was saturable, with a K_m of about 8×10^{-6} M as measured at 37°C in the presence of sodium succinate as the energy source. Growth on glutamate resulted in a twofold increase in the activity of the glutamate transport system. Glutamate transport in the E. coli K12 glutamate-utilizing mutant CS1 was highly specific: only structural analogs of L-glutamic acid with an intact α-amino and α-carboxyl group and a proper distance between the two carboxyl groups (D-glutamic acid, γ-methyl- and γ-ethyl-esters of L-glutamic acid, β-hydroxyglutamic acid, α-methyl-DL-glutamic acid, and L-glutamine) behaved as strictly competitive inhibitors; α-ketoglutarate, γ-aminobutyrate, and L-aspartate did not compete for the glutamate binding site. However, several amino acids (L-aspartate, L-alanine, L-serine, L-valine, L-methionine, L-proline, and L-histidine) and α-ketoglutarate noncompetitively inhibited glutamate uptake, while γ-aminobutyrate activated the system. Glycine, L-phenylalanine, and DL-methionine sulfoxide were inactive even at a 10,000-fold molar excess (117). The authors suggest that the glutamate permease is an allosteric protein with a binding site for the permeant, which also has affinity for its structural analogs, and an effector site with affinity for a number of amino acids. Binding of an effector to this latter site affects the activity of the transport site. Kinetic experiments in which the effect of different effector pairs at varying concentrations of each member was examined suggest the existence of two effector sites: one for amino acids and one for α-ketoglutarate (unpublished results).

When glutamate uptake was measured in the absence of any added energy source, or in the presence of glycerol or glucose instead of sodium succinate, curvilinear Lineweaver-Burk plots were obtained. Addition of sodium succinate resulted in increased activity and "normalization" of uptake kinetics. The authors interpreted these findings on the assumption that the effector site is also capable of binding glutamate and succinate and that the binding of glutamate to this allosteric site reduces the activity of the permease (117). However, an alternative explanation was suggested by the work of Kahana & Avi-Dor (119) which indicated that Na$^+$ ions may affect glutamate uptake in E. coli. The specific requirement for NaCl for the active transport of L-glutamate by the halophile H. salinarium (120) and Na$^+$-activated transport of α-aminoisobutyric acid by a marine pseudomonad (121–123a) have been described. In fact, a number of reports have recently appeared documenting the dependence of threonine (124), melibiose (125), and citric acid (126, 127) transport in nonhalophilic bacteria on cations. The most direct indication was given in the work of Frank & Hopkins (128), who found that glutamate transport in *E. coli* B was stimulated by Na^+ ions and that the effect was on the affinity of the uptake system for glutamate, rather than on the maximal velocity when the uptake system was saturated with the permeant. It was therefore very likely that the peculiar shape of substrate-saturation curves, in reaction mixtures in which sodium glutamate was the only source of Na^+ ions, was due to the Na^+ requirement of the glutamate uptake system. Reexamination of the problem established that the active transport of glutamate in *E. coli* K12 was dependent on the simultaneous presence of both Na^+ and K^+ ions. Whereas Na^+ affected the affinity of the transport system for glutamate, varying the concentration of K^+ ions affected the capacity for glutamate uptake but had no effect on affinity. In the presence of saturating concentrations of Na^+ and K^+ (15 mM of each), reciprocal plots of rate of uptake vs concentration were always rectilinear regardless of which compound served as the energy source (129).

A similar requirement for Na^+ and K^+ has been described by Thompson & MacLeod for the transport of α-aminoisobutyric acid in whole cells (130) and isolated membrane vesicles of a marine pseudomonad (131). These authors have proposed a model according to which the binding of Na^+ to the carrier results in a conformation with increased affinity for the substrate. The ternary complex traverses the membrane, and in the presence of energy and K^+ the affinity of the Na^+-carrier complex for the substrate is somehow reduced and the substrate is released and accumulates in the cell (130).

Comparative kinetic analysis of a series of independently isolated glutamate-utilizing *E. coli* K12 mutants showed that in all of them the maximum rate and capacity for glutamate uptake increased 4–8-fold as compared to the parent strain, but the affinity of the transport system did not change (115, 118, 132). This suggested the possibility that glutamate transport in wild-type *E. coli* K12 was partially repressed and that the mutations isolated by us were in a control gene of the system, causing derepression of glutamate permease synthesis. This idea was borne out by the isolation of two further types of mutants. One class consisted of three apparent revertants of the glutamate-utilizing mutant *E. coli* K12 CS7. Glutamate uptake experiments revealed that these mutants not only had reduced rates and capacities of glutamate uptake similar to those of wild-type *E. coli* K12 but they also exhibited a more than tenfold decrease in affinity (10^{-4} M in the revertant CS7/50 vs 0.5 $\times$ 10^{-5} M in its glutamate-utilizing parents CS7 and 10^{-5} M in the wild-type strain). Mapping by interrupted mating and by transduction with phage Plkc showed that the original mutation and the reversion were at closely linked but separable loci between the *tna* and *pyrE* genes. The gene determining a structural element of glutamate transport and identified by the reverse mutation was designated *gltS*, and the adjacent control gene, tentatively considered as the operator locus, was designated *gltC*. A third locus was defined by mutants isolated for their ability to utilize glutamate as a carbon source at 42°C but not at 30°C. One such mutant, CS2^{TC}, produced a thermolabile repressor, which could be heat-inactivated in the absence of growth. The regulatory gene, *gltR*, specifying the synthesis of the glutamate permease repressor, maps 6 min to the right of the *gltS-gltC* cluster; it is 3% linked to *metA* by transduction (132).

A glutamate-binding protein released from the periplasmic space of *E. coli* K12 CS7 during the preparation of spheroplasts has been isolated. The spheroplasts showed a decrease in glutamate transport activity to about 50% of that of untreated cells. The activity could be fully restored by the addition of concentrated shock fluid or purified glutamate-binding protein. The in vitro behavior of the glutamate-binding protein is very similar to that of the glutamate transport system in whole cells in regard to specificity, kinetic constants, type of inhibition, and cation requirement (133–135). The essential role of the binding protein in transport is also very strongly indicated by our recent finding that the *gltR*^{TL} mutant CS2^{TC} had about 2.3 times as much glutamate-binding protein when grown at 42°C as when grown at the nonpermissive temperature of 30°C, and the amount of binding protein obtained from strain CS7 was 1.7 times that obtained from the wild-type parent strain (135). However, the glutamate binding protein does not seem to be the only specific structural component of the glutamate transport system. Membrane vesicles of *E. coli* K12 CS8, a glutamate-utilizing mutant, carry out D-lactate stimulated concentrative glutamate transport of comparable activity to that in spheroplasts. Nevertheless, no glutamate-binding protein could be detected in these vesicles. Membrane vesicles of wild-type *E. coli* K12 or of the CS2^{TC} mutant grown at 30°C showed very little if any D-lactate-stimulated glutamate transport (135 and unpublished data).

The Na$^+$ requirement for glutamate transport by intact cells of *E. coli* B (128) was recently examined at the subcellular level (glutamate binding by protein released from the periplasmic space and glutamate transport by isolated membrane vesicles) in wild-type and mutant strains (135a). In contrast to our findings with *E. coli* K12, binding of glutamate by protein released from *E. coli* B was not dependent on Na$^+$. However, shocked wild-type cells retained Na$^+$-stimulated transport. Membrane vesicle preparations of the different strains mimicked the behavior of whole cells in regard to Na$^+$-stimulated transport. The authors conclude that in *E. coli* B the Na$^+$-activated glutamate transport system is located in the cytoplasmic membrane and that releasable binding protein is not intimately involved in transport.

Rapid first-order exit of ^{14}C-glutamate from preloaded cells was observed in *E. coli* K12 (136). The exit reaction was strongly temperature dependent, with a Q$_{10}$ of 2.4 between 27 and 37°C and a rate constant of 0.086 min^{-1} at 30°C under physiological conditions in the presence of glycerol and succinate. Addition of NaN$_3$, 0.01 *M*, accelerated the apparent rate of exit 2.4-fold. Exit was also accelerated 2.4-fold upon addition of cold L-glutamate or any competitive or noncompetitive inhibitor of glutamate uptake. However, uptake of ^{14}C-glutamate was not accelerated by preloading the cells with cold L-glutamate. The acceleration of ^{14}C-glutamate exit by cold L-glutamate and by NaN$_3$ was not additive. In the presence of 25% sucrose the rate of glutamate exit was about one fifth of that in the control, whereas the rate of entry was not affected (137). Unlike glutamate uptake, which was severalfold higher in glutamate-utilizing mutants than in wild-type strains and whose activity was doubled following growth on glutamate, the apparent rate of exit in unpoisoned cells was even some 20% lower in the glutamate-utilizing mutants than in the wild type, and growth on glutamate did not affect it. These data have been used as an argument for the nonidentity of carriers mediating entry and exit (136, 137).

It is not very likely that acceleration of glutamate exit by cold glutamate and by noncompetitive inhibitors of glutamate uptake was due to exchange diffusion and counterflow (138), especially because no symmetrical acceleration of entry by pre-loading was observed. It is therefore suggested (136) that the apparent acceleration of exit was due to prevention of recapture (139). This is supported by the fact that the exit rate in the absence of uptake inhibitors is some 20% faster in the wild type, which has a lower rate of recapture, and its acceleration by cold L-glutamate or azide is 25% less than in the mutant with rapid uptake (136).

From these data and from the lack of additivity of the effects of azide and inhibitors of uptake on the rate of exit, one can hardly escape the notion that in glutamate transport in *E. coli* energy is directly coupled to entry. This is in accord with the recent findings of Koch on the energy expenditure in the downhill transport of galactosides (140), but is contrary to the beliefs of many others (89, 138, 141, 142). The idea of separate carriers for entry and exit has also been suggested by Burrous & DeMoss for tryptophan transport in *E. coli* (77) and by Gryder & Adams for the transport of hydroxyproline in *P. putida* (91).

The transport of glutamate and aspartate in lactobacilli has been studied by Holden and his associates (143–146). The two dicarboxylic acids were taken up by the same transport systems. In *S. faecalis* there are two systems: one with higher affinity, which takes up L-glutamate with a K_m of 2×10^{-5} M and L-aspartate with a K_m of 10^{-5} M, and a second system with lower affinity, with K_m's of 8.3×10^{-3} and 8.3×10^{-4} M for L-glutamate and L-aspartate, respectively. A mutant requiring high concentrations of glutamate for growth had simultaneously lost the high affinity transport of both glutamate and aspartate. The mutant also lost the ability to transport α-methylglutamate and 2-amino-3-phosphonopropionic acid; the transport of other amino acids has not been reduced in the mutant (145, 146).

An inducible glutamate transport system has been described in mycobacteria (147). While uptake of L-glutamate in *M. avium* is saturable, the rate of D-glutamate uptake is proportional to the concentration of the amino acid in the medium. Nevertheless, D-glutamate uptake is sensitive to inhibition by azide, decreases drastically below 4°C, and is strongly inhibited by L-aspartate and L-glutamate, but only very weakly inhibited by L- and D-alanine and by D-aspartate (148). The uptake of L-glutamate in *Mycobacterium smegmatis* is mediated by both an active and a passive process, whereas only a passive component showing diffusion kinetics has been found for the transport of D-glutamate (149). A very similar picture obtains for the transport of the two optical enantiomorphs of aspartic acid in *M. smegmatis*. From inhibition studies the author concludes that L-glutamate and L-aspartate are transported by the same permeation system (150).

Transport of L-aspartate in *E. coli* K12 has been studied by Kay & Kornberg (151–153). Lineweaver-Burk plots were biphasic with K_m's of 3.4 × 10^{-6} *M* and 3.9 × 10^{-5} *M*. In mutants resistant to 3-fluoromalate and showing impaired transport of C$_4$-dicarboxylic acids (*dct* mutants) the reciprocal plot for aspartate was rectilinear with a K_m of 3.5 × 10^{-6} *M* and a V_{max} of 1.5 nmol/mg dry wt/min. In *ast⁻* mutants selected for resistance to inhibition by DL-*threo-β*-hydroxyaspartic acid the Lineweaver-Burk plot for aspartate transport was also a straight line, but with a K_m of 3 × 10^{-5} *M* and a V_{max} of 25 nmol/mg dry wt/min. The high affinity system was inhibited by β-hydroxyaspartate and to a lesser extent by L-glutamate but not by C$_4$-dicarboxylic acids. The low affinity system was sensitive to inhibition by C$_4$-dicarboxylic acids and was induced by them. The specific high affinity system was not induced by aspartate. Membrane vesicles actively transported and accumulated L-aspartate and were stimulated by D-lactate. This activity was missing in vesicles prepared from *ast* mutant cells (153). The *dct* gene was mapped by interrupted mating and placed at about 70 min, near *xyl*, in the order *ilv, xyl, dct* (152).

These studies clearly established the existence of two transport systems facilitating the transport of aspartate in *E. coli* K12.

Asparagine and Glutamine

The discovery of a highly specific asparagine active transport system in *E. coli* K12 has been reported by Willis & Woolfolk (154). Only compounds with a modified amide group, such as diazooxonorvaline (DONV) or β-aspartylhydroxamate, were competitive inhibitors; even L-glutamine did not inhibit L-asparagine uptake. Asparagine uptake greatly stimulated by glucose has been recently described in *Lactobacillus plantarum* and *S. faecalis* by Holden & Bunch (155). Curvilinear Lineweaver Burk plots have been obtained in both organisms, suggesting two catalytic components and a diffusion term, but the data could also be accounted for by assuming diffusion and only one catalytic system. In *L. plantarum* the most effective inhibitors were L-asparagine and L-glutamine, but in *S. faecalis* L-asparagine uptake was most strongly inhibited by L-alanine, L-serine, L-α-aminoisobutyrate, L-cysteine, and L-methionine. It is therefore possible that in spite of the similar kinetics the asparagine transport systems in the two organisms differ significantly.

Active transport of L-glutamine in *E. coli* has been described by Weiner et al (156) and Weiner & Heppel (157). Osmotic shock released glutamine-binding protein into the medium and resulted in a 90% decrease in the rate of transport. A mutant capable of utilizing glutamine as the sole source of carbon showed a fourfold higher initial rate of glutamine uptake and released 3.3 times as much binding protein as did the wild-type parent (156). The purified protein had a K_d of 3 × 10^{-7} *M* for L-glutamine. The K_m for transport was 0.8 × 10^{-7} *M*. Only α-glutamylhydrazide and α-glutamylhydroxamate were competitive inhibitors of glutamine binding and uptake. Binding of L-glutamine to the protein is accompanied by changes in its absorbance and fluorescence (157). Conformational changes upon addition of L-glutamine as reflected in the proton magnetic resonance (PMR) spectrum have also been recently reported by Kreishman et al (158).

AMINO ACID TRANSPORT IN ISOLATED BACTERIAL MEMBRANE VESICLES

It is not my intention to dwell here on the basic and more general aspects of the membrane vesicles system. Nor do I discuss in detail the evidence for the coupling of transport in membrane vesicles to the electron transfer chain and the model proposed to explain its mechanism. These have been amply discussed in recent articles by Kaback, including several reviews (10, 11, 159). In the next few paragraphs I present recent data on amino acid transport in bacterial membrane vesicles stressing the genetic basis where the relevant information is available and compare the specificity and other important features of the amino acid transport systems discussed, to those in intact cells.

One of the first amino acid transport systems studied in membrane vesicles was that of proline in *E. coli* (160, 161). Membrane preparations of *E. coli* W6, a proline auxotroph, catalyzed an energy-dependent concentrative uptake of proline, while membranes from the proline transport-deficient mutant, W157, did not. No other amino acid except for hydroxyproline inhibited proline uptake. The activity of the system was about 50% or more of that observed in spheroplasts. A membrane protein fraction solubilized with Brij 36-T was fractionated on Sephadex G-100 in the presence of the detergent. Three proline-binding fractions were obtained: a relatively low molecular weight fraction with high specific proline-binding activity and two heavier fractions with low specific proline-binding activities and with D-lactate dehydrogenase activity. Proline binding by the lower molecular weight fraction, devoid of D-lactate dehydrogenase activity, was highly specific and was not inhibited by structurally unrelated amino acids. This fraction also bound glycine, serine, lysine, and tyrosine (162). A similar correlation between the transport activity of membrane vesicles and that of intact cells was observed in regard to the uptake of glycine (163, 164). Membrane vesicles from wild-type *E. coli* W exhibited an energy-dependent transport of glycine. The transport system was specific for glycine, DL-alanine, DL-serine, and DL-threonine. Membranes of a D-serine-resistant transport-deficient mutant, WS, were unable to take up glycine or exchange ^{14}C-glycine with the external medium.

A thorough and extensive study of amino acid transport in membrane vesicles of *E. coli* ML308-225 has been recently made by Kaback and co-workers (165, 166). They have shown that the concentrative uptake of 16 amino acids by these vesicles is stimulated by D-lactate and to lesser extents by succinate, L-lactate, DL-α-hydroxybutyrate, and NADH. None of 36 other metabolites and cofactors, including ATP and P-enolpyruvate, has stimulated transport. The five energy sources that have shown activity differ in their relative effectiveness in regard to the uptake of individual amino acids. Anaerobiosis and inhibitors of electron transfer inhibited amino acid uptake, but inhibitors of oxidative phosphorylation showed only mild effects.

The following transport systems have been indicated by competition for entry and exchange and by kinetic studies:

1. Proline—highly specific for proline; even hydroxyproline has no effect on the system.

2. Lysine—highly specific for lysine; arginine and ornithine at ninetyfold molar excess produce very small effects.

3. Glycine and alanine—specific for glycine, alanine, and D-serine; L-serine has no effect. This system is markedly defective in membranes from a D-serine-resistant K12 mutant.

4. Serine and threonine—common for L-serine and threonine; D-serine at a fifty-fold molar excess is only moderately competitive.

5. Glutamic and aspartic acid—common system for the two acidic amino acids.

6. Phenylalanine, tyrosine, and tryptophan—a general transport system for the three aromatic amino acids; it also shows some affinity for histidine.

7. Histidine—specific for histidine; phenylalanine, tyrosine, and tryptophan are moderately competitive at a 100-fold molar excess.

8. Leucine, isoleucine, and valine—possibly two overlapping systems transporting the three branched-chain amino acids: one preferential for leucine and the other more specific for isoleucine and valine. Cysteine can also be transported with lower efficiency by one or both systems. Methionine also competes, although it is not taken up by membrane vesicles.

9. Cysteine—specific for cysteine; serine at a twentyfold excess is only slightly effective.

Although the above picture is very similar to the pattern of amino acid transport activities in whole cells of *E. coli*, a number of significant differences should be pointed out.

1. L-methionine, L-cystine, L-arginine, L-ornithine, L-asparagine, and L-gluta-mine are all actively transported in whole cells, but apparently not in membrane vesicles.

2. The LAO system, catalyzing the uptake of lysine, ornithine, and arginine in whole cells, has not been found in membrane vesicles.

3. The separate systems for the uptake of aspartate, glutamate, phenylalanine, tyrosine, and tryptophan present in intact cells of *E. coli* K12 have not been demonstrated in membrane vesicles of strain ML308–225.

4. The second uptake system for L-alanine, not shared by glycine, described in whole cells of *E. coli* K12 and W, has not been observed in membrane preparations.

Another comment relevant to this discussion concerns the quantitative comparison of transport activities in membranes and in intact cells. With the exception of serine and glutamic acid, the transport activity of membrane vesicles toward several amino acids thus compared was less than 30% of that in *intact* cells, being particularly low in the case of histidine (8%) and leucine (15%) (see Table VII in reference 166).

Concentrative uptake of amino acids has been shown to take place in membrane vesicles from a number of bacterial species in addition to *E. coli* when an artificial electron donor system, ascorbate-phenazine methosulfate, was used. Activity to different amino acids has been found in preparations from *S. typhimurium, P. putida, P. mirabilis, B. megaterium, B. subtilis, S. aureus,* and *Micrococcus denitrifi-*

cans (167). Konings et al have described the following uptake systems in membrane vesicles from *B. subtilis:* glycine-alanine, leucine-isoleucine-valine, serine-threonine, asparagine-glutamine, aspartate-glutamate, cysteine-methionine, proline, phenylalanine-tyrosine, lysine, and arginine (168, 169).

A detailed study of amino acid transport in membrane vesicles from *S. aureus* has been recently published by Kaback and associates (170, 171). The main difference from the *E. coli* system was that in *S. aureus* α-glycerophosphate was the primary physiological electron donor instead of D-lactate. NADH, succinate, and L-lactate showed very little activity, but ascorbate-PMS was a good electron donor. N-ethylmaleimide and *p*-hydroxymercuribenzoate inhibited amino acid transport and α-glycerophosphate oxidation but did not significantly affect α-glycerophosphate dehydrogenase activity with dichlorophenolindophenol as the electron acceptor. Hence, the site of coupling of α-glycerophosphate dehydrogenase to amino acid transport is between the dehydrogenase and the cytochrome chain (170). The following 12 specific systems have been observed: glycine-alanine, leucine-isoleucine-valine, serine-threonine, aspartate-glutamate, asparagine-glutamine, lysine, histidine, arginine, phenylalanine-tyrosine-tryptophan, cysteine, methionine and proline. All the transport systems tested except for those for glycine-alanine, leucine-isoleucine-valine, and glutamate-aspartate are stereo-specific for the L-amino acid isomer (171).

COUPLING OF ENERGY TO AMINO ACID TRANSPORT

Hong & Kaback (172) isolated mutants of *S. typhimurium* unable to grow on D-lactate but growing normally on glucose. Different classes of mutants were obtained. The *dld* mutants defective in D-lactate dehydrogenase activity could grow on D-ribose, glycerol, succinate, fumarate, and malate. Proline, glutamate, and tyrosine uptake in whole cells was normal. In membrane vesicles transport became dependent on succinate, which stimulated uptake even more than in the parent strain, although there was no increase in succinate dehydrogenase activity in the mutant. In a recent study, Reeves, Hong & Kaback (173) have been successful in reconstituting D-lactate-dependent transport of amino acids and lactose and valinomycin-induced rubidium uptake in membrane vesicles from a *dld⁻* mutant by treatment with a 0.75 *M* guanidine extract of wild-type vesicles. The specific activity of D-lactate dehydrogenase of the extract is approximately 3 times higher than that of intact ML308-225 membranes. Reconstituted transport activity is a saturable function of the amount of extract added. Another class of mutants isolated by Hong & Kaback (172) were those defective in electron transfer (*etc*). These could not grow on succinate, fumarate, malate, and D-ribose and grew slowly on glucose and glycerol under aerobic conditions. They had normal activity of D-lactate-dichlorophenolindophenol reductase. The *etc* mutation mapped at 120 min and was 40% cotransduced with *asn1* (asparagine requirement). Amino acid transport was severely affected in these mutants (172). An *E. coli* mutant uncoupled for oxidative phosphorylation and showing a 95% decrease in the activity of Ca^{2+}, Mg^{2+}-activated ATPase, AN120 (*uncA,* map position 73.5 min), isolated by Butlin et al (174), was examined by Prezioso et al (175). Amino acid transport in the mutant was comparable to that in wild type, both in whole cells and in membrane vesicles. Stimulation by electron donors and inhibition by anoxia, cyanide, and 2,4-dinitrophenol were the same in mutant and wild-type preparations. Furthermore, Mg^{2+} ions, which markedly stimulated ATPase activity in wild-type vesicles, and EDTA and dicyclohexylcarbodiimide (DCCD), which inhibited it, did not markedly affect respiration-dependent transport. The authors concluded that oxidative phosphorylation was *not* involved in respiration-linked transport. These results were in contrast to those obtained by Simoni & Shallenberger (176) who found a fourfold decrease in proline and alanine transport in whole cells of an *E. coli* mutant defective in Ca^{2+},Mg^{2+}-activated ATPase. Membrane vesicles from the mutant were totally devoid of transport activity. This discrepancy may very well be due to different ATPase lesions in the two mutants. This is supported by the recent report of Bragg & Hou on two ATPase-negative mutants (177). *E. coli* DL-54 lacking Ca^{2+},Mg^{2+}-ATPase activity was defective in amino acid transport and in energy-dependent transhydrogenase activity. Transhydrogenase activity could be restored in this mutant by coupling factor (ATPase) from wild-type strain. However, restoration of transhydrogenase could not be achieved in another Ca^{2+},Mg^{2+}-ATPase mutant, *E. coli* N_{144}, isolated by Kanner & Gutnick (178). Bragg & Hou (177) suggest that the defective enzyme particle in mutant N_{144} adheres more strongly to its membrane site than in mutant DL-54. A neomycine-resistant mutant of *E. coli*, NR70, lacking membrane-bound Mg^{2+}-ATPase, studied recently by Rosen (179), was also defective in amino acid and sugar transport, both in intact cells and in membrane vesicles. Amino acid transport could be restored by DCCD which was inhibitory in the wild-type strain. The author proposes a model according to which a protein(s) is responsible for the coupling of energy from oxidation-reduction to energy-requiring processes such as transport. The presence of ATPase protein (not in its enzymatic capacity) is required for the activity of the energy-coupling protein. DCCD can cause the same (conformational?) changes in the absence of the ATPase protein. This model could likewise explain the behavior of wild-type and DCCD-resistant strains of *S. faecalis* described by Abrams et al (180). Klein & Boyer studied the uptake of proline, leucine, α-methyl-glucoside, and Rb^+ in whole cells and in membrane vesicles of *E. coli* ML-308-225, under a variety of conditions (181). Proline transport in vesicles in the presence of D-lactate was dependent on O_2, but was unaffected by arsenate. ATP did not stimulate proline transport. Intact cells incubated at high concentrations of arsenate and showing low levels of ATP and PEP had unimpaired aerobic proline uptake but suffered a drastic decrease in proline uptake under anaerobic conditions. Similarly, DCCD, an ATPase inhibitor, had no effect on aerobic transport of proline, while being markedly inhibitory under anaerobiosis. These results show that intact cells can use energy either from oxidation or from phosphorylation to drive active transport. Similar conclusions have been reached by Schairer & Haddock for β-galactoside transport (182) and by Van Thiener & Postma in regard to the transport of serine by *E. coli* K12 (182a). Applying a very similar experimental approach but using simultaneously a wild-type and ATPase defective (DL-54) strain, Berger (183) confirmed Klein & Boyer's results concerning proline transport and concurred with their conclusion. However, energy coupling for glutamine transport required an active ATPase when the oxidative pathway was used (D-lactate as energy donor). The glycolytic component (when glucose was the energy donor) was resistant to cyanide and uncouplers and functioned normally in the ATPase-defective mutant. Arsenate abolished glutamine transport energized by either pathway. To use the author's own words, "the results suggest that proline transport is driven directly by an energy-rich membrane state, which can be generated by either electron transport or ATP hydrolysis. Glutamine uptake, on the other hand, is apparently driven directly by phosphate-bond energy formed by way of oxidative or substrate level phosphorylations." The fact that, in contrast to proline, glutamine is not transported in isolated membrane vesicles of *E. coli* (166) and that there is a glutamine-binding protein (156, 157, 184) but no proline-binding activity (8, 162) in the periplasmic space of these bacteria may be related to the different ways of energization of these two transport systems.

An alternative mechanism for the coupling of respiration to transport based on Mitchell's chemiosmotic hypothesis has been recently proposed by Harold and co-workers (184a). A synthetic lipid-soluble cation, dibenzylmethylammonium ion, was used as an indicator of electrical potential. The authors claim that oxidation of D-lactate by membrane vesicles of *E. coli* generates a membrane potential, vesicle interior-negative, of the order of −100 mV. In the absence of substrate an electrical potential was created by inducing electrogenic efflux of K^+ with the aid of K^+-ionophores. Under these conditions transient accumulation of ^{14}C-proline and other metabolites was observed. Results from Harold's laboratory on amino acid uptake by *S. faecalis* have also been interpreted by the authors in terms of proton symport, according to the chemiosmotic hypothesis (61a). Proton symport as a possible mechanism for active transport has also been suggested by Rosen in his recent paper on β-galactoside transport in wild-type and Mg^{2+}-ATPase-less mutant strains of *E. coli* (184b).

CONCLUDING REMARKS

It seems to me, and I hope to the reader, that considerable and significant progress has been made in the last pentad, toward understanding biological membrane transport. The rising interest among cell biologists, physiologists, biochemists, immunologists, pharmacologists, and medical scientists in the structure and function of biological membranes has been a major contributing factor to this welcome development. The three recent areas of main emphasis and activity in the field of transport, namely binding proteins, transport in membrane vesicles, and energization of transport well illustrate this point.

In regard to the nature and identity of the carrier(s) two points can be made:

(a)   Some binding proteins at least are essential components of amino acid transport systems in Gram-negative bacteria. The most notable example is the histidine-binding J protein of *S. typhimurium* whose participation in histidine transport rests on rather convincing kinetic, biochemical, physiological, and genetic evidence. However, the J protein is not the only structural element of the histidine transport system (99, 100, 102, 185). There is also quite good evidence for the involvement in transport of other amino acid binding proteins [glutamine (156, 157, 184), glutamate (133–135), LIV-protein (58, 68), L-protein (68)], but firm genetic evidence is yet to be forthcoming.

(b)   Membrane vesicles of *E. coli* exhibiting high transport activity for the great majority of amino acids do not seem to contain detectable amounts of periplasmic binding proteins (11, 135, 162, 166).

A number of possibilities suggest themselves to account for the existing data and they probably will be clarified in the next year or two:

1. Different systems may have different carriers, some of them firmly anchored to the plasma membrane, while others enjoy the demimonde existence in the periplasm or may even be loosely attached to the membrane.

2. The "true" carrier sits always in the inner membrane of the Gram-negative bacterium, and the binding proteins only serve as a vehicle for more efficient movement of substrate across the cell envelope (15).

3. A more rigorous version of point 2 is to say that because of the complex structure of the envelope in Gram-negative bacteria the substrate *cannot* reach the carrier in the inner membrane unless carried there by the binding protein. The binding protein-substrate complex interacts directly with the carrier protein molecule. The latter also has affinity for the free substrate (if any is around). In membrane vesicles the carrier being exposed to the medium does not require the good services of the binding protein.

It also seems reasonable to hope that the questions concerning energy requirement and coupling will be solved at the molecular level in the present decade.

ACKNOWLEDGMENT

The preparation of this study was supported by the Special Foreign Currency Program of the National Library of Medicine, National Institutes of Health, Public Health Service, US Department of Health, Education and Welfare, Bethesda, Maryland, under an agreement with the Israel Journal of Medical Sciences, Jerusalem, Israel.

Literature Cited

1. Freeland, J. C., Gale, E. F. 1947. *Biochem. J.* 41:135–38
2. Gale, E. F. 1947. *J. Gen. Microbiol.* 1:53–76
3. Cohen, G. N., Monod, J. 1957. *Bacteriol. Rev.* 21:169–94
4. Britten, R. J., McClure, F. T. 1962. *Bacteriol. Rev.* 26:292–335
5. Holden, J. T. 1962. *Amino Acid Pools,* ed. J. T. Holden, 566–94. New York: Elsevier. 815 pp.
6. Kepes, A., Cohen, G. N. 1962. *The Bacteria,* ed. I. C. Gunsalus, R. Y. Stanier, 4:179–221. New York: Academic. 459 pp.
7. Kepes, A. 1964. *The Cellular Functions of Membrane Transport,* ed. J. F. Hoffman, 155–69. New Jersey: Prentice Hall. 291 pp.
8. Kaback, H. R. 1970. *Ann. Rev. Biochem.* 39:561–98
9. Roseman, S. 1972. *Metabolic Pathways,* ed. L. E. Hokin, 6:41–88. New York: Academic. 704 pp.
10. Kaback, H. R. 1970. *Current Topics in Membranes and Transport,* ed. F. Bronner, H. Kleinzeller, 35–99. New York: Academic. 243 pp.
11. Kaback, H. R. 1972. *Biochim. Biophys. Acta* 265:367–416
12. Pardee, A. B. 1968. *Science* 162:632–37
13. Heppel, L. A. 1969. *J. Gen. Physiol.* 54:95s–109s
14. Lin, E. C. C. 1971. *Structure and Function of Biological Membranes,* ed. L. T.

Rothfield, 286–341. New York: Academic. 486 pp.
15. Oxender, D. L. 1972. *Ann. Rev. Biochem.* 41:777–814
16. Oxender, D. L. 1972. *Metabolic Pathways*, ed. L. E. Hokin, 6:133–85. New York: Academic. 704 pp.
17. Lin, E. C. C. 1970. *Ann. Rev. Genet.* 4:225–62
18. Sussman, A. J., Gilvarg, C. 1971. *Ann. Rev. Biochem.* 40:397–408
19. Payne, J. W., Gilvarg, C. 1971. *Advan. Enzymol.* 35:187–244
20. Grenson, M., Hou, C., Crabeel, M. 1970. *J. Bacteriol.* 103:770–77
21. Grenson, M., Hennaut, C. 1971. *J. Bacteriol.* 105:477–82
22. Grenson, M. 1973. *Genetics of Industrial Microorganisms,* ed. Z. Vanek, Z. Hastalk, J. Cudlin, 179–93. Prague: Academia
23. Lester, G. 1966. *J. Bacteriol.* 91:677–84
24. Jacobson, E. S., Metzenberg, R. L. 1968. *Biochim. Biophys. Acta* 156: 140–47
25. Pall, M. L. 1969. *Biochim. Biophys. Acta* 173:113–27
26. Tisdale, J. H., DeBusk, A. G. 1970. *J. Bacteriol.* 104:689–97
27. Wolfinbarger, L. Jr., DeBusk, A. G. 1971. *Arch. Biochem. Biophys.* 144: 503–11
28. Wolfinbarger, L. Jr., Jervis, H. H., DeBusk, A. G. 1971. *Biochim. Biophys. Acta* 249:63–68
29. Wolfinbarger, L. Jr., DeBusk, A. G. 1972. *Biochim. Biophys. Acta* 290: 355–67
30. Sanchez, S., Martinez, L., Mora, J. 1972. *J. Bacteriol.* 112:276–84
31. Magill, C. W., Nelson, S. O., D'Ambrosio, S. M., Glover, G. I. 1973. *J. Bacteriol.* 113:1320–25
32. Jones, O. T. G., Watson, W. A. 1962. *Nature* 194:947–48
33. Benko, P. V., Wood, T. C., Segel, I. H. 1969. *Arch. Biochem. Biophys.* 129:498–508
34. Sinha, U. 1969. *Genetics* 62:495–505
35. Oxender, D. L., Christensen, H. N. 1963. *J. Biol. Chem.* 238:3686–99
36. Christensen, H. N., Oxender, D. L., Liang, M., Vatz, K. A. 1965. *J. Biol. Chem.* 240:3609–16
37. Inui, Y., Christensen, H. N. 1966. *J. Gen. Physiol.* 50:203–24
38. Christensen, H. N., Liang, M., Archer, E. G. 1967. *J. Biol. Chem.* 242:5237–46
39. Winter, C. G., Christensen, H. N. 1964. *J. Biol. Chem.* 239:872–78
40. Eavenson, E., Christensen, H. N. 1967. *J. Biol. Chem.* 242:5386–96
41. Winter, C. G., Christensen, H. N. 1965. *J. Biol. Chem.* 240:3594–3600
42. Hagahira, H., Wilson, T. H., Lin, E. C. C. 1962. *Am. J. Physiol.* 203:637–40
43. Munck, B. G. 1966. *Biochim. Biophys. Acta* 120:97–103
44. Munck, B. G. 1966. *Biochim. Biophys. Acta* 120:282–91
45. Webber, W. A. 1962. *Am. J. Physiol.* 202:577–83
46. Maas, W. K., Davis, B. D. 1950. *J. Bacteriol.* 60:733–45
47. Schwartz, J. H., Maas, W. K., Simon, E. J. 1959. *Biochim. Biophys. Acta* 32:582–83
48. Lubin, M., Kessel, D. H., Budreau, A., Gross, J. D. 1960. *Biochim. Biophys. Acta* 42:535–38
49. Kessel, D., Lubin, M. 1965. *Biochemistry* 4:561–65
50. Levine, E. M., Simmonds, S. 1960. *J. Biol. Chem.* 235:2902–9
51. Levine, E. M., Simmonds, S. 1962. *J. Bacteriol.* 84:683–93
52. Levine, E. M., Simmonds, S. 1962. *J. Biol. Chem.* 237:3718–24
53. Piperno, J. R., Oxender, D. L. 1968. *J. Biol. Chem.* 243:5914–20
54. Wargel, R. J., Shadur, C. A., Neuhaus, F. C. 1970. *J. Bacteriol.* 103:778–88
55. Wargel, R. J., Shadur, C. A., Neuhaus, F. C. 1971. *J. Bacteriol.* 105:1028–35
56. Russel, R. R. B. 1972. *J. Bacteriol.* 111:622–24
57. Cosloy, S. D. 1973. *J. Bacteriol.* 114:679–84
58. Robbins, J. C., Oxender, D. L. 1973. *J. Bacteriol.* 116:12–18
59. Leach, R. F., Snell, E. E. 1960. *J. Biol. Chem.* 235:3523–31
60. Mora, J., Snell, E. E. 1963. *Biochemistry* 2:136–41
61. Reitz, R. H., Slade, H. D., Neuhaus, F. C. 1967. *Biochemistry* 6:2561–70
61a. Ashgar, S. S., Levin, E., Harold, F. M. 1973. *J. Biol. Chem.* 248:5225–33
62. Marquis, R. E., Gerhardt, P. 1964. *J. Biol. Chem.* 239:3361–71
63. Cohen, G. N., Rickenberg, H. V. 1955. *CR Acad. Sci.* 240:2086–88
64. Cohen, G. N., Rickenberg, H. V. 1956. *Ann. Inst. Pasteur* 91:693–720
65. Smulson, M. E., Rabinovitz, M., Breitman, T. R. 1967. *J. Bacteriol.* 94: 1890–95
66. Rahmanian, M., Oxender, D. L. 1972. *J. Supramolecular Struct.* 1:55–59
67. Tager, H. S., Christensen, H. N. 1971. *J. Biol. Chem.* 246:7572–80
68. Rahmanian, M., Claus, D. R., Oxender, D. L. 1973. *J. Bacteriol.* 116:1258–66

69. Neal, J. L. 1972. *J. Theor. Biol.* 35:113–18
70. Penrose, W. R., Nichoalds, G. E., Piperno, J. R., Oxender, D. L. 1968. *J. Biol. Chem.* 243:5921–28
71. Furlong, C. E., Weiner, J. H. 1970. *Biochem. Biophys. Res. Commun.* 38: 1076–83
72. Guardiola, J., DeFelice, M., Klopotowski, T., Iaccarino, M. 1974. *J. Bacteriol.* 117:383–92
73. Guardiola, J., Iaccarino, M. 1971. *J. Bacteriol.* 108:1034–44
74. Guardiola, J., DeFelice, M., Klopotowski, T., Iaccarino, M. 1974. *J. Bacteriol.* 117:393–405
74a. Templeton, B. A., Savageau, M. A. 1974. *J. Bacteriol.* 117:1002–9
75. Thorne, G. M., Corwin, L. M. 1972. *J. Bacteriol.* 110:784–85
76. Boezi, J. A., DeMoss, R. D. 1961. *Biochim. Biophys. Acta* 49:471–84
77. Burrous, S. E., DeMoss, R. D. 1963. *Biochim. Biophys. Acta* 73:623–37
78. Brown, K. D. 1970. *J. Bacteriol.* 104:177–88
79. Brown, K. D. 1971. *J. Bacteriol.* 106:70–81
80. Ames, G. F. 1964. *Arch. Biochem. Biophys.* 104:1–18
81. Ames, G. F., Roth, J. R. 1968. *J. Bacteriol.* 96:1742–49
82. Kay, W. W., Gronlund, A. F. 1971. *J. Bacteriol.* 105:1039–46
83. Rosenfeld, H., Feigelson, P. 1969. *J. Bacteriol.* 97:705–14
84. Guroff, G., Bromwell, K., Abramowitz, A. 1969. *Arch. Biochem. Biophys.* 131:543–50
85. Guroff, G., Bromwell, K. 1970. *Arch. Biochem. Biophys.* 137:379–87
86. Kuzuya, H., Bromwell, K., Guroff, G. 1971. *J. Biol. Chem.* 246:6371–80
86a. D'Ambrosio, S. M., Glover, G. I., Nelson, S. O., Jensen, R. A. 1973. *J. Bacteriol.* 115:673–81
86b. Glover, G. I., D'Ambrosio, S. M., Jensen, R. A. 1974. *Proc. Nat. Acad. Sci. USA.* In press.
87. Kessel, D., Lubin, M. 1962. *Biochim. Biophys. Acta* 57:32–43
88. Tristram, H., Neale, S. 1968. *J. Gen. Microbiol.* 50:121–36
89. Kay, W. W., Gronlund, A. F. 1969. *Biochim. Biophys. Acta* 193:444–54
90. Kay, W. W., Gronlund, A. F. 1969. *J. Bacteriol.* 98:116–23
91. Gryder, R. M., Adams, E. 1970. *J. Bacteriol.* 101:948–58
92. Behki, R. M. 1967. *Can. J. Biochem.* 45:1819–30
93. Ayling, P. D., Bridgeland, E. S. 1972. *J. Gen. Microbiol.* 73:127–41
93a. Kadner, R. J. 1974. *J. Bacteriol.* 117:232–41
94. Leive, L., Davis, B. D. 1965. *J. Biol. Chem.* 240:4362–69
95. Leive, L., Davis, B. D. 1965. *J. Biol. Chem.* 240:4370–76
96. Berger, E. A., Heppel, L. A. 1972. *J. Biol. Chem.* 247:7684–94
97. Shifrin, S., Ames, B. N., Ames, G. F. 1966. *J. Biol. Chem.* 241:3424–29
98. Krajewska-Grynkiewicz, K., Walczak, W., Klopotowski, T. 1971. *J. Bacteriol.* 105:28–37
99. Ames, G. F., Lever, J. 1970. *Proc. Nat. Acad. Sci. USA* 66:1096–1103
100. Ames, G. F., Lever, J. 1972. *J. Biol. Chem.* 247:4309–16
101. Rosen, B. P., Vasington, F. D. 1971. *J. Biol. Chem.* 246:5351–60
102. Lever, J. 1972. *J. Biol. Chem.* 247:4317–26
103. Kustu, S. G., Ames, G. F. 1973. *J. Bacteriol.* 116:107–13
104. Wilson, O. H., Holden, J. T. 1969. *J. Biol. Chem.* 244:2737–42
105. Wilson, O. H., Holden, J. T. 1969. *J. Biol. Chem.* 244:2743–49
106. Rosen, B. P. 1971. *J. Biol. Chem.* 246:3653–62
107. Rosen, B. P. 1973. *J. Biol. Chem.* 248:1211–18
108. Rosen, B. P. 1973. *J. Bacteriol.* 116:627–35
109. Celis, T. F. R., Rosenfeld, H. J., Maas, W. K. 1973. *J. Bacteriol.* 116:619–26
110. Maas, W. K. 1972. *Mol. Gen. Genet.* 119:1–9
111. Miller, D. L., Rodwell, V. W. 1971. *J. Biol. Chem.* 246:1765–71
112. Fan, C. L., Miller, D. L., Rodwell, V. W. 1972. *J. Biol. Chem.* 247:2283–88
113. Friede, J. D., Gilboe, D. P., Triebwasser, K. C., Henderson, L. M. 1972. *J. Bacteriol.* 109:179–85
114. Halpern, Y. S., Umbarger, H. E. 1961. *J. Gen. Microbiol.* 26:175–83
115. Halpern, Y. S., Lupo, M. 1965. *J. Bacteriol.* 90:1288–90
116. Marcus, M., Halpern, Y. S. 1969. *Biochim. Biophys. Acta* 177:314–20
117. Halpern, Y. S., Even-Shoshan, A. 1967. *J. Bacteriol.* 93:1009–16
118. Marcus, M., Halpern, Y. S. 1967. *Bacteriol.* 93:1409–15
119. Kahana, L., Avi-Dor, Y. 1966. *Israel J. Chem.* 4:59–66
120. Stevenson, J. 1966. *Biochem. J.* 99:257–60
121. Drapeau, G. R., Matula, T. I., MacLeod, R. A. 1966. *J. Bacteriol.* 92:63–71

122. Wong, P. T. S., Thompson, J., Mac-Leod, R. A. 1966. *J. Biol. Chem.* 244:1016–25
123. DeVoe, I. W., Thompson, J., Costerton, J. W., MacLeod, R. A. 1970. *J. Bacteriol.* 101:1014–26
123a. Thompson, J., MacLeod, R. A. 1973. *J. Biol. Chem.* 248:7106–11
124. Shiio, I., Miyajima, R. 1972. *J. Biochem. Tokyo* 72:773–75
125. Stock, J., Roseman, S. 1971. *Biochem. Biophys. Res. Commun.* 44:132–38
126. Eagon, R. G., Wilkerson, L. S. 1972. *Biochem. Biophys. Res. Commun.* 46:1944–50
127. Willecke, K., Gries, E. M., Oehr, P. 1973. *J. Biol. Chem.* 248:807–14
128. Frank, L., Hopkins, I. 1969. *J. Bacteriol.* 100:329–36
129. Halpern, Y. S., Barash, H., Dover, S., Druck, K. 1973. *J. Bacteriol.* 114:53–58
130. Thompson, J., MacLeod, R. 1971. *J. Biol. Chem.* 246:4066–74
131. Sprott, G., MacLeod, R. A. 1972. *Biochem. Biophys. Res. Commun.* 47:838–45
132. Marcus, M., Halpern, Y. S. 1969. *J. Bacteriol.* 97:1118–28
133. Barash, H., Halpern, Y. S. 1971. *Biochem. Biophys. Res. Commun.* 45:681–88
134. Halpern, Y. S., Barash, H. 1972. *Abstr. Commun. Meet. FEBS 8, Amsterdam*
135. Barash, H., Kahane, S., Marcus, M., Halpern, Y. S. 1973. *Proc. 1st Int. Congr. Bacteriol., Jerusalem* 2:84 (Abstr.)
135a. Miner, K. M., Frank, L. 1974. *J. Bacteriol.* 117:1093–98
136. Halpern, Y. S. 1967. *Biochim. Biophys. Acta* 148:718–24
137. Halpern, Y. S., Barash, H., Druck, K. 1973. *J. Bacteriol.* 113:51–57
138. Jaquez, J. A., Sherman, J. H. 1965. *Biochim. Biophys. Acta* 109:128–41
139. Robbie, J. P., Wilson, T. H. 1969. *Biochim. Biophys. Acta* 173:234–44
140. Koch, A. L. 1971. *J. Mol. Biol.* 59:447–59
141. Wilson, T. H., Kusch, M. 1972. *Biochim. Biophys. Acta* 255:786–97
142. Kashket, E. R., Wilson, T. H. 1972. *J. Bacteriol.* 109:784–89
143. Holden, J. T., Holman, J. 1959. *J. Biol. Chem.* 234:865–71
144. Reid, K. G., Utech, N. M., Holden, J. T. 1970. *J. Biol. Chem.* 245:5261–72
145. Utech, N. M., Reid, K. G., Holden, J. T. 1970. *J. Biol. Chem.* 245:5273–80
146. Holden, J. T., van Balgooy, J. N. A., Kittredge, J. S. 1968. *J. Bacteriol.* 96:950–57
147. Lyon, R. H., Rogers, P., Hall, W. H., Lichstein, H. C. 1967. *J. Bacteriol.* 94:92–100
148. Yabu, K. 1967. *Biochim. Biophys. Acta* 135:181–83
149. Yabu, K. 1970. *J. Bacteriol.* 102:6–13
150. Yabu, K. 1971. *Jap. J. Microbiol.* 15:449–56
151. Kay, W. W., Kornberg, H. L. 1971. *Eur. J. Biochem.* 18:274–81
152. Kay, W. W., Kornberg, H. L. 1969. *FEBS Lett.* 3:93–96
153. Kay, W. W. 1971. *J. Biol. Chem.* 246:7373–82
154. Willis, R. C., Woolfolk, C. A. 1970. *Bacteriol. Proc.* p. 127
155. Holden, J. T., Bunch, J. M. 1973. *Biochim. Biophys. Acta* 307:640–55
156. Weiner, J. H., Furlong, C. E., Heppel, L. A. 1971. *Arch. Biochem. Biophys.* 124:715–17
157. Weiner, J. H., Heppel, L. A. 1971. *J. Biol. Chem.* 246:6933–41
158. Kreishman, G. P., Robertson, D. E., Ho, C. 1973. *Biochem. Biophys. Res. Commun.* 53:18–23
159. Kaback, H. R. 1972. *The Molecular Basis of Biological Transport, Miami Winter Symposia,* ed. J. F. Woessner Jr., F. Huijing, 3:291–328. New York: Academic. 328 pp.
160. Kaback, H. R., Stadtman, E. R. 1966. *Proc. Nat. Acad. Sci. USA* 55:920–27
161. Kaback, H. R., Deuel, T. F. 1969. *Arch. Biochem. Biophys.* 132:118–29
162. Gordon, A. S., Lombardi, F. J., Kaback, H. R., 1972. *Proc. Nat. Acad. Sci. USA* 69:358–62
163. Kaback, H. R., Kostellow, A. B. 1968. *J. Biol. Chem.* 243:1384–89
164. Kaback, H. R., Stadtman, E. R. 1968. *J. Biol. Chem.* 243:1390–4000
165. Kaback, H. R., Milner, L. S. 1970. *Proc. Nat. Acad. Sci. USA* 66:1008–15
166. Lombardi, F. J., Kaback, H. R. 1972. *J. Biol. Chem.* 247:7844–57
167. Konings, W. N., Bisschop, A., Daatselaar, M. C. C. 1972. *FEBS Lett.* 24:260–64
168. Konings, W. N., Barnes, E. M., Kaback, H. R. 1971. *J. Biol. Chem.* 246:5857–61
169. Konings, W. N., Freese, E. 1972. *J. Biol. Chem.* 247:2408–18
170. Short, S. A., White, D. C., Kaback, H. R. 1972. *J. Biol. Chem.* 247:298–304
171. Short, S. A., White, D. C., Kaback, H. R. 1972. *J. Biol. Chem.* 247:7452–58
172. Hong, J. S., Kaback, H. R. 1972. *Proc. Nat. Acad. Sci. USA* 69:3336–40
173. Reeves, J. P., Hong, J. S., Kaback,

H. R. 1973. *Proc. Nat. Acad. Sci. USA* 70:1917-21
174. Butlin, J. D., Cox, G. B., Gibson, F. 1971. *Biochem. J.* 124:75-81
175. Prezioso, G., Hong, J. S., Kerwar, G. K., Kaback, H. R. 1973. *Arch. Biochem. Biophys.* 154:575-82
176. Simoni, R. D., Shallenberger, M. K. 1972. *Proc. Nat. Acad. Sci. USA* 69:2663-67
177. Bragg, P. D., Hou, C. 1973. *Biochem. Biophys. Res. Commun.* 50:729-36
178. Kanner, B. I., Gutnick, D. L. 1972. *FEBS Lett.* 22:197-99
179. Rosen, B. P. 1973. *J. Bacteriol.* 116:1124-29
180. Abrams, A., Smith, J. B., Baron, C. 1972. *J. Biol. Chem.* 247:1484-88
181. Klein, W. L., Boyer, P. D. 1972. *J. Biol.*

Chem. 247:7257-65
182. Schairer, H. U., Haddock, B. A. 1972. *Biochem. Biophys. Res. Commun.* 48:544-51
182a. Van Thienen, G., Postma, P. W. 1973. *Biochim. Biophys. Acta* 323:429-40
183. Berger, E. A. 1973. *Proc. Nat. Acad. Sci. USA* 70:1514-18
184. Heppel, L. A., Rosen, B. P., Friedberg, I., Berger, E. A., Weiner, J. H. 1972. See Ref. 159, pp. 133-56
184a. Hirata, H., Altendorf, K., Harold, F. M. 1973. *Proc. Nat. Acad. Sci. USA* 70:1804-8
184b. Rosen, B. P. 1973. *Biochem. Biophys. Res. Commun.* 53:1289-96
185. Ames, G. F. 1972. *Membrane Research*, ed. C. F. Fox, 409-26. New York: Academic. 501 pp.

GENETIC AND ANTIBIOTIC MODIFICATION OF PROTEIN SYNTHESIS

♦3068

David Schlessinger
Department of Microbiology, Washington University School of Medicine,
St. Louis, Missouri 63110

INTRODUCTION

The details and ramifications of protein synthesis are so complex that all the available analytical tools seem barely sufficient. The arsenal of physiological genetics is increasingly invaluable because 1. genetic lesions are the only systematic way—and at least as productive as the most inspired guesses—to identify all the participants in a process, or the relation of one process to others; and 2. the location and dominance relationships of the genes involved often can help us to understand the nature and regulation of the process.

Very complete analyses of major portions of relevant material on tRNA suppressor action (1) and the genetics of bacterial ribosomes (2) have recently appeared. Therefore, this discussion is a survey principally of point one: genetics as a tool for analyzing the physiology of genetic expression, primarily in *Escherichia coli*.

Such studies are, as usual, limited only by the ingenuity and luck of the genetic selection or screening designed. A variety of mutants in regulation, in formation, and in function of components of the protein synthetic machinery have gradually become available. Conditional lethal genetic lesions can have effects that imitate the response of cells to antibiotics that affect protein synthesis. The discussion below treats several cases in which analyses of events in mutants and in antibiotic-treated cells have overlapped or been complementary.

In these analyses, the complexities that cause difficulties also make for much of the interest. Mutants and antibiotics are often used to try to isolate intermediates in blocked physiological processes; however, these same experiments are often used to try to define the mechanism of action of the antibiotics. Fortunately, the results deviate from the expected often enough to prevent circular reasoning.

A specific line of reasoning is followed here, which is that some modification is necessary in the traditional obeisance to two "genetic" principles:

135

1. "The finding of a viable cell that has an amber mutation in a gene, or of a mutation leading to a sharp decrease in the level of an enzyme by a change in regulation or misreading, argues that the gene product is dispensable for the cell." The argument made below is that only a deletion or a drastic frameshift is a safe criterion for dispensability.

2. "Temperature-sensitive lethals (or conditional lethals) are the method of choice (or the "obligate" method) for the genetic analysis of vital processes." Instead, examples are given to suggest that mutations that reduce the functional capacity of a gene toward a low limiting state provide an especially strong alternative.

INFERENCES FROM ABNORMAL PHYSIOLOGY

Millennia of evolution have adapted physiological processes to efficient coordination. As a result, a mutation in a major process is highly pleiotropic. In a sense this is trivial; for example, if mRNA synthesis is blocked, messages decay and protein synthesis stops. But the interactions are not always so straightforward. For example, in an organelle like the ribosome, functional interdependence of its constituent elements can be extreme. A mutation in one element of the structure can negate, mask, or reverse the phenotype caused by another mutation (2, 3). In the elaborated studies of Gorini and his co-workers, the balance of state of the *ram* and *str* ribosomal proteins can finely tune the level of ambiguity seen in the translation of the genetic code (2, 4, 5).

This extensive pleiotropy extends to comparable interactions between elements in different fundamental processes. As an example, the arrest of protein synthesis by amino acid starvation (6) leads to the failure of further accumulation of stable RNA and of certain mRNA species as well ("polarity"; 7, 8). The normal ribosome cycle in protein synthesis thereby shows a "coupling" or functional interdependence of translation and transcription.

Coupling interactions also occur between "synthetic" and "degradative" processes. As examples, 1. Processing of nascent ribosomal RNA (9–12), mRNA (13), and protein (14) have all been reported. The relation of rRNA processing to other processes is just being studied, but already it is clear that failure of rRNA processing can block formation of ribosomes (15, 16). 2. When translation is interrupted, the degradation of mRNA is also inhibited (see section, *Coupling of Processes*, below).

Another example of extreme, unforeseen interactions comes from the pleiotropic effects of mutations in the cell membrane. Poorly defined binding interactions of membranes with various macromolecules have been reported for a decade. Some mutants temperature-sensitive in membrane formation or function show selectivity in the rate of shut-off of various processes (17), and recently the *lon* wall mutation in *E. coli* has been reported to reduce the rate of intracellular protein turnover [as one of a family of pleiotropic effects (18)].

Many problems arise in defining primary and secondary effects of lesions or drugs, but these studies also provide a major source of information about interactions.

SELECTIONS AND ANTIBIOTICS CLASSIFIED BY EFFECT

Affecting Formation of Components of Machinery
for Macromolecule Synthesis

General selections have been reported for the following:
(*a*) Mutants temperature-sensitive in stable RNA accumulation (19, 20). These
include mutants that selectively fail to make rRNA at high temperature and
some that make the "stable" RNA but rapidly degrade it at a very high rate.
(*b*) Mutants in tRNA formation. One of these is specifically temperature-sensitive
in RNase P (21), which is required for the maturation of tRNA (22). Another
comes from a selection of mutants dependent on exogenous tRNA for growth
at 42° and contains many temperature-sensitive tRNAs (23).
(*c*) Mutants in ribosome formation. Assembly-defective mutants have been obtained among cold-sensitive mutants (24–27).

Reported effects of antibiotics on synthetic processes are generally less selective
for the individual types of RNA or protein. But the general blockage of RNA
formation by actinomycin D, proflavine, rifampicin, etc, has been invaluable for
many studies, especially since few mutants in RNA polymerase formation have been
reported (see 28).

An interesting use has been made of ethionine for the selective blockage of
methylating steps in RNA and ribosome formation. In a detailed study, Beaud &
Hayes (29) showed that inactive 30S and 50S particles formed during growth of
E. coli met⁻ in the presence of ethionine contain normal amounts of 5S, 16S, and
23S RNA and complete sets of 30S and 50S proteins in which all methionines are
replaced by ethionines. These particles were also shown to be nonfunctional until
their 16S and 23S RNAs and a small number of their proteins were methylated by
provision of S-adenosylmethionine and appropriate enzyme factors in cells or extracts.

Conceivably, antibiotics could be found for such specific steps, for example, those
active against individual methylases. At least one antibiotic, kasugamycin, interacts
with methylation at a specific site in 30S ribosomal precursors (30). As a result,
*ksg*ʳ mutants lack a corresponding methylase (31).

Affecting Function of Components

The types of mutants available include those changed in the miscoding properties
of the ribosome (32), in the structure of tRNA (33), and in resistance to drugs that
otherwise block protein synthesis (2). Temperature-sensitive mutants have been
isolated in a variety of aminoacyl tRNAs (34) and in the elongation factors Ts (35)
and G (36). Mutants temperature-sensitive in elongation factor G have been reported from both *E. coli* (36) and *Bacillus subtilis* (37).

No usable temperature-sensitive mutants in RNA polymerase have been reported,
and only one in ribosome function, a mutant with a temperature-sensitive *strA*
protein (38).

The mutants in function complement or overlap the known antibiotics in many ways. For example, the antibiotic fusidic acid blocks the action of elongation factor G (EFG), with results comparable in many ways to those observed with temperature-sensitive EFG (39, 40).

Few mutants have been isolated specific for the initiation or termination steps in protein synthesis. A nonsense mutation affects the dissociation of ribosomes (41), but the lesion in the mutant strain has not been chemically characterized. Thus, at initiation and termination steps a variety of antibiotics have made possible most of the inhibition studies. In vitro, aurintricarboxylic acid has proven to be a specific inhibitor of initiation (42). In vitro and in vivo, inhibitors that are more or less specific for initiation include pactamycin (43) and a number of the aminoglycosides. Among that class, kasugamycin has a relatively specific effect at chain initiation. Streptomycin, which has a stronger effect on initiation (45, 46), also affects chain elongation (47) and even chain termination (48). Presumably the structure of the ribosome is such that binding that produces an effect on initiation often affects chain elongation. Antibiotics relatively specific for chain elongation are known, like chloramphenicol, for which the proteins involved in binding the drug to ribosomes are now identified (49).

Affecting Regulation of Component Function

Mutations are usually classified as regulatory when their effects are pleiotropic. In some cases, like the *adc* mutants in adenyl cyclase (50) and *crp* mutants in the cAMP receptor protein (51, 52), they help to define the mechanism by which small molecules (in that instance, 3'-5'cAMP) participate in cellular metabolism. In other cases, the lesion is not known. Each such mutant would require a review of its own. Included are the mutation in the famous *rel* locus, which permits accumulation of rRNA during amino acid starvation (6), and suppressors of polarity (53), which cause the appearance of certain mRNA sequences otherwise missing from points on a long mRNA distal to a point of translation arrest. Another example is the *lon* locus, mutations in which affect the rate of protein turnover as well as produce ultraviolet light sensitive cells that form mucoid colonies (18). Other mutants of complex phenotype include ones that affect the rate of turnover of mRNA (54) or rRNA (55).

SELECTIONS CLASSIFIED BY GENETIC STRATEGY

Since the functional elements required for protein synthesis are indispensable to the cell, mutants containing nonsense mutations, direct deficiencies, and deletions were for some time considered to be inviable, and therefore impossible to obtain. As a result, instead (apart from special selective tools like antibiotics), conditional lethals and especially temperature-sensitive mutants have been sought by physiological geneticists. Here I summarize some extensions of the approach by conditional lethals, some ways introduced to get around the "lethality" of nonsense mutations, and several general methods that ignore both these traditional approaches.

Nonsense Mutations and Deletions in Vital Functions

Several studies have avoided or even capitalized on the indispensability of certain functions. In two cases, partial diploidy has been used to permit, in theory, the selection of frameshift, deletion, insertion, or amber mutations. One case is an analysis of the organization of genes for ribosomal proteins. Recessive antibiotic resistant mutations were installed on one chromosome copy. Their expression was contingent on the incorporation of an inactivating sequence ("μ" DNA) into the corresponding region of the alternative chromosome (56–58).

In an extension of these studies, Nomura et al have made amber mutations in one of diploid copies of r-protein genes, using the amber mutations as an analog to μ insertion (59). A similar principle had been used earlier to obtain a nonsense lesion in an RNA polymerase gene that codes for the protein subunit sensitive to rifampicin. When the sensitive allele is knocked out, the product of a second, rif-resistant allele can be observed (60).

Amber mutants can now also be potentially obtained as conditional amber mutants, starting from a strain that harbors a temperature-sensitive amber suppressor (61, 62). Furthermore, nonsense mutations in a "suppressor-less" strain (see below, 3) are also known.

Natural Variants in the Protein-Synthetic Machinery

By using subfractions of ribosomes or other elements from widely varying species, large scale variants of comparable cell macromolecules can be obtained from nature's stocks. Examples of the power of this approach are the mapping of ribosomal protein and RNA genes by interspecific crosses (63–69) and the successful reconstitution of 30S ribosomes from *E. coli* 16S ribosomal RNA and *B. stearothermophilus* ribosomal proteins (70).

As another example, the finding of MRE600 as a natural *E. coli* strain with very low levels of the ribonuclease RNase I (71) has been used to support the argument that RNase I is dispensable (but see next section).

Conditional Lethals

Temperature-sensitive (ts) mutants may not be possible in all genes. This may be the reason that ts ribosomes, for example, are rare. Nevertheless, many ts mutants in other elements, as well as cold-sensitive ribosomes, have been obtained, as described above. Furthermore, a number of *alternative* conditional lethalities have been developed. A familiar example is the large number of streptomycin-dependent strains (72). A more recent, powerful approach has produced mutants with a conditional dependence on added tRNA (23). Such mutants are isolated in a strain that has been mutated to permit uptake of exogenous tRNA. From that parent, which could take up tRNA, mutants were made that could grow at 42° only in presence of added tRNA. Such mutant strains have included some defective in tRNA synthesis or protein synthesis at the high temperature (like the HAK88 strain which has a ts elongation factor Ts; reference 36).

Limits of Conditional Lethals and Use of Leaky Mutants

Lethality is defined as the loss of ability of a cell or organism to replicate itself. However, total growth limitation in cellular mutants is too severe a criterion and may discard many of interest. The case is very different from the paradigmatic analyses of phage mutants (74) in two respects:

1. Some mutations in vital genes can lead to a drastic reduction in an enzymatic function without limiting growth at all. This is because growth is normally limited by an unusually slow reaction, probably involved in the translocation step of protein synthesis (75), that usually holds the rate of growth proportional to the cellular content of ribosomes (76). In other words, a great many enzymes are present at a 100-fold or greater excess above that needed to support maximal growth. In fact, a similar point has been made with respect to large excesses of a number of phage proteins during T-even phage infection (R. Epstein, personal communication).

2. Even when a mutation lowers the content of an enzyme to the point at which it *does* become growth limiting, the strain can continue to grow at a slow exponential rate, limited by the missing component. In nature, a mutant strain growing at one fifth the rate of its parental type would rapidly be overgrown and lost, but in the laboratory such a mutant would traditionally be scored only as a slow-growing strain, unsuitable for most genetic analysis.

Unfortunately, for many enzymes the level of leakiness of mutations, even amber mutations, can be of a magnitude sufficient for normal growth. The first well-documented case of such an "adequate leakiness" was the famous *polA* mutation. Amber mutants in the *polA* gene retain 1–2% of the wild-type activity which a number of investigators of DNA replication, led by Okazaki and his collaborators (77, 78), now consider very likely to be sufficient to provide an indispensable activity.

A number of comparable deficiencies that affect RNA and protein synthesis are now accumulating. Two that have been intensively studied and are discussed below are a deficiency in RNase III in strain AB105 (79–81) and a deficiency in tRNA nucleotidyltransferase in strain 5C15 (82). A possible example from nature is the existence of unselected strains with a low level of glycyl-tRNA synthetase activity (83).

Thus, the levels of leakiness in missense and nonsense mutants are likely to be sufficient to permit retention of vital functions in many cases. In the case of mutations to the nonsense triplet UGA, levels of leakage of 2% or more are well documented (4). Possibly this is a safety measure for indispensable gene functions, a kind of defense against the deleterious effects of mutations. However, as a result, the frequent argument is clearly weakened that if cells survive an amber mutation in a particular gene, that gene is dispensable. Only a deletion is foolproof.

An example of a case where adequate leakiness might suggest an alternative interpretation is that of mutants in *E. coli* RNase I. Mutants that "lack" RNase I grow normally (85), but such mutants customarily contain 1 to 2% or the wild-type activity. In a recent analysis of a mutant with a high rate of turnover of ribosomes, the possibility has been raised that the low levels of RNase I in deficient mutants may be enough to give full activity in vivo (86). Similar doubts can be proposed about the dispensability of the suppressor A (*suA*) gene product. Mutations in that gene can alleviate the severity of polar effects (87); in that case also, the criterion for dispensability has been the survival of amber *suA* mutants (88).

Some comparable difficulties arise in the analysis of temperature-sensitive "conditional lethals." Often the rate of inactivation of a component is slow, a fraction is resistant by virtue of a bound substrate, or very little of the component is necessary for cell function. As a result, even when a cell component is quite sensitive in a mutant, the strain may show extensive residual growth (linear or even exponential) at the nonpermissive temperature.

A further difficulty arises in the analysis of temperature-sensitive mutants isolated as having lesions in *processes* (rather than in specific elements). Mapping of bacterial mutants remains relatively cumbersome, and the construction of gene catalogs from mutants, analogous to the virus models, is slow. Instead, physiological experiments tend to come first, but in those studies it is often difficult to determine which effects are primary and which are secondary. For example, among randomly isolated temperature-sensitive mutants, 15% showed a rapid shut-off of protein synthesis (89); however, in similar screenings in our laboratory, less than 1/700 showed temperature-sensitive protein synthesis from charged tRNA in cell extracts.

To avoid the unknown consequences of specific lesions, and in an attempt to pursue conditional lethal analysis, Reiner (90) turned to the screening of cells heavily mutagenized with modern chemical mutagens. His rationale was that cells treated with mutagens such as nitrosoguanidine should have mutations in any given gene in about 0.1% of the cells. Thus, the growth in liquid culture and subsequent testing of the order of 1000 random temperature-sensitive colonies should have a reasonable expectation of turning up a mutant in any indispensable function. In accord with such an expectation was the frequency of lesions in polynucleotide phosphorylase and a mutant temperature-sensitive in the exonuclease RNase II (90).

From the discussion thus far, it should be clear that the approach of screening after strong mutagenesis is intrinsically powerful, but that for mutants in specific enzymes, the feature of temperature sensitivity can very likely be dropped. Instead, the physiological analysis of mutants that contain *low levels* of a given enzyme is often adequate. The characterization of processes that become rate limiting with respect to the same processes in the parental strain can often suggest critical details of the function of the enzyme in vivo.

This approach is exemplified by the use of the *polA* mutants by Okazaki et al (77, 78), with the specification of a likely role for DNA polymerase I in cell metabolism. More pertinent to the discussion are the cases of mutations in RNase III and in the tRNA nucleotidyl transferase (or "CCA enzyme").

In the former case, Kindler et al (79) screened for, and found in colony 105, a mutant with low levels of RNase III (<5%). The strain accumulated, transiently, large precursors of rRNA and mRNA in vivo. The extracted precursors were subsequently demonstrated to be processed by RNase III (80, 81).

In the case of the -CCA enzyme (82), Deutscher & Hilderman tested 5000 colonies that survived heavy mutagenesis and found seven colonies with reduced enzymatic activity. One, the most severely affected, showed a normal rate of growth, with a residual activity of the -CCA enzyme 2% that of the parental strain. In vivo analysis showed that 16% of the tRNA in the strain lacked the CCA terminus and, furthermore, that much nascent tRNA appeared transiently in a form lacking -CCA (W. H. McClain and M. Deutsches, personal communication). Thus, a role for the -CCA enzyme in tRNA biosynthesis was proven. In both these cases, "incomplete" lesions were not only adequate but also may represent the most severe lesion attainable by point mutation in those genes.

NEW PROCESSES TURNED UP BY GENERAL SELECTIONS

The previous section recommends analyses by direct screening for mutants in specific enzymes. That approach works only for processes in which the components are well defined and able to be easily assayed by screening procedures. In order to use genetics to saturate the catalog of components in processes, general selections remain indispensable. Relevant examples are obtained from the following:

1. Isolation of mutants in ribosome assembly. A number of cold-sensitive mutants that accumulate subribosomal particles contain lesions at locations distinct from those of ribosomal proteins or RNA (91). By complementation tests some of the lesions define other proteins that aid in the formation of ribosomes from particles accumulated in the mutants. These proteins may be "morphopoietic factors" in ribosome formation, not found in the finished particles. Still another unanticipated feature of ribosome formation was revealed by the "*sad*" mutant analysis. Mutations in a protein of the 30S ribosome can lead to the accumulation of precursors of 50S ribosomes (92, 93). The inferred interaction of 30S and 50S ribosome formation has no detailed basis as yet.

2. *sts* (starvation temperature-sensitive) mutants. This slightly modified selection of temperature-sensitive mutants is based on provision of an additional barrier of ribosome depletion that mutants must overcome at 42° during screening (94). Among the interesting mutants turned up by this procedure are one in peptidyl hydrolase (95) and another in a factor involved in polypeptide chain termination (95). Neither of these elements has a place in standard formulations of protein synthesis, and they perhaps indicate that the present view is incomplete.

3. $su^+ \longrightarrow su^-$ tRNA analyses. Smith's laboratory (34, 97, 98) has exploited a system in which forward and reverse mutations could be generated in a suppressor tRNA. These mutations have revealed a number of the details of tRNA function and also often accumulate precursors of tRNA. The tRNA-dependent selection (see above) also generates mutants in tRNA processing as well as function.

PHYSIOLOGY STUDIED WITH THE COMBINED USE OF MUTANTS AND ANTIBIOTICS

Ribosome Cycle During Translocation Blockage

The normal cycle of ribosome function is recalled in Figure 1. The initial expectation was that blockage at various steps would "freeze" intermediates; for example, a block at the initiation of protein synthesis would lead to the accumulation of blocked

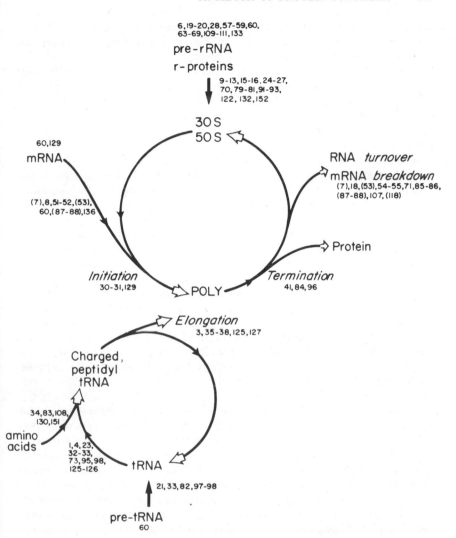

Figure 1   Schematic of the interlocking functional cycles of ribosomes and tRNA in protein synthesis. The movement of ribosomes from a free pool to polyribosomes and back is shown (for a detailed review, see reference 147). Effects on the overall process are distinguished at the formation of precursors to RNA (pre-tRNA, etc), on the entry of components into the cycles, and on steps in the polymerization process. Processes, as distinguished from intermediates, are shown in italic type. The numbers are references in the text that use mutants in that process or a related one. Relevant corresponding uses of antimetabolites or antibiotics are at text locations neighboring the references to mutants; most (42–49) relate to the polymerization process (see text). Because the mechanism(s) of certain polarity phenomena remain incompletely assigned, some of the reference numbers are shown in parentheses.

initiation complexes or free ribosomes. This notion has been considerably refined as the blockage of peptide bond formation or ribosome translocation has been studied using a range of inhibitors and several mutants.

Among the most widely used inhibitors is chloramphenicol, which interferes with tRNA binding (99) and blocks protein synthesis by a reaction with the 50S ribosome, probably at protein L16 (49). In a way, a physiological analog of this process is provided by the specific starvation of amino acid auxotrophs. Starvation again causes an arrest of protein synthesis at the positions of the limiting amino acids (100).

A very useful mutant for comparisons has been the mutant *G1*, which has a nonleaky temperature-sensitive lesion in elongation factor G. Events in the mutant at high temperature can be compared with the effects of the antibiotic fusidic acid, which specifically inhibits EFG function, as well as with the effects of inhibitors like chloramphenicol.

Especially with reversible inhibitors like chloramphenicol and fusidic acid, a situation arises analogous to that in the analysis of leaky mutants in vital functions. As the dose is increased, inhibition increases; but even in presence of maximal levels of drug, amino acid incorporation continues at a level 0.3 to 1% that in growing cells. Short peptides are formed, are apparently terminated prematurely, and accumulate in the cell. On the average, each ribosome forms 1 peptide every 10 minutes in cells treated at 30°.

From these results a model has been suggested in which ribosome movement is linked to continued peptide bond formation. In chloramphenicol-treated cells, ribosomes are released from mRNA after a short peptide is formed (101). A similar model has been suggested for a truncated ribosome cycle in amino acid-starved cells (100).

No studies have reported the use of small variations in temperature with temperature-sensitive mutants to achieve an analog of a "dose-response" curve with drugs. This is difficult because inactivation tends to be all or none at the critical temperature. Mutant *G1* at high temperature is more completely blocked than are cells treated with chloramphenicol, and even low levels of amino acid incorporation are essentially abolished (101). Thus the mutant provides a critical control that the polypeptide formation in chloramphenicol-treated cells is truly translocation dependent. The strain has also been used to show that EFG can inhibit polypeptide chain initiation (S. L. Huang and S. Ochoa, personal communication).

Coupling of Processes

Comparable analyses with antibiotics and temperature-sensitive mutants have produced evidence for the coupling of continued protein synthesis to the accumulation of rRNA and certain mRNA species.

REGULATION OF rRNA ACCUMULATION   Starvation for an amino acid restricts the accumulation of rRNA drastically in normal strains but not in "relaxed" mutants. The mechanism was shown by a genetic experiment of Eidlic & Neidhardt (108) to act through a depletion of charged tRNA. The locus and mechanism are now much better specified, thanks to a combination of genetic and biochemical studies. Production of a guanine tetraphosphate nucleotide is correlated with the restriction of rRNA accumulation (109). Studies with strains *G1* (110) and HAK88 (111), respectively temperature-sensitive in EFG and EFTs, showed that neither EFG (11) nor EFTs (111) is required for synthesis of the nucleotide. Consistent with the studies in vivo, studies in vitro have shown that nucleotide formation seems to be initiated by the absence of charged tRNA at a specific ribosomal site (112, 113). At present it is thought that a phosphorylated guanine nucleotide interacts with another element(s) of the system that functions in the production of rRNA.

COUPLING OF TRANSCRIPTION AND TRANSLOCATION   When translocation is blocked in strain *G1* (102), or in a mutant temperature-sensitive in the *str* ribosomal protein (8), rRNA, and some mRNA species, continue to be made, but other mRNA types, and especially their distal portions (103), tend to be absent from newly formed RNA sequences. Similar "polar" effects are observed distal to translocation blockage by chloramphenicol (54) or distal to nonsense codons in many mRNA species (104). Possible bases for this "coupling" of some mRNA synthesis to translation, and its possible significance, have been extensively discussed (103, 105, 106). Of particular interest are several mutants in which polarity effects are partially alleviated (87, 107), but the mechanism of polarity, and of its alleviation, remains unclear (88, 109; see below).

COUPLING OF TRANSLATION AND mRNA DECAY   Again in mutants like *G1* at high temperature (102), or in strains treated with chloramphenicol (101, 114, 115), both preexisting and some newly formed mRNA chains are substantially stabilized against chemical decay. As in the other cases mentioned above, the mechanism remains unknown, but ribosomes bind transiently to mRNA even in absence of active EFG, and much discussion has dealt with the possibility that nucleases might bind at or behind a ribosome on mRNA (see below).

Use of Coupled Systems In Vitro

The coupling of physiological processes in vivo can be imitated in complex coupled systems in vitro. Because the number of elements involved is very large, the major method available for analysis of the processes is through the preparation of coupled systems lacking one or another component.

For example, the DNA-coupled system for β-galactosidase formation developed by Zubay and his collaborators (116) has been instrumental in analyzing the role of 3',5'-cyclic AMP in the regulation of mRNA formation. Mutants in the nucleotide binding protein (52, 53) provided a complex system specifically deficient in the control element. In these studies, addition of the purified protein to the system confirmed its role in *lac* mRNA formation. Similar experiments have helped to decipher the regulation of *trp* mRNA production (117).

However, in these cases the important results could be obtained with a few components involved only in "pure" transcription. More germane here are the cases in which the analysis of "coupling" effects is attempted. For example, Wetekam &

Ehring (118) have partially analyzed a case of a polar effect on *gal* operon mRNA formation. The coupled system for enzyme production was analyzed in two steps; first, mRNA containing a nonsense codon in a proximal gene was formed, and then the expression of a proximal gene and one distal to the nonsense codon was tested. No polarity was observed in the system from a normal strain, but polar lack of production of the distal enzyme was observed in the clarified lysate in presence of an additional component obtained from membranous debris. Furthermore, the component could be obtained from a wild-type strain, but not from equivalent extracts of an suA strain. (The suA strain shows weakened polar effects in vivo, as mentioned above). Thus these results argue that at least in this case, polar effects can occur on a finished mRNA chain. In other cases, polar effects may well result from effects primarily on transcription of mRNA (e.g. 103).

Several systems have been tried for the investigation of the coupling of translation and degradation (119–121). In all, the inhibition of breakdown of mRNA by chloramphenicol has been observed, and evidence has accumulated that the important nuclease activity is probably in polyribosomes (22, 119, 126), but as yet no mutants have been reported to specify the mechanism further.

THE START OF COMPARABLE ANALYSES OF EUKARYOTIC SYSTEMS

As the movement of techniques and ideas up from bacteria to higher cells begins, it seems a move toward simplicity in one respect, for "coupling" phenomena are less of a problem, with transcription inside the nucleus clearly preceding translation. However, this simplicity may be apparent only to the ignorant. Already, temperature-sensitive mutants at a number of different loci in yeast are all found to inhibit ribosome formation at a very early step (122). Also, the blockage of protein synthesis by puromycin in HeLa cells (123) or starvation for certain amino acids (124) restrict net formation of ribosomal RNA. There is as yet no idea of whether the unknown mechanisms in these cases will be analogous to the partially known mechanisms in bacteria.

Antibiotics that can be used against specific components are available: fusidic acid against the eukaryotic analog of EFG, cycloheximide against polypeptide formation as an analog of chloramphenicol. In yeast, a number of antibiotic-resistant mutants have already been isolated. For example, mutants with ribosomes resistant to cycloheximide (125) and others showing ribosomes modified in cryptopleurine-resistant strains (126) have been reported. Also, a strain of use for selectively obtaining antibiotic-sensitive mutants has been made in *Saccharomyces cerevisiae* (127).

Use of conditional lethals has been seriously pursued in simple eukaryotes, with the isolation of many interesting mutants in *S. cerevisiae* (122, 128, 130) and *Chlamydomonas* (131). The advantages of the complete yeast genetic system are evident, for example in the identification of ten genes that control ribosome formation (132). And while cytogenetics is not one of the major advantages of fungal systems, it has been possible to map the chromosomal location of yeast ribosomal RNA by techniques analogous to those used with *E. coli* (133). (I pass over as too vast for mention the extensive analyses of genetics of mitochondria and other organelles, themselves topics of many reviews.)

By comparison to bacteria and fungi, the analyses of higher cells are still fragmentary and tentative. It is easy to point out difficulties: for example, the failure of many temperature-sensitive isolates from tissue culture cells to breed true after subcloning, the failure to produce any antibiotic-resistant ribosome mutants, the enormous physical labor involved in growth and testing of mammalian cell mutants, the incomplete genetic systems afforded by somatic cell cultures, and the extensive chromosome aberrations and regulatory changes observed in cell lines. However, there are also empirical reasons for optimism:

1. Many of the difficulties are based on fragmentary knowledge of the physiology as well as the genetics of the cells. For example, with respect to the action of antibiotics that affect transcription, α-amanitin specifically inhibits one RNA polymerase in cell extracts (134), but it shows a very different pattern of inhibition of RNA synthesis in vivo (135). In some α-amanitin-resistant mutants, however (135), the specific RNA polymerase is nevertheless altered.

Comparable confusion exists with respect to rifampicin, which does not inhibit purified preparations of RNA polymerase from higher cells (137) but inhibits RNA synthesis in yeast rendered permeable by wall mutation (138) or in yeast or HeLa cells permeabilized with polyene antibiotics (139). Presumably, as the modes of antibiotic action and the relevant physiology become better known, more fruitful selections can be designed.

2. Gradually, many of the requisite genetic procedures are being developed. For example, the frequency of specific lesions in appropriately mutagenized cells, and corresponding changes in enzyme activity, are being determined (136, 140–143). Also, the isolation of conditional lethal mutants as temperature-sensitive survivors of ^{3}H-uridine or ^{3}H-leucine incorporation at high temperatures ("suicide" selections), first applied to *E. coli* by Tocchini-Valentini (36), has now been extended to higher cells (144–150), including Chinese hamster and mouse L cell lines. After considerable initial difficulties, the mutants obtained thus far include one in the leucyl-tRNA synthetase (151) and one in ribosomal RNA maturation (152) in different hamster cell lines.

Among the many topics slighted in a brief review that deals primarily with mutations in cultured cells is that of mutational material from whole organisms. Analysis of leaky mutations in eukaryotic, and especially in human, genetics is of course not new. Some of the clearest examples of regulatory mutations that affect the levels of proteins are the thalassemias, while missense mutations that lower or modify activity to limiting values include many of the unstable hemoglobins, as well as a number of variants of glucose-6-phosphate dehydrogenase. When a mutational lesion leads to clinical illness it is, unsurprisingly, usually a leaky mutation (otherwise, it would most often be lethal). Also, as discussed in standard compilations on such subjects (153), clinical material customarily provides genetic markers in specialized cells and very often in degradative pathways. One might think that mutants in the protein synthetic mechanism are scarcely to be expected in higher organisms; lethality would be too frequent. On the other hand, detailed analyses of mutations in ribosomal RNA specification in *Drosophila* (e.g. 154) indicate that much more such material may exist.

SUMMARY STATEMENT

For the present, the mutations mentioned in the text are assigned in Figure 1 to appropriate points in the cycle of ribosome and tRNA function.

In contrast to physiological regulation, which most often occurs specifically at initiation, mutants and antibiotics can clearly affect any step in the overall process. However, the distribution of references supports the inference that mutations have usually been isolated in the formation of the molecules and primary intermediates in protein synthesis, while the complex events at the ribosome have been much less susceptible to genetic analysis. For those steps, study of the effects of antibiotics (or of genetic modification in drug response) has till now yielded a more extensive literature.

New types of conditional lethals, as well as studies of leaky lesions, promise to have extraordinary power for physiological analysis and to remedy the shortage of genetic material at some of the steps. Perhaps a review in a short time will proceed much further than a catalog of initial steps and kinds of selections.

One can also anticipate that comparable mutants will be as useful for the study of the physiology of the eukaryotic cell. As an example, the single problem of coupling interactions across each cellular membrane is already a formidable challenge: don't we need conditional mutants in the unknown processes by which certain nucleases and polymerases and ribosomal proteins return to the nucleus after their synthesis in the cytoplasm, or in the analogous process by which some ribosomal proteins made in the cytoplasm enter mitochondria while others enter nucleoli?

ACKNOWLEDGMENT

Work from my own laboratory has been sustained in large part by a grant from the National Science Foundation (GB 23052).

Literature Cited

1. Gorini, L. 1970. Informational suppression. *Ann. Rev. Genet.* 4:107–34
2. Davies, J., Nomura, M. 1972. The genetics of bacterial ribosomes. *Ann. Rev. Genet.* 6:203–34
3. Apirion, D., Schlessinger, D. 1969. Functional interdependence of ribosomal components of *Escherichia coli. Proc. Nat. Acad. Sci. USA* 63:794–99
4. Gorini, L. 1969. The contrasting role of *str A* and *ram* gene products in ribosomal functioning. *Cold Spring Harbor Symp. Quant. Biol.* 34:101–9
5. Gorini, L. 1971. Ribosomal discrimination of tRNAs. *Nature New Biol.* 234:261–64
6. Edlin, G., Broda, P. 1968. Physiology and genetics of the ribonucleic acid control locus in *Escherichia coli. Bacteriol. Rev.* 32:206–26
7. Morse, D. E., Guertin, M. 1971. Regulation of mRNA utilization and degradation by amino-acid starvation. *Nature New Biol.* 232:155–69
8. Imamoto, F., Kano, Y. 1971. Inhibition of transcription of the tryptophan operon in *Escherichia coli* by a block in initiation of translation. *Nature New Biol.* 232:169–73
9. Sogin, M., Pace, B., Pace, N. R., Woese, C. R. 1971. Primary structural relationship of p16 to m16 ribosomal RNA. *Nature New Biol.* 232:48–49

10. Brownlee, G. G., Cartwright, E. 1971. Sequence studies on precursor 16S ribosomal RNA of *Escherichia coli*. *Nature New Biol.* 232:50–52

11. Lowry, C. V., Dahlberg, J. E. 1971. Structural differences between the 16S ribosomal RNA of *E. coli* and its precursor. *Nature New Biol.* 232:52–54

12. Hayes, F., Hayes, D., Fellner, P., Ehresmann, C. 1971. Additional nucleotide sequences in precursor 16S ribosomal RNA from *Escherichia coli*. *Nature New Biol.* 232:54–55

13. Dunn, J. J., Studier, F. W. 1972. T7 early RNAs are generated by site-specific cleavages. *Proc. Nat. Acad. Sci. USA* 70:1559–63

14. Hauschild-Rogat, P. 1968. N-formyl-methionine as an N-termial group of *E. coli* ribosomal protein. *Mol. Gen. Genet.* 102:95–101

15. Corte, G., Schlessinger, D., Longo, D., Venkov, P. 1971. Transformation of 17S to 16S ribosomal RNA using RNase II of *Escherichia coli*. *J. Mol. Biol.* 60:325–38

16. Pace, N. R. 1973. The structure and synthesis of the ribosomal RNA of prokaryotes. *Bacteriol. Rev.* 37:562–603

17. Cronan, J. E., Ray, T. K., Vagelos, P. R. 1970. Selection and characterization of an *E. coli* mutant defective in membrane lipid biosynthesis. *Proc. Nat. Acad. Sci. USA* 65:737–44

18. Shineberg, B., Zipser, D. 1973. The *lon* gene and degradation of β-galactosidase nonsense fragments. *J. Bacteriol.* 116:1469–71

19. Laffler, T., Gallant, J., Harada, B. 1974. *spo*T, a new genetic locus involved in the stringent response in *E. coli*. *Cell* 1:27–30

20. Chaney, S. G., Schlessinger, D. 1974. *Escherichia coli* mutants deficient in RNA accumulation at high temperature. In preparation

21. Schedl, P., Primakoff, P. 1973. Mutants of *Escherichia coli* thermosensitive for the synthesis of transfer RNA. *Proc. Nat. Acad. Sci. USA* 70:2091–95

22. Robertson, H. D., Altman, S., Smith, J. D. 1972. Purification and properties of a specific *Escherichia coli* ribonuclease which cleaves a tyrosine transfer ribonucleic acid precursor. *J. Biol. Chem.* 247:5243–51

23. Yamamoto, M., Endo, H., Kuwano, M. 1972. A temperature-sensitive mutation in *Escherichia coli* transfer RNA. *J. Mol. Biol.* 69:387–96

24. Guthrie, C., Nashimoto, H., Nomura, M. 1969. Structure and function of *E. coli* ribosomes. VIII. Cold-sensitive mutants defective in ribosome assembly. *Proc. Nat. Acad. Sci. USA* 63:384–91

25. Nashimoto, H., Held, W., Kaltschmidt, E., Nomura, M. 1971. Structure and function of bacterial ribosomes. XII. Accumulation of 21S particles by some coldsensitive mutants of *Escherichia coli*. *J. Mol. Biol.* 62:121–38

26. Tai, P., Kessler, D. P., Ingraham, J. 1969. Cold-sensitive mutations in *Salmonella typhimurium* which affect ribosome synthesis. *J. Bacteriol.* 97:1298–1304

27. Tyler, B., Ingraham, J. L. 1973. Studies on ribosomal mutants of *Salmonella typhimurium LT-2*. *Mol. Gen. Genet.* 122:197–214

28. Khesin, R. B. et al 1969. Studies on the functions of the RNA polymerase components by means of mutations. *Mol. Gen. Genet.* 105:243–61

29. Beaud, G., Hayes, D. 1971. Proprietés des ribosomes et de l'RNA synthetisés par *Escherichia coli* cultivé en presence d'ethionine. *Eur. J. Biochem.* 19:323–39

30. Helser, T. L., Davies, J. E., Dahlberg, J. E. 1971. Change in methylation of 16S ribosomal RNA associated with mutation to kasugamycin resistance in *Escherichia coli*. *Nature New Biol.* 233:12–14

31. Helser, T. L., Davies, J. E., Dahlberg, J. E. 1971. Mechanism of kasugamycin resistance in *Escherichia coli*. *Nature New Biol.* 235:6–9

32. Biswas, D. K., Gorini, L. 1972. Restriction, derestriction and mistranslation in missense suppression. Ribosomal discrimination of transfer RNA's. *J. Mol. Biol.* 64:119–35

33. Smith, J. D. 1972. Genetics of transfer RNA. *Ann. Rev. Genet.* 6:235–57

34. Taylor, A. L., Trotter, C. D. 1973. Linkage map of *Escherichia coli* strain K12. *Bacteriol. Rev.* 36:504–24

35. Kuwano, M. et al 1973. Elongation factor T altered in a temperature-sensitive *Escherichia coli* mutant. *Nature New Biol.* 244:107–9

36. Tocchini-Valentini, G. P., Mattoccia, E. 1971. A mutant of *E. coli* with an altered supernatant factor. *Proc. Nat. Acad. Sci USA* 61:146–51

37. Aharonowitz, Y., Ron, E. Z. 1973. A temperature-sensitive mutant in *Bacillus subtilis* with an altered elongation factor G. *Mol. Gen. Genet.* 119:131–38

38. Kang, S. 1970. A mutant of *Escherichia coli* with a temperature sensitive streptomycin protein. *Proc. Nat. Acad. Sci. USA* 65:544–50

39. Kinoshita, T., Kawano, G., Tanaka, N. 1968. Association of fusidic acid sensitivity with G factor in a protein-synthesizing system. *Biochem. Biophys. Res. Commun.* 33:769–73

40. Leder, P. et al 1969. Protein biosynthesis studies using synthetic and viral mRNAs. *Cold Spring Harbor Symp. Quant. Biol.* 34:411–17

41. Herzog, A., Ghysen, A., Bollen, A. 1971. Characterization of a nonsense mutation affecting the activity of the dissociation factor of *Escherichia coli* ribosomes. *Mol. Gen. Genet.* 110: 211–17

42. Grollman, A. P., Stewart, M. L. 1968. Inhibition of the attachment of messenger ribonucleic acid to ribosomes. *Proc. Nat. Acad. Sci. USA* 61:719–25

43. Tai, P. C., Wallace, B. J., Davis, B. D. 1972. Actions of aurintricarboxylate, kasugamycin and pactamycin on *Escherichia coli* polysomes. *Biochemistry* 12:616–20

44. Okuyama, A., Machiyama, N., Kinoshita, T., Tanaka, N. 1971. Inhibition by kasugamycin of initiation complex formation on 30S ribosomes. *Biochem. Biophys. Res. Commun.* 43:196–99

45. Wallace, B. J., Davis, B. D. 1973. Cyclic blockade of initiation sites by streptomycin damaged ribosomes in *Escherichia coli:* an explanation for dominance of sensitivity. *J. Mol. Biol.* 75:377–90

46. Luzzatto, L., Apirion, D., Schlessinger, D. 1969. Polyribosome depletion and blockage of the ribosome cycle by streptomycin in *Escherichia coli. J. Mol. Biol.* 42:315–35

47. Modollel, J., Davis, B. D. 1968. Rapid inhibition of polypeptide chain extension by streptomycin. *Proc. Nat. Acad. Sci. USA* 61:1279–86

48. Caskey, C. T., Beaudet, A. L. 1972. Antibiotic inhibitors of peptide chain termination. *Molecular Mechanisms of Antibiotic Action on Protein Biosynthesis and Membranes,* ed. E. Munoz, F. Garcia-Ferrandiz, D. Vasquez, 326–36. New York: Elsevier

49. Pongs, D., Bald, R., Erdmann, V. 1973. Identification of CHL-binding protein in *Escherichia coli* ribosomes by affinity labeling. *Proc. Nat. Acad. Sci. USA* 70:2229–33

50. Yokota, T., Gots, J. S. 1970. Requirements of adenosine 3',5'-cyclic phosphate for flagella formation in *Escherichia coli* and *Salmonella typhimurium. J. Bacteriol.* 103:513–16

51. Emmer, M., deCrombrugghe, B., Pastan, I., Perlman, R. 1970. Cyclic AMP receptor protein of *E. coli:* Its role in the synthesis of inducible enzymes. *Proc. Nat. Acad. Sci. USA* 66:480–87

52. Zubay, G., Schwartz, D., Beckwith, J. 1970. Mechanism of activation of catabolite-sensitive genes: a positive control system. *Proc. Nat. Acad Sci. USA* 66:104–10

53. Morse, D. E. 1971. Polarity induced by chloramphenicol and release by SuA. *J. Mol. Biol.* 55:1113–16

54. Lennette, E., Gorelic, L., Apirion, D. 1971. An *Escherichia coli* mutant with increased messenger ribonuclease activity. *Proc. Nat. Acad. Sci. USA* 68: 3140–44

55. Ohnishi, Y., Schlessinger, D. 1972. Total breakdown of ribosomal and transfer RNA in a mutant of *Escherichia coli. Nature New Biol.* 238:228–31

56. Taylor, A. L. 1963. Bacteriophage-induced mutation in *Escherichia coli. Proc. Nat. Acad. Sci. USA* 50:1043–51

57. Cabezon, T., Faelen, M., Bollen, A. 1972. Results quoted in Ref. 59

58. Nomura, M., Engbaek, F. 1972. Expression of ribosomal protein genes as analyzed by bacteriophage mu-induced mutations. *Proc. Nat. Acad. Sci. USA* 69:1525–30

59. Jaskunas, S. R., Englbaek, F., Nomura, M. 1973. Amber mutants of a ribosomal protein operon in *E. coli.* 1973. Abstracts of Cold Spring Harbor *Ribosomes Meeting.* p. 37

60. Austin, S. J., Tittawella, I. P. B., Hayward, R. S., Scaife, J. G. 1971. Amber mutations of *Escherichia coli* RNA polymerase. *Nature New Biol.* 232: 133–36

61. Beckman, D., Cooper, S. 1973. Temperature-sensitive nonsense mutations in essential genes of *Escherichia coli. J. Bacteriol* 116:1336–42

62. Horiuchi, T. 1973. Temperature-sensitive amber suppressor in *Escherichia coli. Mol. Gen. Gent.* 123:89–106

63. Muto, A., Takata, R., Osawa, S. 1971. Chemical and genetic analysis of 16S ribosomal RNA in *Escherichia coli. Mol. Gen. Genet.* 111:15–21

64. Jarry, B., Rosset, R. 1973. Further mapping of 5S RNA cistrons in *Escherichia coli. Mol. Gen. Genet.* 126: 29–35

65. Matsubara, M., Takato, R., Osawa, S. 1972. Chromosomal loci for 16S riboso-

mal RNA in *Escherichia coli. Mol. Gen. Genet.* 117:311–17
66. O'Neil, D. M., Sypherd, P. S. 1971. Cotransduction of *strA* and ribosomal protein cistrons in *Escherichia coli-Salmonella typhimurium* hybrids. *J. Bacteriol.* 105:947–56
67. Dekio, S., Takata, R., Osawa, S. 1970. Genetic studies of the ribosomal proteins in *Escherichia coli.* VI. Determination of chromosomal loci for several ribosomal protein components using hybrid strains between *Escherichia coli* and *Salmonella typhimurium. Mol. Gen. Genet.* 109:131–41
68. Dekio, S. 1971. Genetic studies of the ribosomal proteins in *Escherichia coli.* VII 1971. Mapping of several ribosomal protein components by transduction experiments between *Shigella dysenteriae* and *Escherichia coli. Mol. Gen. Genet.* 113:20–30
69. Takata, R. 1973. Genetic studies of the ribosomal proteins in *Escherichia coli.* VIII. Mapping of ribosomal protein components by intergeneric mating experiments between *Serratia marcescens* and *Escherichia coli. Mol. Gen. Genet.* 118:363–71
70. Nomura, M., Traub, P., Bechmann, H. 1968. Hybrid 30S ribosomal particles reconstituted from components of different bacterial origins. *Nature* 219:793–99
71. Cammack, K. A., Wade, H. E. 1965. The sedimentation behaviors of ribonuclease-active and -inactive ribosomes from bacteria. *Biochem. J.* 96:671–80
72. Hashimoto, K. 1960. Streptomycin resistance in *Escherichia coli* analyzed by transduction. *Genetics* 45:49–62
73. Yamamoto, M., Ishizawa, M., Endo, H. 1971. Ribonucleic acid-permeable mutant of *Escherichia coli. J. Mol. Biol.* 58:103–15
74. Epstein, R. H. et al 1963. Physiological studies of conditional lethal mutants of bacteriophage T4D. *Cold Spring Harbor Symp. Quant. Biol.* 38:375–94
75. Wilhelm, J., Haselkorn, R. 1970. The chain growth rate of T4 lysozyme *in vitro. Proc. Nat. Acad. Sci. USA* 65:388–94
76. Maaløe, O., Kjeldgaard, N. 0. 1966. *Control of Macromolecular Synthesis.* New York: Benjamin
77. Okazaki, R., Sugimoto, K., Okazaki, T., Imae, Y., Sugino, A. 1970. DNA chain growth: *in vivo* and *in vitro* synthesis in a DNA polymerase-negative mutant of *E. coli. Nature* 228:223–12

78. Okazaki, R., Arisawa, M., Sugino, A. 1971. Slow joining of newly replicated DNA chains in DNA polymerase I-deficient *Escherichia coli* mutants. *Proc. Nat. Acad. Sci. USA* 68:2954–57
79. Kindler, P., Keil, T. U., Hofschneider, P. H. 1973. Isolation and characterization of a Ribonuclease III deficient mutant of *Escherichia coli. Mol. Gen. Genet.* 126:53–60
80. Dunn, J. J., Studier, F. W. 1973. T7 early RNAs and *Escherichia coli* ribosomal RNAs are cut from large precursor RNAs *in vivo* by ribonuclease III 1973. *Proc. Nat. Acad. Sci. USA* 70:3298–3303
81. Nikolaev, N., Silengo, L., Schlessinger, D. 1973. Synthesis of a large precursor to ribosomal RNA in a mutant of *Escherichia coli. Proc. Nat. Acad. Sci. USA* 70:3361–65
82. Deutscher, M., Hilderman, R. 1974. Isolation and partial characterization of *Escherichia coli* mutants with low levels of tRNA nucleotidyltransferase. *J. Bacteriol.* 118:621–27
83. Folk, W. R., Berg, P. 1970. Characterization of altered forms of glycyl transfer ribonucleic acid synthetase and the effects of such alterations on amino acyl transfer ribonucleic acid synthesis *in vivo. J. Bacteriol.* 102:204–12
84. Weiner, A. M., Weber, K. 1973. Single UGA codon as a termination signal in the coliphage Qβ coat protein cistron. *J. Mol. Biol.* 80:837–55
85. Gesteland, K. 1966. Isolation and characterization of ribonuclease I mutants of *Escherichia coli. J. Mol. Biol.* 16:67–84
86. Ohnishi, Y. 1974. Genetic analysis of an *Escherichia coli* mutant with a lesion in stable RNA turnover. *Genetics* 76:185–94
87. Morse, D. E., Primakoff, D. 1970. Relief of polarity in *E. coli.* by "SuA". *Nature* 226:28–31
88. Morse, D. E., Guertin, M. 1972. Amber SuA mutations which relieve polarity. *J. Mol. Biol.* 63:605–8
89. Russell, R. R. B., Pittard, A. J. 1971. Mutants of *Escherichia coli* unable to make protein at 42°C. *J. Bacteriol.* 108:790–98
90. Reiner, A. M. 1969. Isolation and mapping of polynucleotide phosphorylase mutants of *Escherichia coli. J. Bacteriol.* 97:1431–36
91. Bryant, R. E., Sypherd, P. S. 1974. Genetic analysis of cold-sensitive, ribosome maturation mutants of *Escherichia coli. J. Bacteriol.* 117:1082–89

92. Nashimoto, H., Nomura, M. 1970. Structure and function of bacterial ribosomes. XI. Dependence of 50S ribosomal assembly on simultaneous assembly of 30S subunits. *Proc. Nat. Acad. Sci. USA* 67:1440–47

93. Kreider, G., Brownstein, B. L. 1971. A mutation suppressing streptomycin dependence. II. An altered protein on the 30S ribosomal subunit. *J. Mol. Biol.* 61:135–43

94. Phillips, S. L., Schlessinger, D., Apirion, D. 1964. Mutants in *Escherichia coli* ribosomes: a new selection. *Proc. Nat. Acad. Sci. USA* 62:772–77

95. Menninger, J. R., Walker, C., Tan, P. F., Atherly, A. G. 1973. Studies on the metabolic role of peptidyl-tRNA hydrolase. I. Properties of a mutant *E. coli* with temperature-sensitive peptidyl hydrolase. *Mol. Gen. Genet.* 121: 307–15

96. Phillips, S. L. 1971. Termination of messenger RNA translation in a temperature-sensitive mutant of *Escherichia coli. J. Mol. Biol.* 59:461–72

97. Smith, J. D., Anderson, K. W., Cathmore, A. R., Hopper, M. L., Russell, R. L. 1970. Studies on the structure and synthesis of *Escherichia coli* tyrosine transfer RNA. *Cold Spring Harbor Symp. Quant. Biol.* 35:21–29

98. Anderson, K. W., Smith, J. D. 1970. Still more mutant tyrosine transfer ribonucleic acids. *J. Mol. Biol.* 69: 349–57

99. Pestka, S. 1970. Studies on the formation of transfer ribonucleic acid-ribosome complexes. VIII Survey of the effects of antibiotic on N-Acetyl-Phenylalanyl-Puromycin formation: possible mechanism of chloramphenicol action. *Arch. Biochem. Biophys.* 136: 80–88

100. Brunschede, H., Bremer, H. 1971. Synthesis and breakdown of proteins in *Escherichia coli* during amino acid starvation. *J. Mol. Biol.* 57:35–57

101. Cremer, K., Silengo, L., Schlessinger, D. 1974. Polypeptide formation and polyribosomes in *Escherichia coli* treated with chloramphenicol. *J. Bacteriol.* 118:582–89

102. Craig, E. 1972. Messenger RNA metabolism when translocation is blocked. *Genetics* 70:331–33

103. Imamoto, F. 1973. Diversity of regulation of genetic transcription. *J. Mol. Biol.* 74:113–36

104. Martin, R. G. 1969. Control of gene expression. *Ann. Rev. Genet.* 3:181–216

105. Stent, G. 1964. The operon: on its third anniversary. *Science* 144:816–20

106. Schlessinger, D. 1971. Repertoire of genetic control of gene expression in prokaryotes. *Stadler Symp.* 3:25–36

107. Carter, T., Newton, A. 1971. New polarity suppressors in *Escherichia coli:* suppression and messenger RNA stability. *Proc. Nat. Acad. Sci. USA* 68: 2962–66

108. Eidlic, L., Neidhardt, F. C. 1965. Protein and nucleic acid synthesis in two mutants of *Escherichia coli* with temperature-sensitive aminoacyl ribonucleic acid synthetase. *J. Bacteriol.* 89:706–11

109. Cashel, M. 1969. The control of ribonucleic acid synthesis in *Escherichia coli* IV. Relevance of unusual phosphorylated compounds from amino acid starved stringent strains. *J. Biol. Chem.* 244:3133–41

110. Atherly, A. G. 1973. Temperature-sensitive relaxed phenotype in a stringent strain of *Escherichia coli J. Bacteriol.* 113:178–82

111. Glazier, K., Schlessinger, D. 1974. Magic spot metabolism in an *Escherichia coli* mutant temperature-sensitive in EFTs. *J. Bacteriol.* 117:1195–1200

112. Haseltine, W. A., Block, R. 1973. Synthesis of guanosine tetra- and pentaphosphate requires the presence of a codon-specific, uncharged transfer ribonucleic acid in the acceptor site of ribosomes. *Proc Nat. Acad. Sci. USA* 70:1564–68

113. Pedersen, F. S., Lurd, E., Kjeldgaard, N. O. 1973. Codon specific, tRNA dependent *in vitro* synthesis of ppGpp and pppGpp. *Nature New Biol.* 243:13–15

114. Levinthal, C., Fan, D. P., Higa, A., Zimmerman, R. A. 1963. The decay and protection of messenger RNA in bacteria. *Cold Spring Harb. Symp. Quant. Biol.* 28:183–90

115. Gurgo, C., Apirion, D., Schlessinger, D. 1969. Polyribosome metabolism in *Escherichia coli* treated with chloramphenicol, neomycin or tetracycline. *J. Mol. Biol.* 45:205–20

116. Zubay, G., Chambers, D. A., Cheong, L. C. 1970. Cell-free studies on the regulation of the lac operon. In *The Lactose Operon,* ed. J. Beckwith, D. Zipser, p. 375. New York: Cold Spring Harbor Lab.

117. Zubay, G., Morse, D. E., Shrenk, W. J., Miller, J. H. M. 1973. Detection and isolation of the repressor protein for the

tryptophan operon. *Proc. Nat. Acad. Sci. USA* 69:1100–3
118. Wetekam, W., Ehring, R. 1973. A role for the product of gene suA in restoration of polarity *in vitro. Mol. Gen. Genet.* 124:345–58
119. Mangiarotti, G., Schlessinger, D., Kuwano, M. 1971. Initiation of ribosome-dependent breakdown of T4-specific messenger RNA. *J. Mol. Biol.* 60:441–52
120. Mangiarotti, G., Turco, E. 1973. Ribonuclease activity in *Escherichia coli* poly-ribosomes. *Eur. J. Biochem.* 38:507–15
121. Cremer, K., Schlessinger, D. 1974. Ca²⁺ ions inhibit mRNA degradation, but permit mRNA transcription and translation in DNA-coupled systems from *Escherichia coli. J. Biol. Chem.* 249:In press
122. Warner, J. R., Udem, S. A. 1972. Temperature sensitive mutations affecting ribosome synthesis in *Saccharomyces cerevisiae. J. Mol. Biol.* 65:243–59
123. Latham, H., Darnell, J. E. 1965. Entrance of mRNA into HeLa cell cytoplasm in puromycin-treated cells. *J. Mol. Biol.* 14:13–22
124. Vaughan, M. H., Warner, J. R., Darnell, J. E. 1967. Ribosomal precursor particles in the HeLa cell nucleus. *J. Mol. Biol.* 25:235–51
125. Jiminez, A., Littlewood, B., Davies, J. 1972. Inhibition of protein synthesis in yeast. In *Symp. Mol. Mech. Antibiot. Action Protein Biosyn. Membranes, Granada, 1971,* p. 192. New York: Elsevier
126. Skogerson, L., McLaughlin, C., Wakatoma, E. 1973. Modification of ribosomes in cryptopleurine-resistant mutants of yeast. *J. Bacteriol.* 116:818–22
127. Littlewood, B. S. 1972. A method for obtaining antibiotic-sensitive mutants in Saccharomyces cerevisiae. *Genetics* 71:305–8
128. Hartwell, L. H., McLaughlin, C. S. 1968. Temperature-sensitive mutants of yeast exhibiting a rapid inhibition of protein synthesis. *J. Bacteriol.* 96:1664–71
129. Hartwell, L. H., Hutchison, H. T., Holland, T. M., McLaughlin, C. S. 1970. The effect of cycloheximide upon polyribosome stability in two yeast mutants defective respectively in the initiation of polypeptide chains and in messenger RNA synthesis. *Mol. Gen. Genet.* 106:347–61

130. McLaughlin, C. S., Hartwell, L. H. 1969. A mutant of yeast with a defective methionyl-tRNA synthetase. *Genetics* 61:557–66
131. McMahon, D. 1971. The isolation of mutants conditionally defective in protein synthesis in *Chlamydomonas reinhardii. Mol. Gen. Genet.* 112:80–86
132. Hartwell, L. H., McLaughlin, C. S., Warner, J. R. 1970. Identification of ten genes that control ribosome formation in yeast. *Mol. Gen. Genet.* 109:42–55
133. Finkelstein, D. B., Blamire, J., Marmur, J. 1972. Location of ribosomal RNA cistrons in yeast. *Nature New Biol.* 240:279–81
134. Stirpe, F., Fiume, L. 1967. Studies on the pathogenesis of liver necrosis by α-amanitin. Effect of α-amanitin on ribonucleic acid polymerase in mouse liver nuclei. *Biochem. J.* 105:779–82
135. Tata, J. R., Hamilton, M. J., Shields, D. 1972. Effects of α-amanitin *in vivo* on RNA polymerase and nuclear RNA synthesis. *Nature New Biol.* 238:161–64
136. Chan, V. L., Whitmore, G. F., Siminovitch, L. 1972. Mammalian cells with altered forms of RNA polymerase II. *Proc. Nat. Acad. Sci. USA* 69:3119–23
137. Wehrli, W. 1971. Actions of the rifamycins. *Bacteriol. Rev.* 35:290–303
138. Venkov, P., Hadjiolov, A., Battaner, E., Schlessinger, D. 1974. *Saccharomyces cerevisiae:* sorbitol-dependent fragile mutants. *Biochem. Biophys. Res. Commun.* 56:599–604
139. Medoff, G., Kwan, C. N., Schlessinger, D., Kobayashi, G. S. 1973. Potentiation of rifampicin, rifampicin analogues and tetracycline against animal cells by amphotericin B and polymyxin B. *Cancer Res.* 33:1146–49
140. Sharp, J. D., Capecchi, N. E., Capecchi, M. R. 1973. Altered enzymes in drug-resistant variants of mammalian tissue culture cells. *Proc. Nat. Acad. Sci. USA* 70:3145–49
141. Beaudet, A. L., Roufa, D., Caskey, C. T. 1973. Mutations affecting the structure of hypoxanthine: guanine phosphoribosyltransferase in cultured chinese hamster cells. *Proc. Nat. Acad. Sci. USA* 70:320–24
142. Roufa, D. J., Sadow, B. N., Caskey, C. T. 1973. Derivation of TK⁻ clones from revertant TK⁺ mammalian cells. *Genetics* 75:515–30
143. Albrecht, A. M., Biedler, J. L., Hutchinson, D. J. 1972. Two different species of dihydrofolate reductase in mammalian cells differentially resistant to

amethopterin and methasquin. *Cancer Res.* 32:1539–46

144. Thompson, L. H. et al 1970. Isolation of temperature-sensitive mutants of L-cells. *Proc. Nat. Acad. Sci. USA* 66: 377–84

145. Thompson, L. H. et al 1971. Selective and nonselective isolation of temperature-sensitive mutants of mouse L-cells and their characterization. *J. Cell Physiol.* 78:431–40

146. Smith, D. B., Chu, E. H. Y. 1973. Isolation and characterization of mutants temperature-sensitive for cytokinesis. *J. Cell Physiol.* 80:253–60

147. Kaempfer, R. 1974. The ribosome cycle. In *Ribosomes,* ed. M. Nomura, P. Lengyel, A. Tissieres. New York: Cold Spring Harbor Lab. In press

148. Meiss, H. K., Basilico, C. 1972. Temperature-sensitive mutants of BHK 21 cells. *Nature New Biol.* 239:66–68

149. Naha, P. M. 1973. Early functional mutants of mammalian cells. *Nature New Biol.* 241:13–16

150. Scheffler, I. E., Buttin, G. 1973. Conditionally lethal mutations in chinese hamster cells. I-Isolation of a temperature-sensitive line and its investigation by cell cycle studies. *J. Cell Physiol.* 81:199–216

151. Thompson, L. H., Harkins, J. L., Stanners, C. P. 1973. A mammalian cell mutant with a temperature-sensitive leucyl-transfer RNA synthetase. *Proc. Nat. Acad. Sci. USA* 69:3094–98

152. Toniolo, D., Meiss, H. K., Basilico, C. 1973. A temperature-sensitive mutation affecting 28S ribosomal RNA production in mammalian cells. *Proc. Nat. Acad. Sci. USA* 70:1273–77

153. Stanbury, J. B., Wyngaarden, J. B., Frederickson, D. S. 1972. *The Metabolic Basis of Inherited Disease.* New York: McGraw. 1778 pp.

154. Ritossa, F. M., Scala, G. 1969. Equilibrium variations in the redundancy of rDNA in *Drosophila melanogaster. Genetics* 61:305–17

ON THE ORIGIN OF RNA TUMOR VIRUSES

♦3069

Relationships among RNA tumor viruses and between RNA tumor viruses and cells: implications for the origin of viruses, the evolution of vertebrates, and cellular genetic systems

Howard M. Temin

McArdle Laboratory for Cancer Research, University of Wisconsin, Madison, Wisconsin 53706

INTRODUCTION

RNA tumor viruses are enveloped animal viruses whose virions contain an RNA genome and a DNA polymerase. RNA tumor viruses have been isolated from reptiles, birds, and many orders of mammals, including primates. However, none has been isolated from man. In susceptible animals, RNA tumor viruses cause leukemias, sarcomas, carcinomas, rapid lethal infections with anemia and necrosis, slow infections of the nervous system and lung, and, most frequently, no disease. (The nonpathogenic RNA tumor viruses are assayed by the formation of virion components.)

The name *RNA tumor viruses* describes the oncogenicity of some of the members of the virus group. It does not mean that all of the viruses that belong to this group cause tumors. RNA tumor viruses are also called leukoviruses, rousviruses, oncornaviruses, retraviruses, retroviruses, ambiviruses, rnadnaviruses, oncoviruses, and ribodeoxyviruses. I prefer the name *ribodeoxyviruses.*

RNA tumor viruses are very popular organisms for study for several reasons: 1. all RNA tumor viruses replicate using RNA to DNA to RNA information transfer, a previously unknown mode of information transfer discovered in this group of viruses; 2. some RNA tumor viruses are the apparent cause of much "natural" cancer in chickens and in mice; and 3. some RNA tumor viruses, for example the strongly transforming Rous sarcoma virus, are the most efficient carcinogenic agents known in animals or in cell culture. (Transformation is used in animal cell culture work to describe a stable, heritable change in the morphology and growth properties of cells.)

In this article, I do not discuss the relationship of RNA tumor viruses to cancer, the emerging genetics of RNA tumor viruses themselves as reviewed in Vogt (1), or the replication of RNA tumor viruses as reviewed in Tooze (2) or Temin (3). I discuss the relationships of representative RNA tumor viruses to each other and to uninfected cells. The discussion is not an exhaustive compilation of these relationships, but uses them to illustrate points of possible evolutionary or general genetic significance. In particular, I suggest that RNA tumor viruses repeatedly evolved from a normal cellular genetic system important in differentiation and evolution, and that the relationships among RNA tumor viruses and between RNA tumor viruses and cells reflect this origin from similar normal cellular processes.

DESCRIPTION OF RNA TUMOR VIRUSES

This section provides general descriptions of the RNA tumor virus virion, the classification of RNA tumor viruses, and the major features of RNA tumor virus replication. It is a summary of longer descriptions which can be found in Tooze (2) and Temin (3). References are not cited when present in these review articles, and evidence for the conclusions is not presented here.

RNA Tumor Virus Virions

GENOME RNA   The genome of RNA tumor viruses is composed of two to four 35S single-stranded RNAs. Each 35S RNA has a molecular mass of approximately 3×10^6 daltons. There is still uncertainty whether or not each of these 35S RNAs has a different sequence (Vogt 4). Most have a poly(A) stretch of about 100–200 nucleotides. [The function of the poly(A) stretch is not known. It may relate to the messenger role of these molecules.] The 35S RNAs are normally found in a 60–70S complex with some associated small RNAs.

SMALL RNAs   There are a number of small RNAs in the virion. Some are associated with the 35S RNAs in the 60–70S aggregate, and others are free. These small RNAs in the virion are primarily host tRNAs, but they are probably not a random sample of the host tRNAs. They may include primer molecules for the viral RNA-directed DNA polymerase activity (Dahlberg et al 5).

DNA POLYMERASE   The virion DNA polymerase is associated with the virus 60–70S RNA which acts as an endogenous template-primer for synthesis of the viral DNA intermediate, the DNA provirus. This DNA polymerase is virus coded and is necessary for infection. It is approximately 70,000 daltons in mammalian C-type viruses and in reticuloendotheliosis viruses and approximately 150,000 daltons in avian leukosis-sarcoma viruses.

INTERNAL PROTEINS   The viral 60–70S RNA and DNA polymerase are found associated with a few small basic proteins in a ribonucleoprotein particle. These small basic proteins have molecular masses of 10,000–35,000 daltons (see Fleissner

& Tress 6). This ribonucleoprotein particle is surrounded by another protein called the core shell, matrix, or membrane protein. The ribonucleoprotein particle plus the core shell is called the core (see Figure 1).

ENVELOPE GLYCOPROTEINS   The core is surrounded by a lipid-containing envelope with glycoprotein molecules on the outside. The glycoproteins are apparently not necessary for the structural integrity of the virion, but they are necessary for infectivity. They are often present in morphologically visible spikes.

VIRION   An RNA tumor virus virion has a distinctive structural morphology (Figure 1). A ribonucleoprotein of no clearly observable symmetry and containing 60–70S RNA, basic protein(s), and a DNA polymerase is surrounded by a core shell. The core is surrounded by a lipid bilayer with external glycoproteins usually present in visible spikes. The diameter of the virion is about 100–150 nm.

Classification of RNA Tumor Viruses

All RNA tumor viruses (ribodeoxyviruses) have a virion of the type described above. The ribodeoxyviruses are divided into the following groups: 1. avian leukosis-sarcoma (ALV), 2. mouse C-type (MLV), 3. feline C-type, 4. rat C-type, 5. hamster C-type, 6. simian C-type, 7. RD-114-like C-type, 8. murine, feline, and simian sarcoma, 9. viper C-type, 10. mouse mammary tumor, 11. Mason-Pfizer monkey, 12. visna, 13. syncytium-forming, and 14. reticuloendotheliosis viruses. (As discussed below, there are some relationships between these groups.) In addition, new groups are being described at frequent intervals.

These groups were originally separated on the basis of the species of origin and the host range of the viruses. The classification is now primarily based on serological considerations, especially the relationships of the virion internal proteins and DNA polymerases. This classification based on group-specific serological relationships of the virion proteins has been confirmed by nucleic acid hybridization experiments with the viral RNAs and complementary DNAs and with the viral RNAs and uninfected cell DNAs.

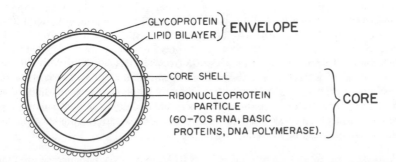

Figure 1   Cross section of a typical ribodeoxyvirus virion. The diameter of the virion is about 100–150 nm. The diameter of the ribonucleoprotein particle is about 50 nm.

In addition, noninfectious particles with morphology or other features resembling ribodeoxyvirus virions have been widely described (see references in 3, 7–10). Although some of these virus-like particles contain proteins related to those of known ribodeoxyviruses, generally they do not. They may be thought of as stages in the evolution of ribodeoxyviruses (see below).

Major Features of Replication of RNA Tumor Viruses

PRODUCTIVE INFECTION (INFECTION OF SENSITIVE CELLS LEADING TO PRODUCTION OF PROGENY VIRUSES)

Entrance   Specific receptors are required for entrance of virus into cells. In chicken cells, these receptors are subgroup specific and are controlled by four genes. Susceptibility, that is, the presence of receptors, is dominant. The presence of these receptors is a necessary condition for a cell to be sensitive to infection. The chemical nature of the receptors and the actual mode of entrance of virus into sensitive cells are unknown or disputed (see Dales 11). The site of uncoating of the virion is also unknown or disputed.

Formation of the DNA provirus   The virion DNA polymerase synthesizes a DNA copy of the virion 35S RNA molecules, perhaps using some of the small RNAs associated with the 60–70S complex as a primer. The kinetics of this synthesis and its location in the cell are unknown or disputed.

Integration   The viral DNA provirus is apparently inserted into and covalently bound to host-cell nuclear DNA (Varmus et al 12). It is not known whether this integration is in a unique site(s) and whether the 35S RNAs are integrated in series or in separate sites in the host-cell DNA. There are only a small number, five or less, DNA proviruses per infected cell.

Activation of transcription   The DNA provirus is not transcribed until the infected cell passes through a normal replicative cell cycle (Humphries & Temin 13). Therefore, some cellular process(es) are needed to activate the newly formed DNA provirus.

Transcription of viral RNA   Viral RNA is transcribed from the integrated DNA provirus. It is not known whether this transcription utilizes a host DNA-directed RNA polymerase, a modified host RNA polymerase, or a virus-coded RNA polymerase. The rate of RNA transcription from the DNA provirus is high compared with that of RNA transcription from most cellular DNA. The stable RNA transcript is probably 35S (Fan & Baltimore 14, Schincariol & Joklik 15). No information on the possible existence of processing of viral RNA in the nucleus has been established, that is, whether the viral RNA is formed from a higher molecular weight precursor. Viral RNA is transcribed from only one of the strands of the double-stranded DNA provirus.

Translation   The virion RNA and the viral mRNAs are apparently of the same size and nucleic acid sequence. The virion proteins may be translated as a large precursor from the 35S mRNA and then be cleaved after translation (post-translational cleavage) (Vogt & Eisenman 16).

Assembly   The 35S RNA associates with the small basic protein(s) to form the ribonucleoprotein particles which then form a core or nucleoid. Concurrently or later an envelope is formed by budding from the host-cell plasma membrane. The relative formation time of the core and the envelope varies from virus to virus.

Fate of Infected Cells   In productive infections, the sensitive cells may be transformed or killed, or may manifest no effects of the virus infection. The type of response apparently depends primarily on the virus genotype. For example, specific viral genes are known that control transformation. Mutation in these genes leads to changes in the type of transformation, temperature sensitivity of the transformation, or loss of the ability to transform. Viruses are known that kill cells in an acute infection and have no apparent effects in a chronic infection (Temin & Kassner 17). The genetically same virus causes both the acute and the chronic infections. In addition, the differentiated state of the cell may affect whether or not transformation occurs.

NONPRODUCTIVE INFECTION (INFECTION BY A VIRUS THAT DOES NOT RESULT IN THE PRODUCTION OF PROGENY VIRUSES)

Defective virus in permissive cells   Genomes of defective viruses [viruses containing a deletion or other lethal mutation(s)] are often able to infect permissive cells as a result of phenotypic mixing with a nondefective helper virus. In addition, defective virus can sometimes be introduced into a cell nonspecifically by fusion mediated by inactivated Sendai virus (a paramyxovirus apparently able to fuse cell and viral membranes). The replication of the defective virus is then the same as that of nondefective viruses except that no virions are produced or virions are produced that lack some protein essential for infectivity.

Another type of nonproductive infection is the result of infection of a nonpermissive cell by an infectious virus.

Early steps in infection of nonpermissive cells   Usually in infection of nonpermissive cells, the virus does not enter through specific receptors. A less efficient mode of entry, perhaps "fusion," is used. The formation of the DNA provirus is apparently the same in nonpermissive cells as it is in productive infections (Varmus et al 12).

Integration in nonpermissive cells   It may be that in nonproductive infections of nonpermissive cells, the integration is not at the same chromosomal site(s) as in productive infections. However, the integration is apparently nuclear in both types of infection.

Transcription, translation, and assembly in nonpermissive cells   One or more of these steps is partially blocked in nonpermissive infections. Which step is affected differs in different nonproductive infections.

Effects on infected cells   The infected cells are either transformed or have no visible effects.

RELATIONSHIPS AMONG RNA TUMOR VIRUSES (RIBODEOXYVIRUSES)

The ribodeoxyviruses are subdivided into the 14 groups listed above. The larger groups, for example, ALV and MLV, have been divided into subgroups, and the larger of these subgroups have been divided into types. The classification of a particular strain of ribodeoxyvirus is therefore group, subgroup, and type.

Specificities of relationships among RNA tumor viruses are seen at three different levels. At the lowest level, type-specific and subgroup-specific antigenic determinants or nucleic acid sequences are present only in some members of a group of ribodeoxyviruses. At the next level, group-specific antigenic determinants or nucleic acid sequences are present in all members of a group of ribodeoxyviruses and are not present in members of other groups of ribodeoxyviruses. At the third level, intergroup-specific antigenic determinants or nucleic acid sequences are present in members of more than one group of ribodeoxyviruses. When the intergroup-specific antigenic determinants or nucleic acid sequences are also found in proteins or nucleic acids of uninfected cells of animals in several different orders, they are also called class-specific.

Genome RNA (35S RNA)

Each infectious ribodeoxyvirus virion contains two to four 35S RNAs. The sequences of the RNAs of different viruses have been compared with each other primarily by nucleic acid hybridization. In addition, a few preliminary nucleic acid fingerprints, resolving only a few percent of the RNA, have appeared (Lai et al 18). Three types of nucleic acid hybridization experiments have been performed. The most common approach, in which labeled DNA complementary to virion RNA has been annealed with excess viral RNA, suffers from the facts that the labeled DNA does not usually contain all of the sequences of the virion RNA and the sequences in the labeled DNA are present in different abundances. However, this method requires the least amount of material. In another approach, labeled virion RNA has been annealed to excess DNA complementary to virion RNA. Here, all of the sequences of the virion RNA are studied, but the sequences are present in different abundances in the DNA, and much more DNA is required than with the first method. Wright & Neiman (19) used a third approach, in which labeled virion RNA was annealed to excess cellular DNA containing proviruses of that virus. This method has been combined with competition with unlabeled virion RNAs.

INTRAGROUP RELATIONSHIPS   Only the avian leukosis-sarcoma viruses, the mouse C-type ribodeoxyviruses, and the RD-114-like viruses have been studied extensively.

In general, it appears that members of a group of ribodeoxyviruses have at least approximately two thirds of their nucleic acid sequences in common. This homology has been shown so far for the following viruses:[1] 1. ALV group: Prague Rous sarcoma virus, Rous-associated virus-0, and Rous-associated virus-7 (Neiman et al 20); Schmidt-Ruppin Rous sarcoma virus and Rous-associated virus-2 (Hayward & Hanafusa 21); B77 virus, Rous sarcoma virus—Rous-associated virus-0, I-induced leukosis virus, Schmidt-Ruppin Rous sarcoma virus, and Rous-associated virus-61 (Kang & Temin 22). 2. Reticuloendotheliosis group: four members tested (Kang & Temin 22). 3. RD-114 group: RD-114 virus and Crandell cat cell virus (Quintrell et al 23, East et al 24). These nucleic acid sequence homologies were found under stringent conditions of hybridization. The hybrid molecules had melting temperatures near those of the homologous nucleic acids for the following virus pairs of the ALV group: Prague Rous sarcoma virus and Rous-associated virus-0 (Neiman 25); and I-induced leukosis virus and Rous-associated virus-61 (Kang & Temin 22). Hybrid molecules from the following viruses had slightly reduced melting temperatures: MLV group: Moloney, Rauscher, and AKR MLV (Haapala & Fischinger 26); ALV group: I-induced leukosis virus and Schmidt-Ruppin Rous sarcoma virus (Kang & Temin 22).

In addition, it appears that viruses in one group of ribodeoxyviruses differ in the sequences of the remaining portion of the viral RNA as shown for the following viruses: B77 virus and Rous-associated virus 61 (Kang & Temin 22); Rous-associated virus-0, Rous-associated virus-7, Prague Rous sarcoma virus, and transformation-defective Prague Rous sarcoma virus (Neiman et al 20); B77 virus and transformation-defective B77 virus (Lai et al 18); Moloney and SCRF MLV (Fan & Baltimore 14); AKR and NIH-MLV (Chattopadhyay et al 27); RD-114 virus and Crandell cat virus (Quintrell et al 23, East et al 24). These regions of different nucleic acid sequences may be correlated with different biological properties of the viruses, for example, whether the viruses can cause transformation of fibroblasts in culture and whether they cause leukemia or no disease in animals (Wyke & Linial 28, Wyke 29).

INTERGROUP RELATIONSHIPS   In general, little or no intergroup nucleic acid sequence homology has been reported. In most cases, these negative experiments were carried out with labeled virus DNA hybridized to excess RNA, so only an unknown fraction, usually small, of the nucleic acid sequences were studied. However, using labeled viral RNA and excess DNA, which examines all viral RNA sequences, Kang & Temin (22) compared ALV and reticuloendotheliosis viruses, and Quintrell et al (23) compared MLV, RD-114 viruses, feline C-type viruses, and visna virus. Again, little or no nucleic acid homology was found between different groups of viruses.

However, several reports of small percentages of intergroup nucleic acid hybridization have appeared; for example, between RD-114 viruses and feline leukemia viruses and between simian sarcoma virus and Moloney MLV (Okabe et al 30, East et al 24, Gallo et al 31). These intergroup nucleic acid hybridizations involved

---

[1]Here and later where lists of different viruses are given, these are most of the data published before the end of 1973. They are assumed to be representative of all such comparisons that could be made.

hybridization of 5–10% of DNA probes containing 30–60% of viral RNA sequences. Therefore, they might indicate a very small percentage of nucleic acid sequence homology. Further work will have to be done to see whether these findings represent regions of real nucleic acid sequence homology. Haapala & Fischinger (26) also reported more annealing between nucleic acids of mouse and feline C-type viruses when less stringent hybridization conditions were used.

Virion DNA Polymerase

The virion DNA polymerase apparently represents 7–15% of the information in the viral RNA (70,000 to 150,000 daltons of protein and 10^7 daltons of RNA) assuming 3 nucleotides of average molecular mass 330 daltons code for 1 amino acid of average molecular mass 100 daltons [or three times as much if the virus is triploid (Vogt 4)] (32). Serological methods have been used to determine possible antigenic relationships. Primarily, neutralization of DNA polymerase activity has been studied. The specificity of the neutralization has not appeared to depend upon the nature of the templates used. In addition, Mizutani & Temin (33) have found more relationships using a more sensitive antibody blocking test. Both tests probably examine only a small portion of the DNA polymerase molecule near the active site of the enzyme.

INTRAGROUP RELATIONSHIPS   Antibody prepared against a partially purified DNA polymerase of one virus seems to neutralize with the same kinetics the activities of DNA polymerases of all members of that group (Nowinski et al 34, Mizutani & Temin 33, 35). In particular, Mizutani & Temin (33, 35) showed that antibodies prepared against the partially purified DNA polymerases of avian myeloblastosis virus and of Rous sarcoma virus—Rous-associated virus-0 neutralized with the same kinetics the activities of DNA polymerases from strongly, weakly, and non-transforming avian leukosis-sarcoma viruses; from avian leukosis-sarcoma viruses passed horizontally or by parental infection of progeny (vertical transmission). However, this similarity may only reflect the insensitivity of the neutralization test used.

INTERGROUP RELATIONSHIPS   All of the DNA polymerases of the mammalian C-type ribodeoxyviruses appear to share some intergroup specificity (see references in 32). For example, antibody prepared against a partially purified DNA polymerase from a mouse C-type ribodeoxyvirus neutralized the activities of DNA polymerases of rat, hamster, and feline C-type ribodeoxyviruses. In all cases, the neutralization was less than in the homologous system. Similarly, antibodies prepared against other mammalian C-type ribodeoxyvirus DNA polymerases neutralized the activity of DNA polymerases from viruses in a variety of groups of mammalian C-type ribodeoxyviruses. These relationships were expected because of earlier described intergroup specificities in the virion internal proteins of mammalian C-type ribodeoxyviruses (see below).

Similar intergroup relationships have been found between the DNA polymerases of the two groups of avian C-type ribodeoxyviruses—the ALV and the reticuloendotheliosis viruses (Mizutani & Temin 33). This antigenic relationship is apparently somewhat less close than that seen among the mammalian C-type ribodeoxyviruses. It was clearly demonstrated only in antibody-blocking tests.

Internal Proteins

There are two to five major internal proteins in the ribodeoxyvirus virion. These proteins probably represent about 10–30% of the information in the viral RNA. Most studies have been of a major protein P28 or P30 which is apparently the core shell protein (Figure 1). Serological methods have been used primarily to determine possible antigenic relationships among these internal proteins, but some amino acid sequencing has begun. Either double-diffusion tests or radioimmunoassays have been used in the serological studies. In general, the radioimmunoassays have picked up smaller differences in antigenic specificities.

INTRAGROUP RELATIONSHIPS   The internal proteins of all members of any group of ribodeoxyviruses seem to be very similar in serological tests using complement fixation or double-diffusion. They have therefore been called group-specific antigens or gs-1 antigens. However, in the two published comparisons using radioimmunoassays, the mouse C-type virus P12, another internal protein different from P30, was different in three different strains of mouse C-type virus (Tronick et al 36), as was P30 in four different MLVs (Strand & August 37).

INTERGROUP RELATIONSHIPS   All of the mammalian C-type ribodeoxyviruses share a common antigenic specificity called gs-3 or interspecies antigen. This antigenic determinant is on the same molecule (the core shell protein, P30) as the group-specific, gs-1, antigen (the core shell protein, P30). However, this intergroup specificity is not identical in different groups. Radioimmunoassays clearly show that there is similarity, but not identity among the P30s of mouse, feline, primate, rat, and RD-114 C-type viruses (Stephenson & Aaronson 38, Parks & Scolnick 39, Strand & August 37). This relationship is similar to that discussed above for the mammalian C-type virus virion DNA polymerases. Furthermore, the P30s in different groups of mammalian C-type viruses differ in their isoelectric points (see references in 3).

Oroszlan et al (40) have published the sequences of 15 amino acids at the amino termini of P30s from mouse, feline, and RD-114-like C-type ribodeoxyviruses. As predicted from the serological tests, there were similarities in these amino acid sequences, but they were not identical. Oroszlan et al (40) estimate that these amino acid differences could reflect at least eight nucleotide differences in the nucleic acid sequences coding for these 15 amino acids.

Envelope Glycoproteins

There are usually two different glycoproteins in ribodeoxyvirus virions: one about 60,000–70,000 daltons and one smaller, 15,000–35,000 daltons. The relative sizes of the carbohydrate and the protein portions in these glycoproteins are not known. The glycoproteins seem to carry the neutralizing antigens and to control the viral host range when it is determined by specific cell surface receptors. (There are also other determinants of host range involving later steps in infection.)

INTRAGROUP RELATIONSHIPS In contradistinction to the virion structural components discussed so far, there is considerable variability among viruses within a group in the specificity of neutralization of infectivity, the receptor specificity, and the size of the glycoproteins. The virion glycoproteins are thus said to have subgroup and type specificities.

In the avian leukosis-sarcoma virus group, there are A, B, C, D, E, and F subgroup specificities defined by the relative efficiency of infection of genetically different chicken cells and interference between viruses in each subgroup. Each subgroup contains viruses with different glycoproteins, which are different either in antigenic specificity, as seen in neutralization tests, or in size (Lai & Duesberg 41).

Such extensive intragroup (subgroup) variation has not been reported for other ribodeoxyviruses. The mouse C-type viruses are separated into only two subgroups on the basis of neutralization (Aoki et al 42), the feline C-type viruses into three on the basis of interference (Sarma & Log 43); the reticuloendotheliosis viruses are not subdivided on the basis of interference (Temin & Kassner, unpublished).

Strand & August (37, 44) have reported purification of the native GP70 from a mouse C-type ribodeoxyvirus. Antibody prepared against this protein did have group specificity as well as type specificity in radioimmunoassays.

INTERGROUP RELATIONSHIPS Tests using whole virus (neutralization or interference tests) did not show any intergroup specificities in the envelope glycoproteins. However, Strand & August (37, 44), using radioimmunoassay, found intergroup specificity between mouse and feline C-type virus GP70s. No relationship of the mouse GP70 to RD-114 or primate C-type virus proteins was found. These results indicate that the intergroup specificity in GP70 resides in that portion of the molecule attached to or buried in the lipid bilayer of the virion envelope or in a portion not important for infection. Table 1 presents a summary of these relationships for ALV.

RELATIONSHIPS OF RIBODEOXYVIRUSES TO SPECIFIC CELLULAR GENES AND PRODUCTS

Genome RNA (35S RNA)

SPECIES OF ORIGIN AND ENDOGENOUS VIRUSES In many cases, a group of ribodeoxyviruses has been isolated from only one species and outside of experimental situations is transmitted only vertically (from parent to progeny), although it can be transmitted horizontally in laboratory situations. In addition, viruses representative of some ribodeoxyvirus groups can be "activated" from uninfected cells of only one species. In these cases, that one species has been called the species of origin of that virus, for example, avian leukosis-sarcoma viruses from chickens, mouse C-type ribodeoxyviruses from mice, mouse mammary tumor virus from mice, and RD-114-like viruses from cats.

Table 1 Distribution of avian leukosis virus genes[a]

Genes For	In Viruses	In Cells
DNA polymerase	ALV, (REV)	Ch, (Ph), (Du)
Viral RNA	ALV	Ch, (Ph), (Q)
Internal proteins	ALV	CH, (Ph)
Envelope glycoprotein	(ALV)	(Ch)

[a]Abbreviations: ALV = Avian leukosis viruses, REV = reticuloendotheliosis viruses, Ch = chicken, Ph = pheasant, Du = duck, Q = quail, ( ) = related but not identical or only in some members.

Nucleic acid hybridization has been used to determine whether there are RNAs or DNAs in uninfected cells related to ribodeoxyvirus RNAs. Labeled virion 60–70S RNA has been annealed to excess cellular DNA, or labeled DNA complementary to virion RNA has been annealed to excess cellular RNA or DNA. In addition, older work looked at repetitive sequences in cellular DNA related to viral RNA by annealing excess labeled virion RNA to cellular DNA. When nucleic acids homologous to viral RNA are found in uninfected cells of the species of origin, there is said to be endogenous virus or endogenous virus-related genes present in these cells.

DNA   DNA in uninfected cells of the species of origin which is homologous to viral RNA from the endogenous virus has been detected in several cases. In chickens, mice, and cats, uninfected cells contain DNA homologous to many of the sequences of RNA from virions of viruses of the corresponding virus group. The DNA appears to represent a majority of the nucleic acid sequences in a nontransforming representative of the virus group; for example, chickens contain DNA homologous to over 70% of the RNA of Rous-associated virus-0, but contain DNA homologous only to those portions of Rous sarcoma virus RNA that are homologous to Rous-associated virus-0 RNA (Neiman et al 20). It is not known whether all or only most of the sequences of the RNA of the endogenous virus are present in the DNA of the uninfected cells of the species of origin; for example, one can ask whether or not all of the sequences of Rous-associated virus-0 RNA are present in the DNA of uninfected chickens.

Some of the endogenous sequences are apparently in the repetitive fraction of cell DNA (Ruprecht et al 45, Gillespie et al 46, Baluda & Roy-Burman 47, Fujinaga et al 48). It is not known whether this is a special fraction of viral DNA per se or whether the RNA transcript of repetitive DNA enters virions, perhaps because it contains promoters for transcription of viral RNA (Britten & Davidson 49).

In mice and cats, there is DNA homologous to two groups of ribodeoxyviruses. Cats have DNA related to RD-114 virus RNA (Ruprecht et al 45, Gillespie et al 46, Neiman 50, Fujinaga et al 48) and apparently also to feline C-type virus RNA (Quintrell et al 23). Mice have DNA related to both mouse mammary tumor virus RNA (Varmus et al 51, Parks et al 52) and mouse C-type virus RNA (Gelb et al 53). Furthermore, one strain of inbred mouse can be activated to produce more than one kind of mouse C-type virus (Aoki & Todaro 54, Aaronson & Stephenson 55).

NONENDOGENOUS VIRUSES   In addition to those virus groups with large amounts of endogeneous sequences in DNA of uninfected cells, there are apparently other groups of ribodeoxyviruses with little or no endogenous nucleic acid sequences in uninfected cells of the species from which they were isolated. These groups of viruses are biologically distinguished from the groups with endogeneous nucleic acid sequences in the following ways: 1. They have been isolated from more than one species, for example, reticuloendotheliosis viruses have been isolated from turkeys, chickens, and Pekin ducks, and Mason-Pfizer monkey viruses have been isolated from Rhesus monkey and human cells. 2. They are horizontally transmitted in nonlaboratory situations, for example, visna virus of sheep. 3. They have been isolated only a few times, as is perhaps true for some primate C-type ribodeoxyviruses. In all of these cases, few or no sequences homologous to viral RNA have been found in the DNA of the species from which the viruses were isolated. Most reticuloendotheliosis virus RNA sequences are absent in chickens, ring-necked pheasants, turkeys, and Pekin ducks (Kang & Temin 56). Mason-Pfizer monkey virus RNA sequences are apparently absent in human and rhesus monkey cells (Parks et al 57). Visna virus RNA sequences are apparently absent in sheep cells (Haase & Varmus 58), and primate C-type ribodeoxyvirus RNA sequences are apparently absent in primate cells (Scolnick et al 59). (However, these experiments have not looked at all of the sequences in viral RNA and so may be misleading.)

Similarly, the endogenous nucleic acid sequences are not homologous to all of the nucleic acid sequences of the pathogenic members of the virus groups; for example, the parts of Rous sarcoma virus RNA that are different from Rous-associated virus-0 RNA are not found in uninfected chicken cells (Neiman et al 20).

OTHER SPECIES   For those viruses with an apparent species of origin, it has also been asked whether nucleic acid sequences homologous to virion RNA are found in related species. About 10% of ALV RNA sequences are found in ring-necked pheasants, Japanese quail, and turkeys (Neiman 60, Kang & Temin 56), and 0% in ducks (Varmus et al 12, Kang & Temin 56). Mouse C-type and mouse mammary tumor virus RNA sequences are not found in rats (Gelb et al 61, Varmus et al 51), and RD-114 virus RNA sequences are not found in ocelets (Quintrell et al 23).

Therefore, the distribution in related species of endogenous ribodeoxyvirus nucleic acid sequences is narrow, like that of the distribution of the infectious viruses.

RNA   The RNA in uninfected cells usually represents only a part of the virion RNA sequences. The amount of viral-specific RNA in uninfected cells varies and is apparently genetically controlled (Hayward & Hanafusa 21, Varmus et al 62). In some cases the RNA is translated (see later). In other cases it is not. For example, RD-114 virus RNA is found in many normal cat cells, and mouse mammary tumor virus RNA is found in many normal mouse cells without any viral products (antigens) having been found (Okabe et al 30, Varmus et al 62). Interestingly, the mouse mammary tumor virus RNA in different mouse strains is different, as shown by the lowered melting temperature of the nucleic acid hybrids with mouse mammary tumor virus DNA (Varmus et al 62).

Small RNAs

The virion small RNAs are apparently host coded and apparently do not share large, or any, portions of their sequences with viral 35S RNA. They appear to be primarily homologous to host tRNAs (Sawyer & Dahlberg 63, Sawyer et al 64, Elder & Smith 65). One of the major virion small RNAs in Rous sarcoma virus can act as a primer for in vitro viral RNA-directed DNA synthesis and is also found in uninfected chicken cells (Dahlberg et al 5, Sawyer et al 64). In addition, other host RNAs, for example, ribosomal RNAs, 7S RNA, and 5S RNA (Erikson et al 66, Faras et al 67), and numerous host enzymes can be found in purified RNA tumor virus virions (see references in 68).

DNA Polymerase

Few studies have been performed to determine whether there are relationships between ribodeoxyvirus virion DNA polymerases and host cell DNA polymerases. Partially, this reflects the lack of knowledge of host cell DNA polymerases. Antibodies to mouse C-type or avian leukosis virus DNA polymerases do not neutralize the two major DNA polymerases of mouse and chicken cells, respectively (Ross et al 69, Mizutani & Temin 35). However, use of a more sensitive blocking test has shown serological cross reactions between DNA polymerases of avian leukosis-sarcoma viruses, reticuloendotheliosis viruses, and the large and small DNA polymerases of chickens, pheasants, turkeys, and ducks (Mizutani & Temin 33). Thus, these avian DNA polymerases, both viral and cellular, have class specificity [or higher vertebrate specificity (33, 70)]. However, DNA polymerases identical with virion DNA polymerases have not been found in uninfected cells of these avian species.

Kang & Temin (71, 72) reported endogenous RNA-directed DNA polymerase activity in uninfected chicken embryo fibroblasts, chicken amnion cells, and chicken embryos. The DNA polymerase of this activity was more closely related to the small DNA polymerase in chicken cells than to that in avian leukosis-sarcoma virus (Mizutani & Temin 35). The template of the chicken embryo endogenous DNA polymerase activity also did not have nucleic acid sequences that are homologous to RNAs of avian leukosis-sarcoma or reticuloendotheliosis viruses (Kang & Temin 22).

A possibly similar activity has been found in some human acute leukemic cells (Sarngadharan et al 73, Baxt et al 74). However, the template of this endogenous DNA polymerase activity from human leukemic cells appears to have some sequences in common with the RNA of a primate C-type virus (Gallo et al 31, Baxt et al 74). Furthermore, the DNA polymerase purified from this human endogenous activity is antigenically related to that of primate C-type virus (Todaro & Gallo 75).

Internal Proteins

Group-specific proteins for avian leukosis-sarcoma virus and mouse C-type virus are found in most or all uninfected chickens and mice, respectively (Stephenson & Aaronson 38, Stephenson et al 76, Chen & Hanafusa 77). Results with sensitive radioimmunoassay tests indicate that almost all uninfected cells of these species contain this antigen(s). The antigens in the uninfected chicken cells have the same mobilities in sodium dodecyl sulfate/polyacrylamide gel electrophoresis as P27 and P22 from the avian leukosis-sarcoma virus virion (Fleissner & Tress 78). The amount of these antigens varies greatly from individual to individual. This difference is controlled by a single Mendelian gene; more antigen present is dominant (Payne & Chubb 79, Hanafusa et al 80). In spite of the widespread occurrence of these antigens in cells of embryo and adult chickens and mice, there is no immunological tolerance to them (Ruoslahti et al 81, Oldstone et al 82).

ALV group-specific antigen(s) has also been found in a few pheasants and jungle fowl, but not in ducks or quail (Weiss & Payne 83, Weiss et al 84, Friis 85). Mouse C-type group-specific antigen has not been found in uninfected cells of any mammalian species other than the mouse (Stephenson & Aaronson 38). No mouse C-type virus P12 has been found in uninfected mouse cells (Tronick et al 36).

In addition, rat C-type, feline C-type, and RD-114 virus group-specific antigens have not been found in uninfected rats and cats, respectively (Stephenson & Aaronson 38).

Therefore, viral group-specific antigens have been detected in uninfected cells of a species less frequently than the nucleic acid sequences of endogenous viral RNA. Furthermore, the distribution of group-specific antigen(s) is narrow like that of the endogenous viral nucleic acid sequences and infectious viruses (see above).

Envelope Glycoproteins

Some uninfected chicken cells contain viral genetic material, called chick helper factor. Chick helper factor appears to be the structural gene for a virion glycoprotein which supplies by phenotypic mixing glycoprotein of E subgroup to defective Rous sarcoma viruses. The uninfected chicken cells apparently contain the avian-leukosis sarcoma virus GP70 molecule of E subgroup specificity (Hanafusa et al 86). The GP70 molecule is not present in all chicken cells. Its presence appears to be controlled by a single Mendelian gene (Weiss & Payne 83).

Table 1 presents a summary of these relationships for ALV and cells.

A large number of apparent mouse C-type virus envelope glycoproteins have been found on the surfaces of uninfected mouse cells (see description and references in Nowinski & Peters 87), and MLV GP70 was present in most mouse strains (Strand, Lilly & August 87a). In addition, virus-specific (having the same antigenicity as antigens appearing after virus infection), but not virion, antigens have been found on the surfaces of uninfected mouse cells. These antigens are not found on all uninfected mouse cells, and their presence appears to depend upon the differentiated state of the cells and on genetic factors. The genetics originally was thought to be fairly simple but now appears to be complex, with the genetic factors located on several chromosomes.

Virion Receptors

Chickens have several genes controlling the presence of receptors for sensitivity to infection by avian leukosis-sarcoma viruses (see references in 3). These receptors are subgroup-specific. Each gene has a sensitive and a resistant allelle. Sensitivity is dominant. The chemical nature of the receptors is unknown.

Inhibitor or Resistance Genes

In addition to specific resistance resulting from the lack of a specific receptor, two other types of resistance have been described (see references in 3). One is the *Fv-1* locus in mice, which controls a dominant resistance and has two allelles: *n* and *b*. Mouse C-type viruses are N-tropic, B-tropic, NB-tropic, or xenotropic (Levy 88). N-tropic viruses are less efficient at infecting *Fv-1^b* cells than are B-tropic viruses. The kinetics of infection of B-type cells by N-tropic viruses are one-hit, and the kinetics of infection of B-type cells by N-tropic viruses are two-hit. Viruses can mutate to NB-tropisim, and they then grow equally well in both types of cells.

In chickens, there is also a dominant gene for resistance to subgroup E viruses which maps separately from the receptor genes. This gene may be the same as that for the subgroup E glycoprotein (Crittenden et al 89).

Activation Genes

Some mice and chickens produce ribodeoxyviruses spontaneously or after treatment with halogenated pyrimidines (see references in 3, Stephenson & Aaronson 90, Rowe 91). The mechanism of this activation is not known. The activation is controlled by dominant Mendelian genes. In different strains of inbred mice, the number of these genes varies from two to over five. The viruses produced in response to different genes are different in their host range, perhaps indicating that each gene represents the presence or absence of different inactive or defective proviruses. The genes are located on different chromosomes in different strains of inbred mice (Rowe 91).

IMPLICATIONS OF THESE RELATIONSHIPS

Comparison of Ribodeoxyvirus Relationships with Those of Other Animal Viruses

Ribodeoxyviruses differ from the other 15 or so defined groups of animal viruses in two respects: only ribodeoxyviruses are known to use RNA to DNA information transfer and only ribodeoxyviruses have genes related to specific cellular genes and products.

There are few known relationships between animal viruses other than ribodeoxyviruses and specific cellular genes and products. Host genes are known which affect disease production by animal viruses, both ribodeoxyviruses and other animal viruses (see Fenner 92). However, specific cellular genes acting at the cellular level on virus replication or cellular genes for endogenous virus products are apparently not known for animal viruses other than ribodeoxyviruses.

Shubak-Sharpe (93) has suggested, based on the frequency of nucleotide pairs, that papova-, parvo-, and picornaviruses evolved from animal cell DNA. However, no nucleic acid sequence homology apparently exists between these viruses and uninfected cells (2).

Defective proviruses of temperate bacteriophage are found in many strains of bacteria. However, in all cases, these proviruses are of DNA viruses, not of ribodeoxyviruses (see Hershey 94). Thus, ribodeoxyviruses appear to have a unique kind of relationship to genomes of uninfected cells.

Cellular Genes for Virion Components

In chickens and mice, there are cellular genes for some virion components (Table 1). For example, all chickens and mice apparently possess the gene for the core shell protein (group-specific antigen), and some of the protein is made in almost all chickens and mice. There is, however, polymorphism in the concentration of this protein in cells of genetically different chickens and mice. The almost universal presence of this protein seems to indicate that it has some function giving selective advantage to its continued presence.

In chickens, mice, and cats there are inactive or defective proviruses of ribodeoxyviruses. Included in these inactive proviruses are genes needed for the horizontal transmission of these "endogenous" viruses. These endogenous viruses are usually nonpathogenic. Their wide distribution in uninfected, normal animals seems to indicate that their presence confers some selective advantage on the host animals or reflects something in the cell genome that confers a selective advantage on the host.

Furthermore, there are wide intergroup antigenic relationships among some ribodeoxyvirus virion structural proteins. For example, there appears to be a core of conserved antigenic relationships among some ribodeoxyvirus proteins [DNA polymerase, one internal protein (P30), and perhaps an envelope glycoprotein (GP70)]. However, there are little or no intergroup nucleic acid sequence homologies among ribodeoxyviruses, even among those viruses with proteins antigenically related, for example, murine and feline leukemia viruses. There are also antigenic relationships between viral proteins and cellular proteins [DNA polymerase, one virion internal protein (P30), and perhaps an envelope glycoprotein (GP70)], and nucleic acid sequence homologies between viral RNA and cellular DNA. The antigenic relationships are called intergroup-specific. They could equally be called class-specific since they involve viral and cellular gene products, for example, DNA polymerases of avian ribodeoxyviruses and avian cells. The nucleic acid sequence homologies are species-specific, for example, mouse leukemia virus RNA has homology to mouse DNA, but not to rat DNA.

The two questions posed by these relationships are 1. why do these relationships between groups of ribodeoxyviruses and between ribodeoxyviruses and cells and these potentials to form ribodeoxyviruses exist so widely in higher vertebrates? and 2. how did these relationships originate?

The Protovirus Hypothesis

I suggest an explanation for these relationships based upon an expansion of my earlier protovirus hypothesis (Temin 3, 95–97). The protovirus hypothesis states that RNA-directed DNA polymerase activity exists in normal animals, that it plays a role in normal cellular processes like differentiation, that RNA tumor viruses (ribodeoxyviruses) evolved from this activity, and that cancer arises from variational events in the functioning of this activity.

The core of conserved antigenic relationships among viral and cellular proteins, discussed above, reflects the same kind of genetic persistence and antigenic and amino acid sequence relationships seen in other essential proteins of similar function, for example, cytochrome c and hemoglobin. However, the nucleic acid sequences coding for these proteins in different ribodeoxyviruses have varied so they are not seen to be homologous in nucleic acid hybridization experiments using different ribodeoxyviruses RNAs, or, alternatively, the extent of nucleic acid homology reflected in these antigenic relationships is too small a percentage of the nucleic acid sequences to be recognized in nucleic acid hybridization experiments.

Thus, we find related (homologous) structural genes for these proteins in cells or in viruses even though the viral nucleic acid sequences are mostly distinct. Therefore, to explain the presence of genes for functional homologous proteins in viruses of distinct nucleic acid sequences, there must have been considerable evolution with change and perhaps generation of nucleic acid sequences.

I suggest that ribodeoxyviruses and the potential to become ribodeoxyviruses evolved along with the host organisms. The proviruses and potential proviruses (protoviruses) acted as agents for much of the expansion and alteration of cellular nucleic acid sequences seen in evolutionary times, as for example in making gene duplications, and perhaps in the differentiation of an organism.

This mechanism of gene expansion would be in addition to polyploidization or duplication of whole chromosomes and single gene duplication through errors in crossing over. The advantages of the protovirus mechanism are that it can involve any length of DNA from part of a gene to many genes, that no deletion of the original DNA is required, and that the duplicated DNA can be inserted anywhere in the cell genome.

This type of information transfer would be needed by all organisms. Therefore, the potential for it would have been conserved in all organisms. However, protoviruses would have undergone independent but parallel evolution in different organisms. Possibly, the closer to an active provirus this system became, the more efficient it was at its primary cellular functions of providing genetic substratums for embryonic development and for evolution, including transfer of information between cells.

Stated more concisely, ribodeoxyviruses evolved to independent genetic status from a normal cellular system of DNA to RNA to DNA information transfer. The relationships between viruses and cells, that is, the species-specific and class-specific relationships discussed above, reflect this origin. The group-specific and intergroup-specific relationships between viruses reflect their parallel evolution from cellular components which derived from common evolutionary precursors.

An Alternate Hypothesis

Another possibility is that the viruses with these relationships had another origin, unspecified as are most virus origins, and by horizontal transmission these viruses infected the germ lines of the ancestors of all surviving organisms (Huebner &

Todaro 98, Gilden 99, Gross 100). In this hypothesis, the present-day relationships among C-type viruses are reflections of this common origin in one ancestral virus.

The major problems with this model are (a) it does not approach the question of origin; (b) it requires 100% efficient initial infections of the germ lines of all ancestors of present-day chickens, cats, and mice; (c) it requires many different infections to explain the different types of endogenous ribodeoxyviruses and the multiple, distinguishable, endogenous representatives of some of them; and (d) it has no mechanism for the maintenance and evolution in these inactive proviruses of structural genes for many functional virion structural proteins, for example, a DNA polymerase distinct from host-cell DNA polymerases. (The viral DNA polymerase is needed only for horizontal transmission of proviral or protoviral information. The other cellular DNA polymerases take care of vertical transmission of the provirus.)

There frequently is seen cellular host range restriction against horizontal transmission of the endogenous ribodeoxyvirus. In the germ-line infection hypothesis, this restriction must be assumed to reflect a change in organisms after the initial infection, as no repression analogous to that seen with temperate phage is known for ribodeoxyviruses. Furthermore, if the mammalian C-type virus relationships are relics from an original germ-line infection by a single type of virus, another hypothesis is needed to explain the separate evolution of mammalian C-type viruses which have been passed as inactive proviruses. There would have been no chance for selection acting on these inactive proviruses, but they have evolved into different potentially infectious ribodeoxyviruses, for example, the different types of endogenous mouse ribodeoxyviruses. This problem remains even if there was a selective advantage to the retention of the original provirus.

In contrast, the protovirus hypothesis requires a functioning DNA polymerase for the physiological role which I suggest insures the maintenance of this protovirus information transfer system. As the protovirus hypothesis provides a mechanism for the origin and variation of new nucleic acid sequences and requires a functioning information transfer machinery, it resolves these problems inherent in the ancestor germ-line infection hypothesis.

CONCLUSIONS

The protovirus hypothesis states that cells of ancestral organisms had a genetic system for DNA to RNA to DNA information transfer. The enzymes and other proteins involved in this transfer were purely cellular. This information transfer system was originally used for differentiation, that is, information transfer in somatic cells. If this type of information transfer occurred in a germ line by DNA to RNA to DNA transfer in the germ line or by transfer of RNA from somatic cells . to the germ line, there would be a change in progeny organisms. This change could provide a substratum for evolution. As this information transfer system became more efficient, it might have evolved DNA polymerases and structural proteins separate from those for other cellular information transfer. Finally, this system could be capable of horizontal transmission and would have become a ribodeoxyvirus.

Some idea of the variation that could be introduced into organisms by this system is seen in the numerous groups of ribodeoxyviruses and the types in each group. For example, mouse cells can have DNA sequences for proviruses of two different groups of ribodeoxyviruses and from two to over five different types of viruses of one of the groups, with the inactive proviruses probably located on separate chromosomes.

As this protovirus system would provide expanded and altered genomes for evolution, it would have been present in the ancestors of present-day species. Therefore, the wide occurrences of ribodeoxyviruses and the relic relationships among them could be explained.

If duplication and evolution happened only in somatic cells, viruses or virus-like particles could be formed with new nucleic acid sequences not present in the DNA of other organisms of the same species. The "spontaneous" origin of ribodeoxyviruses with few endogenous nucleic acid sequences would be a rare and rapid example of this evolution.

The origin of strongly transforming ribodeoxyviruses, for example, Rous sarcoma virus, would involve evolution and recombination of protoviruses. The origin of multiple endogenous viruses would involve duplication and evolution. The "induction" of inactive endogenous proviruses would represent the last step in the evolutionary origin of a ribodeoxyvirus.

The protovirus hypothesis predicts amino acid sequence homologies between viral structural proteins and cellular proteins. The DNA polymerases of viruses and cells, and perhaps virion envelope proteins and cellular antibodies or histocompatability antigens, could be homologous. The protovirus hypothesis also predicts the existence of DNA sequences altered in amount or location in differentiated cells and a role for the ribodeoxyvirus group-specific antigen in normal cellular processes.

SUMMARY

RNA tumor viruses (ribodeoxyviruses) are viruses with a special type of virion. There are at least 14 groups of ribodeoxyviruses. There are some intergroup antigenic and nucleic acid sequence relationships. There are also antigenic and nucleic acid sequence relationships between ribodeoxyviruses and uninfected cells.

The protovirus hypothesis states that ribodeoxyviruses repeatedly evolved from normal cellular processes that were similar in related animals because of their common ancestry. The normal cellular processes of DNA to RNA to DNA information transfer could be important in differentiation and evolution.

ACKNOWLEDGMENTS

I wish to thank Drs. G. M. Cooper, J. Crow, R. DeMars, E. Fritsch, E. Humphries, and D. Zarling for their useful comments on this manuscript. The research in my laboratory is supported by Program Project Grant CA-07175 from the National

Cancer Institute and research grant VC-7 from the American Cancer Society. I hold Research Career Development Award CA-8182 from the National Cancer Institute.

Literature Cited

1. Vogt, P. K. 1972. Guest Editorial: The emerging genetics of RNA tumor viruses. *J. Nat. Cancer Inst.* 42:3–8
2. Tooze, J., Ed. 1973. *The Molecular Biology of Tumor Viruses.* Cold Spring Harbor, N.Y.: Cold Spring Harbor Lab. 743 pp.
3. Temin, H. M. 1974. The cellular and molecular biology of RNA tumor viruses, especially avian leukosis-sarcoma viruses, and their relatives. *Advan. Cancer Res.* 19:47–104
4. Vogt, P. K. 1973. The genome of avian RNA tumor viruses: a discussion of four models. In *Possible Episomes in Eukaryotes,* ed. L. G. Silvestri, 35–41. Amsterdam: North-Holland
5. Dahlberg, J. E. et al 1974. Transcription of DNA from the 70S RNA of Rous sarcoma virus I. Identification of a specific 4S RNA which serves as primer. *J. Virol.* 13:1126–33
6. Fleissner, E., Tress, E. 1973. Isolation of a ribonucleoprotein structure from oncornaviruses. *J. Virol.* 12:1612–15
7. Volkman, L. E., Krueger, R. G. 1973. Characterization of C-type particles produced by tissue culture-adapted murine myeloma. *J. Virol.* 12:1589–97
8. Nayak, D. P., Murray, P. R. 1973. Induction of type-C viruses in cultured guinea pig cells. *J. Virol.* 12:177–87
9. Lieber, M. M., Benveniste, E., Livingston, D., Todaro, G. J. 1973. Mammalian cells in culture frequently release type C virus. *Science* 182:57–59
10. Schidlovsky, G., Ahmad, M. 1973. C-type virus particles in placentas and fetal tissue of rhesus monkeys. *J. Nat. Cancer Inst.* 51:225–33
11. Dales, S. 1973. Early events in cell-animal virus interactions. *Bacteriol. Rev.* 37:103–35
12. Varmus, H. E., Vogt, P. K., Bishop, J. M. 1973. Integration of deoxyribonucleic acids specific for Rous sarcoma virus after infection of permissive and non-permissive hosts. *Proc. Nat. Acad. Sci. USA* 70:3067–71
13. Humphries, E., Temin, H. M. 1974. Requirement for cell division for initiation of transcription of Rous sarcoma virus RNA. *J. Virol.* 14:In press
14. Fan, H., Baltimore, D. 1973. RNA metabolism of murine leukemia virus: detection of virus-specific RNA sequences in infected and uninfected cells and identification of virus-specific messenger RNA. *J. Mol. Biol.* 80:93–117
15. Schincariol, A. L., Joklik, W. K. 1973. Early synthesis of virus-specific RNA and DNA in cells rapidly transformed with Rous sarcoma virus. *Virology* 56:532–48
16. Vogt, V. M., Eisenman, R. 1973. Identification of a large polypeptide precursor of avian oncornavirus proteins. *Proc. Nat. Acad. Sci. USA* 70:1734–38
17. Temin, H. M., Kassner, V. K. 1974. Replication of reticuloendotheliosis viruses in cell culture: acute infection. 13:291–97
18. Lai, M. M. C., Duesberg, P. H., Horst, J., Vogt, P. K. 1973. Avian tumor virus RNA: a comparison of three sarcoma viruses and their transformation-defective derivatives by oligonucleotide fingerprinting and DNA-RNA hybridization. *Proc. Nat. Acad. Sci. USA* 70:2266–70
19. Wright, S. E., Neiman, P. E. 1974. Base sequence relationships between avian RNA endogenous and sarcoma viruses assayed by competitive RNA-DNA hybridization. *Biochemistry* 13:1549–54
20. Neiman, P. E., Wright, S. E., McMillin, C., MacDonnell, D. 1974. Nucleotide sequence relationships of avian RNA tumor viruses: measurement of the deletion in a transformation defective mutant of Rous sarcoma virus. *J. Virol.* 13:837–46
21. Hayward, W. S., Hanafusa, H. 1973. Detection of avian tumor virus RNA in uninfected chicken embryos cells. *J. Virol.* 11:157–67
22. Kang, C. -Y., Temin, H. M. 1973. Lack of sequence homology among RNA's of avian leukosis-sarcoma viruses, reticuloendotheliosis viruses, and chicken endogenous RNA-directed DNA polymerase activity. *J. Virol.* 12:1314–24
23. Quintrell, N., Varmus, H. E., Bishop, J. M., Nicolson, M. O., McAllister, R. M. 1974. Homologies among the nucleotide sequences of the genomes of C-type viruses. *Virology* 58:568–75
24. East, J. L., Knesek, J. E., Allen, P. T., Dmochowski, L. 1973. Structural characteristics and nucleotide sequence analysis of genomic RNA from RD-114

virus and feline RNA tumor viruses. *J. Virol.* 12:1085–91

25. Neiman, P. E. 1972. Rous sarcoma virus nucleotide sequences in cellular DNA: measurement by RNA-DNA hybridization. *Science* 178:750–52

26. Haapala, D. K., Fischinger, P. J. 1973. Molecular relatedness of mammalian RNA tumor viruses as determined by DNA-RNA hybridization. *Science* 180:972–74

27. Chattopadhyay, S. K., Lowry, D. R., Teich, N., Levine, A. S., Rowe, W. P. 1974. Evidence that AKR murine leukemia virus genome is complete in DNA of the high virus AKR mouse and incomplete in the DNA of the "virus-negative" NIH mouse. *Proc. Nat. Acad. Sci. USA* 71:167–71

28. Wyke, J. A., Linial, M. 1973. Temperature-sensitive avian sarcoma viruses: physiological comparison of twenty mutants. *Virology* 53:152–61

29. Wyke, J. A. 1973. Complementation of transforming functions of temperature-sensitive mutants of avian sarcoma virus. *Virology* 54:28–36

30. Okabe, H., Gilden, R. V., Hatanaka, M. 1973. RD-114 virus-specific sequences in feline cellular RNA: detection and characterization. *J. Virol.* 12:984–94

31. Gallo, R. C., Miller, N. R., Saxinger, W. C., Gillespie, D. 1973. Primate RNA tumor virus-like DNA synthesized endogenously by RNA-dependent DNA polymerase in virus-like particles from fresh human acute leukemic blood cells. *Proc. Nat. Acad. Sci. USA* 70: 3219–24

32. Temin, H. M., Mizutani, S. 1974. RNA tumor virus DNA polymerases. In *The Enzymes*, ed. P. D. Boyer, 10:211–35. New York: Academic

33. Mizutani, S., Temin, H. M. 1974. Specific serological relationships among partially purified DNA polymerases of avian leukosis-sarcoma viruses, reticuloendotheliosis viruses, and avian cells. *J. Virol.* 13:1020–29

34. Nowinski, R. C., Watson, K. F., Yaniv, A., Spiegelman, S. 1972. Serological analysis of the deoxyribonucleic acid polymerase of avian oncornaviruses. II. Comparison of avian deoxyribonucleic acid polymerases. *J. Virol.* 10:959–64

35. Mizutani, S., Temin, H. M. 1973. Lack of serological relationships among DNA polymerases of avian leukosis-sarcoma viruses, reticuloendotheliosis viruses, and chicken cells. *J. Virol.* 12:440–46

36. Tronick, S. R., Stephenson, J. R., Aaronson, S. A. 1973. Immunological characterization of a low molecular weight polypeptide of murine leukemia virus. *Virology* 54:199–206

37. Strand, M., August, J. T. 1974. Structural proteins of mammalian oncogenic RNA viruses: multiple antigenic determinants of the major internal protein and envelope glycoproteins. *J. Virol.* 13:171–80

38. Stephenson, J. R., Aaronson, S. A. 1973. Expression of endogenous RNA C-type virus group-specific antigens in mammalian cells. *J. Virol.* 12:564–69

39. Parks, W. P., Scolnick, E. M. 1972. Radioimmunoassay of mammalian C-type viral proteins: interspecies antigenic reactivities of the major internal polypeptides. *Proc. Nat. Acad. Sci. USA* 69:1766–70

40. Oroszlan, S., Copeland, T., Summers, M. R., Gilden, R. V. 1973. Feline leukemia and RD-114 virus group-specific proteins: comparison of amino terminal sequences. *Science* 181:454–56

41. Lai, M. M. C., Duesberg, P. H. 1972. Differences between the envelope glycoproteins and glycopeptides of avian tumor viruses released from transformed and from nontransformed cells. *Virology* 50:359–72

42. Aoki, T., Old, L. J., Boyse, E. A. 1966. Serological analysis of leukemia antigens of the mouse. *Nat. Cancer Inst. Monogr.* 22:449–57

43. Sarma, P. S., Log, G. 1973. Subgroup classification of feline leukemia and sarcoma viruses by viral interference and neutralization tests. *Virology* 54:160–69

44. Strand, M., August, J. T. 1973. Structural proteins of oncogenic ribonucleic acid viruses. Interspec II, a new interspecies antigen. *J. Biol. Chem.* 248:5627–33

45. Ruprecht, T. R. M., Goodman, N. C., Spiegelman, S. 1973. Determination of natural hosts and taxonomy of RNA tumor viruses by molecular hybridization: application to RD-114, a candidate human virus. *Proc. Nat. Acad. Sci. USA* 70:1437–41

46. Gillespie, D., Gillespie, S., Gallo, R. C., East, J. L., Dmochowski, L. 1973. Genetic origin of RD-114 and other RNA tumor viruses assayed by molecular hybridization. *Nature New Biol.* 244: 51–54

47. Baluda, M. A., Roy-Burman, P. 1973. Partial characterization of RD-114 virus by DNA-RNA hybridization studies. *Nature New Biol.* 244:59–62

48. Fujinaga, K. et al 1973. RD-114 virus: analysis of viral gene sequences in feline and human cells by DNA-DNA reassociation kinetics and RNA-DNA hybridization. *Virology* 56:484–95

49. Britten, R. J., Davidson, E. H. 1969. Gene regulation for higher cells: a theory. *Science* 165:349–57

50. Neiman, P. E. 1973. Measurement of RD-114 virus nucleotide sequences in feline cellular DNA. *Nature New Biol.* 244:62–64

51. Varmus, H. E., Bishop, J. M., Nowinski, R., Sarkar, N. H. 1972. Detection of mouse mammary tumor virus-specific nucleotide sequences in mouse strains with high and low incidence of spontaneous tumors. *Nature New Biol.* 238:189–91

52. Parks, W. P., Scolnick, E. M. 1973. Mouse mammary tumor cell clones with varying degrees of virus expression. *Virology* 55:163–73

53. Gelb, L. D., Milstein, J. B., Martin, M. A., Aaronson, S. A. 1973. Characterization of murine leukemia virus-specific DNA present in normal mouse cells. *Nature New Biol.* 44:76–79

54. Aoki, T., Todaro, G. J. 1973. Antigenic properties of endogenous type-C viruses from spontaneously transformed clones of BALB/3T3. *Proc. Nat. Acad. Sci. USA* 70:1598–1602

55. Aaronson, S. A., Stephenson, J. R. 1973. Independent segregation of loci for activation of biologically distinguishable RNA C-type viruses in mouse cells. *Proc. Nat. Acad. Sci. USA* 70:2055–58

56. Kang, C. -Y., Temin, H. M. 1974. Hybridization of reticuloendotheliosis virus RNA and DNA with nucleic acids of cells. *J. Virol.* 14:In press

57. Parks, W. P. et al 1973. Mason-Pfizer virus characterization: a similar virus in a human amniotic cell line. *J. Virol.* 12:1540–47

58. Haase, A. T., Varmus, H. E. 1973. Demonstration of a DNA provirus in the lytic cycle of visna virus. *Nature New Biol.* 245:237–39

59. Scolnick, E. M. et al 1974. Primate and murine type-C viral nucleic acid association kinetics: analysis of model systems and natural tissues. *J. Virol.* 13: 363–69

60. Neiman, P. E. 1973. Measurement of endogeneous leukosis virus nucleotide sequences in the DNA of normal avian embryos by RNA-DNA hybridization. *Virology* 53:193–204

61. Gelb, L. D., Aaronson, S. A., Martin, M. A. 1971. Heterogeneity of murine leukemia virus in vitro DNA; detection of viral DNA in mammalian cells. *Science* 172:1353–55

62. Varmus, H. E. et al 1973. Transcription of mouse mammary tumor virus genes in tissues from high and low tumor incidence mouse strains. *J. Mol. Biol.* 79:663–79

63. Sawyer, R. C., Dahlberg, J. E. 1973. Small RNA's of Rous sarcoma virus: characterization by two-dimensional polyacrylamide gel electrophoresis and fingerprinting analysis. *J. Virol.* 12: 1226–37

64. Sawyer, R. C., Harada, F., Dahlberg, J. E. 1974. A virion-associated RNA primer for Rous sarcoma virus DNA synthesis: isolation from uninfected cells. *J. Virol.* 13:1302–11

65. Elder, A. T., Smith, A. E. 1973. Methionine transfer ribonucleic acids in avian myeloblastosis virus. *Proc. Nat. Acad. Sci. USA* 70:2823–26

66. Erikson, E., Erikson, R. L., Henry, B., Pace, N. R. 1973. Comparison of oligonucleotides produced by RNase T1 digestion of 7S RNA from avian and murine oncornaviruses and from uninfected cells. *Virology* 53:40–46

67. Faras, A. J., Garapin, A. C., Levinson, W. E., Bishop, J. M., Goodman, H. M. 1973. Characterization of the low-molecular-weight RNAs associated with the 70S RNA of Rous sarcoma virus. *J. Virol.* 12:334–42

68. Temin, H. M., Baltimore, D. 1972. RNA-directed DNA synthesis and RNA tumor viruses. *Advan. Virus Res.* 17:128–86

69. Ross, J., Scolnick, E. M., Todaro, G. J., Aaronson, S. A. 1971. Separation of murine cellular and murine leukaemia virus DNA polymerases. *Nature New Biol.* 231:163–69

70. Chang, L. M. S., Bollum, F. J. 1972. Antigenic relationships in mammalian DNA polymerase. *Science* 175: 1116–17

71. Kang, C. -Y., Temin, H. M. 1972. Endogenous RNA-directed DNA polymerase activity in uninfected chicken embryos. *Proc. Nat. Acad. Sci. USA* 69:5050–54

72. Kang, C. -Y., Temin, H. M. 1973. Early DNA-RNA complex from endogenous RNA-directed DNA polymerase activity of uninfected chicken embryos. *Nature New Biol.* 242:206–8

73. Sarngadharan, M. G., Sarin, P. S., Reitz, M. S., Gallo, R. C. 1972. Reverse transcriptase activity of human acute leukaemic cells: purification of the enzyme, response to AMV 70s RNA, and characterization of the DNA product. *Nature New Biol.* 240:67–72

74. Baxt, W., Helmann, R., Spiegelman, S. 1972. Human leukaemic cells contained reverse transcriptase associated with a high molecular weight virus-related RNA. *Nature New Biol.* 240:72–75

75. Todaro, G. J., Gallo, R. C. 1973. Immunological relationship of DNA polymerase from human acute leukaemic cells and primate and mouse leukaemia virus reverse transcriptases. *Nature* 244:206–8

76. Stephenson, J. R., Wilsnack, R. E. P., Aaronson, S. A. 1973. Radioimmunoassay for avian C-type virus group-specific antigen: detection in normal and virus-transformed cells. *J. Virol.* 11:893–99

77. Chen, J. H., Hanafusa, H. 1974. Detection of a protein of avian leukoviruses in uninfected chick cells by radioimmunoassay. *J. Virol.* 13:340–46

78. Fleissner, E., Tress, E. 1973. Chromatographic and electrophoretic analysis of viral proteins from hamster and chicken cells transformed by Rous sarcoma virus. *J. Virol.* 11:250–62

79. Payne, L. N., Chubb, R. C. 1968. Studies on the nature and genetic control of an antigen in normal chicken embryos which reacts in a COFAL test. *J. Gen. Virol.* 3:379–91

80. Hanafusa, T., Hayward, W. S., Hanafusa; H. 1973. Tumor virus genetic information in avian cells. In *Virus Research. 2nd ICN-UCLA Symp. Mol. Biol.*, ed. C. F. Fox, W. S. Robinson, 387–402. New York: Academic

81. Ruoslahti, E., Vaheri, A., Estola, T., Sandelin, K. 1973. Antibodies to avian gs antigen in chickens infected naturally and experimentally with RNA tumor viruses. *Int. J. Cancer* 11:595–603

82. Oldstone, M. B. A., Aoki, T., Dixon, F. J. 1972. The antibody response of mice to murine leukemia virus in spontaneous infection: absence of classical immunological tolerance. *Proc. Nat. Acad. Sci. USA* 69:134–38

83. Weiss, R. A., Payne, L. N. 1971. The heritable nature of the factor in chicken cells which acts as a helper virus for Rous sarcoma virus. *Virology* 45:508–15

84. Weiss, R. A., Friis, R. R., Katz, E., Vogt, P. K. 1971. Induction of avian tumor viruses in normal cells by physical and chemical carcinogens. *Virology* 46:920–38

85. Friis, R. R. 1973. Genetic interactions of avian RNA tumor viruses and their host cells. *Am. J. Clin. Pathol.* 60:25–30

86. Hanafusa, H., Aoki, T., Kawai, S., Miyamoto, T., Wilsnack, R. E. 1973. Presence of antigens common to avian tumor viral envelope antigen in normal chicken embryo cells. *Virology* 56:22–32

87. Nowinski, R. C., Peters, E. 1973. Cell-surface antigens associated with murine leukemia virus: definition of the G_l and G_t antigenic systems. *J. Virol.* 12:1104–17

87a. Strand, M. S., Lilly, F., August, J. T. 1974. Host control of endogenous leukemia virus gene expression: concentrations of viral proteins in high and/or leukemia mouse strains. *Proc. Nat. Acad. Sci. USA.* In press

88. Levy, J. A. 1973. Xentropic viruses: murine leukemia viruses associated with Swiss, NZB, and other mouse strains. *Science* 182:1151–53

89. Crittenden, L. B., Wendel, E. J., Motta, J. Z. 1973. Interaction of genes controlling resistance to RSV (RAV-O). *Virology* 52:373–84

90. Stephenson, J. R., Aaronson, S. A. 1973. Segregation of loci for C-type virus induction in strains of mice with high and low incidence of leukemia. *Science* 180:865–66

91. Rowe, W. P. 1973. Genetic factors in the natural history of murine leukemia virus infection. GHA Clowes Memorial Lecture. *Cancer Res.* 33:3061–68

92. Fenner, F. 1968. *The Biology of Animal Viruses.* New York: Academic

93. Shubak-Sharpe. 1969. The doublet pattern of the nucleic acid in relation to the origin of viruses. In *Handbook of Molecular Cytology,* ed. A. Lima-de-Faria, 67–87. Amsterdam: North-Holland

94. Hershey, A. D. Ed. 1971. *The Bacteriophage Lambda.* Cold Spring Harbor, NY: Cold Spring Harbor Lab.

95. Temin, H. M. 1970. Malignant transformation of cells by viruses. *Perspect. Biol. Med.* 14:11–26

96. Temin, H. M. 1971. The protovirus hypothesis. *J. Nat. Cancer Inst.* 46:III–VIII

97. Temin, H. M. 1972. The protovirus hypothesis and cancer. In *RNA Viruses and Host Genomes in Oncogenesis,* ed. P. Emmelot, P. Bentvelzen, 351–63. Amsterdam/London: North Holland

98. Huebner, R. J., Todaro, G. 1969. The Oncogene Hypothesis. *Proc. Nat. Acad. Sci. USA* 64:1087–91

99. Gilden, R. V. 1972. Immunochemical studies of the major group-specific antigens of mammalian C-type viruses. See Ref. 97, pp. 183–96

100. Gross, L. 1974. Facts and theories on viruses causing cancer and leukemia. *Proc. Nat. Acad. Sci. USA* 71:2013–17

PHEROMONES AS A MEANS OF ❖3070
GENETIC CONTROL OF BEHAVIOR

Jack E. Leonard and Lee Ehrman
Division of Natural Sciences, State University of New York, College at Purchase,
Purchase, New York 10577

Anita Pruzan
Hunter College of the City University of New York, New York, NY 10021

INTRODUCTION

In reading this review you are unlikely to get much information by sniffing the paper
and ink. This is not due to technological limitations, as it would be easy to encapsu-
late odors and print them with the words. This failure reflects human genetic and
environmental endowments; with few exceptions, we use our chemical senses
crudely. This crudeness in using our chemical senses of taste and smell to map our
environment sets us apart even from many of our closest biological relations. Lack
of awareness of the richness of the olfactory environment probably accounts in part
for the scientific neglect of the olfactory cues which animals employ to influence one
another's behavior. With improvements in chemical analytical tools and an increasing interest in the biology of behavior, a number of research groups have initiated
studies on the behavioral effects of chemical signs. In recent reviews of the topic (1,
2) Regnier has proposed the word *semiochemicals* (from the Greek, *semeion,* a mark
or signal) to describe such behaviorally active chemical signs.

By far the most intensively studied class of semiochemicals to date are the
pheromones: compounds transmitted by one organism which affect the behavior of
other organisms of the same species. In particular the pheromones of insects have
been investigated by several research groups and are the subject of thorough reviews
(2–6). Several significant factors contribute to this concentration on insects. They
are small and easily cultivated in large numbers; they are behaviorally simple
compared with vertebrates, making changes in responses due to pheromones easier
to monitor; perhaps most significant, pheromones may provide a species-specific
tool for monitoring and controlling insect pest populations (4, 8–10). For the purpose of this review, there is one additional factor of great importance: several species
of insects are well characterized genetically. Although the pheromones of higher

179

organisms have also come under scrutiny, we confine our attention to insects. We do not attempt to provide comprehensive literature coverage of insect pheromones, since recent commendable reviews already exist which do that (2, 4, 5, 7). Instead we attempt to provide two kinds of information: 1. a schematic view of the processes of pheromonal production, release, reception, and effects, and 2. specific information about the few existing studies in the genetics of pheromone-mediated behavior. At present there are few such studies, especially if only genetically well-characterized organisms are considered. For this reason we also include several investigations of the taxonomic or evolutionary significance of pheromones. We include areas where genetic studies might be fruitfully conducted. We hope to demonstrate thereby the potential power of pheromonal studies in expanding our now limited knowledge of the genetics of behavior.

Karlson (11, 12) originally proposed the term pheromone (from the Greek, *pherein,* to carry, plus *horman,* to excite) to replace the earlier term "ectohormone" and in this way to emphasize the action of these substances *between* rather than solely within organisms. To structure this review we use the temporal sequence of this organismal interaction, rather than treat pheromones primarily from the viewpoint of the taxonomy of the producing organism. Our sequence is 1. pheromone production, 2. pheromone release, 3. pheromone reception, and 4. pheromone-mediated behavioral responses. Finally, we discuss factors complicating the simple model developed in our general discussion, including environmental effects and multicomponent pheromone systems.

PHEROMONE PRODUCTION

The processes involved in the production of chemical compounds that make up a pheromone system are poorly understood. Production is apparently carried out in specialized glands which are then also active in the release of the pheromone; more is said about them in the following section.

Little direct evidence has accumulated about the biochemical pathways by which pheromones are synthesized. Tabulations of structures, such as that found in the review by Law & Regnier (2), clearly show that a large fraction of identified sex pheromones are related to natural fatty acids. That is, they are esters, alcohols, or hydrocarbons related to the saturated or unsaturated carboxylic acids of straight chain hydrocarbons. The predominance of this type of compound is probably an indication that the origin of the sex pheromone synthesis pathway was a specialization of existing pathways for fatty acid anabolism. Elucidation of the biochemical genetics of pheromone systems requires more discriminating examination of the actual pathways in specific organisms. Although the techniques and organisms with which such studies could be conducted are available, no reports of such studies have appeared.

Work has been done, however, on the rate of pheromone production by various organisms. Shorey & Gaston have investigated the production rate and evaporation rate of the pheromone of female *Trichoplusia ni,* a noctuid moth species. During the sexually active period each female contains approximately 1 μg (13). In *T. ni,* no pheromone can be found in female pupae, but production begins immediately upon emergence and builds to a maximum steady-state pheromone level within a day. In other noctuid species pupal pheromone production has been found (14). By manually everting the pheromone gland of an intact female, Sower, Shorey & Gaston showed that the passive release rate (evaporation) from the gland is independent of diurnal rhythm but strongly dependent on wind velocity (15). Both whole gland content and evaporation rate were dependent on female age and on population factors which could not be explained. This variability between different laboratory populations of the same organism has important methodological implications for laboratory and field assays, since it can lead to wholly spurious differences in attractancy unless controlled for or assayed independently (for example, by chemical means).

Studies on the control of sex pheromone production in cockroaches have been conducted. The mating behavior of the cockroach involves a volatile pheromone produced by the female which serves to attract the male. As the couple approach each other, the male releases a pheromone which in turn elicits feeding and mounting behavior in the female. Barth showed that the female sex pheromone production is regulated by the corporea allata in the species *Byrsotria fumigata* (17). Allatectomy led to a total loss of pheromone production, while reimplantation led to its restoration. Hartman & Suda showed that in the species *Nauphoeta cinerea,* male pheromone is not under the regulation of the corpora allata, since neither reproductive maturity nor rate of mating are altered by allatectomy (16).

Additional studies of the importance of hormonal vs direct neural control of the release and production of pheromones are clearly in order. The paucity of information on the regulation and biochemistry of pheromone production, so evident above, will hopefully be overcome in the next few years.

PHEROMONE RELEASE

Pheromone release is touched on briefly in the last section. In this section, we discuss the anatomy of the pheromone glands and environmental factors influencing release behavior. The experimental situation in this area is similar to that of pheromone production: a paucity of studies, primarily on the Lepidoptera, but a tantalizing variety of results.

Pheromone release is controlled by exocrine glands whose structure and function appear to be highly specialized. The sexual scent organs of Lepidoptera females are typically located on the abdomen near the genital opening (18). They originate from an intersegmental fold and are usually situated between the eighth and ninth abdominal segments. It would be reasonable, therefore, to inquire whether there are differences between families and species of Lepidoptera in the structure of these female sex pheromone glands and, if such exist, to relate them to evolutionary processes. While comparing the morphology of the female sex pheromone glands of eight species of the family Noctuidae, Jefferson, Shorey & Rubin (19) found differences among but not within subfamily lines. The glands of *Autographa californica, Pseudoplusia includens,* and *Rachiplusia ou,* all belonging to the subfamily

Plusiinae, have a dorsal-eversible sac or fold and are similar histologically. The glandular epithelium consists of tall columnar cells containing many vacuoles; these cells rest on a relatively thick (1–1.5 μm) basement membrane. The segments are devoid of scales and hairs. Several reports present the structure of the pheromone-releasing gland of *Trichoplusia ni,* also belonging to the same subfamily. Other than some small differences such as the thickness of the basement membrane (2.5–3 μm) and that of the other apical membrane (3.6 μm for *T. ni* vs 0.5 μm for the other species), the shape of the cells and the location of the pheromone gland justify the inclusion of *T. ni* within the subfamily Plusiinae (21, 22).

The presence of the pheromone gland in *T. ni* can be recognized 5 days before emergence. In the succeeding days the size of the individual cells and the size and number of intracellular vacuoles undergo changes correlated with the secretory activity of these cells, initially in the deposition of the cuticle and later in pheromone secretion. The maximum in size and number of vacuoles is reached 1½ days post-emergence; this time corresponds to the time when pheromone levels reach their steady-state value. Histological studies indicate that sex pheromone is continuously secreted by *T. ni* females from 1–7 days of age (20). As noted in the previous section, similar production patterns are found in other noctuids, but there are clear inter-specific differences (14). It would be useful to know whether such differences corre-late with observable differences in mating rhythm.

In the subfamily Amphipyrinae, on the other hand, a sac-shaped gland is situated ventrally in the intersegmental membrane between the eighth and ninth segments (19, 23). These segments are densely covered with long hairs. The glandular cells are tall, uneven, and columnar, and rest on a basement membrane approximately 1 μm thick. Individual cells appear to contain fewer vacuoles than do those of Plusiinae.

Further morphological and histological differences within this order were found within the subfamily Heliothinae (19). The sex pheromone gland forms a complete ring situated between the eighth and the ninth abdominal segments. In the species examined the ring is more fully developed ventrally and is devoid of scales and hairs, but minute spines do cover membranous areas of the eighth and ninth segments. The cells within the glands of Heliothinae appear cuboidal to slightly columnar, rest on a basement membrane approximately 0.5 μm thick, and contain fewer vacuoles than do the cells of Plusiinae.

Studies similar to the above were performed in the lepidopterous family Tor-tricidae (24), again demonstrating subfamilial differences in the structure of the pheromone sex glands. It was found that, in general, the pheromone-producing glands in species of Tortricidae possess many large folds in the invaginated area, whereas the glands found in species of *Olethreutinae* appear as a thick compact layer following the general outline of the invagination. Interestingly, strawberry leaf roller females *Ancylis comptana fragariae,* have either an atrophied or an underdeveloped gland. Since their habitat is low-growing strawberry plants, Roelofs & Feng (24) suspect that their apparently atrophied pheromone gland evolved through natural selection in response to need for short-range attraction between sexes and thus a greater reliance on other distance cues.

Since our intent is not so much to present a comprehensive review of the literature, but rather to point out further questions of potential interest, the reader is directed to the studies of White et al (25, 26) and the comprehensive review by Jacobson (4) for discussions of anatomy and physiology of the sex pheromone glands. With such information at hand one might inquire, for example, how the observed differences in subfamilies evolved. What is the relationship between the habitats occupied by members of different subfamilies and the structure of the individual pheromone glands? What is the relationship between mating patterns and the location of the gland?

Blum has discussed the pheromone glands of social insects (6). Within the family Formicidae a variety of trail pheromone glands and ducts are found. These structures are clearly different in anatomical origin, involving such diverse structures as the hindgut, the stinger, and leg tarsi. Apparently there is a high selective value to trail laying in social insects, resulting in a polyphyletic origin for this ability.

One striking feature revealed by studies investigating pheromone release is the temporal coordination between conspecific females and males in terms of sexual readiness and maximal sex pheromone release and use (2). In noctuid moths, for example, there is evidence that time of mating is controlled by the female through her release of sex pheromones and that this correlates with circadian rhythms of male responsiveness (27–29). As a general rule, the two major environmental influences on circadian rhythms are photoperiods and the temperature changes which normally accompany them (30, 31), both of which may result in the setting of biological clocks. These and other factors may also have a phase-shifting effect, that is, changing the time of expression of the circadian rhythm.

Factors such as length of light–dark period, temperature and wind velocity, and the effect of each on the onset and frequency of episodes of pheromone release were investigated by Shorey's group in *T. ni* females (14, 15, 32–35). When exposed to a 12:12 (light:dark) photoperiod during the first 3 days post-emergence, females persist in their release rhythm for up to 60 hr of continuous darkness. Within a 12:12 photoperiod, maximum release occurs 8 to 11 hr into the dark period and decreases to zero before the light period, at constant ambient temperature of 24°C. When this temperature regime was changed to 12 hr at 30°C in the light and 12 hr at 18°C in the dark (roughly approximating natural field conditions), there was a marked phase shift resulting in earlier onset of pheromone release. A mechanism such as this which shifts mating activity in response to temperature variation presumably has adaptive significance to these insects. It could be that, as the season progresses, cooler night temperatures force the insects to complete all mating activity earlier in the evening. Behavioral rhythms adapted to this situation would seem to be of survival value to the population.

Another factor influencing sex pheromone release was found to be air velocity. Females of *T. ni* spent more time releasing pheromone when exposed to velocities ranging from 0.3–1.0 m/sec than they did at either higher or lower velocities. In addition, these females appear to be able to adapt the length of each episode of pheromone release to the prevailing wind velocity, increasing the time spent releasing with decreasing wind velocity. This finding is interesting in light of the fact that the amount of material evaporating from the exposed gland is directly proportional to the wind velocity (15). Thus, the behavioral mechanism may insure that the total quantity released is independent of the wind speed.

Sex pheromone release time can also be affected by ambient temperature during the development of silkmoth *Antherae pernyi* females (36). For pupae exposed to 25°C throughout their development, calling behavior (episodes of active release of pheromone by the female) began about 6 hr after lights off, whereas for female pupae reared at 12°C, the onset of calling behavior was advanced to about 2 hr after lights off. Corresponding shifts in flight activity were noted for *A. pernyi* males, again illustrating the temporal coordination between the sexes.

PHEROMONE RECEPTION AND RESPONSE

One striking feature of insect pheromones is the high degree of specificity and sensitivity which characterizes their reception and response by their target organism. Specificity can be related to production, recognition, or both. To date there is very little experimental evidence supporting the view that specific production is important. Indeed, in the genus *Dendroctonous,* Renwick found that the aggregation pheromones of these bark beetles are characterized by a considerable overlap of compounds actually produced by the various species, coupled with a marked specificity of action in each species (37). As we discuss in the next section, multicomponent pheromone systems may be the rule rather than the exception.

Recognition specificity can be produced either at the receptor level or at the behavioral level, and examples of both are known. Chemoreceptors in insects are primarily concentrated on the surfaces of the legs and antennae, and species may be observed which utilize each of these receptor systems for pheromone recognition. Males of the mosquito subgenus *Stegomyia* recognize the female pheromone via the tarsi (38), whereas extensive studies of the Lepidoptera have shown that these organisms rely primarily on their antennae (39–46). Detailed information about the receptor mechanism has been obtained by the indirect method of anatomical examination (46, 47), by behavioral testing on altered organisms (48), and by determining the sensory response to various substances utilizing recording electroantennograms (39–47, 49). The electroantennogram allows the investigator to determine the effect of even relatively minor changes in structure, so it is a powerful tool for elucidating specificity (see, for example, 41). To date this technique has not been exploited as a tool in genetic studies.

A recent study has cast some light on the molecular mechanism of pheromone reception (50). The alarm pheromone of the leaf-cutting ant *Atta texana* contains S–(+)–4–methyl–3–heptatone. This optically active compound and its enantiomer have been synthesized and its activity tested. The enantiomeric compound possesses at most 1% of the activity of the pheromone itself. The differentiation between these two requires a chiral receptor site, possibly proteinaceous, of some complexity.

The specificity of pheromones generally refers to their action at the behavioral level. A pheromone is said to be specific because in a particular species it produces a defined behavioral response which is not elicited (or is more weakly elicited) by related compounds. The overtness of the behavior forms the basis for the sensitive and specific assay needed to work with material extracted from small organisms. Thus, the most common form of classifying pheromones has been by their evoked behavior, for example, sex pheromones, aggregation pheromones, trail pheromones, alarm pheromones.

Because of the relative simplicity of the assay involved and the possible significance for controlling populations of economically important pests, most pheromone work has centered around sex pheromones (see Jacobson, 4). In addition, a large part of this work has concerned itself with the Lepidoptera. Though there are significant exceptions, we treat the major ideas which have evolved from lepidopteran studies and then present data on other orders where needed to augment this foundation.

In a striking study, Shorey and his co-workers were able to show that there is a separation between reception and response sensitivity. In *T. ni* they showed that age, light, and time of day all affected the *behavior* of male moths in the presence of the female pheromone, yet none of these environmental factors influenced the *sensitivity of reception* as measured by recording electroantennograms (45). The degree of response sensitivity can be quite high: in this same species, the behavioral response could be evoked even with extracts equivalent to 10^{-6} female moths (51). Traynier has also succeeded in demonstrating the effect of age, photoperiod, and light intensity on the pheromonal response of *Anagasta (Ephestia) kuhniella* (52). A variety of studies have shown that the responsiveness of male moths to natural or synthetic pheromones can be modulated by seasonal and diurnal rhythms (53–55). It has been found in the cabbage looper, *T. ni,* that the synthetic chemosterilant tepa not only produces abnormal copulatory behavior and shortens the male lifespan, but also that tepa-fed males are less responsive to the female sex attractant than are controls (56). Tepa-sprayed males, however, are attracted to the females equally with the controls, and the effect of tepa on these noctuid moths is dependent not only on method of application and dosage but also on the age of the males at the time of treatment. Shorey's group has demonstrated that pheromonal responsiveness (together with mating ability) is the final stage in the sexual maturation of noctuid moths (57).

The modulation of response may have important evolutionary consequences, as it is clear that a number of closely related species may be sympatric and share the same sex pheromone (or the same major component). Field testing has revealed that males of more than one species can be attracted to pheromones obtained from *Homomelina nigricans* (58), *T. ni* (59), *Grapholitha molesta* (60), and *Choristoneura fumiferana* (61). Detailed study of the attraction of *T. ni* and *Autographa californica* to the same pheromone shows that interspecific mating is attempted but cannot be completed, indicating that other isolating mechanisms are operative (62). Roelofs has reviewed the literature of lepidopteran sex attractants and discussed the relationship of identified pheromones to other taxonomic information available (63).

Instances of the opposite case have also been found: that is, species whose only apparent difference lies in their pheromone responses. In two closely related sympatric gelechiid species Roelofs & Comeau found that one species was attracted exclusively to *cis*-9-tetradecenyl acetate and the other attracted exclusively to the *trans* isomer (64). The evolutionary mechanism by which this could arise is not clear. However, as is seen in the next section, such isolation could derive from an initially multicomponent pheromone system. Roelofs & Comeau propose the wider use of such pheromonal distinctions for taxonomic purposes (64). But, in later work from Roelofs' laboratory, results were obtained which cast doubt on the reliability of pheromonal distinctions alone. In field tests of *trans*-11 tetradecenyl acetate with the European cornborer, Roelofs' group discovered that males in flights early in the summer (late June) were attracted to the material, but moths flying in late summer (August) were not (65).

An intraspecific difference was also found in the territorial-marking scent of the palearctic bumblebee *Bombus lucorum,* in which one form utilizes ethyl dodecanoate and the other uses ethyl *cis*-9-tetradecenoate. In this instance there is also morphological differentiation between the two forms, as well as indications that the pheromone may be multicomponent (66). Intraspecific variability in the pheromone, coupled with multicomponent systems, may prove an obstacle to the long-term success of the much publicized technique of controlling pest populations by permeation of the air with a synthetic pheromone or its inhibitors leading to disorientation of the males present (67–69). To date no report has appeared in which this theoretical difficulty has proved a practical obstacle.

The social insects provide several examples of intraspecific specialization of response (5, 6). Topoff and his co-workers found that in the army ant genus *Eciton,* workers of different sizes show different pheromone responsiveness (70). Workers of all sizes respond equally to the trail pheromones, yet only the major workers respond to the alarm pheromone.

Lanier analyzed the genetics of pheromone-mediated sexual isolation directly in the spruce-infesting bark beetle, *Ips amiskwiensis,* and in two species with which it is sympatric over portions of its range, *I. borealis* and *I. pilifrons* (71). Although hybrids between *I. amiskwiensis* and the other two species can be obtained in the laboratory, such hybrids have never been found in nature. The components of the pheromone systems of these beetles have not been isolated, but tests on the separate species show significantly more intraspecific than interspecific attraction. Furthermore, hybrids between the three species show intermediate responses and are recognized as different by each nonhybrid parental species. The discrimination involved is greater at actual mate selection, copulation, than it is for premating attraction of the females to the males.

Discrimination leading to mating preferences is usually considered an essential function of sex pheromone systems. However, recent results in our laboratory (unpublished data) and studies by Averhoff & Richardson (72) show that, at least in some Drosophila species, pheromonal information can serve as the basis for negative assortative mating—outbreeding—in intraspecific crosses. Averhoff & Richardson found that in *D. melanogaster* the behavioral basis for the preference is apparently a reduced level of male courtship behavior in the exclusive presence of females from the same inbred strain as the males.

The behavior exhibited by *D. pseudoobscura* in the studies in our laboratory is more complex. Ehrman, and others, have shown in earlier studies that females of several *Drosophila* species exhibit a mate preference which depends on the relative frequencies of the various strains of male present (73–75). This frequency-dependent mating behavior has been shown to be mediated by olfactory cues in whole organism studies (75). Working with the Chiricahua (CH) and Arrowhead (AR) strains, differing by a third chromosome inversion, success has been achieved in isolating the behaviorally active substance. In both CH and AR males there is a chemically extractable volatile material which can lead females of either strain to exhibit frequency-dependent mate selection, even though the two strains of males are present in equal abundance. Although the active fraction has been identified in neither strain, these two closely related strains obviously are identified by chemically very distinct substances. In the AR strain the active fraction is acetone soluble (76) and hexane insoluble. The opposite solubility is observed in the CH male extracts (see note added in proof at end). The females apparently respond to the extracts just as they would to the presence of the males, because the mating preference can be varied continuously by using successively higher concentrations of the extract.

While evidence is lacking concerning the relevance of these behaviors to Drosophila in the wild, these studies give clear indication that the recognition information provided by pheromones can serve as the basis for very complex evolutionary mechanisms by favoring outbreeding or the maintenance of genetic polymorphisms.

COMPLICATIONS TO THE SIMPLE ONE-PHEROMONE/ONE-RESPONSE MODEL

One of the most rapidly developing segments of pheromone research has been the study of pheromones composed of several different compounds. When the authoritative review of Law & Regnier was published in 1971, no discussion of the topic was included, although they did tabulate a few such compounds among their lists of identified pheromonal compounds (2). Blum, in his 1970 discussion of social insect pheromones, considered the role of multicomponent pheromones and mentioned their value as a source of species specificity (5). However, it was known that trail and alarm pheromones possess lower sensitivity and specificity (2, 5), so the significance of these findings for other pheromone systems was not apparent.

Improvement in analytical techniques has led to the discovery that even previously studied pheromones are multicomponent (77, 78). In some cases, an additional component of the pheromone appears to have an augmentative effect (77), perhaps providing recognition information at close range, though this has not been substantiated. In other cases, the additional components are clearly synergistic, providing more attractivity in combination than individually (78, 79), although either or both components evoke the response when tested alone. Finally, the pheromone may be a truly interrelated multi-component system, with no individual component behaviorally active (80–83). Of course, some multicomponent systems exist in which the action of individual components is not known (84, 85). All of the studies cited (77–85) are of lepidopteran sex attractants. This specialization is an indication of the great volume of research activity in this area, rather than an indication that the phenomenon does not occur in other orders. The existence of the interrelated type of multicomponent systems provides a clear methodological caveat: purification can lead to loss of the assay.

Although the exact nature of such various multicomponent systems is only now being elucidated, they were clearly foreshadowed in the development of synergists and inhibitors for use with synthetic pheromones. The usefulness of these synthetic additives has been confirmed both by field tests and by electroantennograms (86–93).

Even before distinct identification of the components of any particular multicomponent pheromone system, Ganyard & Brady had proposed that such systems could account for the sexual isolation of related species of Lepidoptera without requiring an immense number of different pheromonal substances (94). Minks et al (95) found what may be evidence of this with their study of two tortricid species: *Adoxophyes orana*, the summer fruit moth, and *Clepsis spectrana*, the leaf roller. These two moths are sympatric in the wild, and their diurnal activity is almost entirely overlapping. The *A. orana* pheromone is apparently a 9:1 mixture of *cis*-9-tetradecenyl acetate (*cis*-9-TDA) and *cis*-11-TDA. Preliminary investigations of the *C. spectrana* pheromone suggests that the same two compounds are present, with the 11-unsaturated ester predominating. The electroantennogram responses to the two compounds are very similar in the two species, though there is clearly a selective response toward the isomer predominating in the respective pheromone of each. The field response to various mixtures of the isomers clearly distinguishes the two species. Maximum response in *A. orana* is exhibited at a 90:10 (9–:11-*cis*-TDA) ratio of homologues, while the *C. spectrana* response peaks at a 25:75 ratio.

Such concentration isolation is not limited to interspecific interactions. Intraspecific isolation also occurs, but in existing studies it is related to geographic isolation; whether it is important in sympatric populations is yet to be established. In 1972 Lanier and his co-workers showed that three populations of the ubiquitous pine engraver beetle (California, Idaho, and New York) can be distinguished on the basis of their response to each other's pheromones (96). The distinction between the eastern and western populations is much greater than that between the two western populations, and, interestingly, the two populations (east/west) can also be distinguished by a predator *Enoclerus lecontei* and a parasitoid *Tomicobia tibialis*. However, it was not established that the differences itemized would actually foster sexual isolation between the populations. Also, as the tests were carried out using the females to attract the males, it was not established that this effect is related to variable ratios of several chemical components. However, one important 1973 study has definitely shown such ratio specificity (82). Three moth species use *trans*-11-TDA with admixtures of *cis*-11-TDA as their sex pheromone: *Ostrinia nubilalis*, the European corn borer; *Argyrotaenia velutinana*, the redbanded leaf roller; and *Ostrinia obumbratalis*, the smartweed borer (82). The attraction of the smartweed borer was a maximum at a 50:50 *trans-cis* ratio, while the leaf roller showed maximum preference for a 7:93 ratio. For the European corn borer, though, its Iowa population preferred a 3:97 mixture, while its New York population preferred almost pure *trans*. Thus, both inter- and intraspecific differentiation occurs in the same pheromone system. As mentioned in the previous section, such extreme values for a multicomponent pheromone may serve as a preliminary stage in the behavioral isolation of populations that may lead to speciation.

A related but distinguishable complication to the assumption that there is a simple connection between a molecule and an associated behavioral response lies in the existence of threshold effects. Although we discussed the very high sensitivity exhibited towards pheromones, there is always a concentration threshold below which the behavioral response does not occur. Indeed, Bartell & Shorey found that in the light-brown apple moth *Epiphyas postvittana* there is a sequence of behaviors related to the mating ritual, and each step has a characteristic pheromonal threshold (97). Schwink showed a similarly stepwise response to female sex attractant in the behavior of male *Bombyx mori* (98). Wilson has shown that the alarm pheromone of *Pogonomyrmex badius* functions as an attractant at low concentrations but as an alarm substance at higher ones (99). Blum has discussed the importance of threshold concentrations as one possible basis for pheromone differentiation in social insects (5).

Using the chemically well-characterized pheromone, *cis*-7-dodecenyl acetate, Kaae, Shorey & Gaston showed that the threshold concentration was a significant factor in the behavioral isolation of Lepidoptera in the field (100). *T. ni,* the cabbage looper, and *Autographa californica,* the alfalfa looper, frequent the same host plants and overlap in both their geographic distribution and seasonal cycles. Their mating rhythms are distinguishable but appear to be insufficient to isolate them. By using an evaporator to control pheromone release, it was found that at low concentrations *A. californica* males were attracted almost exclusively. As the release rate was increased, however, fewer *A. californica* males appeared and the number of *T. ni* males attracted increased. As expected, this type of species isolation is not complete, and the incompleteness is observable in the field. When the traps are baited with females of one species, males of both species are attracted, although conspecific males predominate. The generality of threshold isolation has yet to be determined. Furthermore, since the isomeric purity of the synthetic pheromone used in the above study was not reported, there is a possibility that the response observed is actually a threshold response to a trace of an impurity, such as the *trans* isomer, rather than to the reported compound.

THE CHEMICAL AND PHYSICAL ENVIRONMENT

The chemical and physical environment separates the transmitting organism from the target organism, and it is within this environment that the evoked behavior will finally be expressed. In the laboratory this environment can be carefully controlled —if indeed one knows what to control—but in the field it must simply be measured where possible, and accepted, measurable or not.

The most thoroughly studied aspect of the environmental influence on phero-mones is the atmosphere's ability to transport pheromone molecules. The pivotal paper in this regard was coauthored by Bossert & Wilson in 1963 (101). Assuming a molecular diffusion mechanism for transport, they showed that the essential variables are the emission rate Q and the threshold concentration K. The Q/K ratio provides information about the transmission radius for a given molecule in still air. However, it is not clear that a simple concentration gradient is sufficient to allow the target organism to orient accurately toward the transmitting organism. For behaviors such as alarm pheromones, such orientation is simply not necessary, and for trail pheromones laid on the ground, the Bossert-Wilson analysis may not provide an accurate picture of the gradient involved. However, for airborne trail pheromones and sex attractants, some orientation capability is apparently essential. A current of pheromone on the wind could provide orientational information if the organism is able to orient toward air currents (anemotaxis). Although anemotaxis has been demonstrated in *Drosophila melanogaster* (102) and the yellow fever mosquito, *Aedes aegypti* (103), it is believed to be vision-dependent and, at any rate, its generality has not been established. In the lepidopteran species *Pectinophora gossypiella*, Farkas & Shorey found that all that is necessary for orientation is a structured plume of pheromonal substance in air, even in the absence of an air current (104). Additional behavioral information is needed to decide which mechanisms allow insects to orient accurately in the presence of pheromones. Furthermore, a mechanism for the transmission of multicomponent pheromones without loss of information is needed, since in such systems the diffusion rate of the individual components are not necessarily identical.

Studies on the emission rate and its effect on the target organism have been important in developing pest control systems (105, 106). The relevance of this rate to threshold behavior is illuminated by evidence that in *T. ni* a plot of release rate vs the attractancy of the synthetic pheromone goes through a maximum (107).

As we have mentioned above (45–56), pheromonal responses are partially under the influence of light intensity, diurnal rhythm, and even seasonal variation. In *T. ni,* Shorey found that light intensity affected both the timing and occurrence of mating, and that the pheromone response is also susceptible to a temperature effect (108). Environmental factors may serve to enhance the specificity of pheromones in the perfection of behavioral-isolating mechanisms.

The environment is more than a passive mediator of these processes, however, as species other than the transmitter and target are involved. This can lead to unusual chemical interactions. The classic case is the silkmoth *Antherea polyphemus:* release of the female sex pheromone is stimulated by *trans*-2-hexenal from the leaves of host hardwood plants (109). Host-plant odor produced measurable changes in the behavior of gravid females of the cabbage root fly *Erioischeria brassicae* (110). Furthermore, mimics of the sex pheromone of the American cockroach can be isolated from spruce and fir needles (111); the field significance of this result has not been determined.

For a fledgling field of research, pheromone studies have yielded a wealth of new information about animal communication, behavior, and evolutionary mechanisms. As our understanding of individual pheromone systems is augmented, new information about the interaction between populations of organisms and their environment may bring new insight into the nature of evolution.

ACKNOWLEDGMENTS

The compilation of this chapter was generously supported by an American Association of University Women Grant to L. E., and by a City University of New York dissertation year fellowship to A. P. The pheromonal studies at SUNY, Purchase were supported by USPHS grant GM18907. L. E. is the recipient of a USPHS research career award 5K03 #H09033-10. We would like to thank Professor H. H. Shorey and his research group for critically reading the manuscript. They made several helpful changes. Any remaining errors, however, are our responsibility.

Note added in proof: Further studies on the frequency-dependent mating behavior in *Drosophila pseudoobscura* (see page 187 above) have been carried out. The behavior can be duplicated with whole body extracts as the only recognition cues. Furthermore, this extract-induced behavior can be used as the basis for a quantitative bioassay of the CH pheromone (112). Chromatographic separation and the use of synthetic pheromonal materials indicate that a synergistic multicomponent pheromone system is involved (unpublished data).

Literature Cited

1. Regnier, F. E. 1971. *Biol. Reprod.* 4:309
2. Law, J. H., Regnier, F. E. 1971. *Ann. Rev. Biochem.* 40:533
3. Regnier, F. E., Law, J. H. 1968. *J. Lipid Res.* 9:541
4. Jacobson, M. 1972. *Insect Sex Pheromones.* New York:Academic
5. Blum, M. S. 1970. In *Chemicals Controlling Insect Behavior*, ed. M. Beroza. New York:Academic
6. Blum, M. S. 1972. *Am. Zool.* 12:553
7. Dean, R. W., Roelofs, W. L. 1970. *J. Econ. Entomol.* 63:684
8. Roelofs, W. L., Gloss, E. H., Tette, J., Comeau, A. 1970. *J. Econ. Entomol.* 63:1162
9. Shorey, H. H. 1972. *Proc. N. Cent. Br. Entomol. Soc. Am.* 27:30
10. Marx, J. 1973. *Science* 181:736
11. Karlson, P., Luscher, M. 1959. *Nature* 183:55
12. Karlson, P., Butenandt, A. 1959. *Ann. Rev. Entomol.* 4:39
13. Shorey, H. H., Gaston, L. K. 1965. *Ann. Entomol. Soc. Am.* 58:604
14. Shorey, H. H., McFarland, S. U., Gaston, L. K. 1968. *Ann. Entomol. Soc. Am.* 61:372
15. Sower, L. L., Shorey, H. H., Gaston, L. K. 1972. *Ann. Entomol. Soc. Am.* 65:954
16. Hartman, H. B., Suda, M. 1973. *J. Insect Physiol.* 19:1417
17. Barth, R. H. 1962. *Gen. Comp. Endocrinol.* 2:53
18. Götz, B. 1951. *Experientia* 7:406
19. Jefferson, R. N., Shorey, H. H., Rubin, R. E. 1968. *Ann. Entomol. Soc. Am.* 61:861
20. Jefferson, R. N., Shorey, H. H., Gaston, L. K. 1966. *Ann. Entomol. Soc. Am.* 59:1166
21. Miller, T., Jefferson, R. N., Thomson, W. W. 1967. *Ann. Entomol. Soc. Am.* 60:707
22. Jefferson, R. N., Rubin, R. E. 1973. *Ann. Entomol. Soc. Am.* 66:277
23. Jefferson, R. N., Rubin, R. E. 1970. *Ann. Entomol. Soc. Am.* 63:431
24. Roelofs, W. L., Feng, K. C. 1968. *Ann. Entomol. Soc. Am.* 61:312
25. White, M. R., Amborski, R. L., Hammond, A. M., Amborski, G. F. 1972. *In Vitro* 8:30
26. White, M. R., Amborski, R. L., Hammond, A. M., Amborski, G. F. 1973. *J. Insect Physiol.* 19:1933
27. Callahan, P. S. 1958. *Ann. Entomol. Soc. Am.* 51:271
28. Shorey, H. H. 1964. *Ann. Entomol. Soc. Am.* 57:371
29. Shorey, H. H., Gaston, L. K. 1964. *Ann. Entomol. Soc. Am.* 57:775
30. Beck, S. D. 1968. *Insect Photoperiodism.* New York:Academic
31. Danilevsky, A. S., Gouphin, N. I., Tyschenko, V. P. 1970. *Ann. Rev. Entomol.* 15:201
32. Sower, L. L., Gaston, L. K., Shorey, H. H. 1971. *Ann. Entomol. Soc. Am.* 64:1448

33. Sower, L. L., Shorey, H. H., Gaston, L. K. 1971. *Ann. Entomol. Soc. Am.* 64:1488
34. Sower, L. L., Shorey, H. H., Gaston, L. K. 1970. *Ann. Entomol. Soc. Am.* 63:1090
35. Kaae, R. S., Shorey, H. H. 1972. *Ann. Entomol. Soc. Am.* 65:436
36. Truman, J. W. 1973. *Science* 182:727
37. Renwick, J. A. A. 1972. *Abstr. 14th Int. Congr. Entomol.* 42
38. Riddiford, L. M. 1970. *J. Insect Physiol.* 16:653
39. Roelofs, W. L., Comeau, A., Hill, A., Milicevic, G. 1971. *Science* 174:297
40. Roelofs, W. L., Carde, R., Benz, G., von Salis, G. 1971. *Experientia* 27:1438
41. Roelofs, W. L., Comeau, A. 1971. *J. Insect Physiol.* 17:1969
42. Roelofs, W. L., Tette, J. P., Taschenberg, E. F., Comeau, A. 1971. *J. Insect Physiol.* 17:2235
43. Roelofs, W. L. 1971. *NY Food Life Sci. Quart.* 4:7
44. Payne, T. L., Shorey, H. H., Gaston, L. K. 1970. *J. Insect Physiol.* 16:1043
45. Payne, T. L., Shorey, H. H., Gaston, L. K. 1973. *Ann. Entomol. Soc.* 66:703
46. Roelofs, W. L. 1972. *Abstr. 14th Int. Congr. Entomol.* 43
47. Priesner, E. 1972. *Abstr. 14th Int. Congr. Entomol.* 43
48. Bell, W. J., Burk, T., Sams, G. R. 1973. *Behav. Biol.* 9:251
49. Jefferson, R. N., Rubin, R. E., McFarland, S. U., Shorey, H. H. 1970. *Ann. Entomol. Soc. Am.* 63:1227
50. Riley, R. G., Silverstein, R. M., Moser, J. C. 1974. *Science* 183:760
51. Shorey, H. H., Gaston, L. K., Fukuto, T. R. 1964. *J. Econ. Entomol.* 57:252
52. Traynier, R. M. M. 1970. *Can. Entomol.* 102:534
53. Shorey, H. H., Gaston, L. K. 1965. *Ann. Entomol. Soc. Am.* 58:597
54. Saario, C. A., Shorey, H. H., Gaston, L. K. 1970. *Ann. Entomol. Soc. Am.* 63:667
55. Sharma, R. K., Rice, R. E., Reynolds, H. T., Shorey, H. H. 1971. *Ann. Entomol. Soc. Am.* 64:102
56. Henneberry, T. J., Shorey, H. H., Kishaba, A. N. 1966. *J. Econ. Entomol.* 59:573
57. Shorey, H. H., Morin, K. L., Gaston, L. K. 1968. *Ann. Entomol. Soc. Am.* 61:857
58. Roelofs, W. L., Carde, R. T. 1971. *Science* 171:684
59. Kaae, R. S., Shorey, H. H., McFarland, S. U., Gaston, L. K. 1973. *Ann. Entomol. Soc. Am.* 66:444
60. Roelofs, W. L., Comeau, A., Selle, R. 1969. *Nature* 224:723
61. Weatherston, J., Roelofs, W., Comeau, A., Sanders, C. J. 1971. *Can. Entomol.* 103:1741
62. Shorey, H. H., Gaston, L. K., Roberts, J. S. 1965. *Ann. Entomol. Soc. Am.* 58:600
63. Roelofs, W. L. 1971. *Proc. 2nd Int. Congr. Pestic. Chem.* 91
64. Roelofs, W. L., Comeau, A. 1969. *Science* 165:398
65. Roelofs, W. L., Carde, R. T., Bartell, R. J., Tierney, P. G. 1972. *Environ. Entomol.* 1:606
66. Bergstrom, G., Kullenberg, B., Stallberg-Stenhagen, S. 1973. *Chem. Scr.* 2:1
67. Shorey, H. H., Gaston, L. K., Saario, C. A. 1967. *J. Econ. Entomol.* 60:1541
68. Shorey, H. H., Kaae, R. S., Gaston, L. K., McLaughlin, J. R. 1972. *Environ. Entomol.* 1:641
69. Kaae, R. S., McLaughlin, J. R., Shorey, H. H., Gaston, L. K., 1972. *Environ. Entomol.* 1:651
70. Topoff, H., Lawson, K., Richards, P. 1973. *Ann. Entomol. Soc. Am.* 66:109
71. Lanier, G. N. 1970. *Science* 169:71
72. Averhoff, W. W., Richardson, R. H. 1974. *Behav. Genet.* 4:203
73. Ehrman, L. 1969. *Evolution* 23:59
74. Pruzan, A., Ehrman, L. 1974. *Behav. Genet.* 4:159
75. Ehrman, L. 1972. *Behav. Genet.* 2:69
76. Ehrman, L., Wissner, A., Meinwald, J. 1973. *Genetics* 74:S69
77. Brady, W. E. 1973. *Life Sci.* 13:227
78. Leyrer, R. L., Monroe, R. E. 1973. *J. Insect Physiol.* 19:2267
79. Young, J. C., Silverstein, R. M., Birch, M. C. 1973. *J. Insect Physiol.* 19:2273
80. Meijer, G. M., Ritter, F. J., Persoons, C. J., Minks, A. K., Voerman, S. 1972. *Science* 175:1469
81. Roelofs, W. L. et al 1973. *Nature New Biol.* 244:149
82. Klun, J. A. et al 1973. *Science* 181:661
83. Hummel, H. E., Gaston, L. K., Shorey, H. H., Kaae, R. S., Byrne, K. J., Silverstein, R. M. 1973. *Science* 181:873
84. George, D. A., McDonough, L. M. 1972. *Nature* 239:109
85. Mody, N. V., Miles, D. H., Neel, W. W., Hedin, P. A., Thompson, A. C., Gueldner, R. C. 1973. *J. Insect Physiol.* 19:2063
86. Shorey, H. H., Gaston, L. K. 1967. *Ann. Entomol. Soc. Am.* 60:847
87. Roelofs, W. L., Comeau, A. 1968. *Nature* 220:600
88. Roelofs, W. L., Tette, J. P. 1970. *Nature* 226:1172

89. Wright, R. H., Chambers, D. L., Keiser, I. 1971. *Can. Entomol.* 103:267
90. Roelofs, W. L., Comeau, A. 1971. *J. Insect Physiol.* 17:435
91. Roelofs, W. L., Bartell, R. J., Hill, A. S., Carde, R. T., Waters, I. H. 1972. *J. Econ. Entomol.* 65:1276
92. McLaughlin, J. R., Shorey, H. H., Gaston, L. K., Kaae, R. S., Stewart, F. D. 1972. *Environ. Entomol.* 1:647
93. McLaughlin, J. R., Gaston, L. K., Shorey, H. H., Hummel, H. E., Stewart, F. D. 1972. *J. Econ. Entomol.* 65:1592
94. Ganyard, M. C., Brady, U. E. 1971. *Nature* 234:415
95. Minks, A. K., Roelofs, W. L., Ritter, F. J., Persoons, C. J. 1973. *Science* 180:1073
96. Lanier, G. N., Birch, M. C., Schmitz, R. F., Furniss, M. M. 1972. *Can. Entomol.* 104:1917
97. Bartell, R. J., Shorey, H. H. 1969. *Ann. Entomol. Soc. Am.* 62:1206
98. Schwink, I. 1958. *Proc. 10th Int. Congr. Entomol.* 2:577
99. Wilson, E. O. 1958. *Psyche* 65:41
100. Kaae, R. S., Shorey, H. H., Gaston, L. K. 1973. *Science* 179:487
101. Bossert, W. H., Wilson, E. O. 1963. *J. Theor. Biol.* 5:443
102. Kellog, F. E., Frizel, D. E., Wright, R. H. 1962. *Can. Entomol.* 94:884
103. Wright, R. H. 1962. *World Rev. Pest Contr.* 1:2
104. Farkas, S. R., Shorey, H. H. 1972. *Science* 178:67
105. Kuhr, R. J., Comeau, A., Roelofs, W. L. 1972. *Environ. Entomol.* 1:625
106. Sower, L. L., Kaae, R. S., Shorey, H. H. 1973. *Ann. Entomol. Soc. Am.* 66:1121
107. Gaston, L. K., Shorey, H. H., Saario, C. A. 1971. *Ann. Entomol. Soc. Am.* 64:381
108. Shorey, H. H. 1966. *Ann. Entomol. Soc. Am.* 59:502
109. Riddiford, L. M. 1967. *Science* 158:139
110. Traynier, R. M. M. 1967. *Entomol. Exp. Appl.* 10:321
111. Bowers, W. S., Bodenstein, W. G. 1971. *Nature* 232–59
112. Leonard, J. E., Ehrman, L., Schorsch, M. E. 1974. *Nature* 250:261–62

GENE EXPRESSION IN SOMATIC CELL HYBRIDS

❖3071

Richard L. Davidson
Genetics Division, Children's Hospital Medical Center and Department of Microbiology and Molecular Genetics, Harvard Medical School, Boston, Massachusetts 02115

INTRODUCTION

Regulation of gene expression in cells of higher organisms is basic to the orderly development of the embryo and to the maintenance of the stable condition of the adult organism. Only a fraction of the total genetic potential of a differentiated cell is ever expressed, and the mechanisms that regulate selective expression of the genetic material are not understood. This article focuses on one approach presently being used to study gene regulation in mammalian cells: somatic cell hybridization.

Experiments on the regulation of gene expression in microorganisms had demonstrated the potential of recombinational systems for the study of genetic control mechanisms. Over the past decade, there has been an increasing effort to elucidate the mechanisms of gene regulation in mammalian cells by means of recombination between genomes of somatic cells in vitro. The basic idea of these experiments is to combine within a single nucleus the genomes of two cells that differ in their phenotypes and to examine the factors underlying the phenotypic differences by analyzing the interactions between the genomes.

Fusion between cells of different types can lead to the formation of either heterokaryons (in which the nuclei of the parental cells exist in a common cytoplasm but as separate entities) or hybrids (in which the genomes of the parental cells are combined within a single nucleus). Fused cells have been involved in experiments on topics as diverse as gene regulation, gene mapping, cell membranes, virus rescue, and malignancy. The field of somatic cell hybridization has been reviewed recently (Davidson 1) and has been covered in depth in the proceedings of a recent symposium (*Somatic Cell Hybridization,* 2). Most studies on gene regulation in fused cells have involved hybrids rather than heterokaryons, and this article is limited to studies with hybrid cells.

Viable hybrid cells can be formed from the fusion between mammalian cells of essentially any type, e.g. between cells of the same or different species, between malignant and normal cells, and between differentiated and undifferentiated cells. The hybrids are generally rapidly dividing cells and have an apparently unlimited potential for proliferation (if one of the parental cells is from a permanent line). In hybrids between cells of the same species or between cells of different rodent species (e.g. rat X mouse or hamster X mouse),.the majority of the chromosomes of the two parental cells are retained in the hybrids. In contrast, in hybrids between rodent cells and human cells (e.g. mouse X human or rat X human), the chromosomes of the human parent are preferentially lost. This characteristic makes rodent X human hybrids especially useful for studies on gene mapping.

Most experiments on gene expression in hybrids have involved the fusion of cells of different species. Interspecific hybrids are advantageous in such experiments because the cells of different species have numerous genetic markers, such as electrophoretic differences between homologous enzymes, which make it possible to monitor the activity of the genetic material of the two parental cells in the hybrids. While there are important advantages to the use of interspecific hybrids, the interpretation of the results with such hybrids must take into account the fact that such combinations of genomes would never occur in nature. Also, for all of the hybrids described in this article, at least one of the parental cells was from a permanent line. While it has never been demonstrated that the use of cells of different species or of permanent lines has an effect on the results of hybridization experiments, this possibility must be kept in mind.

SHARED FUNCTIONS AND ISOZYMES IN HYBRID CELLS

In early studies on cell hybridization, it was a matter of theoretical interest to determine whether, and to what extent, the genomes of both parental cells were expressed in the hybrids. These investigations involved the study of functions expressed by both parental cells prior to fusion. Recently, the expression in hybrids of functions common to both parental cells has been used more to provide evidence for the specificity of gene regulation or to indicate the continued presence of specific chromosomes in the hybrids.

Shared Functions in Intraspecific Hybrids

One of the earliest studies to demonstrate the continued expression of both parental genomes in hybrids involved cell surface antigens. Mouse fibroblasts characterized by different isoantigens were hybridized with each other, and it was shown that the isoantigens of both parental cells were codominantly expressed in the hybrids (Spencer et al 3). In another study fibroblasts derived from strains of mice that differed in the heat stability of the enzyme β-glucuronidase were hybridized with each other (Ganshow 4). The results of heat inactivation tests suggested that both parental forms of the enzyme were produced in the hybrids. These experiments showed that at least some of the genes of both parental cells continue to function in intraspecific hybrids.

In some cases, hybrids have been isolated between cells in which homologous molecules were not distinguishable but were produced at different rates. For example, cells of two mouse fibroblast lines that synthesized collagen at different rates were hybridized with each other (Green et al 5). In the hybrids, collagen synthesis occurred at an intermediate rate. Similarly, human fibroblasts characterized by different levels of alkaline phosphatase activity were hybridized with each other (Simoni et al 6). The hybrids were found to have an intermediate level of alkaline phosphatase activity. Results of such experiments to date have provided no evidence for interactions in hybrids between genomes that differ in the levels at which specific functions are carried out.

Isozymes in Interspecific Hybrids

With the isolation of hybrids between cells of different species, studies on the expression of the parental genomes in hybrid cells became much more feasible. With interspecific hybrids, there are many enzymes for which the forms produced by the parental cells can be readily distinguished, e.g. by means of electrophoretic mobility, isoelectric point, or heat stability. These studies have provided evidence for the continued activity of the genomes of both species in interspecific hybrids and have also provided information on molecular homology between species.

The NAD-dependent enzyme malate dehydrogenase (MDH) in Syrian hamster X mouse hybrids provides a good example (Davidson, Ephrussi & Yamamoto 7). MDH in the hamster and mouse cells exists in a single form, and the hamster and mouse forms can be resolved electrophoretically. In the hybrid cells, both forms of MDH are observed, indicating the continued presence and activity of the MDH genes of both species. There is also a third form of MDH in the hybrids, with an electrophoretic mobility intermediate between those of hamster and mouse MDH. These results suggest that in the parental cells MDH is a dimer made up of two identical subunits and that the intermediate form of MDH in the hybrid cells represents an enzymatically active hybrid molecule, composed of one hamster subunit and one mouse subunit.

A similar but slightly more complex example is provided by lactate dehydrogenase (LDH). This enzyme is a tetramer, with two different types of subunits (A and B) which can associate in all possible combinations to give five isozymes. LDH-5 (composed of four A-type subunits) from rat and mouse can be resolved by electrophoresis, and the enzyme was studied in rat X mouse hybrids (Weiss & Ephrussi 8). In the hybrids, both rat and mouse LDH-5 were produced as well as three forms with electrophoretic mobility intermediate between those of rat and mouse LDH. These intermediate forms were hybrid enzymes, composed of rat and mouse LDH-A subunits associated in all possible combinations.

These two cases illustrate a general rule in hybrid cells: when the two parental cells produce an enzyme before fusion, both genomes continue to produce the enzyme in the hybrids (assuming no chromosome loss). This rule has been observed to hold in a wide variety of hybrids, involving fusion of cells from a number of different rodent and primate species. Results of these studies indicate that cell hybridization itself does not alter the expression of the parental genomes in the hybrids, even when the parental cells are of different species.

It has also been observed as a general rule that enzymatically active hybrid molecules are formed when the enzyme in the parental cells is made up of subunits. A list of enzymes for which hybrid molecules have been observed is presented in reference 1. The existence of the hybrid enzymes indicates a degree of molecular homology between species. However, it is not known whether the hybrid molecules are of functional significance.

Ribosomes and Mitochondria

The studies described above on the expression in hybrid cells of functions common to both parental cells were based in large part on the ability to distinguish the forms of homologous enzymes produced by cells of different species. It is also possible to resolve certain types of nucleic acids, e.g. ribosomal RNA (rRNA) and mitochondrial DNA (mDNA), on the basis of species differences. This has made it possible to study these nucleic acids in interspecific hybrids.

Ribosomal RNA was studied in hybrids between mouse fibroblasts and Syrian hamster fibroblasts (Stanners, Eliceiri & Green 9). The hybrids were found to produce both mouse and hamster types of 28S rRNA. In contrast, hybrids between mouse fibroblasts and human fibroblasts were found to contain no human 28S RNA, even though some of the hybrids had as many as 35 human chromosomes (Eliceiri & Green 10).

Mitochondrial nucleic acids have also been examined in hybrid cells. In hybrids between mouse fibroblasts and Syrian hamster fibroblasts, the mitochondrial RNAs of both species were produced, demonstrating (indirectly) the presence of the mDNAs of both species in the hybrids (Eliceiri 11). In contrast, no human mDNA could be detected (by direct assay) in hybrids between mouse fibroblasts and human fibroblasts or lymphocytes, even in hybrids that had more than 20 human chromosomes (Clayton et al 12, Attardi & Attardi 13). In a recent study, it was shown that even though mouse X human hybrids contained no human mDNA, the hybrids could still produce human mitochondrial enzymes coded for (presumably) by nuclear genes (Van Heyningen, Craig & Bodmer 14). At present, it is not known why human X mouse hybrids that have lost some human chromosomes but have retained at least half of the human genome do not contain human mDNA or rRNA.

In hybrids between human and mouse cells, the general rule has been that the human chromosomes are lost (Weiss & Green 15). It has recently been shown, however, that hybrids between freshly explanted mouse cells and human cells of permanent lines often show the opposite pattern of segregation, i.e. the loss of the mouse chromosomes (Coon, Horak & Dawid 16, Minna & Gilman 17). In those hybrids that had lost mouse chromosomes, the human mDNA was retained while the mouse mDNA was lost (Coon, Horak & Dawid 16).

EXPRESSION OF MUTANT PHENOTYPES

Analysis of the mechanisms of genetic regulation requires, in addition to recombinational systems, the existence of cells that exhibit stable phenotypic differences. With microorganisms, such differences are achieved through the use of mutant strains. With mammalian cells in vitro, it also has been possible to isolate variant populations exhibiting stable phenotypic differences. (As will be seen below, differentiated cells can also be used to provide the phenotypic differences.) These variants will be referred to as "mutants," because of the heritability of the altered characteristics, even though it has not been demonstrated that true genetic changes (at the DNA level) have occurred. A number of different types of mutant cells have been studied in hybridization experiments.

Drug-Resistant Mutants

The mutants most widely used in hybridization experiments have been those isolated because of their resistance to the drugs 8-azaguanine (AG) or 5-bromodeoxyuridine (BUDR). The widespread use of these mutants has been associated with their role in the HAT system for the biochemical selection of hybrid cells (Littlefield 18), but more recent studies have focused on the mutations themselves. Cells lacking hypoxanthine-guanine phosphoribosyl transferase (HGPRT) activity or thymidine kinase (TK) activity can be isolated because of their ability to grow in the presence of AG or BUDR respectively. Such cells cannot grow in HAT medium, which contains hypoxanthine, aminopterin, and thymidine, because aminopterin blocks the cells' de novo synthesis of hypoxanthine and thymidine, and, in the absence of HGPRT or TK activity, the cells cannot utilize the exogenous compounds (Littlefield 18, Szybalski & Szybalska 19). When AG-resistant (HGPRT$^-$/TK$^+$) and BUDR-resistant (TK$^-$/HGPRT$^+$) mouse fibroblasts were mixed in HAT medium, cells of neither parental line survived, but hybrids between the two lines were able to grow (Littlefield 18). Measurements of HGPRT and TK activity revealed that the levels of enzyme activity in the hybrids were intermediate between those of the enzyme-deficient cell and the enzyme-producing cell. These results suggested that the mutations leading to the enzyme deficiencies were recessive, in the sense that hybrids between enzyme-producing and enzyme-deficient cells were able to synthesize the enzyme.

Many HGPRT$^-$ and TK$^-$ cells have been isolated (by selection in AG or BUDR) from lines of mouse, rat, Syrian hamster, Chinese hamster, and human cells, and these lines have been used in a variety of hybridization experiments. In no case has an enzyme deficiency in one parental cell prevented the appearance of enzyme activity in the hybrids. [In some cases, a decrease below 50% in activity has been observed in hybrids. For example, hybrids between TK$^-$ Chinese hamster cells and TK$^+$ Armenian hamster cells were found to have TK activity that is only 20% as high as the TK$^+$ parental cell (Sonnenschein, Roberts & Yerganian 20). However, the significance of such results is not clear at present.] The interpretation of these results is complicated by the fact that the selective system used to isolate the hybrids would prevent the growth of any hybrids that lacked either HGPRT or TK activity.

The HAT selective system described above was involved in the demonstration that human X mouse hybrids, which preferentially lose human chromosomes, could be used to map the human genome. In hybrids between TK$^+$ human cells and TK$^-$ mouse cells, it was possible to correlate the presence of the human TK with the presence of the human chromosome number 17 (Weiss & Green 15, Miller et al 21). In one case, hybrids that had lost all recognizable human chromosomes still produced human TK (Migeon, Smith & Leddy 22). These results suggested that a small piece of human genetic information (possibly a chromosome fragment) had been stably incorporated into the genome of another species (mouse).

There has been much interest in HGPRT⁻ cells, partly because the Lesch-Nyhan syndrome in man (a type of cerebral palsy) is characterized by the absence of HGPRT activity. HGPRT⁻ mouse fibroblasts were fused with chick erythrocytes (which retain their nuclei, unlike mammalian erythrocytes), and cells able to grow in HAT were isolated (Schwartz, Cook & Harris 23). The cells were found to have HGPRT activity, and the enzyme resembled (in electrophoretic mobility) chick rather than mouse HGPRT, even though no chick chromosomes could be observed in the cells. These results suggested that a small piece of chick genetic information had been stably inserted into the mouse genome. This cross between HGPRT⁻ mouse cells and chick erythrocytes was recently repeated, and the same results were obtained (Klinger & Shin 24). It was also shown that maintenance of the chick HGPRT activity in the "hybrids" was dependent upon the presence of aminopterin in the medium. It was suggested that the chick HGPRT activity, which is constitutive in chick fibroblasts, was behaving like an inducible function in the hybrids.

Recently the apparent "reactivation" of mutant HGPRT genes has been observed in hybrid cells. In one experiment HGPRT⁻ mouse cells were fused with HGPRT⁺ human cells (Watson et al 25). Out of five lines selected in HAT, three lines contained no human chromosomes. These three lines had HGPRT activity but the activity resembled mouse rather than human HGPRT. In control experiments no revertants of the HGPRT⁻ mouse line could be isolated in HAT. Similar results were obtained in two other recent experiments, involving the fusion of HGPRT⁻ mouse fibroblasts with HGPRT⁺ chick fibroblasts (Bakay et al 26) and HGPRT⁻ rat fibroblasts with HGPRT⁺ human fibroblasts (Croce et al 26).

The reappearance of rodent HGPRT activity in these experiments is difficult to interpret. However, the results raise questions about the interpretation of the HGPRT⁻ mutation in these cells as being a structural gene alteration. Such an interpretation is consistent with the previously mentioned observation that hybridization of HGPRT⁻ and HGPRT⁺ cells does not lead to loss of HGPRT activity, but it can not easily explain the reactivation of HGPRT just described.

In some recent experiments HGPRT⁻ cells were fused with other HGPRT⁻ cells. In one experiment, four independent HGPRT⁻ Chinese hamster fibroblast lines were isolated and fused with each other in various combinations (Sekiguchi & Sekiguchi 27). From certain combinations, viable hybrids with HGPRT activity could be selected in HAT. The results suggested the restoration of enzyme activity by complementation between different mutants. Complementation between mutants of different species was observed in a cross between HGPRT⁻ mouse fibroblasts and HGPRT⁻ human (Lesch-Nyhan) fibroblasts (Roy & Ruddle 28). Hybrid cells were isolated in HAT, and the hybrids were found to have HGPRT activity with an isoelectric point intermediate between those of mouse and human HGPRT.

Drug-resistant mutants in addition to the AG- and BUDR-resistant cells described above have been used in hybridization experiments. Fluoroadenine-resistant mouse fibroblasts that lacked adenine phosphoribosyl transferase (APRT) activity were isolated (Kusano, Long & Green 29). The APRT⁻ mouse cells were fused with APRT⁺ human fibroblasts, and hybrids were selected in medium containing adenine plus alanosine. (Alanosine prevents the formation of AMP from IMP, and APRT⁻ cells cannot survive in adenine-alanosine medium.) The hybrids had APRT activity, which was shown by isoelectric focusing to be of human origin. The results indicated that the APRT⁻ mutation, like the HGPRT⁻ and TK⁻ mutations, was recessive.

Drug-resistant cells characterized by overproduction of an enzyme rather than by an enzyme deficiency were studied in hybridization experiments. Aminopterin-resistant Syrian hamster cells, which had greatly elevated levels of folate reductase activity, were hybridized with aminopterin-sensitive Syrian hamster cells (Littlefield 30). The hybrids were found to have levels of folate reductase activities intermediate between those of the parental cells. The results provided no evidence for interactions in hybrids between the genome of a mutant that overproduced an enzyme and the genome of a nonmutant cell.

X-Linked Mutants and Human Disorders

Cells derived from humans with various disorders have been used in hybridization studies. In one experiment, cells of two human diploid strains, each carrying a different X-linked mutation, were hybridized with each other (Siniscalo et al 31). One strain was derived from a male patient with the Lesch-Nyhan syndrome (HGPRT⁻) and the other was derived from a male patient with congenital non-spherocytic anemia [glucose-6-phosphate dehydrogenase (G6PDH) deficient]. Tetraploid cells were observed, and it was suggested that they were hybrids because they had both HGPRT and G6PDH activity. The results are difficult to interpret because no pure hybrid lines were isolated, but they suggested that the two X chromosomes, one from each parental cell, remained active in the tetraploid hybrids.

In another study, HGPRT⁻ human cells from a permanent line were fused with cells from an individual with the condition orotic aciduria, characterized by the absence of the enzymes orotic acid (OMP) decarboxylase, and pyrophosphorylase (Silagi, Darlington & Bruce 32). Hybrid cells were isolated and it was seen that the enzymes lacking in both parental cells were all present in the hybrids. In addition, the activity of the parental X chromosomes could be followed in the hybrids, since the parental cells produced different forms of G6PDH. Both parental forms of G6PDH were observed in the hybrids, indicating the continued activity of the X chromosomes from both parental cells.

It may be recalled that in diploid female cells with two X chromosomes one of the X chromosomes is inactivated, whereas in tetraploid male cells with two X chromosomes both of the X's remain active. [The mechanisms of X chromosome inactivation are discussed by Lyon (33).] The results of the hybridization experiments are consistent with the idea that it is not the absolute number of X chromosomes per cell that determines how many X's remain active, but rather the balance between the number of X's and the ploidy that is the critical factor. Experiments have been performed to determine whether cell hybridization can induce the activity of an inactivated human X chromosome (Migeon 34). Cells of a clone of diploid human fibroblasts from a female heterozygous for G6PDH (A/B) were hybridized with mouse fibroblasts. The particular clone of human cells hybridized expressed only the B form of G6PDH, and no indication of the reactivation of the A allele of G6PDH in the hybrids was observed.

Membrane-Associated Mutants

Some recent studies have involved mutations associated with membrane properties. Actinomycin D-resistant Chinese hamster cells were selected and fused with actino-mycin D-sensitive Chinese hamster cells (Sobel et al 35). The hybrids were found to be almost (but not quite) as resistant to actinomycin D as were the drug-resistant parental cells. Resistance to actinomycin D is thought to be due to a decreased permeability of the cell membrane to the drug (Biedler & Riehm 36). If this is so, the altered permeability is acting as a dominant characteristic in the hybrids, in the sense that the permeability of the membrane of the drug-sensitive parental cell to actinomycin D is not being expressed in the hybrids. Similar results were observed with vinblastine-resistant cells (Harris 37). Like actinomycin D resistance, vinblas-tine resistance may be due to a decreased permeability of the cell membrane to the drug, and, like actinomycin D resistance, vinblastine resistance acted like a domi-nant trait in the hybrids. It is of interest to note that vinblastine resistance has been observed in cells isolated for resistance to actinomycin D (Biedler & Riehm 36). [In the cross involving vinblastine resistance, the vinblastine-sensitive cells were from a line isolated for resistance to cytosine arabinoside (*araC*). The hybrids were found to be sensitive to *araC*.]

Another membrane-associated mutation which has been studied in hybrid cells is resistance to the drug ouabain, which inhibits the Na^+/K^+-activated membrane ATPase (Baker et al 38). In resistant cells, the enzyme has an increased resistance to inhibition by ouabain. Ouabain-resistant Chinese hamster cells were selected and fused with ouabain-sensitive Chinese hamster cells. The hybrids were found to be much more resistant to ouabain than were the sensitive parental cells, but not quite as resistant as the resistant parental cells. Thus ouabain resistance behaved as a dominant or codominant trait in the hybrids.

Although the drug-resistant phenotype continued to be expressed in the three cases cited above in which resistance was associated with membrane alterations, it should not be assumed that membrane-associated resistance to extrinsic agents is always expressed in hybrids between sensitive and resistant cells. For example, human cells are sensitive to polio virus, whereas mouse cells are naturally resistant, presumably because of the absence of a membrane receptor for the virus. Hybrids between human and mouse cells have been found to be sensitive to polio virus (Wang et al 39).

Nutritional Mutants

Nutritional mutants have been studied in hybridization experiments involving a series of auxotrophic mutants of Chinese hamster cells isolated with the BUDR-light technique. Mutants were isolated that required various nutrients for growth, e.g. glycine, adenine, or proline. The mutants were hybridized with each other in all combinations, and in all cases it was found that hybrids between cells with different requirements could be isolated in medium lacking the nutrients required by the parental cells (Kao, Johnson & Puck 40, Kao & Puck 41). For example, hybrids between glycine and adenine-requiring mutants could be isolated in the absence of glycine and adenine. These results indicated that all of the mutations in these auxotrophs were recessive in hybrids.

Studies were also carried out on the complementation between mutants of a given type. Thirteen glycine⁻ mutants were isolated, and the mutants were fused with each other in all possible combinations in the absence of glycine. It was found that there were four complementation groups (Kao, Chasin & Puck 42). Mutants of one group could be fused with mutants of any other group to give hybrids able to grow in the absence of glycine. These results suggested that there were four different loci at which defects could lead to the requirement for glycine. One of the defects may involve serine hydroxymethylase activity. Similar studies involving 11 adenine-requiring mutants revealed that there were two complementation groups: called ade A and ade B (Kao & Puck 41). The adenine⁻ mutants were fused with chick erythrocytes, and viable hybrids were isolated in the absence of adenine (Kao 43). Hybrids with the ade A mutants retained one specific chick chromosome while hybrids with the ade B mutants retained a different chick chromosome. The results suggested that genes on different chick chromosomes complemented the ade A and B defects.

Temperature-Sensitive Mutants

Temperature-sensitive (ts) mutants have also been studied in hybridization experiments. In one study, a series of ts Syrian hamster lines, unable to grow at 39°C, were isolated (Meiss & Basilico 44). Mutants of the different lines were fused in various combinations, and hybrids were selected at the nonpermissive temperature. The fact that hybrids grew at 39°C was taken to mean that the ts mutations were recessive and that mutants of different types could complement each other.

Two other types of ts mutations were found to be recessive in hybrids. In one mutant there was a ts block at a step late in G1 required for DNA synthesis (Smith & Wigglesworth 45). In the other mutant, there was a ts block to azaguanine and hypoxanthine uptake (Harris & Whitmore 46).

All of the ts mutations studied to date have been found to be recessive in hybrids. However, in two of the three studies described above (Meiss & Basilico 44, Harris & Whitmore 46) the hybrids were selected at the nonpermissive temperature. Thus any hybrids that exhibited the ts characteristics would have been killed by the selection procedure.

Summary

A summary of the expression of mutant phenotypes in hybrid cells is presented in Table 1. The nature of the mutations remains to be elucidated. However, in none of the cases described above (with the possible exceptions of actinomycin D and

Table 1 Expression of mutant phenotypes in hybrid cells

Phenotype of Mutant[a]	Expression in Hybrids[b]
Azaguanine R; HGPRT⁻	–
Bromodeoxyuridine R; TK⁻	–
Fluoroadenine R; APRT⁻	–
Aminopterin R; excess folate reductase	±
Cytosine arabinoside R	–
Actinomycin R	+
Vinblastine R	+
Ouabain R	+
G6PDH⁻	–
OMP decarboxylase and pyrophosphorylase⁻	–
Auxotrophic (several types)	–
Temperature sensitive (several types)	–

[a] The symbol R indicates resistance to the drug.

[b] The mutant cells were fused with cells that did not exhibit the mutant phenotype. The + symbol indicates that the mutant phenotype was expressed in the hybrids, ± that it was expressed at an intermediate level, – that it was not expressed.

vinblastine resistance) was there any evidence for interactions between the genomes of the mutant and nonmutant cells in the hybrids. To date, studies on the hybridization of mutant cells have provided little specific information on the mechanisms of gene regulation in mammalian cells. However, it is reasonable to expect that the desired information will be forthcoming as more and more mutant types are isolated and characterized genetically in hybridization experiments.

REGULATION OF DIFFERENTIATED FUNCTIONS

Somatic cell hybridization has provided a new approach to the problem of the genetic regulation of differentiation. Stated simply, the problem is as follows: What are the mechanisms that regulate the expression of the genetic material in differentiated cells such that only a specific small part of the total genetic potential is expressed in carrying out the tissue-specific functions? A wide variety of cell hybridization experiments have been carried out in recent years in attempts to provide information on this problem. The basic approach has been to hybridize cells that differ in the expression of at least one differentiated function and to follow the expression of that function in the hybrids in order to obtain evidence on the mechanisms that control the function. These experiments are formally analogous to those described in the previous section, in which enzyme-producing cells were hybridized with enzyme-deficient mutant cells. However, in the experiments discussed below, it is assumed that both parental cells have the genetic information necessary to carry out the differentiated function even though only one of the cells is actually expressing it, i.e. the difference in phenotype between the parental cells is assumed to involve a change in gene expression, rather than a change in the genetic information as may be the case with the mutants described above.

One of the central questions in the experiments on the regulation of differentiated functions in hybrids is whether there are soluble regulatory molecules in mammalian cells that control gene expression. As seen below, the results of the hybridization experiments suggest that mammalian cells do use soluble regulatory molecules. Most of the experiments on the regulation of differentiation in hybrid cells can be classified in one of the following areas: pigment synthesis, brain-specific functions, liver-specific functions, and immunologically related functions. The experiments that best illustrate the mechanisms of gene regulation are discussed below.

Pigment Synthesis

The regulation of melanin synthesis has been extensively studied in a series of experiments on hybrids between pigmented Syrian hamster melanoma cells and unpigmented mouse fibroblasts. Over 100 hybrid lines have been observed, and all were unpigmented (Davidson, Ephrussi & Yamamoto 7, 47). The melanoma cells had high Dopa oxidase activity (the only enzyme activity necessary to synthesize melanin from tyrosine), whereas the hybrids, like the fibroblast parents, lacked Dopa oxidase activity. The results of further studies with the hybrids suggested that the absence of pigmentation was not due to the presence of an inhibitor of Dopa oxidase activity (Davidson & Yamamoto 48), to nonspecific suppression of the entire melanoma genome, to loss of the structural genes necessary for pigment synthesis, or to the fusion of fibroblasts with unpigmented variants in the melanoma population (Davidson 49). On the basis of these results, it was hypothesized that the genome of a fibroblast produces a diffusible regulator substance that blocks the expression of the pigment-forming genes, possibly by preventing the synthesis of the enzyme Dopa oxidase.

Gene dosage experiments were carried out in order to study the quantitative aspects of regulation (Davidson 50, Fougere, Ruiz & Ephrussi 51). Hybrids containing two pigment cell genomes and one fibroblast genome were isolated (These hybrids will be referred to as P/P/F hybrids, indicating that they contain two pigment cell genomes, in contrast to the hybrids described above that contained only one pigment cell genome and which will be referred to as P/F hybrids.) Approximately half of the P/P/F hybrids were darkly pigmented and had high Dopa oxidase activities. With prolonged growth, the pigmented hybrids showed a tendency to segregate off unpigmented cells (Davidson 50).

The findings that some but not all of the P/P/F hybrids were pigmented is difficult to interpret. However, the observed segregation of unpigmented from pigmented hybrids raises the possibility that all of the P/P/F hybrids may have been initially pigmented. If so, the difference in pigmentation between the P/F and P/P/F hybrids could be explained on the basis of the genome ratios in the two types of hybrids. For example, if the absence of melanin in P/F hybrids were due to the presence of a soluble regulator substance produced by the fibroblast genome and which prevented pigmentation, it could be hypothesized that there was not a sufficient amount of this regulator substance to prevent melanin synthesis when a second melanoma genome was added in the P/P/F hybrids. It could just as easily be hypothesized that a substance produced by the melanoma genome and necessary for pigment synthesis did not attain a sufficient concentration to allow pigmentation in P/F hybrids but did reach such a concentration in P/P/F hybrids. In either case, the results of these experiments suggest that there is a quantitative aspect to regulation, and that the factors governing the expression of a differentiated function may not be produced in large excess.

In addition to providing evidence on gene regulation, observation of pigmented P/P/F hybrids demonstrated that neither the act of cell fusion itself, nor the subsequent increase in cell size and chromosome number, nor the combination of the genomes of different species within a single nucleus was the reason for the absence of pigmentation in the P/F hybrids.

Brain-Specific Functions

Rat glioma cells that exhibited a number of differentiated functions have been used in several hybridization experiments. Since the glial cells expressed more than one function, it was possible to obtain information on the coordination of the multiple functions characteristic of a given cell type.

In one set of experiments, the glial cell functions were analyzed in hybrids containing either one rat glial cell genome and one mouse fibroblast genome (G/F hybrids) or two glial cell genomes and one mouse fibroblast genome (G/G/F hybrids). The parental glial cells produced S-100, a brain-specific protein whose function is unknown. In the hybrids (both G/F and G/G/F) there was at least a 90% reduction in the S-100 activity, as measured immunologically (Benda & Davidson 52). The hybrids contained a small amount of material that reacted with anti-S-100 antibody, but this material did not react with the antibody (in terms of complement fixation) in exactly the same way as did S-100 protein in the glial cells. It is not known at present whether the activity in the hybrids was due to a cross-reacting (non-S-100) material or to a modified S-100.

The parental glial cells were also characterized by a high level of glycerol phosphate dehydrogenase (GPDH) activity and the inducibility (2–3-fold) on GPDH by hydrocortisone (HC). The baseline level of GPDH in the fibroblasts was less than 10% of that of the glial cells, and there was no increase in enzyme activity in the presence of HC. In the G/F hybrids, the baseline level of GPDH was reduced approximately 75% relative to the glial cells, and there was no enzyme inducibility (Davidson & Benda 53). In the G/G/F hybrids, the baseline GPDH level was within 15% of that of the 1s glial cells, and there was a slight increase (approximately 25%) in GPDH activity in the presence of HC. The fact that the G/G/F hybrids had a baseline level of GPDH almost as high as the 1s glial cells and yet were much less inducible than the 1s glial cells suggested that the baseline level and inducibility of GPDH are controlled by separate mechanisms.

It was mentioned above that the G/F and G/G/F hybrids had the same S-100 activity. In contrast, the G/G/F hybrids had a GPDH baseline level fourfold higher than that of the G/F hybrids. This comparison suggested that two proteins characteristic of the differentiated state of glial cells are controlled not coordinately but by independent mechanisms.

The rat glial cells were also hybridized with human leukocytes (Horn & Davidson 54). Like other human X rodent hybrids, these hybrids preferentially lost human chromosomes. The hybrids were analyzed to see what effect the loss of chromosomes of the noninducible parent would have on enzyme inducibility. Some of the hybrid lines showed GPDH inducibility, while others did not. From one of the noninducible hybrids, two subclones that had regained GPDH inducibility were isolated. It has not yet been shown that a specific human chromosome was absent in the hybrids that regained enzyme inducibility. However, the results suggested that a human gene produced a regulator substance that prevented enzyme inducibility in the hybrids and that inducibility returned when the chromosome carrying this gene was lost. By analogy with the results of this experiment, it may be hypothesized that the genome of the fibroblast in the G/F (rat X mouse) hybrids described above produced a similar regulator substance that prevented the expression of the inducibility of the glial cell genome but that the amount of the regulator substance produced was not sufficient to control the activity of the two glial cell genomes in the G/G/F hybrids. It is not known whether the block in inducibility involves a decrease in the amount of hydrocortisone receptor in the cells or whether the block occurs at some level closer to the GPDH gene.

The control of cAMP levels has been investigated in hybrids between rat glial cells and mouse fibroblasts (Gilman & Minna 55, Minna & Gilman 56). The cAMP level in the glial cells increased greatly when the cells were treated with catecholamines (e.g. isoproterenol) but not when treated with prostaglandin E_1 (PGE_1). In contrast, the cAMP level in the fibroblasts responded to PGE_1 but not to isoproterenol. The hybrids showed very little response to catecholamines but showed a marked response to PGE_1. Thus the hybrids resembled the fibroblasts rather than the glial cells with respect to both responses. The catecholamine response was lost in hybrids between responsive and nonresponsive cells whereas the PGE_1 response was maintained in such hybrids.

A number of neuron-specific functions have been examined in hybrids between mouse neuroblastoma cells and mouse fibroblasts. The neuroblastoma cells exhibited several neural characteristics, including electrically excitable membranes, neurite formation, acetyl cholinesterase (AChE) activity, acetylcholine (ACh) sensitivity, steroid sulfatase activity, and protein 14-3-2 activity (protein 14-3-2 is a nerve-specific protein assayed immunologically whose function is not known). It was found that most of the nerve-specific functions were expressed in at least some of the hybrids. For example, about half of the hybrid cells had high levels of AChE activity (Minna, Glazer & Nirenberg 57). Similarly, some hybrids exhibited excitable membranes (Minna et al 58), or neurite formation (Minna et al 58), or ACh sensitivity (Peacock, McMorris & Nelson 59), or protein 14-3-2 activity (McMorris et al 60). Hybrids that exhibited one of these functions did not necessarily exhibit others. In contrast to the above functions, steroid sulfatase activity was consistently absent in the hybrids (McMorris et al 60).

Hybrids were also isolated between mouse neuroblastoma cells and human fibroblasts. In one hybrid line, the activity of choline acetyl transferase (ChAT) was detected, even though neither parental cell exhibited that activity (McMorris & Ruddle 61). This new activity could be due to the activation of a gene not expressed in the parental cells. (It is also possible that the hybrid with ChAT activity resulted from the fusion of a fibroblast with a rare cell in the neuroblastoma population that did possess ChAT activity.)

The variable expression of the neural functions in hybrids is difficult to interpret. In one study, it was suggested that there may be a type of sequential control, i.e. that function B can be expressed only if function A is expressed, etc (Minna et al 57). In another study, it was suggested that there is a tendency for coordinate expression of neural functions in the hybrids, i.e. that hybrids expressing one neural function at a high level have a tendency to express other neural functions also (McMorris & Ruddle 61).

In comparing the hybrids between fibroblasts and two types of cells derived from neural tissue (glial cells and neuroblastoma cells) it is seen that the glial-specific functions are not expressed in the hybrids, whereas the neuroblastoma-specific functions generally are expressed in the hybrids. The significance of this is not clear. One possible (but as yet untested) explanation for the difference between the two types of hybrids involves gene dosage effects. The glial cells used were near diploid, whereas the neuroblastoma cells were near tetraploid, and it was shown above (for hybrids with both glial cells and melanoma cells) that changing the number of parental genomes in the hybrids can affect the expression of differentiated functions. The differences between the glial cell and neuroblastoma cell hybrids may also reflect fundamental differences in the mechanisms that regulate the differentiated functions of the two cell types.

Liver-Specific Functions

Hepatoma cells, expressing a variety of liver-specific functions, have been extensively studied in hybridization experiments. The specific functions that have been studied include the production of serum proteins, the induction of a number of different enzymes by steroid hormones, and the production of liver-specific forms of certain enzymes.

Rat hepatoma cells characterized by high levels of tyrosine aminotransferase (TAT) and the inducibility of the enzyme by hydrocortisone were hybridized with mouse fibroblasts that had very low levels of TAT and in which TAT was not inducible (Schneider & Weiss 62). In the hybrids TAT activity was reduced to a very low level and the enzyme was not inducible. The rat hepatoma cells were also fused with rat epithelial cells, and the hybrids initially resembled the hepatoma X fibroblast hybrids in terms of TAT level and inducibility (Weiss & Chaplain 63). However, in the hepatoma X epithelial cell hybrids there was an anomalous loss of chromosomes. One hybrid that had lost 30–40% of the chromosomes initially present was found to have regained TAT inducibility, even though the noninduced level of TAT in this hybrid remained as low as in the noninducible hybrids. The reappearance of TAT inducibility was interpreted as evidence that the presence of specific chromosomes of the noninducible parent was required to prevent enzyme inducibility in the hybrids.

In another study on chromosome segregation and TAT inducibility, HGPRT⁻ rat hepatoma cells were hybridized with HGPRT⁺ human fibroblasts. As long as the hybrids were maintained in HAT medium, survival in which required the continued presence of the (X-linked) human HGPRT gene, there was no TAT inducibility (Croce, Litwack & Koprowski 64). The hybrids were then backselected in azaguanine in order to isolate HGPRT⁻ cells, and it was found that these hybrids, which had lost the human X chromosome, had regained TAT inducibility. These results suggested that a gene located on the human X chromosome produces a regulatory factor that prevents TAT inducibility. It was shown that the absence of TAT inducibility in the hybrids was not due to a loss of steroid receptor activity. This suggested that the regulatory substance was acting at some level specifically related to TAT induction and not related to hormone responsiveness in general.

The hybrids between rat hepatoma cells and rat epithelial cells described above have been analyzed for a number of liver-specific functions in addition to TAT inducibility. These functions included synthesis of aldolase B (Bertolotti & Weiss 65, 66) and liver alcohol dehydrogenase (ADH) (Bertolotti & Weiss 67), production of high levels of alanine aminotransferase (AAT), and induction of AAT by steroid hormones (Sparkes & Weiss 68), none of which functions was exhibited by the parental epithelial cells. In hybrids that did not show significant chromosome segregation, none of these liver-specific functions was exhibited. However, in those hybrids that had lost many chromosomes, some liver-specific activities could be observed. It was suggested that the absence of the differentiated functions was due to the presence of certain chromosomes of the epithelial cell and that the expression of the functions was blocked only as long as those chomosomes were maintained in the hybrids. In the hybrids that reexpressed some liver-specific activities, it was observed that different subclones could express different activities. For example one hybrid subclone was inducible for TAT but not AAT, while another subclone produced high levels of aldolase B but not liver ADH. These results therefore suggested that the multiple differentiated functions characteristic of liver cells are not coordinately controlled in hybrid cells but are instead controlled by independent regulatory systems. (The interpretation of the results of these experiments is slightly complicated, because the parental epithelial cells, which did not express any liver-specific activities, were themselves ultimately derived from liver. These cells thus may have been initially differentiated liver cells that dedifferentiated in vitro. It is not certain that the same mechanism is involved in the absence of a function due to embryonic differentiation and the loss of a function due to dedifferentiation.)

Other liver-specific functions that have been examined in hybrids between rat hepatoma cells and mouse fibroblasts include catalase production and production of the second component of complement (Levisohn & Thompson 69). In the hybrids both of these functions were found to be maintained at intermediate levels. In another cross, cells of a permanent mouse liver line (not established from a hepatoma) were hybridized with mouse fibroblasts (Rintoul, Colofiore & Morrow 70). The liver cells were characterized by the accumulation of glycogen granules and by the inducibility of tryptophan pyrrolase by steroid hormones. Neither of these functions was expressed in the hybrids.

A number of experiments have been performed on the production of serum proteins in hybrids. These experiments involved lines of rodent hepatoma cells that produced albumin. In one series of experiments, hybrids containing either one rat hepatoma genome and one mouse fibroblast genome (L/F hybrids) or two rat hepatoma genomes and one mouse fibroblast genome (L/L/F hybrids) were isolated, and the hybrids were analyzed for albumin production (Peterson & Weiss 71). In L/F hybrids, albumin production was maintained at a low level, 5–30% as high as in the hepatoma cells. Immunological tests showed that the albumin being produced was only of the rat (hepatoma) type. In contrast to these results, synthesis of mouse albumin was detected in three out of five L/L/F hybrids (only one of which also synthesized rat albumin). These results suggested that albumin synthesis is regulated by a soluble substance that acts to elicit its production. The results also clearly demonstrated the maintenance of the genes for albumin synthesis in a cell type which, as a result of differentiation, did not produce albumin. The activation of mouse albumin production in L/L/F hybrids but not L/F hybrids could be explained on the basis of changes in gene dosage, analogous to the explanation proposed for expression of differentiated functions in hybrids containing two pigment cell genomes or two glial cell genomes.

More recently albumin production was analyzed in hybrids between rat hepatoma cells and mouse lymphoblast cells that did not produce albumin (Malawista & Weiss 72). In 8 out of 9 hybrids containing one hepatoma genome and one lymphoblast genome, activation of mouse albumin synthesis occurred. At first glance these results seem to differ from those of the previous cross, in which production of only rat albumin was detected in L/F hybrids. However, the fibroblasts used in the previous cross contained almost twice as many chromosomes as the lymphoblast cells used in this cross. In terms of parental chromosome input, the ratio of rat to mouse chromosomes in the hepatoma X lymphoblast hybrids resembled the ratio in the L/L/F hybrids rather than that in the L/F hybrids. Thus these results also are consistent with the idea that gene dosage may play a role in determining the expression of differentiated functions in hybrid cells.

Activation of human albumin production has been observed in hybrids between mouse hepatoma cells and human leukocytes that do not produce albumin (Darlington 73). All of the hybrids produced approximately as much mouse albumin as the hepatoma parent, and two of four hybrid clones also produced a small amount of human albumin. No intact human chromosomes could be identified in the hybrids that produced human albumin.

Activation of a serum protein other than albumin was observed in another cross (Dannies & Tashjian 74). Hybrids were isolated between rat hepatoma cells and mouse fibroblasts. Two out of four hybrids produced rat albumin and none produced mouse albumin, but three out of four produced a mouse serum protein that was not albumin. These results again demonstrated the activation in hybrids of a gene for a serum protein. In this case, however, it is not possible to propose any mechanism of activation because it is not yet known whether the hepatoma cells produced the equivalent, nonalbumin serum protein.

A comparison of the expression of liver-specific functions in hybrids reveals that, in general, synthesis of tissue-specific enzymes is not observed in the hybrids, whereas synthesis of serum proteins is. Whether or not this apparent distinction has any significance cannot be said at present.

Immunologically Related Functions

Hybrids between immunoglobulin (Ig) producing myeloma cells and non-Ig-producing cells have been analyzed in studies on the control of antibody synthesis. In two studies with hybrids between Ig-producing mouse myeloma cells and mouse fibroblasts, it was seen that there was complete, or almost complete, absence of Ig in the hybrids (Periman 75, Coffino et al 76). In another study, mouse myeloma cells that produced IgG as well as free kappa light chains were hybridized with mouse lymphoma cells that produced neither (Mohit 77). A mass population of hybrid cells, presumably representing a mixture of a number of independent hybrid lines, was isolated, and it was found that this population continued to produce kappa chains but no IgG. The mass population was cloned, and two types of hybrid cell clones were observed. Approximately 85% of the hybrids produced only kappa chains, and 15% produced both kappa chains and IgG. The production of IgG by these hybrids is difficult to interpret because of the uncertain origin of the different types of hybrids. In another cross, human lymphoblasts that produced lambda light chains were hybridized with mouse fibroblasts, and the hybrids were found to maintain the production of human lambda chains (Orkin et al 78).

The activation of Ig production was observed in hybrids between mouse myeloma cells that produced IgA and human peripheral lymphocytes that produced no Ig (Schwaber & Cohen 79). In the hybrids, production of mouse Ig continued and, in addition, the synthesis of human Ig was initiated. While the results demonstrated the activation of the human genes for Ig production, the hybrids contained only very few human chromosomes and synthesized human Ig from the time they were isolated. Thus it is not possible to say what mechanism led to the activation of human Ig synthesis.

The serum complement system, which has an effect on antigen-antibody reactions, also has been studied in hybrid cells. Freshly isolated macrophages from a guinea pig genetically unable to produce the fourth component of complement (C4) were hybridized with human cells of a permanent line (HeLa) that also did not make C4 (Colten & Parkman 80). In the hybrids C4 production occurred, and it was shown that the C4 was of the human type. In subsequent studies, it was shown that the C4-deficient guinea pig macrophages release into the medium a substance capable of eliciting C4 production by HeLa cells (Colten 81). It thus seems that the control of C4 production involves a soluble substance that acts in some way to cause C4 production. Thus far it has not been possible to demonstrate that this substance has an effect on cells other than HeLa cells.

Other Differentiated Functions

Much of the work on regulation of differentiation in hybrids has been performed with a relatively small number of different cell types, as described above. However, other cell types have been used in hybridization experiments, and the findings of some such studies are described below.

In one experiment, where rat growth hormone-producing cells from a pituitary tumor were hybridized with mouse fibroblasts (Sonnenschein, Richardson & Tashjian 82), no growth hormone could be detected in the hybrids. In another experiment, mouse renal adenocarcinoma cells that produced the kidney-associated esterase ES-2 were hybridized with human fibroblasts (Klebe, Chen & Ruddle 83), and seven out of eight hybrids exhibited ES-2 activity. The hybrid that did not make ES-2 was cloned, and a subclone was isolated that had regained ES-2 activity. [Initially there was thought to be a correlation between the presence of human chromosome number 10 and the absence of ES-2 activity. However, this correlation was not confirmed in more recent experiments (G. Darlington, P. Bernhard, and F. Ruddle, personal communication).] It was suggested that a human regulator gene was responsible for the absence of ES-2 activity and that ES-2 activity reappeared following the loss of the human chromosome carrying this gene.

A new activity exhibited by neither parental cell was observed in hybrids between adenine requiring Chinese hamster fibroblasts and human fibroblasts (Kao & Puck 84). As long as the hybrids were grown in medium lacking adenine, they retained a human B group chromosome and produced three esterases not observed in either parental cell. When the hybrids were grown in medium containing adenine, the human B group chomosome was lost and the three esterases disappeared. Comparative tests suggested that the three new esterases in the hybrids corresponded to three esterases found in ovarian tissue of Chinese hamster. The results suggested that a gene on the human B group chromosome was responsible for the activation of the hamster esterase genes in the hybrids. Interpretation of these results in terms of differentiation is complicated by the fact that the hamster cells were derived from ovarian tissue that produced the three esterases, and therefore the absence of the esterases in the parental hamster cells may have occurred as a result of cultivation in vitro and not from differentiation per se.

Mouse teratoma cells which produce a wide variety of differentiated structures (bone, cartilage, neurotubules, striated muscle, etc) when injected into mice have been used in hybridization experiments (Finch & Ephrussi 85, Jami, Failly & Ritz 86). The teratoma cells were fused with mouse fibroblasts, and the hybrids were injected into mice. The hybrids formed tumors that lacked all the differentiated elements of the teratoma parent. It should be noted that the teratoma cells, unlike the differentiated cells used in the crosses described above, express their differentiated functions only in vivo, i.e. they are cells with a potential for differentiation not expressed at the time of fusion. It is therefore not clear whether the absence of differentiation in the hybrid tumors is due to an effect on the expression of the differentiated functions of the teratoma genome or an effect on the potential of the teratoma genome for differentiation.

Concluding Remarks

A summary of the results of the experiments on the expression of differentiated functions in hybrids is presented in Table 2. The results are divided into three

Table 2 Expression of differentiated functions in hybrid cells

Function[a]	Differentiated Cell Type	Comments[b]
1. Loss of activity		
Melanin	melanoma	dosage effect
S-100 protein	glioma	
GPDH and inducibility	glioma	dosage effect, segregation
cAMP response to catecholamines	glioma	
Steroid sulfatase	neuroblastoma	
TAT and inducibility	hepatoma	segregation
AAT and inducibility	hepatoma	segregation
Aldolase B	hepatoma	segregation
Liver alcohol dehydrogenase	hepatoma	segregation
TP and inducibility	hepatoma	
Glycogen accumulation	hepatoma	
Immunoglobulin	myeloma	segregation?
Growth hormone	pituitary tumor	
ES-2	kidney tumor	segregation
Multipotency	teratoma	
2. Maintenance of activity		
cAMP response to PGE_1	fibroblast	
Excitable membranes	neuroblastoma	
Acetyl cholinesterase	neuroblastoma	
Acetylcholine sensitivity	neuroblastoma	
14-3-2 protein	neuroblastoma	
Catalase	hepatoma	
C2	hepatoma	
Immunoglobin light chain	myeloma	
3. Appearance of new activity		
Choline acetyl transferase	neuroblastoma	
Albumin	hepatoma	dosage effect
Mouse serum protein	hepatoma	
C4	macrophage	
Chinese hamster esterases I, II, III	fibroblast	

[a] Abbreviations used for differentiated functions: GPDH = glycerol phosphate dehydrogenase, TAT = tyrosine aminotransferase, AAT = alanine aminotransferase, TP = tryptophan pyrrolase, C2 and C4 = second and fourth components of complement, PGE_1 = prostaglandin E_1.

[b] The comments indicate whether a given result is altered by a loss of chromosomes or by a change in genome ratios.

categories: loss of activity, maintenance of activity, and appearance of new activity. The majority of the differentiated functions were suppressed in hybrids between cells that expressed the function and cells that did not. (The loss of a differentiated function will be referred to as "suppression" in this discussion. The use of this term is not meant to imply any mechanism whatsoever.) For the purpose of constructing a simple model, it is assumed that all of the functions suppressed are controlled by the same type of regulatory system. (This is obviously an oversimplification.) Suppression of differentiated functions in the hybrids suggests that control of the functions involves diffusible regulatory signals. In terms of control systems involving only a single regulator substance, the absence of a differentiated function could be due to (*a*) a regulator substance produced by the genome of the undifferentiated parent which acts to prevent the expression of the function in the hybrids, or (*b*) a regulator substance produced by the genome of the differentiated parent which acts to induce expression of the function but which does not reach its threshold concentration in the hybrids. The observation that differentiated functions can be reexpressed in hybrids following loss of some of the chromosomes of the undifferentiated parent (in hybrids with glial, liver, and kidney cells) makes the first of these two possibilities more likely, especially in those cases in which reexpression can be correlated with loss of a single specific chromosome.

Reexpression of differentiated functions following chromosome segregation suggests that suppression of the differentiated function requires the continuous presence of the genes that produce the proposed regulator substances. Furthermore, the fact that the hybrids can maintain the ability to carry out specific functions in the absence of expression suggests that the regulator substances affect expression of the differentiated functions rather than the potential of the cells to express those functions. The results of the gene dosage experiments (in hybrids with pigment and glial cells) suggest that the regulator substances are produced in limited quantities, in the sense that a twofold change in the ratio of differentiated:undifferentiated cell genomes in the hybrids can affect the expression of the differentiated functions. Results of experiments on the coordination of multiple differentiated functions (in hybrids with glial and liver cells) suggest that each differentiated function may be controlled by its own specific regulatory substance.

On the basis of the above discussion, it can be hypothesized that mammalian cells are continually producing a large number of specific regulator substances that act to suppress the expression of the majority of the differentiated functions that the cells do not exhibit. At present no evidence exists to indicate whether there is an additional level of regulation to control the potential of a cell to differentiate, or whether the loss of any regulator substance from any cell type is sufficient to result in the expression of the corresponding function. (Given the embryonic distinction between determination and overt differentiation, the former seems more likely.) Furthermore there is as yet no evidence on the level at which the proposed regulator substances operate. As pointed out in a recent review (Davis & Adelberg 87), regulator substances need not act directly on the expression of the relevant structural genes but may operate indirectly through a series of regulatory steps. Obviously, very complicated regulatory circuits could be constructed. However, for simplicity, and in the absence of evidence to the contrary, a one-step system seems preferable as a model.

The above model was proposed for differentiated functions suppressed in hybrids. However, a similar model could be proposed to explain the activation of differentiated functions, e.g. albumin or C4 synthesis, in hybrids. In these cases the results again support the idea that regulation of the differentiated functions involves diffusible regulator substances continually produced in limited quantities. For these functions, however, a regulator substance that induces differentiation would be substituted in the hypothesis for the regulator substance proposed above which suppresses differentiation. Finally, reactivation of a function demonstrates that the relevant structural genes are maintained in cells that, as a result of differentiation, do not express the function.

The above model is proposed as a working hypothesis. While the results discussed in this review are consistent with the model, they clearly do not prove its validity, and other mechanisms consistent with the same results obviously could be proposed. In addition, the cautions stated in the introduction concerning hybrids in which the parental cells belong to different species or permanent lines should be recalled. Further genetic experiments, involving the hybridization of other types of differentiated cells, embryonic cells, and, if available, mutants of differentiated cells, should help to elucidate the mechanisms of gene regulation. Final proof of the mechanisms of regulation may have to await the isolation of regulator substances from mammalian cells and studies of their mode of action on purified mammalian genes.

ACKNOWLEDGMENT

The author would like to acknowledge the significant contribution of his wife Jalane, who helped in writing this article.

Literature Cited

1. Davidson, R. 1973. *Somatic Cell Hybridization: Studies on Genetics and Development.* Reading, Mass.: Addison Wesley
2. Davidson, R., de la Cruz, F., Eds. 1974. *Somatic Cell Hybridization.* New York: Raven
3. Spencer, R., Hauschka, T., Amos, D., Ephrussi, B. 1964. Codominance of isoantigens in somatic hybrids of murine cells grown in vitro. *J. Nat. Cancer Inst.* 33:893–903
4. Ganschow, R. 1966. Glucuronidase gene expression in somatic hybrids. *Science* 153:84–85
5. Green, H., Ephrussi, B., Yoshida, M., Hamerman, D. 1966. Synthesis of collagen and hyaluronic acid by fibroblast hybrids. *Proc. Nat. Acad. Sci. USA* 55:41–44
6. Simoni, G., Balacco, S., Nuzzo, F., Larizza, L., DeCarli, L. 1970. A hybrid line between two clonal derivates of the EUE line with different levels of alkaline phosphatase. *Atti Ass. Genet. Ital.* 15:131–44
7. Davidson, R., Ephrussi, B., Yamamoto, K. 1968. Regulation of melanin synthesis in mammalian cells, as studied by somatic hybridization. I. Evidence for negative control. *J. Cell. Physiol.* 72: 115–28
8. Weiss, M., Ephrussi, B. 1966. Studies of interspecific (rat X mouse) somatic hybrids. II. Lactate dehydrogenase and beta-glucuronidase. *Genetics* 54: 1111–22
9. Stanners, C., Eliceiri, G., Green, H. 1971 Two types of ribosomes in mouse-hamster hybrid cells. *Nature* 230:52–54
10. Eliceiri, G., Green, H. 1969. Ribosomal RNA synthesis in human-mouse hybrid cells. *J. Mol. Biol.* 41:253–60
11. Eliceiri, G. 1973. Synthesis of mitochondrial RNA in hamster-mouse hybrid cells. *Nature* 241:233–34
12. Clayton, D., Teplitz, R., Nabholz, M., Dovey, H., Bodmer, W. 1971. Mitochondrial DNA of human-mouse cell hybrids. *Nature* 234:560–62
13. Attardi, B., Attardi, G. 1972. Fate of mitochondrial DNA in human-mouse somatic cell hybrids. *Proc. Nat. Acad. Sci. USA* 69:129–33
14. Van Heyningen, V., Craig, I., Bodmer, W. 1973. Genetic control of mitochondrial enzymes in human-mouse somatic cell hybrids. *Nature* 242:509–12

15. Weiss, M., Green, H. 1967. Human-mouse hybrid cell lines containing partial complements of human chromosomes and functioning human genes. *Proc. Nat. Acad. Sci. USA* 58:1104–11
16. Coon, H., Horak, I., Dawid, I. 1974. The fate of mitochondrial DNA's in mouse-human and rat-human hybrid cells. See Ref. 2, pp. 59–64
17. Minna, J., Gilman, A. 1974. Genetic analysis of the nervous system, using somatic cell hybrids. See Ref. 2, pp. 191–96
18. Littlefield, J. 1964. Selection of hybrids from mating of fibroblasts in vitro and their presumed recombinants. *Science* 145:709
19. Szybalski, W., Szybalska, E. 1962. Drug sensitivity as a genetic marker for human cell lines. *Univ. Mich. Med. Bull.* 28:277–93
20. Sonnenschein, C., Roberts, D., Yerganian, G. 1969. Karyotypic and enzymatic characteristics of a somatic hybrid cell line originating from dwarf hamsters. *Genetics* 62:379–92
21. Miller, O., Allderdice, P., Miller, D. 1971. Human thymidine kinase gene locus: assignment to chromosome 17 in a hybrid of man and mouse cells. *Science* 173:244–45
22. Migeon, B., Smith, S., Leddy, C. 1969. The nature of thymidine kinase in the human-mouse hybrid cell. *Biochem. Genet.* 3:583–90
23. Schwartz, A., Cook, P., Harris, H. 1971. Correction of a genetic defect in a mammalian cell. *Nature* 230:5–8
24. Klinger, H., Shin, S. 1974. Modulation of the activity of an avian gene transferred into a mammalian cell by cell fusion. *Proc. Nat. Acad. Sci. USA* 71:1398–1402
25. Watson, B., Gormley, I., Gardiner, S., Evans, H., Harris, H. 1972. Reappearance of murine hypoxanthine guanine phosphoribosyl transferase activity in A9 cells after attempted hybridization with human cell lines. *Exp. Cell Res.* 75:401–9
26. Bakay, B., Croce, C., Koprowski, H., Nyhan, W. 1973. Restoration of hypoxanthine phosphoribosyl transferase activity in mouse IR cells after fusion with chick-embryo fibroblasts. *Proc. Nat. Acad. Sci. USA* 70:1998–2002; Croce, C., Bakay, B., Nyhan, W., Koprowski, H. 1973. Reexpression of the rat hypoxanthine phosphoribosyl transferase gene in rat-human hybrids. *Proc. Nat. Acad. Sci. USA* 70:2590–94

27. Sekiguchi, T., Sekiguchi, F. 1973. Interallelic complementation in hybrid cells derived from Chinese hamster diploid clones deficient in hypoxanthine-guanine phosphoribosyl transferase activity. *Exp. Cell Res.* 77:391–403
28. Roy, K., Ruddle, F. 1973. Microscale isoelectric focusing studies of mouse and human hypoxanthine-guanine phosphoribosyl transferases. *Biochem. Genet.* 9:175–85
29. Kusano, T., Long, C., Green, H. 1971. A new reduced human-mouse somatic cell hybrid containing the human gene for adenine phosphoribosyl transferase. *Proc. Nat. Acad. Sci. USA* 68:82–86
30. Littlefield, J. 1969. Hybridization of hamster cells with high and low folate reductase activity. *Proc. Nat. Acad. Sci. USA* 62:88–95
31. Siniscalo, M. et al 1969. Evidence for intragenic complementation in hybrid cells derived from two human diploid strains each carrying an X-linked mutation. *Proc. Nat. Acad. Sci. USA* 62:793–99
32. Silagi, S., Darlington, G., Bruce, S. 1969. Hybridization of two biochemically marked cell lines. *Proc. Nat. Acad. Sci. USA* 62:1085–92
33. Lyon, M. F. 1971. Possible mechanisms of X chromosome inactivation. *Nature* 232:229–32
34. Migeon, B. 1972. Stability of X chromosomal inactivation in human somatic cells. *Nature* 239:87–89
35. Sobel, J., Albrecht, A., Riehm, H., Biedler, J. 1971. Hybridization of actinomycin D- and amethopterin-resistant Chinese hamster cells *in vitro. Cancer Res.* 31:297–307
36. Biedler, J., Riehm, H. 1970. Cellular resistance to actinomycin D in Chinese hamster cells *in vitro:* cross resistance, radioautographic, and cytogenetic studies. *Cancer Res.* 30:1174–84
37. Harris, M. 1973. Phenotypic expression of drug resistance in hybrid cells. *J. Nat. Cancer Inst.* 50:423–29
38. Baker, R. et al 1974. Ouabain-resistant mutants of mouse and hamster cells in culture. *Cell* 1:9–21
39. Wang, R., Pollack, R., Kusano, T., Green, H. 1970. Human-mouse hybrid cell line and susceptibility to polio virus. *J. Virol.* 5:677–81
40. Kao, F., Johnson, R., Puck, T. 1969. Complementation analysis on virus-fused Chinese hamster cells with nutritional markers. *Science* 164:312–14
41. Kao, F., Puck, T. 1972. Genetics of somatic mammalian cells: XIV Genetic

analysis *in vitro* of auxotrophic mutants. *J. Cell. Physiol.* 80:41–50

42. Kao, F., Chasin, L., Puck, T. 1969. Genetics of somatic cells. X. Complementation analysis of glycine requiring mutants. *Proc. Nat. Acad. Sci. USA* 64: 1284–91

43. Kao, F. 1973. Identification of chick chromosomes in cell hybrids formed between chick erythrocytes and adenine-requiring mutants of Chinese hamster cells. *Proc. Nat. Acad. Sci. USA* 70: 2893–98

44. Meiss, H., Basilico, C. 1972. Temperature sensitive mutants of BHK 21 cells. *Nature New Biol.* 239:66–68

45. Smith, B., Wigglesworth, N. 1973. A temperature-sensitive function in a Chinese hamster line affecting DNA synthesis. *J. Cell Physiol.* 82:339–48

46. Harris, J., Whitmore, G. 1974. Chinese hamster cells exhibiting a temperature dependent alteration in purine transport. *J. Cell Physiol.* 83:43–52

47. Davidson, R., Ephrussi, B., Yamamoto, K. 1966. Regulation of pigment synthesis in mammalian cells, as studied by somatic hybridization. *Proc. Nat. Acad. Sci. USA* 56:1437–40

48. Davidson, R., Yamamoto, K. 1968. Regulation of melanin synthesis in mammalian cells, as studied by somatic hybridization. II. The level of regulation of 3,4-dihydroxyphenylalanine oxidase. *Proc. Nat. Acad. Sci. USA* 60:894–901

49. Davidson, R. 1969. Regulation of melanin synthesis in mammalian cells, as studied by somatic hybridization. III. A method of increasing the frequency of cell fusion. *Exp. Cell Res.* 55:424–26

50. Davidson, R. 1972. Regulation of melanin synthesis in mammalian cells: effect of gene dosage on the expression of differentiation. *Proc. Nat. Acad. Sci. USA* 69:951–55

51. Fougere, C., Ruiz, F., Ephrussi, B. 1972. Gene dosage dependence of pigment synthesis in melanoma x fibroblast hybrids. *Proc. Nat. Acad. Sci. USA* 69:330–34

52. Benda, P., Davidson, R. 1971. Regulation of specific functions of glial cells in somatic hybrids. I. Control of S-100 protein. *J. Cell. Physiol.* 78:209–16

53. Davidson, R., Benda, P. 1970. Regulation of specific functions of glial cells in somatic hybrids. II. Control of inducibility of glycerol phosphate dehydrogenase. *Proc. Nat. Acad. Sci. USA* 67: 1870–77

54. Horn, D., Davidson, R. 1974. Reexpression of glycerol phosphate dehydrogenase inducibility in rat glial cell-human leukocyte hybrids. In preparation

55. Gilman, A., Minna, J. 1973. Expression of genes for metabolism of cyclic adenosine 3':5'-monophosphate in somatic cells. I. Responses to catecholamines in parental and hybrid cells. *J. Biol. Chem.* 248:6610–17

56. Minna, J., Gilman, A. 1973. Expression of genes for metabolism of cyclic adenosine 3':5'-monophosphate in somatic cells. II. Effects of prostaglandin E_1 and theophylline on parental and hybrid cells. *J. Biol. Chem.* 248: 6618–25

57. Minna, J., Glazer, D., Nirenberg, M. 1972. Genetic dissection of neural properties using somatic cell hybrids. *Nature* 235:225–31

58. Minna, J., Nelson, P., Peacock, J., Glazer, D., Nirenberg, M. 1971. Genes for neuronal properties expressed in neuroblastoma X L cell hybrids. *Proc. Nat. Acad. Sci. USA* 68:234–39

59. Peacock, J., McMorris, F., Nelson, P. 1973. Electrical excitability and chemosensitivity of mouse neuroblastoma X mouse or human fibroblast hybrids. *Exp. Cell Res.* 79:199–212

60. McMorris, F., Kolber, A., Moore, B., Perumal, A. 1974. Differentiated expression in neuroblastoma cell hybrids: III. Expression of the neuron-specific protein, 14-3-2, and steroid sulfatase. In press

61. McMorris, F., Ruddle, F. 1974. Expression of neuronal phenotypes in neuroblastoma cell hybrids. In press

62. Schneider, J., Weiss, M. 1971. Expression of differentiated functions in hepatoma cell hybrids. I. Tyrosine aminotransferase in hepatoma-fibroblast hybrids. *Proc. Nat. Acad. Sci. USA* 68:127–31

63. Weiss, M., Chaplain, M. 1971. Expression of differentiated functions in hepatoma cell hybrids: reappearance of tyrosine aminotransferase inducibility after the loss of chromosomes. *Proc. Nat. Acad. Sci. USA* 68:3026–30

64. Croce, C., Litwack, G., Koprowski, H. 1973. Human regulatory gene for inducible tyrosine aminotransferase in rat-human hybrids. *Proc. Nat. Acad. Sci. USA* 70:1268–72

65. Bertolotti, R., Weiss, M. 1972. Expression of differentiated functions in hepatoma cell hybrids. II. Aldolase. *J. Cell. Physiol.* 79:211–24

66. Bertolotti, R., Weiss, M. 1972. Aldolases in hepatoma cell hybrids: extinction of the hepatic form and its re-expression following loss of chromosomes. In *Cell Differentiation*, ed. R. Harris, D. Viza, 202–5. Copenhagen: Munksgaard

67. Bertolotti, R., Weiss, M. 1972. Expression of differentiated functions in hepatoma cell hybrids. VI. Extinction and re-expression of liver alcohol dehydrogenase. *Biochimie* 54:195–201

68. Sparkes, R., Weiss, M. 1973. Expression of differentiated functions in hepatoma cell hybrids: alanine amino transferase. *Proc. Nat. Acad. Sci. USA* 70:377–81

69. Levisohn, S., Thompson, E. 1973. Contact inhibition and gene expression in HTC/L cell hybrid lines. *J. Cell. Physiol.* 81:225–32

70. Rintoul, D., Colofiore, J., Morrow, J. 1973. Expression of differentiated properties in fetal liver cells and their somatic cell hybrids. *Exp. Cell Res.* 78:414–22

71. Peterson, J., Weiss, M. 1972. Expression of differentiated functions in hepatoma cell hybrids: induction of mouse albumin production in rat hepatoma-mouse fibroblast hybrids. *Proc. Nat. Acad. Sci. USA* 69:571–75

72. Malawista, S., Weiss, M. 1974. Expression of differentiated functions in hepatoma cell hybrids: high frequency of induction of mouse albumin production in rat hepatoma-mouse lymphoblast hybrids. *Proc. Nat. Acad. Sci. USA* 71:927–31

73. Darlington, G. 1974. Production of human albumin in mouse hepatoma-human leukocyte hybrids. See Ref. 2, 159–62

74. Dannies, P., Tashjian, A. 1974. Unexpected production of a mouse serum protein by rat-mouse somatic cell hybrids. See Ref. 2, 163–72

75. Periman, P. 1970. IgG synthesis in hybrid cells from an antibody-producing mouse myeloma and an L cell substrain. *Nature* 228:1086–87

76. Coffino, P., Knowles, B., Nathenson, S., Scharff, M. 1971. Suppression of immunoglobulin synthesis by cellular hybridization. *Nature* 231:87–90

77. Mohit, B. 1971. Immunoglobulin G and free kappa chain synthesis in different clones of a hybrid cell line. *Proc. Nat. Acad. Sci. USA* 68:3045–48

78. Orkin, S., Buchanan, P., Yount, W., Reisner, H., Littlefield, J. 1973. Lambda-chain production in human lymphoblast-mouse fibroblast hybrids. *Proc. Nat. Acad. Sci. USA* 70:2401–5

79. Schwaber, J., Cohen, E. 1973. Human X mouse somatic cell hybrid clone secreting immunoglobulins of both parental types. *Nature* 244:444–47

80. Colten, H., Parkman, R. 1972. Biosynthesis of C4 (fourth component of complement) by hybrids of C4-deficient guinea pig cells and HeLa cells. *Science* 176:1029–31

81. Colten, H. 1972. *In vitro* synthesis of a regulator of mammalian gene expression. *Proc. Nat. Acad. Sci. USA* 69:2233–36

82. Sonnenschein, C., Richardson, U., Tashjian, A. 1971. Loss of growth hormone production following hybridization of a functional rat pituitary cell strain with a mouse fibroblast line. *Exp. Cell Res.* 69:336–44

83. Klebe, R., Chen, T., Ruddle, F. 1970. Mapping of a human regulator element by somatic cell genetic analysis. *Proc. Nat. Acad. Sci. USA* 66:1220–27

84. Kao, F., Puck, T. 1972. Genetics of somatic mammalian cells: demonstration of a human esterase activator gene linked to the ade B gene. *Proc. Nat. Acad. Sci. USA* 69:3273–77

85. Finch, B., Ephrussi, B. 1967. Retention of multiple developmental potentialities by cells of a mouse testicular teratocarcinoma during prolonged culture *in vitro* and their extinction upon hybridization with cells of permanent lines. *Proc. Nat. Acad. Sci. USA* 57:615–21

86. Jami, J., Failly, C., Ritz, E. 1973. Lack of expression of differentiation in mouse teratoma-fibroblast somatic cell hybrids. *Exp. Cell Res.* 76:191–99

87. Davis, F., Adelberg, E. 1973. Use of somatic cell hybrids for analysis of the differentiated state. *Bacteriol. Rev.* 37: 197–214

REGULATION: POSITIVE CONTROL

❖3072

Ellis Englesberg and Gary Wilcox

Section of Biochemistry and Molecular Biology, Department of Biological Sciences, University of California, Santa Barbara, California 93106

INTRODUCTION

The operon model of gene regulation as originally conceived and developed by Jacob & Monod and their collaborators (1–4) was a scheme of regulation based exclusively on a system of negative control. This concept of negative control was essentially derived from experiments with the *lac* operon in *Escherichia coli* K12. However, it was analysis of the data then available for the tryptophan operon and bacteriophage λ (2, 3), data which fitted so easily into a negative control model, that led Jacob & Monod to propose that all gene regulation was under negative control.

The discovery of regulatory genes and the enunciation of the operon model had a profound influence on biology, as we all realize today. However, at the time, the concept and experimental evidence supporting the operon model were difficult for many to follow. Because there were still many skeptics that needed convincing of the reality of regulatory genes, operons, and negative control, it appears that it was too much to ask that the workers in the field take time to consider the possibility that negative control was not the answer to *all* problems of gene regulation; there were so many biologists that still had to be convinced of the reality of these new concepts.

It was against this background of the ascendancy of the operon model with negative control as the integral part, that the discovery and elaboration of positive control of gene regulation took place. This concept, developed in the L-arabinose system, led to an elaboration of an operon model incorporating both negative and positive control (5–16). Although there was some indication that the repression of alkaline phosphatase synthesis in *E. coli* might be under positive control (18–22), the evidence was contradictory and these studies had little impact on the generalized operon-negative control model. Finally, recognition of the significance of positive control led Epstein & Beckwith in 1968 (17) in a review article to redefine the word operon. Their definition of the operon and its controlling sites is in common usage today. It represents a slight but most important modification of the original definition proposed by Jacob & Monod. Epstein & Beckwith defined an operon "as a

219

group of contiguous structural genes showing coordinate expression and their closely associated controlling sites." Controlling sites are defined "as elements which determine the expression of only those genes to which they are attached, i.e., they have 'cis dominant' effects." The importance of this definition is that it does not presume any particular mechanism of regulation, positive or negative.

This review is concerned mainly with an analysis of systems of regulation in bacteria where there is substantial evidence of positive control. We also summarize some of the work with eukaryotes where there is some presumptive evidence for positive control.

The study of regulation of gene expression in bacteriophage λ has led to the elucidation of at least two positive control systems involving genes N and Q. Gene regulation in λ has been the subject of recent reviews (23–27) and is not covered here because of space limitations.

Because we contrast positive and negative control it is best that we define these terms rigorously. In a negative control system the controlling sites are so coded that transcription occurs in the absence of any cytoplasmic regulatory component specific for the particular system. The operator controlling site is coded to recognize a *repressor,* a specific regulator for the particular operon. The repressor prevents the otherwise free expression of the operon. In a positive control system the coding at the initiator controlling site prevents expression of the operon in the absence of any specific regulatory product. A product from a specific regulatory gene, an *activator,* is required to turn on the operon at the initiator site. Both inducible and repressible systems may be under positive or negative control, depending on the state of the regulator and the controlling site (31).

GENETIC CRITERIA FOR A POSITIVE CONTROL REGULATORY GENE

In this section we summarize the genetic criteria that led to the establishment of a positive control system. These criteria apply equally to inducible or repressible systems.

1. *The occurrence of pleiotropic-negative mutants at a high frequency similar to that found in a structural gene may be taken as presumptive evidence for a positive regulator.* This high frequency is what one might expect of mutations in a structural gene where a negative phenotype is the result of nonspecific, general inactivation of the resulting gene product. A priori, one would expect many positions in a gene producing a positive regulator which would change the amino acid sequence of the gene product so that it could not function. In a purely negative control system the frequency of occurrence of pleiotropic-negative mutations in the regulatory gene, for example, $laci^s$ mutations, is quite low. This is to be expected because only limited, specific changes in the repressor molecule can occur so that it still retains its active site for attachment to the operator but still has a modified effector binding site. This severely restricts the positions in the gene where such i^s mutations can occur.

2. *The isolation of a deletion mutation (or a few different nonsense mutations) in the proposed regulatory gene whose phenotypic expression results in the failure to*

induce or derepress structural genes in one or more operons (pleiotropic-negative mutants, R-). A deletion or nonsense mutation in a negative control system leads to a constitutive phenotype. Thus, the isolation of a deletion or nonsense mutation in a regulatory gene that produces a pleiotropic-negative phenotype under conditions where possible polarity effects in the operon in question are eliminated is the most definitive evidence for positive control. A deletion is crucial in tests for dominance to rule out the possibility of subunit interaction (30). Also, if mutations can be isolated that allow expression of the structural genes in a strain containing a deletion of the regulatory gene, e.g. fusion to another operon or controlling site constitutives, then the possibility that the regulatory gene produces a common subunit for the structural gene products is eliminated.

3. *Demonstration that the regulatory gene is not part of the operon(s) that it is proposed to control.* This is done by demonstrating that nonsense mutations in gene *R* have no polar effect on the operon(s) in question and by mapping experiments that show the regulatory gene is not contiguous with the structural genes it controls. This is essential in order to eliminate the possibility that the effect of deletion or nonsense mutations is the result of producing a strong polar effect which shuts off the concerned operon.

4. *Isolation of constitutive mutants in the regulatory gene (R^c).* It should be possible to isolate mutations in the presumed regulatory gene so that the gene product functions in the absence of the normal effector. In a repressible system, the gene product would no longer recognize the corepressor so that the system would no longer be repressible. In an inducible system, the gene product assumes the activator form in the absence of the effector.

5. *Test for dominance: R^+ (wild-type, inducible allele) and R^c alleles should be dominant to R^-.* The demonstration of the dominance of R^+ and R^c to R^- serves to confirm the nature of the R^- mutation. (The $R^+ \times R^c$ test for dominance may be difficult to interpret because of the possibility of subunit interaction or because of the presence of a negative element in the control.) R^- alleles should complement mutations in the structural genes under question.

6. *The demonstration of cis-dominant mutations in the operon(s) that affect the expression of the R^c and R^+ alleles of the regulatory gene.* The characterization of such mutants serves as presumptive evidence that the gene in question is a regulatory gene.

7. *The finding of only cis-dominant mutations in the controlled operons as "revertants" of a deletion mutation in the regulatory gene.* This serves to eliminate alternative models according to which the real regulatory gene remains to be found. For instance, one alternative model states that the proposed regulatory gene is actually a structural gene for an enzyme which converts the external inducer to the real inducer. The real inducer then is supposed to neutralize a repressor produced by a still undiscovered regulatory gene. If one proposes that the function of this regulatory gene is vital to the organism, it would be necessary to look for heat-labile revertants of the R^- (deletion) mutants or to analyze in an in vitro system the nature of the proposed regulatory gene. This of course is the final and crucial criterion for establishment of the nature of the control exercised by the regulatory gene.

THE L-ARABINOSE GENE-ENZYME COMPLEX

General Aspects and Model

The L-arabinose gene-enzyme complex is a regulon consisting of at least three operons. We are mainly concerned with operon *araBAD* for which we have the most information (Figure 1). The regulatory gene *araC* for the regulon is located between the controlling sites for *araBAD* and *leu* (5, 7, 8, 12, 13, 32–34). Genes *araA, araB,* and *araD* are the structural genes for L-arabinose isomerase, L-ribulokinase, and L-ribulose-5-phosphate-4-epimerase, respectively, the first three enzymes involved in L-arabinose metabolism. The enzymes have been purified to homogeneity and characterized (31). There are at least two operons governing the transport of L-arabinose. One of these, consisting of gene *araE,* is closely linked to thymine *(thyA)* (31). Mutants in gene *araE* give a typical permease-less cryptic response (31, 35, 36). An L-arabinose-binding protein coded for by gene *araF* has been purified and directly implicated in L-arabinose transport (37–40). The *araF* gene has been tentatively mapped at 38–39 min on the circular *E. coli* map close to the structural gene for the galactose-binding protein (41, 42; R. W. Hogg, personal communication).

The regulon is unique in that it is under positive as well as negative control. L-arabinose serves not only as the inducer-effector of this system but also as a source of catabolite effectors (repressors) which reduce by approximately 50% the maximal rate of expression of the operon in a mineral L-arabinose medium (43). Thus, the steady-state rate of expression of the operon is determined not only by the concentration of L-arabinose as the inducer-effector of the operon, but also by one or more intermediates produced from L-arabinose which serve to dampen the expression of the operon (43, 44). The latter effect of self-catabolite deactivation (repression) is a general phenomenon observable in all inducible catabolic pathways studied (43). Because the phenomenon previously called catabolite repression is actually not repression of operon expression (no repressor is involved) but rather an effect on activation, the term catabolite deactivation (29), which we use in this review, more closely describes what is taking place.

Genetic evidence supports the following model of gene regulation of the *araBAD* operon. The regulatory product of gene *araC* is a protein molecule that can exist in two different functionally active configurations: that of a repressor (P1) or that of an activator (P2) of gene expression. The transition from one state to the other is mediated by the effector, L-arabinose. In the absence of L-arabinose, the equilibrium between P1 and P2 is mainly in the direction of P1 with P1 attached to the operator (*araO*) but with small but detectable amounts of P2 present. In the presence of L-arabinose, P1 is removed from *araO* and the equilibrium is shifted toward P2. P2 functions at the initiator site (*araI*) to facilitate expression of the operon. In this system, it must be emphasized that the "absence" of P1 is a necessary but not sufficient condition for the full expression of the operon. P2 must be produced from P1 and act at *araI* before the maximum expression of this gene complex can occur (13–16, 31–34). In addition to the two functions of repression and activation that occur in the *ara*-controlling site region, it has been shown that the segment previously designated *araI* also contains the site for "promoter" function, that is

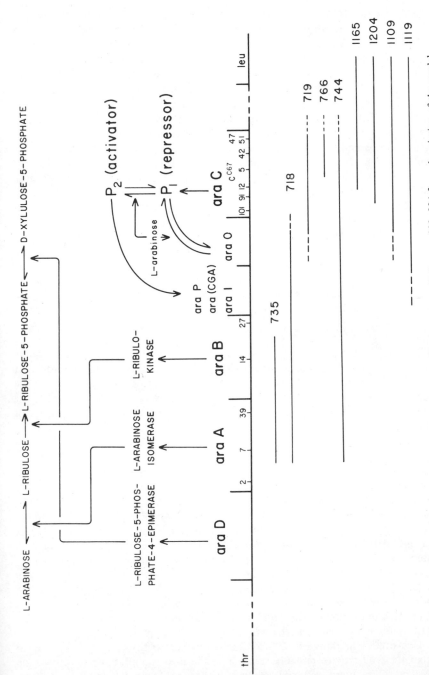

Figure 1 The *araBAD* operon with its controlling sites and the regulatory gene *araC*. See pp. 222, 224 for a description of the model.

the site for the RNA polymerase to bind to the DNA and initiate transcription of the operon (31–34). Moreover, evidence is consistent with placing the site of action of the catabolite gene activator protein, cAMP complex (CGA-cAMP) within the same region (45). To allow for the possibility that all three functions have separate sites of action and so as to focus attention on the structure-function relationships in this region, it has been proposed that individual sites be referred to as follows: *araI* (initiator, site for activator), *ara(CGA)* (catabolite gene activator site, site for action of CGA protein-cAMP), and *araP* (promoter-RNA polymerase site) (45). It should be pointed out that separate and distinct sites may not exist and, for instance, the activator and the CGA protein-cAMP complex may just prepare a site for polymerase binding and initiation of transcription. It has been shown that site *araO* is the first site in the operon (27–29), but the order of the *araI, araP, ara(CGA)* sites, if there are separable sites, remains unknown.

Why the "double lock" of both a repressor and an activator in the regulation of the *ara* regulon? It is possible that under some growth conditions *E. coli* may come in contact with compounds that resemble L-arabinose and thus be able to remove the repressor from the operator site without being able to convert it into the activator form. The requirement that the effector remove the repressor from the operator site and convert it into activator may severely reduce the degrees of freedom in the structure of the effector molecule, so that only L-arabinose has the required structure.

The Regulatory Gene araC

There are three different sets of alleles in the *araC* gene: 1. *araC$^+$*, the inducible wild-type allele—the isomerase, kinase, and epimerase and an L-arabinose transport system are inducible by L-arabinose. 2. *araC$^-$*, the pleiotropic L-arabinose-negative allele—the above products are no longer inducible by L-arabinose. 3. *araC^c*, the pleiotropic constitutive allele—the above products are present in the absence of inducer.

araC$^-$ mutants have been ordered by reciprocal three-factor crosses and deletion mapping (5, 13, 31, 46). There is no complementation in crosses between 15 *araC$^-$* mutants, indicating that the *araC* gene is composed of one cistron (47). Four different *araC$^-$* mutants, mapping in different positions, are the result of nonsense mutations. Several deletions have been characterized that excise different portions of the *araC* gene (Figure 1). These deletions produce a pleiotropic L-arabinose-negative phenotype, similar to that of *araC$^-$* nonsense or missense mutations (14–16, 32–34). All the *araC$^-$* mutants so far tested (nonsense, missense, and deletion mutants) are recessive to *araC$^+$* (*cis* and *trans*), are complemented by *araA$^-$*, *araB$^-$*, and *araD$^-$* alleles, and the pleiotropic-negative phenotype is not the result of a polarity effect on the *araBAD* operon. The results offer conclusive proof that the *araC$^+$* allele produces a protein in the presence of L-arabinose necessary for the expression of the L-arabinose regulon. *araC$^-$* mutants constituted one fourth of all *Ara$^-$* mutants isolated from the wild-type strain (see pp. 220–21).

Constitutive mutants (*araC^c*) for the L-arabinose pathway have been isolated from the wild type as mutants resistant to the D-fucose inhibition of growth on mineral L-arabinose D-fucose medium and from D-ribulokinase-less mutants (dK⁻) as double mutants (dK⁻$araC^c$) able to grow on mineral D-arabinose medium (13, 31). The $araC^c$ mutants were mapped by recombination frequencies and by deletion mapping. $araC^c$ mutant sites are not randomly distributed throughout the $araC$ gene, but are concentrated in the left half of the gene. There does not seem to be any pattern yet discernible in the distribution of $araC^c$ fucose-sensitive alleles among the $araC^c$ fucose-resistant alleles (13, 26). $araC^c$ strains have different but coordinate constitutive levels of the enzymes and permease. This is of some interest as the operons for the L-arabinose permease system are not linked to the $araBAD$ operon. Those strains with constitutive levels below the wild-type inducible level are further inducible by L-arabinose (13). Some are now inducible by D-fucose (48).

$araC^c$ is dominant to $araC^-$ (14, 15, 47). This fits the model and is explained on the basis that the $araC^c$ allele produces an activator in the absence of L-arabinose which is able to turn on the operon both cis and trans to the $araC^c$ allele. Thus, dominance of both $araC^+$ and $araC^c$ alleles to $araC^-$ is consistent with the fact that $araC^-$ mutants can be the result of deletions and nonsense mutations.

$araC^+$ is dominant to $araC^c$. In merodiploids of the type $F'A^+B^-C^+/A^-B^+C^c67$ the constitutive levels of both isomerase and kinase that one would have anticipated on the basis of the haploid constitutive level of the $F^-A^+B^+C^c67$ strain are reduced about fifteenfold in the merodiploid. With the addition of L-arabinose to the growth medium, induction occurs and isomerase levels, for instance, are increased in most cases to wild-type induced levels (13). This experiment presented the first bit of evidence suggesting negative control in this operon although there were alternative models considered at the time. Supporting the conclusion that $araC$ produces a protein that can act as both a repressor and activator depending on the presence of the effector is the finding that both repressor and activator function are heat-labile in temperature-sensitive $araC^-$ mutants (49). On the basis of subsequently defining an operator site, these results are best interpreted on the basis that P1 in the absence of L-arabinose acts as a repressor essentially preventing the expression of the operon by the $araC^c$ activator. When P1 is attached to the operator site, transcription of the operon for the most part ceases even in the presence of the $araC^c$ activator.

Not only is the constitutive phenotype of $araC^c$ mutants isolated as resistant to D-fucose inhibition recessive to $araC^+$, but the fucose-resistant phenotype is also recessive. Several dominant trans-acting constitutive derivatives of $araC^c67$ (a constitutive isolated as resistant to D-fucose) have recently been isolated (50; T. Su and N. Lee, unpublished). The mutants fall into two classes: 1. dominant in both constitutivity and D-fucose resistance and 2. dominant in D-fucose resistance only. Both classes are inducible by D-fucose. We presume that this dominance is the result of subunit interaction with the $araC^+$ gene product. Three of the mutants isolated by Su and Lee (unpublished) appear to be CGA protein-cAMP-independent (see section on controlling sites).

Ara⁺ revertants were isolated from a double mutant deficient in CGA protein and adenylcyclase (L. Hefferman and E. Englesberg, unpublished data; M. Casadaban and J. Beckwith, unpublished data). Studies with pertinent merodiploids indicate that the product is trans-acting and that the mutation maps in $araC$. A complication in the isolation of *cis*-acting Ara$^+$ revertants in the *araBAD*-controlling sites is the fact that the permease genes are not in the *araBAD* operon. Because the permease is strongly affected by catabolite deactivation, selection for Ara$^+$ revertants of the CGA protein adenylcyclase-deficient double mutant forces selection for a *trans*-acting product that would make the entire regulon inducible in the absence of CGA protein and cAMP. The finding of an alteration in the *araC* gene suggests that P2 can partially replace the CGA protein-cAMP complex indicating a similarity in function for the specific and general positive regulators.

The Controlling Sites of the araBAD Operon

Polarity of the *araBAD* operon is in the direction *araB, araA, araD*, suggesting that the controlling elements for this operon must be at the end of the *araB* gene farther from gene *araA* (31). Nonsense mutations in gene *araC* have no *cis* effect on the *araBAD* operon as determined with merodiploids of the type $F'A^-C^+/A^+C^-$ (nonsense) (47), indicating *araC* is not in the *araBAD* operon.

Deletions that extend from the *araB* gene into the *leu* operon remove the remaining structural genes of the *araBAD* operon from the control of L-arabinose and a *araC$^+$* allele in the *trans* position (in merodiploids), and place these genes under leucine control (13). With deletions that extend from before or beyond the *leu* operon into but not beyond the *araC* gene, the *araBAD* operon remains under L-arabinose control (34). Thus, the direction of polarity in the operon, the finding that *araC* is not in the *araBAD* operon, and the observed differences between strains containing deletions of the type *araB araC-leu* on the one hand, and *araC* or *araC-leu* on the other, offer conclusive proof that the controlling site region of the *araBAD* operon is located between genes *araB* and *araC*. A large number of deletions in the *araBAD* operon have recently been isolated and mapped in *E. coli* K12 but have not been further characterized (51).

Further analysis of several deletions that cut into the *araC* gene and the controlling site area has led to the discovery of an operator site (*araO*) and its ordering as the first controlling site of the *araBAD* operon. The initial observations were based on experiments with deletions *766* and *719* (see Figure 1) (16, 32, 33). Merodiploids of the type $F'A^-C^+/\Delta 719$ have a thirtyfold increase in isomerase activity over the haploid deletion *719* strain in the absence of inducer. In contrast, similar merodiploids constructed with deletion *766* have at most a twofold increase in isomerase activity over the haploid strain. The increase in the isomerase activity observed with the deletion *719* merodiploids was explained by assuming that deletion *719* excises all or part of a proposed *araO* site. Strains lacking *araO*, e.g. containing deletion *719*, are sensitive to the small quantities of P2 that exist in equilibrium with P1 (supplied by the episome), causing genes *cis* to the deletion to be partially "expressed." Merodiploids containing deletion *766*, with both operator and initiator sites intact, have little or no increase in isomerase activity (33). In addition, deletions *1165* and *1204* beginning in the leucine operon but excising much more in the *araC* gene than deletion *766*, give results similar to deletion *766* (34, 52). Similar results as that obtained with deletion *719* were also obtained with deletion *1109* (34).

Further characterization of the *araO* site and the location of the *araI* site is based upon complementation analysis with initiator constitutive mutants (I^c) isolated as "revertants" of deletion mutants *719* and *766* subsequent to mutagenesis with diethyl sulfate (32) and 2-amino purine (52), respectively. No spontaneously occurring *araI^c*'s were detectable down to a spontaneous mutation frequency of less than 10^{-10}. All have constitutive levels of L-arabinose isomerase and L-ribulokinase (epimerase activity was not measured) ranging up to 33% of the induced wild type and are not further inducible in the *araC* deletion background.

The *araI^c* mutant sites were mapped to the left of *araO* and closely linked to deletion *719* and deletion *1109* (32, 52). *AraI^c* has been shown to be *cis*-dominant and *trans*-recessive to the *araI^+* allele. In merodiploids of the type $F'A^-C^-/A^+I^c$ $\Delta719$, in the absence of inducer, the isomerase level of the merodiploid is similar to that of the corresponding $F^-A^+I^c\Delta719$ strain. When an *araA^-C^+* allele is substituted for the *araC^-* allele on the episome in these merodiploids, isomerase levels (in the absence of inducer) in several cases increase significantly, although in a few they remained the same as that of the F^- haploid strain (or that found in the merodiploid $F'A^-C^-/A^+\Delta719$). With the *araC^+* allele in the *trans* position the isomerase is further inducible although to different extents depending on the *araI^c* mutation. The change in sensitivity to activator in different *araI^cΔ719* strains containing an *araA^-C^+* episome, in the absence and presence of inducer, and the fact that this type of mutation can be induced by 2AP is good presumptive evidence that the effect results from a change in the activator-sensitive site and is probably not the result of the insertion of a new promoter.

In contrast to the results obtained with the various *araI^cΔ719* mutants, *araI^c* $\Delta766$ strains containing the same *araI^c* mutations have severely repressed isomerase levels in the merodiploids of the type $F'A^-I^+C^+/A^+I^c\Delta766$. This repression is alleviated when a *araC^-* allele is substituted for the *araC^+* allele on the episome. Thus, the *araC^+* allele and not the *araC^-* allele exerts a severe epistatic effect on the expression of the *araI^c* alleles when it is *cis* to deletion *766* but not when *cis* to deletion *719*. Similarly, when the 2-AP-induced mutants isolated in a strain containing deletion *766* are converted to *araI^cC^+* by transduction, the *araC^+* allele exerts a similar repressive effect on the expression of the *araI^c* phenotype observed with *araC^+* *trans* to the *araI^cΔ766.

These results support the observation made with the *araC^+* allele *trans* to deletions *719* and *766*, and they can be best explained on the basis that the *araC^+* allele in the absence of arabinose produces a repressor which recognizes an operator site present in deletion strain *766* but excised in deletion strain *719*. The presence of P1 repressor interferes with the expression of the *cis*-acting constitutive *araI^c* allele. This evidence, together with the observed dominance of *araC^+* to *araC^c*, as mentioned above, demonstrates the repressor function of the *araC* gene product in the absence of L-arabinose. The addition of L-arabinose to a merodiploid $F'A^-C^+/A^+I^c\Delta766$ or to an $F^-I^cC^+$ induces the operon *cis* to the *araI^c* mutations, demonstrating the conversion of the repressor to activator. The finding that there is a low frequency of "reversion" of an *araC* deletion to Ara^+ and that all 29 independently isolated revertants map in the controlling site region of the *araBAD* operon argues against an alternative model to positive control based upon the production of an internal inducer by the *araC* gene that in turn inactivates a repressor produced by a yet undiscovered regulator gene (see pp. 220–21).

On the basis of what is known about the mode of action of the CGA protein-cAMP complex, it has been proposed that the site of action of this complex in the *araBAD* operon is the identified controlling site region (28). Several experimental approaches have been undertaken to test this assumption.

A deletion analysis involving the construction and analysis of reciprocal merodiploids in which alternately the episome and the chromosome in the merodiploid pair served as an internal control for catabolite deactivation is consistent with placing the site for catabolite deactivation in the segment between *araB* and *araO*. It also appears that in a *leu-ara* deletion, if transcription begins from the *leu* operon, even though there is a site remaining that makes the *ara* system subject to catabolite deactivation, *leu* promotion is immune to this phenomenon. Thus, catabolite deactivation is not involved in the elongation of the message.

araI^{c}Δ766 mutants were cycled in mineral arabinose medium, and *araI^{c}Δ1165* mutants in a dK$^-$ background (D-ribulokinase-deficient) were cycled in D-arabinose medium; strains that fermented L-arabinose rapidly were isolated on EMB-L-arabinose plates. Among dozens of revertants, some linked to *araBAD* and some unlinked, several were found that mapped in the region between *araO* and *araB*. One such mutant *(araI^{c}I$^{(c)}$)*, which contains two mutations in the region between *araO* and *araB*, is *cis* dominant, has a higher constitutive level of expression of *araBAD* operon, and is less sensitive than the wild-type or *araC^c* mutant to catabolite deactivation. In the presence of an *araA$^-$C$^+$* episome an epistatic effect is shown, but upon the addition of L-arabinose, L-arabinose isomerase is fully induced. Thus, the double mutant is similar to the *araI^c* mutations but, in addition, is less sensitive to catabolite deactivation and partially independent of CGA protein-cAMP control (J. Colomé and E. Englesberg, unpublished data).

These experiments support the conclusions based upon the deletion analysis and offer evidence that the *ara(CGA)* site does reside in the area between *araO* and *araB*. Nineteen other *araI^c* mutants failed to show any major differences in sensitivity to catabolite deactivation. Thus, there seems to be just a partial overlapping of the *araI* and *ara(CGA)* controlling sites. Perhaps, however, a more careful analysis of the 19 *araI*'s in a double mutant deficient in CGA protein and adenylcyclase may reveal some significant effects of these *araI^c* mutations on catabolite deactivation.

Sheppard and his collaborators (53) have isolated several spontaneous, pleiotropic arabinose-negative, *cis*-dominant, *trans*-recessive, revertible mutants that map between *araO* and the right-most point mutation in *araB*. Evidence suggests that some of these mutations may reside in the controlling site region, but which site(s) is involved has not as yet been determined.

Molecular Basis of Regulation of the araBAD Operon

Regulation of the L-arabinose operon *BAD* in vitro has been studied using cell-free protein synthesizing preparations. Activation of the *araBAD* operon, as measured by the synthesis of either L-arabinose isomerase or L-ribulokinase, requires a DNA template containing the *araBAD* operon and the addition of extracts containing *araC* protein (54–56). As expected, the small molecule effectors L-arabinose and cAMP (54–57, 122) are required to obtain expression of the operon in vitro. In addition, D-fucose, an anti-inducer in vivo, prevents expression of the *araBAD* operon in the presence of L-arabinose in vitro (54, 55). *araC* ^{c} protein extracted from a strain resistant to D-fucose inhibition is insensitive to D-fucose in vitro (55). Thus, the expression of the *araBAD* operon in vitro is under the positive control of the *araC* gene product, as predicted on the basis of in vivo genetic experiments.

If *araI* ^{c} *I* $^{(c)}$ Δ *766* DNA (see pp. 226–28 for the phenotype of *araI* ^{c} *I* $^{(c)}$ mutations) is used to program the cell-free system, the L-arabinose operon is expressed in the absence of either *araC* protein or L-arabinose, demonstrating in vitro the constitutive nature of the *araI* ^{c} *I* $^{(c)}$ mutations. If *araC* protein is added to the cell-free system in the absence of L-arabinose, the rate of expression of the operon is decreased, demonstrating the repressor activity of *araC* protein. Furthermore, if L-arabinose and *araC* protein are added together, the *araBAD* operon *cis* to the *araI* ^{c} *I* $^{(c)}$ mutations is further activated to a level consistent with the levels observed in vivo. In addition the *araI* ^{c} *I* $^{(c)}$ mutations confer partial independence from the CGA protein-cAMP system in vivo. Similarly, in the cell-free system programmed with *araI* ^{c} *I* $^{(c)}$ DNA one can detect expression of the *araBAD* operon in the absence of cAMP (G. Wilcox, unpublished data). Thus, both forms of *araC* protein are observable in vitro, dependent on the presence of L-arabinose and sensitive to the addition of D-fucose. The CGA protein-cAMP complex interacts with the *araI* region, and in the case of *araI* ^{c} *I* $^{(c)}$ mutations this control is relaxed and expression of the operon can occur in the absence of cAMP in the cell-free system. All of the macromolecular interactions thought to occur in vivo in the region of the *ara* controlling sites are observed in the cell-free system and confirm the predictions of the model (54).

araC protein can also be synthesized in the cell-free system. If the system is programmed with an *araA* $^+$ *B* $^+$ *C* $^+$ DNA template, the operon is expressed in the absence of added *araC* protein as indicated by the synthesis of L-ribulokinase (54, 57) and L-arabinose isomerase (54). Neither an *araC* $^+$ *A* $^-$ nor an *araC* $^-$ *A* $^+$ template is capable by itself of stimulating the synthesis of L-arabinose isomerase, but a mixture of the two DNAs behaves like *araA* $^+$ *B* $^+$ *C* $^+$ DNA (54). This demonstrates that the *araC* $^+$ template is able to provide a product, the *araC* protein, which is required for the expression of the *araBAD* operon in the *trans* position.

The *araC* protein produced in vitro has been compared with *araC* protein purified from whole cells. The apparent affinities for L-arabinose and D-fucose are the same for *araC* protein from either source as measured by the concentration dependence of activation of the *araBAD* operon in the cell-free system. The results suggest that the two products are similar, if not identical. The *araBAD* operon is expressed coordinately in the cell-free system, and the appearance of L-ribulokinase, the promoter proximal gene product, precedes the appearance of the more distal gene product, L-arabinose isomerase. The appearance of L-arabinose isomerase activity is delayed in the system programmed with *araA* $^+$ *B* $^+$ *C* $^+$ DNA when compared to the system to which *araC* protein purified from whole cells has been added. This delay represents the time required to synthesize and assemble an amount of *araC* protein required to turn on the operon (54).

The direction of transcription of the *araC* gene has been determined. $\lambda paraC$ phage has been isolated which carries an intact *araC* gene and approximately one half of the *araB* gene. An *araC3* derivative of this phage has been constructed by P1 transduction. *araB* message hybridizes to the heavy but not the light strand, indicating that the sense strand of the *araBAD* operon would be the heavy strand. Hybrid DNA molecules between the *araC⁺* and *araC3* phage were made by reannealing the separated strands and used to program the in vitro system along with $\lambda h80daraC^-(\Delta 766)$ DNA. *araC* protein is synthesized in the in vitro system only when the light strand comes from the $\lambda paraC^+$ phage, indicating that the sense strand for the *araC* gene is on the light strand. Thus, the *araC* gene is transcribed in the opposite direction from the *araBAD* operon (59).

For the purification of the *araC* protein, assays have been developed for both the repressor (P1) and activator (P2) forms (29, 54, 58). *araC* protein has been purified by affinity chromatography on Sepharose 4B to which an anti-inducer of the L-arabinose operon has been covalently attached (29, 58). The purified *araC* protein has both repressor (29, 58) and activator (54) activity. Further purification on salmon-sperm DNA cellulose yields a product that is greater than 90% *araC* protein and possesses both activator (unpublished data) and repressor (60) activities, thus providing further evidence for the proposed dual function of *araC* protein.

The binding of *araC* protein to *ara* DNA in the absence of L-arabinose has been studied using the nitrocellulose membrane filter technique. *araC* protein used in these studies is purified to apparent homogeneity in the absence of L-arabinose. Equilibrium competition experiments using *ara* DNAs containing different deletions of the controlling site region have demonstrated that *araC* repressor binds specifically to the *ara* operator. The apparent K_m of the interaction as determined in equilibrium conditions is $10^{-12}M$. The rates of association and dissociation of the P1-*araO* complex have been determined. A K_a of $2 \times 10^9 M^{-1}sec^{-1}$ and a K_d of $4 \times 10^{-3}sec^{-1}$ are calculated, assuming binding to a single operator site. The equilibrium constant calculated from the ratio of K_a to K_d, $2 \times 10^{-12}M$ is in good agreement with the equilibrium determination and suggests that the kinetic studies are providing true rate constants (60).

L-arabinose does not affect the binding of *araC* protein to *araO* under the conditions of the filter binding assay. This could be due to a partial denaturation of the *araC* protein or to the lack of a necessary component in the binding buffer. It has been suggested that perhaps CGA protein or RNA polymerase are required along with L-arabinose for *araC* protein to dissociate from *araO* and slide down the DNA to *araI* where it is involved in the initiation of transcription (60).

Dithiothreitol, EDTA, glycerol, L-arabinose, and phenylmethyl sulfonyl flouride (PMSF) are effective stabilizing agents for *araC* protein (61). PMSF is a well-known inhibitor of serine-proteases, suggesting that *araC* protein may be susceptible to protease attack (54, 57). The inclusion of the above compounds in the purification buffers confers stability on *araC* protein in crude extracts, but as *araC* protein is further purified, it becomes progressively more unstable at 37°C (61). However, preparations containing 30% P2 can be stored for 3–4 months in liquid nitrogen without apparent loss of activity.

Our best estimates place the minimum number of monomers of *araC* protein per haploid cell at about 20 (62). The presumed susceptibility of *araC* protein to attack by proteases may be a way in which the *araC* protein concentration is controlled in vivo. In the absence of L-arabinose the *araC* protein is made but broken down rapidly by protease action. The presence of L-arabinose may protect the otherwise labile *araC* protein from degradation by the protease(s) and may allow concentrations of *araC* protein required for full induction of the L-arabinose regulon to be realized.

The subunit molecular weight of *araC* protein is 28,000 based on electrophoresis in SDS polyacrylamide gels (60). Recently, it has been possible to cut out the 28,000 dalton band from an SDS gel, reconstitute *araC* protein, and recover the activator form as assayed in the cell-free system, eliminating the possibility that *araC* protein was a minor component of the SDS gels (63).

AraC activator has a sedimentation coefficient of 4S on either 5–20% sucrose or glycerol gradients (61). The 4S form is observed at low concentrations of *araC* protein and over a wide range of concentrations in the presence of L-arabinose. However, in the absence of L-arabinose and at high concentrations, an association to a faster sedimenting form occurs as indicated by an increased variance and faster sedimenting peak. The same phenomenon occurs in the presence of D-fucose. If *araC* protein is a globular protein, the 4S form corresponds to a dimer. As it is the predominant form in the presence of L-arabinose it may be that P2 is a dimer. The presence of higher molecular weight forms in the absence of L-arabinose or in the presence of D-fucose suggests that the repressor may be of higher molecular weight, perhaps a tetramer. These results also provide evidence that L-arabinose interacts directly with *araC* protein.

The affinity of L-arabinose and D-fucose for *araC* protein has been measured by indirect methods both in vivo (64–66) and in vitro (54, 55), and the values obtained are on the order of 10^{-3} to $10^{-2} M$. The in vivo studies are based on the assumption that the internal concentration of L-arabinose is accurately known, and both in vivo and in vitro estimates assume that induction of the operon is proportional to the binding of L-arabinose to *araC* protein. The interaction of L-arabinose and D-fucose with purified *araC* protein has been studied using spectrofluorimetric methods (66). *araC* protein has a fluorescence spectrum typical of tryptophan-containing proteins. The fluorescence is quenched by L-arabinose, suggesting that L-arabinose causes at least one conformational change in *araC* protein. A quenching of fluorescence does not occur in the presence of D-fucose. If one assumes the quenching reflects the binding of L-arabinose, the apparent K_m of the interaction with L-arabinose is 3 × $10^{-3} M$ and the K_i of the interaction with D-fucose is 6 × $10^{-3} M$. The "conformational" change induced by L-arabinose may reflect the transition from P1 to P2. The interaction with D-fucose does not result in the conformational change which yields P2, consistent with the anti-inducer activity of this analog of L-arabinose.

L-arabinose is considered to be the inducer and effector of the L-arabinose regulon because *araA⁻* strains have fully induced L-ribulokinase levels when grown in the presence of L-arabinose. Further support for this conclusion comes from the isolation of *araC^c* strains which were further inducible by D-fucose, a nonmetabolizable analog of L-arabinose. The difference in sedimentation properties of *araC* protein dependent on the presence of L-arabinose and the direct demonstration in the spectrofluorimeter of a quenching of fluorescence caused by L-arabinose proves that *araC* protein interacts directly with L-arabinose.

Several in vivo studies have been directed toward determining whether repression and activation occur primarily at the transcriptional or translational level. *araBAD* mRNA, as detected by hybridization to *λh80dara* DNA, is inducible (67–70). *araBAD* mRNA is present in the absence of L-arabinose in an *araC^c* strain (69, 70). In a merodiploid of genotype *F'araC⁺B⁻/araC^del araO⁺araI^c*, the repressor activity of *araC* protein (P1) was shown to decrease the levels of *araBAD* mRNA present in the cell (69). The half-life of *ara* mRNA is 3 min at 30°C in the presence or absence of L-arabinose. However, a lag before *ara* mRNA decays has been observed in cells treated with rifampin but not in cells exposed to D-fucose, which could indicate that *araC* protein protects *ara* mRNA from an initial step in degradation (70). The conclusion that P1 and P2 act on the transcription process depends on the validity of the assumption that the hybridization technique can detect *ara* mRNA independent of translation.

Recently, there has been direct evidence that the activator form of *araC* protein, P2, is required for transcription of the *ara* operon (63). In a system employing purified components (*λh80daraBAD* DNA, purified RNA polymerase holoenzyme, and CGA protein), synthesis of *ara* mRNA shows an absolute requirement for the presence of *araC* protein. The synthesis of *ara* mRNA in vitro requires the CGA system, consistent with the finding that arabinose enzymes are sensitive to catabolite deactivation in vivo and that an *ara⁺crp⁻cya⁻* cell has an arabinose-negative phenotype. D-fucose, a potent anti-inducer for the *ara* operon, also inhibits the synthesis of *ara* mRNA.

THE MALTOSE GENE-ENZYME COMPLEX

General Aspects

The first two enzymes involved in the metabolism of maltose, the components of the maltose permease system, and the receptor for phage λ are all induced by maltose and controlled by regulatory gene *malT* in *E. coli* K12 (71–77). The maltose gene-enzyme complex is a regulon comprising two unlinked regions on the *E. coli* chromosome, *malA* and *malB*. The *malA* region (Figure 2) is situated between *glpR* and *aroB* at 65 min on the genetic map and closely linked to the former (78, 79). It consists of operon *malPQ*, which contains genes *malP* and *malQ*, the structural genes for the enzymes maltodextrin phosphorylase and amylomaltase, respectively. Amylomaltase converts maltose to glucose and to maltodextrin, and the phosphorylase converts maltodextrin to glucose-1-phosphate. *malT*, the regulatory gene for this entire complex, is located adjacent to the controlling sites for this operon

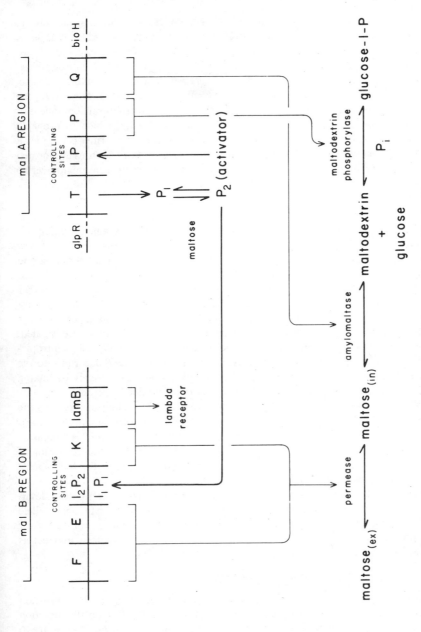

Figure 2   The maltose gene enzyme-complex. See pp. 232–36 for a description of the model.

(75–77). The *malB* region contains all the known genes involved specifically in maltose transport and phage λ adsorption (74). The genes are located between *metA* and *purA* (79–82). Recent experiments suggest that *malB* is composed of two operons. One operon includes *malK*, and *lamB*, cistrons involved in maltose permeation and λ receptor synthesis, respectively. *lamB* appears to be monocistronic, although there is evidence for intracistronic complementation (83). The *lamB* gene product is presumably the protein recently shown to be a structural component of the λ receptor (84). The other operon contains *malE* and *malF*, which are probably two different cistrons also involved in maltose permeation, the former coding for a periplasmic maltose protein (85). The controlling sites for both operons are located between *malE* and *malD*, and evidence suggests that the two operons might overlap in the region of their promoters (82).

The Regulatory Gene malT

The *malT* gene is defined by pleiotropic-negative mutations among which are several nonsense, deletion, and missense mutations. These mutations result in a very low, noninducible level of phosphorylase, amylomaltase, and maltose permease and result in resistance to phage λ by preventing the induction of the phage receptor protein (75–77). These mutations occur at a relatively high frequency and genes for permease and phage receptor are not in *malPQ* operon (see pp. 220–21). Beginning with a pleiotropic-negative mutation in the *glpR* gene, presumably a super-repressor type of mutation, a series of *glp⁺*, *glp*-constitutive deletion mutants were isolated. These deletions terminating within the *malT* gene all have the typical pleiotropic-negative phenotype described above. Deletions that excise *malT* and run into the *malPQ* operon give rise to various levels of amylomaltase, presumably by linking the *malPQ* operon to the *glp* operon or to some unidentified operon.

   MalT⁺ has been shown to be dominant to several *malT⁻* mutants, supporting the conclusion that the *malT* gene product is required for the expression of the *malPQ* operon and the operons in the *malB* region (78). It is clear that *malT* is a separate operon from *malPQ*. Nonsense mutations in *malT* do not affect the expression of the *malPG* operon (77, 78). Furthermore, the controlling sites for the *malPQ* operon are between *malP* and *malT* (78, 83).

   Constitutive mutants have not been isolated in *malT*, and therefore the dominance relationship between the possible constitutive alleles and the inducible allele cannot at present be determined. Thus we are deprived of one means of testing for the possibility that the gene *malT* product, in the absence of maltose, may act as a repressor as in the *ara* system.

The Controlling Sites of the malPQ Operon

A study of the effect of nonsense mutations in *malP* and *malQ* on the expression of the phosphorylase and amylomaltase indicates that polarity is in the direction *malP* to *malQ*, suggesting that the controlling sites for the *malPQ* operon are to the right of *malP* (see Figure 2) (77). *malT* is not in the *malPQ* operon because nonsense mutations in *malT* have no *cis* effect on the expression of the adjacent *malP* and *malQ* structural genes. Studies with deletions that go from *glpR* into *malT* together with the above evidence offer conclusive proof that the controlling sites for operon *malPQ* lie to the right of gene *malP*, between *malP* and the regulatory gene *malT*. No deletion so far has been found that removes a putative operator site, such as those found in the *ara* system. However, out of hundreds of *ara*⁻ deletions isolated and characterized (31; E. Englesberg, unpublished data), only two have been found that yield this repressor-independent phenotype. Thus, there is still the possibility that such a deletion may be isolated in the maltose system.

As we have pointed out above, *malT*⁻ mutations result in the noninducibility of the *malPQ* operon as well as the operons located in the *malB* region. *malT*⁻ cells have approximately 1–2% of the level of expression of the *malPQ* operon and are devoid of maltose permease and λ receptor (and are resistant to phage λ). *mal*⁺ revertants were sought from strains carrying deletions originating in *glpR*, (or in some unidentified operon), and excising various portions of *malT*, or excising *malT* and portions of the controlling site region, and some that continue into the structural genes of the *malPQ* operon (86). Revertants were found, however, only with strains carrying two specific deletions (deletions 3 and 5). Both of these deletions terminated between the *malT* mutation mapping farthest to the right and the *malP* mutation mapping farthest to the left.

Complementation analysis indicated that the strain containing deletion 3 is complemented by a *malT*⁺*P*⁺*Q*⁻ episome and amylomaltase is fully inducible, indicating that the controlling site region that responds to the *malT* gene product is still present in this deletion mutant. On the other hand, the deletion 5 mutant is not complemented by the same episome. Thus, deletion 5 would appear to have excised some of the controlling elements of the *malPQ* operon, either the promoter or the initiator, or both.

Revertants isolated directly from these deletion mutants acquire the ability to transport maltose but are still resistant to λ and have the same constitutive levels of amylomaltase and phosphorylase as the parent strain. The secondary mutations characterize a gene called *bymA*, unlinked to the *malA* and *malB* regions of the chromosome and located between *malB* and *leu*. The mutations appear to have uncovered or modified an existing permease gene, which is now able to transport maltose. There are apparently sufficient constitutive amylomaltase and phosphorylase to permit growth on maltose in the presence of this "new" maltose permease function.

By constructing strains carrying the above short deletions ending in *malT* and the gene *bymA*, another set of "revertants" were isolated. In some respects, these mutations resemble phenotypically the initiator constitutive mutations in *ara* and are called *malI*ᶜ₄. Genetic crosses locate the *malI*ᶜ₄ mutations close to *malT*. However, it cannot be excluded that some or all of the *malI*ᶜ may be in the right-most segment of the *malP* gene, although this does not seem likely. The *malI*ᶜ₄ mutants have been separated from the deletions, and the results of their analysis can best be presented by comparing their characteristics with that of the *araI*ᶜ's.

1. The degree of constitutively resulting from the *malI*ᶜ₄ mutations, in general, is similar to that resulting from *araI*ᶜ mutations.

2. Only some of the $malI^c_A$ mutants, 5 out of 31, are still subject to induction by maltose, that is, are under the control of the $malT$ gene, while all of the 29 $araI^c$ mutants are fully inducible by L-arabinose. It seems likely that when control by $malT$ is lost, the $malI^c_A$ mutations could be due to the insertion of a new promoter into the normal controlling sites of the mal operon. It would be of interest to know whether there is any change in sensitivity to catabolite deactivation in the $malI^c_A$ mutants.

3. In no case is the constitutive phenotype of a $malI^c_A$ mutation repressed by a $malT^+$ allele. However, because only five out of all I^c_A's are still under maltose control, these five would be the only ones that could possibly be expected to respond to the control of the $malT$ gene product and therefore to be repressed by the $malT^+$ allele, in the absence of maltose—that is, of course, if the product produced by the $malT$ gene under these conditions does exist as a repressor.

Although the data do suggest that the $malT$ product is not a repressor in the absence of maltose, the possibility exists that the five inducible $malI^c_A$'s also originated by an insertion of foreign DNA which could have destroyed operator function but still conserved the original initiator site. Under these conditions the $malI^c_A$ would still respond to the activator form of the $malT$ product but not respond to the repressor. The failure so far to isolate constitutive mutants in $malT$, which would have provided a means for detecting repressor function, and the fact that the I^c_A's may have originated by means of an insertion, place some doubt on the conclusion that the system may be one of pure positive control.

OTHER SYSTEMS

Three genes, which seem to be structural genes for L-rhamnose isomerase ($rhaA$), L-rhamnulokinase ($rhaB$), and L-rhamnulose 1-phosphate aldolase ($rhaD$), have been identified and mapped in a cluster near the $metB1$ marker on the $E.$ $coli$ K12 chromosome (87). A presumptive regulatory gene, $rhaC$, is linked to $rhaB$. The following evidence suggests that $rhaC$ produces a positive regulator for the $rhaBAD$ operon: 1. the high frequency of isolation of pleiotropic-negative mutants in the $rhaC$ gene—about one fourth of all Rha^- mutants map in the $rhaC$ gene—and 2. the $rhaC^+$ allele is dominant to $rhaC^-$. Without further analysis of the phenotypic effect of deletion mutants in the $rhaC$ gene, any conclusion with regard to the mode of regulation would be premature.

There is presumptive evidence that cysteine biosynthesis in $E.$ $coli,$ and probably in $Salmonella$ $typhimurium,$ is under positive control. This work has been reviewed recently by Gross (88). The $cysB$ gene apparently produces a product necessary for the function of several structural genes. Constitutive mutants have been isolated in this gene in $S.$ $typhimurium.$ The effect of deletion or nonsense mutations on the phenotype has not been determined.

THE CGA PROTEIN-cAMP SYSTEM

The catabolite deactivation of inducible enzyme synthesis by glucose and other carbon sources has been known for many years (89), and its self-regulatory nature has recently been emphasized (43). The study of catabolite deactivation has led to the delineation of a general positive control system sensitive to the intracellular and extracellular concentrations of cAMP. Glucose lowers the intracellular concentration of cAMP in *E. coli* (90). The addition of cAMP to cells growing in glucose increases the rate of synthesis of B-galactosidase (91, 92), L-arabinose isomerase (43), and *lac* mRNA (93). The presence of cAMP and its effect on the synthesis of inducible enzymes appears to occur throughout the prokaryotes (28).

The initial step in determining how cAMP activates catabolite-sensitive operons was the isolation of mutants with a pleiotropic-negative phenotype for catabolite-sensitive genes. This was accomplished by selection of a number of mutants with apparent defects in two catabolite-sensitive operons (94, 95). Such mutants produce low levels of most catabolite-sensitive enzymes, suggesting they have a defect in a general system which is required for the expression of catabolite-sensitive genes. The mutants can be divided into two classes. One class is deficient in cAMP as a result of a mutation in the structural gene *(cya)* for the enzyme, adenyl cyclase. Gene *cya* maps between *ilv* and *metE* (96) and is not essential in *E. coli* as evidenced by the isolation of deletions of this gene (97). The phenotypic effect of the mutation can be reversed by the addition of cAMP. The second class of mutants map between *strA* and *aroB* (98) and are not phenotypically reversed by cAMP. These mutants with lesions in the *crp* locus have been found to lack a protein factor which is essential for the expression of catabolite-sensitive genes (99, 100). This protein has been referred to as CAP (catabolite activator protein) (99), CRP (cAMP receptor protein) (100), and CGA (catabolite gene activator) protein (101). Although given different names, the protein that was isolated by various workers is identical and in this review is termed CGA protein. The high frequency of pleiotropic-negative mutations in the *crp* gene suggests that the product would be a positive regulator. However, proof of a positive control system has come only from in vitro studies.

CGA protein stimulates the DNA-directed cell-free synthesis of the enzymes of the *lac* operon (99, 100), but the S30 used in these studies is prepared from a strain with an undefined lesion in the *crp* gene; one cannot rule out the possibility that subunit exchange occurs between the defective and wild-type CGA protein, obscuring the control mechanism involved. However, highly purified transcription systems have given proof that CGA protein is a general positive regulator required for the transcription of the *lac* (102, 103), *gal* (104), and *ara* (63) operons.

Studies with altered promoters of the *lac* operon show that certain properties of an intact promoter are needed for the effective action of CGA protein (105). In the *ara* operon, the region between *araO* and *araB* has been shown by deletion analysis to be required for catabolite deactivation (45), and mutants have been isolated that map in or near the *ara* initiator and promoter region that make expression of the *ara* operon independent of the CGA protein and cAMP (see pp. 226–28). In vitro studies confirm that expression of the *ara* operon occurs in the absence of cAMP and CGA protein in these mutants. The above results suggest that the binding of CGA protein in the absence of cAMP to a site at or near the promoter region is required for the expression of catabolite-sensitive operons.

Recently, it has been shown that the galactose promoter has the capacity to bind 6 RNA polymerase molecules in the presence of CGA protein and cAMP (106). In the absence of CGA protein or cAMP, only one polymerase is bound at the *gal* promoter site. The effect is apparently specific for catabolite-sensitive operons because there is no effect on λ promoters tested in the same system. Thus, CGA protein in the presence of cAMP makes a larger region of the promoter available for the binding of RNA polymerases.

It has been proposed that cAMP induces a conformational change in CGA protein (107). Evidence supporting a conformational change has been obtained in studies showing that CGA protein is more sensitive to proteases in the presence of cAMP (108). Spectrofluorimetric studies have demonstrated directly that CGA protein undergoes at least one conformational change in the presence of cAMP (109).

In the presence of cAMP, the CGA protein could interact with RNA polymerase, DNA, or both. No evidence has been published for a direct interaction with RNA polymerase. However, CGA protein has been shown to interact with DNA. The affinity of CGA protein for DNA is increased by cAMP but greatly reduced by cGMP, a competitive inhibitor of cAMP binding (101, 110). No specificity of binding to a controlling site region of a catabolite-sensitive operon has been observed (101, 110), even when using fragments containing *lac* operator of one million molecular weight. (A. Riggs, personal communication).

One might generalize the results obtained with the *gal* operon and propose that CGA protein allows more RNA polymerase molecules to bind at the promoter regions of all catabolite-sensitive operons. The mechanism by which CGA protein carries out this function is obscure. The half-life of CGA protein complexed with either duplex or heat-denatured DNA is the same, and temperature does not have a pronounced effect on the rate of dissociation (A. Riggs, personal communication). These results are not compatible with models proposing that CGA protein is involved with strand separation. Perhaps CGA protein alters the pitch of the DNA helix so that RNA polymerase can recognize more binding sites on the DNA.

REGULATION IN FUNGI

We refer the reader to Gross' review article (88) for a detailed discussion of nitrate reduction in *Aspergillus* and in *Neurospora,* galactose metabolism in yeast, purine degradation in *Aspergillus,* and leucine biosynthesis in *Neurospora*—systems in which there is presumptive evidence for positive control. We cover briefly the regulation of sulfate metabolism and aromatic biosynthesis in *Neurospora* for which there have been recent advances.

Sulfur Metabolism in Neurospora crassa

Genes that appear to be the structural genes for aryl sulfatase, sulfate permease, and the regulator gene *cys-3* are in different linkage groups. Gene *ars* characterized by mutants deficient in just aryl sulfatase is in linkage group VII (116). Gene *cys-3* is located in linkage group II. *Cys-13* and *cys-14* control two different sulfate permeases and are probably the structural genes for these permeases (114). Neither *cys-13* nor *cys-14* is linked to *cys-3*, or to one another, or to *ars* (112). The gene for choline sulfatase has not been located.

A mutant *cys-3* of *Neurospora crassa* lacks the three enzymes involved in sulfur metabolism: sulfate permease, aryl sulfatase, and choline sulfatase (111). There is sufficient presumptive evidence to suggest that this pleiotropic-negative effect is the result of a mutation in a regulatory gene preventing the synthesis of a biologically active positive regulator.

Several alternative hypotheses have been eliminated; we present a few of the relevant ones. For instance, revertants of *cys3⁻* regain all three activities (115). The revertants differ in the extent to which all three activities are recovered. It appears that the reversions occur near, or probably within, the *cys-3* cistron. Among the revertants are some heat-labile strains that fail to produce the enzymes at 37°C but do so at 25°C. An analysis of the physiochemical properties of aryl sulfatase including heat lability in vitro and in vivo show no change from the wild type. Properties of choline sulfatase and sulfate permease also remain the same in the different revertants tested. These experiments seem to rule out the possibility that the *cys-3* gene codes for a polypeptide that is common to all three activities. It also suggests that the original mutation is not likely to be a large deletion excising the structural genes for the three enzymes. The results also cannot be explained on the assumption that the *cys-3* mutation is an extreme polar mutation or a promoter gene mutation. Although aryl and choline sulfatases are generally increased or decreased in a parallel fashion, they are not coordinate. Furthermore, reversion of *cys-3* yields strains with ratios that depart drastically in both directions from that of the wild type. These results could be explained on the basis that the promoters for the different genes are different and the modification of the positive regulator is recognized differently by the individual promoters.

The mutation does not seem to be the result of the production of a super-repressor. If this were the case, revertants should be mainly strains that would be nonrepressible by the relevant corepressor as a result of a further mutation in the *cys-3* gene, resulting in the production of an inactive product. Instead, all revertants contain measurable levels of sulfatase and are completely repressible by methionine and/or inorganic sulfate. In addition, inducibility is dominant to noninducibility in heterokaryons, which is contrary to the super-repressor model. The phenotypes of the temperature conditional mutants also argue against this alternative model (113). It has been considered that the phenotype of *cys-3* mutants is the result of producing a super-repressor and that the super-repressor is confined to the nucleus in which it is produced, so that it is impossible to conduct a reliable test for dominance. In addition, completely derepressed strains may be lethal. To determine whether or not there is any *trans* effect, heterokaryons were constructed with two alleles of the *ars* locus that code for different forms of aryl sulfatase. One of the alleles was introduced along with *cys-3⁻*, the other with *cys-3⁺*. The heterokaryons produced both forms of the enzyme. The super-repressor hypothesis can therefore be discarded. Here again, what is lacking for conclusive proof of positive control is the demonstration that a deletion or nonsense mutation in gene *cys-3* produces a pleiotropic-negative phenotype and that mutations in one or more controlling sites of the three operons alter the effect of the positive regulator.

Quinate Metabolism in Neurospora crassa

In *Neurospora crassa* the conversion of quinate to protocatechuate is catalyzed by three enzymes that are coordinately induced in the presence of quinate (117–119). These enzymes are specified by three closely linked structural genes (*qa-2, qa-3,* and *qa-4*) which code for a 5-dehydroquinate dehydratase, quinate:NAD oxidoreductase, and 5-dehydroshikimate dehydrase, respectively (120). A fourth gene, *qa-1,* which is closely linked to the structural genes, may code for a positive regulatory protein (121). A noninducible pleiotropic-negative allele of the *qa-1* gene has been isolated which is recessive in heterokaryons which also contain a *qa-1⁺* allele (118). Further support for positive control has been provided by the isolation and identification of a constitutive allele (*qa-1ᶜ*) which produces high levels of the *qa-2,3,4* gene products in the absence of inducer. Heterokaryons between *qa-1ᶜ* and *qa-1⁻* show that some *qa-1ᶜ* alleles are dominant to the wild-type allele. This suggests a pure positive control system similar to that found in the maltose operon of *E. coli.* However, proof that the *qa-2, qa-3,* and *qa-4* genes require the *qa-1* gene product for expression must await the isolation of nonsense or deletion mutants in the *qa-1* gene.

Acknowledgments

We thank the following individuals for making available to us unpublished manuscripts of their work: A. Riggs, M. Schwartz, J. P. Thirion, M. Hofnung, D. Hatfield, R. W. Hogg, R. Metzenberg, and N. Giles. We especially thank N. Lee for her critique of the manuscript. Unpublished work of the authors is supported by NSF grant GB24093.

Literature Cited

1. Jacob, F., Perrin, D., Sanchez, C., Monod, J. 1960. *C. R. Acad. Sci.* 250:1727
2. Jacob, F., Monod, J. 1961. *Cold Spring Harbor Symp. Quant. Biol.* 26:193
3. Jacob, F., Monod, J. 1961. *J. Mol. Biol.* 3:318
4. Willson, C., Perrin, D., Cohn, M., Jacob, F., Monod, J. 1964. *J. Mol. Biol.* 8:582
5. Gross, J., Englesberg, E. 1959. *Virology* 9:314
6. Englesberg, E., Killeen, N. 1959. *Genetics* 44:508
7. Englesberg, E. 1961. *J. Bacteriol.* 81:996
8. Englesberg, E. et al 1962. *J. Bacteriol.* 84:137
9. Lee, N., Englesberg, E. 1962. *Proc. Nat. Acad. Sci. USA* 48:335
10. Lee, N., Englesberg, E. 1963. *Proc. Nat. Acad. Sci. USA* 50:696
11. Helling, R. B., Weinberg, R. 1963. *Genetics* 48:1397
12. Cribbs, R., Englesberg, E. 1964. *Genetics* 49:95
13. Englesberg, E., Irr, J., Power, J., Lee, N. 1965. *J. Bacteriol.* 90:946
14. Sheppard, D., Englesberg, E. 1966. *Cold Spring Harbor Symp. Quant. Biol.* 31:345
15. Sheppard, D., Englesberg, E. 1967. *J. Mol. Biol.* 25:443
16. Englesberg, E., Squires, C. 1968. *Proc. 12th Int. Congr. Genet.* Sect. 2, p. 47
17. Epstein, W., Beckwith, J. R. 1968. *Ann. Rev. Biochem.* 37:411
18. Echols, H., Garen, A., Garen, S., Torriani, A. 1961. *J. Mol. Biol.* 3:425
19. Garen, A., Otsujii, N. 1964. *J. Mol. Biol.* 8:841
20. Garen, A., Echols, H. 1962. *J. Bacteriol.* 83:297

21. Garen, A., Echols, H. 1962. *Proc. Nat. Acad. Sci. USA* 48:1398
22. Garen, A., Garen, S. 1963. *J. Mol. Biol.* 6:433
23. Echols, H. 1972. *Ann. Rev. Genet.* 6:157
24. Echols, H. 1971. *Ann. Rev. Biochem.* 40:827
25. Hershey, A. D., Ed. 1971. *The Bacteriophage Lambda.* Cold Spring Harbor, NY: Cold Spring Harbor Lab. 792 pp.
26. Thomas, R. 1971. *Curr. Topics Microbiol. Immunol.* 56:13
27. Herskowitz, I. 1973. *Ann. Rev. Genet.* 7:289
28. Pastan, I., Perlman, R. 1970. *Science* 169:339
29. Wilcox, G., Clemetson, K., Santi, D., Englesberg, E. 1971. *Proc. Nat. Acad. Sci. USA* 68:2145
30. Markovitz, A., Rosenbaum, N. 1965. *Proc. Nat. Acad. Sci. USA* 54:1084
31. Englesberg, E. 1971. In *Metabolic Pathways*, Vol. V, *Metabolic Regulation*, ed. H. J. Vogel, 257–96. New York: Academic
32. Englesberg, E., Sheppard, D., Squires, C., Meronk, F. Jr. 1969. *J. Mol. Biol.* 43:281
33. Englesberg, E., Squires, C., Meronk, F. Jr. 1969. *Proc. Nat. Acad. Sci. USA* 62:1100
34. Kessler, D., Englesberg, E. 1969. *J. Bacteriol.* 98:1159
35. Isaacson, D., Englesberg, E. 1964. *Bacteriol. Proc.* 64:113
36. Isaacson, D. 1964. *L-arabinose inhibition and the isolation and characterization of L-arabinose permeaseless, cryptic mutants of Escherichia coli B/r.* PhD thesis. Univ. Pittsburgh, Pittsburgh, Pa.
37. Hogg, R. W., Englesberg, E. 1969. *J. Bacteriol.* 100:423
38. Schleif, R. 1969. *J. Mol. Biol.* 46:185
39. Brown, C. E., Hogg, R. W. 1972. *J. Bacteriol.* 111:606–13
40. Parsons, R. G., Hogg, R. W. 1974. *J. Biol. Chem.* 249:3602
41. Parsons, R. G., Hogg, R. W. 1974. *J. Biol. Chem.* 249:3608
42. Boos, W. 1972. *J. Biol. Chem.* 247–5414
43. Katz, L., Englesberg, E. 1971. *J. Bacteriol.* 107:34
44. Sheppard, D. E., Walker, D. A. 1969. *J. Bacteriol.* 100:715
45. Bass, R., Heffernan, L., Sweadner, K., Colome, J., Englesberg, E. 1974. *J. Bacteriol.* Submitted
46. Cribbs, R. 1962. *Problems of mapping L-arabinose negative mutants in the L-ribulokinase structural gene of Escherichia coli B/r.* PhD thesis. Univ. Pittsburgh, Pittsburgh, Pa.
47. Irr, J., Englesberg, E. 1970. *Genetics* 65:27
48. Beverin, S., Sheppard, D., Park, S. 1971. *J. Bacteriol.* 107:79
49. Irr, J., Englesberg, E. 1971. *J. Bacteriol.* 105:136–41
50. Nathanson, N., Schleif, R. 1973. *J. Bacteriol.* 115:711
51. Schleif, R. 1972. *Proc. Nat. Acad. Sci. USA* 69:3479
52. Gielow, L., Largen, M., Englesberg, E. 1971. *Genetics* 69:289
53. Eleuterio, M., Griffin, B., Sheppard, D. E. 1972. *J. Bacteriol.* 111:383
54. Wilcox, G., Meuris, P., Bass, R., Englesberg, E. 1974. *J. Biol. Chem.* 249:2946
55. Greenblatt, J., Schleif, R. 1971. *Nature New Biol.* 233:166
56. Zubay, D., Gielow, L., Englesberg, E. 1971. *Nature New Biol.* 233:164
57. Yang, H., Zubay, G. 1973. *Mol. Gen. Genet.* 122:131
58. Wilcox, G., Clemetson, K. 1974. *Methods in Enzymology.* New York: Academic. In press
59. Wilcox, G., Boulter, J., Lee, N. 1974. *Proc. Nat. Acad. Sci. USA.* In press
60. Wilcox, G., Clemetson, K., Cleary, P., Englesberg, E. 1974. *J. Mol. Biol.* 85:589
61. Wilcox, G., Meuris, P. 1974. *J. Biol. Chem.* Submitted
62. Wilcox, G. 1972. *araC protein from Escherichia coli.* PhD thesis. Univ. California, Santa Barbara, Calif.
63. Lee, N. et al 1974. *Proc. Nat. Acad. Sci. USA* 71:634
64. Schleif, R. 1969. *J. Mol. Biol.* 46:197
65. Doyle, M., Brown, C., Hogg, R., Helling, R. 1972. *J. Bacteriol.* 110:56
66. Wilcox, G. 1974. *J. Biol. Chem.* In press
67. Wilcox, G., Singer, J., Heffernan, L. 1971. *J. Bacteriol.* 108:1
68. Schleif, R. 1971. *J. Mol. Biol.* 61:275
69. Power, J., Irr, J. 1973. *J. Biol. Chem.* 248:7806
70. Cleary, P., Englesberg, E. 1974. *J. Bacteriol.* 118:121
71. Monod, J., Torriani, A. M. 1950. *Ann. Inst. Pasteur* 78:65
72. Wiesmeyer, H., Cohn, M. 1960. *Biochim. Biophys. Acta* 39:417
73. Schwartz, M., Hofnung, M. 1967. *Eur. J. Biochem.* 2:132
74. Schwartz, M. 1967. *Ann. Inst. Pasteur* 113:685

75. Hatfield, D., Hofnung, M., Schwartz, M. 1969. *J. Bacteriol.* 98:559
76. Schwartz, M. 1967. *Ann. Inst. Pasteur* 112:673
77. Hatfield, D., Hofnung, M., Schwartz, M. 1969. *J. Bacteriol.* 100:1311
78. Hofnung, M., Schwartz, M., Hatfield, D. 1971. *J. Mol. Biol.* 61:681
79. Schwartz, M. 1966. *J. Bacteriol.* 92:1083
80. Taylor, A. L. 1970. *Bacteriol. Rev.* 34:155
81. Hofnung, M., Hatfield, D., Schwartz, M. 1974. *J. Bacteriol.* 117:40
82. Hofnung, M. 1974. Submitted
83. Thirion, J. P., Hofnung, M. 1972. *Genetics* 71:207
84. Randall-Hazelbauer, L., Schwartz, M. 1973. *J. Bacteriol.* 116:1434
85. Kellerman, D., Szmelcman, S. 1974. *Eur. J. Biochem.* Submitted
86. Hofnung, M., Schwartz, M. 1973. *Mol. Gen. Genet.* 112:117
87. Power, J. 1967. *Genetics* 55:557
88. Gross, S. 1969. *Ann. Rev. Genet.* 3:395
89. Magasanik, B. 1970. In *The Lactose Operon,* ed. D. Zipser, J. Beckwith. Cold Spring Harbor, NY: Cold Spring Harbor Lab. p. 189
90. Makman, R., Sutherland, E. 1965. *J. Biol. Chem.* 240:1309
91. Perlman, R., Pastan, I. 1968. *J. Biol. Chem.* 243:542
92. Ullman, A., Monod, J. 1968. *FEBS Lett.* 2:57
93. Varmus, H., Perlman, R., Pastan, I. 1970. *J. Biol. Chem.* 245:2259
94. Schwartz, D., Beckwith, J. 1970. See Ref. 89, p. 417
95. Perlman, R., Pastan, I. 1969. *Biochem. Biophys. Res. Commun.* 37:151
96. Yokota, T., Gots, J. 1970. *J. Bacteriol.* 103:513
97. Brickman, E., Soll, L., Beckwith, J. 1973. *J. Bacteriol.* 116:582
98. Epstein, W., Kim, B. S. 1971. *J. Bacteriol.* 108:639
99. Zubay, G., Schwartz, D., Beckwith, J. 1970. *Proc. Nat. Acad. Sci. USA* 66:104
100. Emmer, M., De Crombrugghe, B., Pastan, I., Perlman, R. 1970. *Proc. Nat. Acad. Sci. USA* 66:480
101. Riggs, A., Reiness, G., Zubay, G. 1971. *Proc. Nat. Acad. Sci. USA* 68:1222
102. De Crombrugghe, B. et al 1971. *Nature New Biol.* 231:139
103. Eron, L., Block, R. 1971. *Proc. Nat. Acad. Sci. USA* 68:1828
104. Nissley, P., Anderson, W., Gottesman, M., Perlman, R., Pastan, I. 1971. *J. Biol. Chem.* 246:4671
105. Beckwith, J., Grodzicker, T., Arditti, R. 1972. *J. Mol. Biol.* 69:155
106. Willmund, R., Kneser, H. 1973. *Mol. Gen. Genet.* 126:165
107. Anderson, W., Perlman, R., Pastan, I. 1972. *J. Biol. Chem* 247:2717
108. Krakow, J., Pastan, I. 1973. *Proc. Nat. Acad. Sci. USA* 70:2529
109. Wilcox, G. 1974. Submitted
110. Nissley, P., Anderson, W., Gallo, M., Pastan, I. 1972. *J. Biol. Chem.* 247:4264
111. Metzenberg, R., Parson, J. 1966. *Proc. Nat. Acad. Sci. USA* 55:629
112. Metzenberg, R., Ahlgren, S. 1970. *Genetics* 64:409
113. Metzenberg, R., Chen, G., Ahlgren, S. 1971. *Genetics* 68:359
114. Marzluf, G. 1970. *Arch. Biochem. Biophys.* 138:254
115. Marzluf, G., Metzenberg, R. 1968. *J. Mol. Biol.* 33:423
116. Metzenberg, R., Ahlgren, S. 1971. *Genetics* 68:369
117. Giles, N., Case, M., Partridge, C., Ahmed, S. 1967. *Proc. Nat. Acad. Sci. USA* 58:1453
118. Rines, H. 1969. *Genetical and biochemical studies on the inducible quinic acid catabolic pathway in Neurospora crassa.* PhD thesis. Yale Univ., New Haven, Conn.
119. Chaleff, R. 1974. Submitted
120. Giles, N., Case, M., Jacobson, J. 1973. *Oak Ridge Symposium on Molecular Cytogenetics.* New York: Plenum
121. Valone, J., Case, M., Giles, N. 1971. *Proc. Nat. Acad. Sci. USA* 68:1555
122. Lis, J., Schleif, R. 1973. *J. Mol. Biol.* 79:149

ACCESSORY CHROMOSOMES ♦3073

Arne Müntzing
Institute of Genetics, University of Lund, Sweden

INTRODUCTION

In most cases the genome represents a delicately balanced unit of cooperating chromosomes, each segment of which must be present in order to secure normal viability. It has long been known, however, that there is a difference between the genetically active and highly differentiated euchromatic segments of the chromosomes and the heterochromatic segments that seem to be less active or even quite inert. The genetic inertness or partial inertness of heterochromatin is especially well known with regard to the Y chromosome in *Drosophila* and the corresponding sex chromosome in some other organisms.

The genetic inertness of the heterochromatin, especially in the sex chromosomes, forms a connecting link between the normal, regular, and active chromosomes and another chromosome category that Håkansson & Müntzing in 1945 (1, 2) proposed to call *accessory chromosomes*.

This denomination was earlier used to indicate a heterochromatic chromosome associated with sex determination in the male reproductive cells of insects (3), which today coincides with what we call X chromosome in an XO individual. For the chromosome category treated in this survey, "accessory" seems to be adequate since the word (according to *Webster's Dictionary*) means "extra, additional, helping in a secondary or subordinate way."

Several workers prefer the denomination *supernumerary chromosome,* but this term is somewhat ambiguous because it would also include trisomics, tetrasomics, triploids, and many other numerical deviations involving members of the normal chromosome complement. *B chromosome* is another handy label for distinguishing the deviating chromosomes now under consideration from *A chromosomes,* which represent the members of the regular chromosome complement. These denominations, also including the letters C, D, E, and F, were introduced by Randolph (4) to indicate different size classes of chromosomes in maize. It was gradually realized that only the A chromosomes were regular and that the chromosome types C to F were smaller derivatives of the B-type chromosomes. Although the letter denomina-

243

tions A and B have no intrinsic meaning other than size classification, these labels are often used (5) and are handy especially in comparisons between the normal chromosomes and the deviating ones.

In this paper the terms A and B chromosomes (A's and B's) will sometimes be used, but in order to indicate the deviating category of chromosomes, the term accessory chromosome (sometimes abbreviated to acc. chromosome and accessory) seems to be desirable. This view is indirectly supported by McClintock in the statement, referring to maize, that "The B-type chromosome is an accessory chromosome containing no genetic components that alter in any yet recognizable way either the appearance or the physiology of a plant" (6).

This statement evidently indicates that accessory chromosomes such as the B types in maize, in contrast to the A chromosomes, have neither morphological nor physiological effects. However, additional data from a wide range of other species carrying accessories have made it possible to characterize this chromosome category in approximately the following way:

1. Acc. chromosomes are not necessary for the life of the organisms and are present only in part of the individuals of a population. They occur at highly variable frequencies in different populations, the percentages of plants carrying such chromosomes ranging from 0 to almost 100.

2. In general the acc. chromosomes are subinert and have weak genetic effects. Therefore, many chromosomes of this kind (sometimes more than 30) may be present in the same individual.

3. In many species there are special mechanisms leading to a numerical increase of the acc. chromosomes. This is accomplished in several different ways in different groups of material.

4. The acc. chromosomes are frequently heterochromatic and are generally smaller than the other chromosomes. They show no cytological homology to the members of the regular chromosome complement.

5. In most cases the acc. chromosomes are probably ancient and their ultimate origin unknown or obscure.

6. In species in which the effects of acc. chromosomes on their carriers have been studied in detail there is generally a pronounced negative correlation between number of accessories and fertility. As a rule vigor is also reduced but this effect of the accessories is less pronounced.

7. Search for positive effects of acc. chromosomes has given conflicting results, but in some cases the acc. chromosomes may be ecologically favorable. There is also fairly clear evidence that chiasma formation and genetic recombination is influenced and often favored by the presence of acc. chromosomes.

Before detailing recent research achievements, reference should be made to previous summarizing articles (7–11) and especially to the very comprehensive review published by Battaglia in 1964 (5). Thanks to that review, the present report can be limited mainly to the last decade, though some degree of overlapping is unavoidable in order to clarify the development of certain lines of investigation in the expanding field of research on accessory chromosomes.

OCCURRENCE OF ACCESSORY CHROMOSOMES IN DIFFERENT BIOLOGICAL GROUPS

In Battaglia's review (5) the occurrence of acc. chromosomes was reported for approximately 160 species of higher plants belonging to 18 families. Among these families, Graminaceae represented the largest number of species, followed by Liliaceae and Compositae. References were also given to many papers on mosses and liverworts in which numerous species have been found to carry acc. chromosomes. Recent data from this field are given by Wigh (140).

During the last ten years a large number of new species with accessories have become known, but only a few of the most interesting cases can be referred to in this paper.

In the animal kingdom, insects from many genera and various families and orders are known to carry acc. chromosomes in addition to the normal chromosome complement (7), and in the new edition of White's book on animal cytology and evolution (12) the number of such species is close to one hundred. The grasshopper family, Acrididae, is an especially rich field for investigations of this kind, and some of the ingenious roles played by acc. chromosomes in grasshoppers and other insects are considered below.

In animals other than insects, relatively few species are known to carry accessories. In White's recent list (12) only 13 species are mentioned, representing Platyhelmintes, Mollusca, Crustacea, and Vertebrata. Among these groups the Vertebrata also include two mammals: *Schoinobates volans* (13) and *Reithrodontomys megalotis* (14). The first is a marsupial from Australia, and the second is a harvest mouse from the United States. The list of mammals can now be supplemented by another marsupial, *Echymipera kalabu* (15), the field mouse, *Apodemus giliacus* (16), and the Malayan house rat, *Rattus rattus diardii* (17). Finally, there are some indications that acc. chromosomes may even occur in our own species, *Homo sapiens*. In one case (25) a small supernumerary metacentric chromosome in phenotypically normal women was observed in three members of the same family. It is not yet known whether this small chromosome represents a transient phenomenon without deeper significance, but the possibility may still be open that it corresponds to the acc. chromosomes in other organisms. Anyhow, the occurrence of such chromosomes in mammals is now quite certain and deserves further comment.

Several investigations have been carried out (18–24) concerning the occurrence of accessory chromosomes in foxes. The main facts and conclusions are as follows: in domesticated silver foxes (18–20) as well as in wild red foxes from Europe and America (21–24) $2n = 34$ evidently represents the basic complement to which a variable number of microchromosomes may be added. From 1 to 6 such microchromosomes may occur, representing a total variation in chromosome number from 34 to 40. The microchromosomes are approximately one fourth the size of the smallest autosomes. They are probably heterochromatic, generally acrocentric, and similar to the accessory chromosomes known from invertebrates and plants. In male foxes the single Y chromosome is as small as the acc. chromosomes, which makes the distinction between Y and accessories difficult. Another complication is a certain amount of chromosomal variation within individuals owing to mitotic irregularities among the acc. chromosomes.

In the Malayan house rat (17) three widely separated populations were found to comprise individuals with 0 to 4 acc. chromosomes. The incidence of such chromosomes did not differ significantly between populations and was also similar in the two sexes. In the papers concerning the two mouse species referred to above, more comprehensive and detailed information was reported. In the harvest mouse (14) Shellhammer found that the accessories, which are small and dot-like, ranged from 0 to 7, causing a variation in diploid number from 42 to 49 with the modal class at 44. The relative frequency of different numbers varied between populations from different states in the US. However, in the total material examined, as much as 87.1% were found to carry accessories, and they occurred in every part of the geographic range of the species sampled. No correlation could be established between chromosome number and morphological traits.

In the field mouse, which was studied by Hayata (16), extreme variations were found in number and morphology of the chromosomes. The diploid numbers ranged from 48 to 61, but the number of acrocentric elements was consistently 48. In contrast, the number and constitution of several biarmed elements and microchromosomes were highly variable and hence responsible for the polymorphism. The variable elements were accessory chromosomes whereas the 48 acrocentrics seemed to represent regular autosomes and sex chromosomes. Meiotic behavior of the acc. chromosomes was rather irregular, and most accessories did not pair at first metaphase. At least some of the accessories must be heterochromatic because they were more fluorescent than the normal acrocentric chromosomes.

In the remaining two mammals with acc. chromosomes the evidence gathered by Hayman & Martin in *Schoinobates volans* (13) was obtained from observations on 5 females and 2 males. The females had the somatic chromosome numbers 24, 26, 27 (twice), and of the males, one had 25 and the other 26 chromosomes. All the females had 22 large chromosomes and the males 21 large chromosomes. The additional chromosomes, ranging in number from 2 to 6, were very small and uniform in size. They were constant from cell to cell and had the same number in different organs. Since they also had no influence on morphology and sex determination the authors concluded that these chromosomes are closely analogous to the supernumerary (accessory) chromosomes found in plants.

In the other marsupial, *Echymipera kalabu,* examined four years later by Hayman, Martin & Waller (15), the chromosome conditions are more complicated. There is a parallel mosaicism of accessory chromosomes and sex chromosomes, and the species does not have the full chromosome complement in all its adult somatic tissues. The chromosomes missing are the Y chromosome in the male and an X chromosome in the female. The full chromosome complement was found to be present in the corneal epithelium and the reproductive tissue. A parallel mosaicism was observed with regard to small acc. chromosomes that occur in certain individuals of the species. These acessories are highly variable in number and not related to sex determination.

DIFFERENT MECHANISMS OF NUMERICAL INCREASE

In many cases the accessory chromosomes of plants as well as animals have a remarkable ability of increasing their number from generation to generation. This result is achieved by means of several different mechanisms. The first investigations of such phenomena were carried out in rye and maize and then extended to many other members of the family Graminaceae as well as to other families of flowering plants.

Postmeiotic Nondisjunction

The most characteristic process leading to numerical increase of acc. chromosomes in the grass family is chromosome doubling by means of postmeiotic nondisjunction. Rye, *Secale cereale,* has 14 chromosomes in the somatic cells, but during the period 1923 to 1939 fifteen cytologists reported that plants of rye may also have 16 chromosomes (see 26 for references). In 1941 Müntzing & Prakken studied meiosis in a fairly large number of plants belonging to a Swedish variety of rye (26). In one of these plants there were 2 smaller chromosomes besides the 7 regular bivalents. The smaller chromosomes were either unpaired or formed an extra, small bivalent. Offspring after open pollination was raised and expected to have $2n = 15$ but most of the daughter plants had other numbers, ranging from 14 to 20. In the plants with additional chromosomes, these were of the same type: a chromosome about half the size of a normal rye chromosome and with a subterminal centromere, as previously described by other workers (27, 29, 30).

In crosses between plants with 14 and 16 chromosomes both reciprocal combinations gave an excess of 16 chromosomes, and very few plants had the expected number 15 (2, 27). As first observed by Hasegawa (30), the cause of the unexpected chromosome numbers is irregular behavior of the specific extra chromosome at the first pollen mitosis. At this stage the two chromatids often do not separate but are in most cases included in the generative nucleus. As shown by the crosses $(14 + 2) \times 14$, a similar nondisjunction must also occur on the female side (2). Embryological examination, carried out by Håkansson (31), showed this to be indeed true. At the first mitosis in the embryo sac there is a directed nondisjunction, which implies that the daughter chromatids of the accessory chromosome are included in the micropylar nucleus, which by two further divisions gives rise to the egg cell. At these divisions in the embryo sac the accessories divide in a normal way.

In the pollen the corresponding directed nondisjunction process was observed in detail by Müntzing (32). At early anaphase of the first pollen mitosis the accessory chromosome lags between the anaphase groups and looks like a bivalent with two free arms and two localized chiasmata near the centromere. This structure arises because the centromere has divided and the daughter centromeres try to pull the two chromatids apart. Resistance to this separation is offered in a special segment of the long arm not far from the centromere. The chromatids also stick together at the end of the short arm. As anaphase proceeds and the spindle stretches, the nonseparating chromatids of the acc. chromosome in most cases pass to the generative pole where they separate during interphase between the first and second mitosis.

The behavior of the acc. chromosome in the first pollen mitosis and at the exactly corresponding stage in the ovule is remarkable. The process of nondisjunction occurs only a single time during each life cycle, the mitotic divisions of the acc. chromosomes otherwise proceeding quite regularly, as is evident from the generally strict correspondence between the number of accessories in the root tips and in meiosis. As the process of nondisjunction is performed only by the acc. chromosomes and not by the regular chromosomes, the accessories must have some special property leading to this differential behavior. This property is most likely a delayed reproduction of a special chromosome segment near the centromere, but this delay occurs only in the special environments of the pollen grain as well as of the embryo sac.

The nondisjunction process in rye was further elucidated by Müntzing and Lima-de-Faria by studying the fine structure of the accessories at the pachytene stage of meiosis and by comparing the structure and nondisjunction mechanism of the standard type of acc. chromosome with three new derivatives of this chromosome (2, 33–39). Two of these derivatives were isochromosomes, representing twice the long arm and twice the short arm of the standard acc. chromosome, respectively. The third derivative represented a deficiency, lacking the terminal half of the long arm.

The pachytene analysis revealed a very detailed structure of the standard accessory chromosome, the most conspicuous features being a subterminal knob, surrounded by a terminal appendage with very small chromomeres and a proximal, thin "neck region." The other parts of the arms had characteristic and in part recognizable sequences of chromomeres. The subterminal centromere was almost unstained but easily distinguished.

Chromosome counts in progenies of the structurally changed acc. chromosomes showed that the small iso B and the deficiency B had completely lost the ability to carry out nondisjunction, whereas the large iso B performs this trick as regularly as the standard B chromosomes. This sharp difference between the various kinds of accessories was also verified by observations of the mode of pollen mitosis. Whereas the small iso B and the deficiency B divided as regularly as the normal A chromosomes, the large iso B performed nondisjunction as perfectly as the standard B. At metaphase of the first pollen mitosis the four identical arms are extended laterally—two to each side—and are attached to a central ring or rhombus with the centromere divided in an effort to separate the chromatids. They stick together, however, at precisely the same point as in the standard B. In all respects the large iso B represents the long arm of the standard B and its mirror image.

It is probable that the nondisjunction process is conditioned by the presence of the rather large knob near the end of the long arm. At any rate it is striking that this knob is present in the two types of acc. chromosomes that can perform the nondisjunction trick completely, whereas it is absent in the other two types that have lost this ability entirely. This is only true when the small, deficient accessories are alone in the pollen grain (together with the A chromosomes). If the same pollen grain also contains a standard B chromosome, this chromosome will induce the small iso B or deficiency B to perform nondisjunction at the same time as the standard B (40, 41). This was verified by cytological examination of the progeny plants produced and is a phenomenon also known to occur in *Festuca pratensis* where it was first discovered by Bosemark (42). In these cases the best explanation is probably that a chemical substance is secreted from the standard B into the cytoplasm of the pollen grain, where it will reach and affect the chromatids of the defective B's in such a way that nondisjunction is performed. In rye there is reason to assume that this substance is produced by the knob present in the standard B as well as in the large iso B chromosomes.

Rye is unique by its ability to perform the nondisjunction process in the pollen as well as in the embryo sac. Many other grass species have been found to carry acc. chromosomes that undergo nondisjunction in the pollen but are unable to do so on the female side. A few years after the analysis in rye an exactly corresponding behavior of accessories was observed by Östergren in *Anthoxanthum aristatum* (43) and by Bosemark in *Festuca pratensis* (44–47). Later, the same mechanism has been found in other species of grasses belonging to the genera *Festuca, Alopecurus, Briza,* and *Holcus* (48); *Dactylis* (49, 50); *Phleum* (51); *Deschampsia* (52); *Agropyron* (53); and *Aegilops* (54).

Nondisjunction of acc. chromosomes is also known to occur in *Zea mays,* but in this species the mechanism does not operate in the first pollen mitosis but in the second one. In 1941 Randolph (55) published a report about the genetic characteristics of the B chromosomes in maize, and on the basis of their different modes of inheritance in reciprocal crosses he was able to predict that this might be due to some kind of anomalous behavior of the B's during the division of the generative nucleus. This was verified six years later by Roman (56) and further elaborated in later papers (57–59). Roman realized that the production of 2 B gametes from 1 B microspore is balanced by the production of an equivalent class of 0 B gametes. This could be studied by attaching an A-chromosome segment carrying known genes to a B-type segment and thus tracing the behavior of the latter. It should also be possible, by a comparison of various B segments produced by breakage at different points in the chromosome, to locate the element within it that is responsible for its aberrant performance.

Mature pollen from plants with 4 to 10 B's was treated with X rays and applied to silks of plants devoid of B chromosomes. By special screening methods, 8 different A–B interchanges could be isolated. With the aid of marker genes showing their effects in the endosperm it could be shown that one of the interchange chromsomes (B^4) undergoes nondisjunction in the second division of the microspore. The B^4 chromosome possesses the centromere and proximal third of the B chromosome. The interchange chromosome bearing the centromere of chromosome 4, on the contrary, disjoins normally, and a corresponding situation was also met with in several other A–B interchanges tested with the same method. The B^4 chromosome is relatively stable in the endosperm development and undergoes nondisjunction rarely, if ever, in the development of the egg. Hence, the unique behavior of the intact B-type chromosome may also be accounted for on the basis of nondisjunction in the second microspore division.

This result was supplemented in an important way by the next paper in the series (57), in which crosses between normal seed parents without B chromosomes and pollen parents homozygous for an A–B translocation (TB-4a) are described. In such crosses three kinds of seeds arise: 1. seeds with a deficient embryo and a hyperploid endosperm, 2. seeds with a hyperploid embryo and a deficient endosperm, 3. seeds derived from normal disjunction in the pollen parent. The embryo in such cases is heterozygous for the exchange, and the endosperm also carries a single B^4 chromosome. The three classes of seeds could readily be identified through use of the marker gene su, when the B^4 chromosome carries Su.

Without going into further details it can be stated that fertilization was not at random. The union of the egg with the hyperploid gamete was preferred to the reverse order of fertilization, and use of other translocations between B's and various A chromosomes gave similar results. The whole body of facts clearly demonstrates that the B chromosomes in maize (as well as translocation chromosomes with B centromere) undergo nondisjunction at the second pollen mitosis and that this mechanism is combined with *selective fertilization* leading to an increase in B-chromosome number or at least to preservation of this chromosome type in the population. Meiotic loss or fragmentation would otherwise tend to eliminate the B chromosomes from the maize populations (60).

The B chromosome of maize is telocentric (76) or acrocentric (55, 79) and highly heterochromatic. From the proximal to distal ends it contains a centromere, adjacent heterochromatic knob, euchromatic region, and several large blocks of heterochromatin. As pointed out by Carlson (77, 78) the B^4 chromosome carries the "sticky" region of the B which causes nondisjunction. However, the distal heterochromatin of the B also contributes to this process, and the B^4 cannot undergo nondisjunction in the absence of 4^B (78). Similar results were obtained with other translocations between the B and various A chromosomes. Hence, it is clear that nondisjunction of the B in maize is controlled by at least two separable regions in this chromosome. Recently this mechanism has been studied in still more detail by the use of deletions, rearrangements, and isochromosomes derived from B chromosomes of standard type (79). The isochromosome was found to carry all genes necessary for nondisjunction.

Other Mechanisms of Numerical Increase in Plants

In this survey of different mechanisms of numerical increase of acc. chromosomes we have so far dealt only with the nondisjunction occurring at the first or the second postmeiotic division—mechanisms that seem to be entirely limited to the grass family with rye and maize as typical examples. In other plant species acc. chromosomes are, however, known to be accumulated in still other ways, which are less thoroughly known but which should be briefly indicated.

1. The first of these various mechanisms was discovered in 1956, independently by Rutishauser in *Trillium grandiflorum* (61) and by Kayano in *Lilium callosum* (62, 63). This mechanism implies that univalent accessories at the first meiotic division in the macrospore mother cells preferably pass to that dyad from which the embryo sac is developed. This phenomenon, called "directed distribution" by Rutishauser and "preferential segregation" by Kayano, is now known as *meiotic preferential distribution* (5). As described by Fröst (64) it is also characteristic of the acc. chromosomes in *Plantago serraria* and *Phleum nodosum*. In the latter species it was demonstrated that the mechanism works only when the accessories are single or else unpaired at first metaphase of meiosis. For instance, in plants of *Phleum nodosum* with 2 accessories, forming an extra bivalent at meiosis, the numerical increase of accessories in the progenies does not occur. Since this kind of numerical increase has been found in widely different families, many other such cases will probably be discovered eventually.

2. Another mechanism of numerical increase now fairly well known is somatic nondisjunction, which may be exemplified by observations made in *Crepis capillaris* by Rutishauser (65) and Röthlisberger (66). The number of acc. chromosomes is constant in the vegetative parts but begins to vary in the inflorescence at the time of flowering. In addition to cells with a doubled number of accessories, cells arise in which the acc. chromosomes are missing. The effects of the probable nondisjunction process are evident in the pollen as well as in the ovules, and nondisjunction evidently may occur more than once because some cells in the generative region were found to have more than the double number of chromosomes. It is suggested that the instability of the acc. chromosomes in the apex meristem could be the first visible effect of the production of a flowering hormone.

Somatic nondisjunction has been reported in several other genera. In *Miscanthus japonicus* (67) the number of acc. chromosomes in the pollen mother cells (PMC) ranged from 0 to 10. Variance analysis of this numerical variation suggested two major cycles of somatic nondisjunction: one responsible for significant differences between anthers of the same spikelet and the other for the variation among PMC's of the same anther. In the hexaploid species *Panicum nehruense* (68) numerical variation of accessories was attributed to somatic elimination and/or somatic nondisjunction during the premeiotic mitotic stages. In two species of *Haplopappus* (69, 70) there was also evidence of somatic nondisjunction, especially in the male germ line. Another case in *Achillea*, reported by Ehrendorfer (71), is interesting because examples of accumulation as well as reduction in the somatic number of accessory chromosomes are given. In premeiotic mitosis in the male, and especially in the female archespor there is sometimes a strong tendency toward accumulation of accessories in favor of future PMC's and embryo sac mother cells (EMCs). This is probably achieved by nondisjunction and polarized distribution. On the other hand, in somatic tissues with increasing primary numbers of accessories (up to 6) there is a growing tendency to intra-individual reduction, and only individuals with 2 accessories are relatively stable. This reduction is probably due to nondisjunction of accessories and cell competition.

3. In *Crepis conyzaefolia* and probably also *C. pannonica* Fröst has obtained evidence (72–74) that there is a very effective mechanism of numerical increase on the male as well as the female side. Thus, plants with 1, 2, and 3 accessories in the root tips and in somatic tissues of the immature flower heads always had twice the number of accessories in their PMC's. It seems most probable that this doubling is caused by an endomitotic reduplication taking place just before meiosis. This would imply that the accessories reproduce twice in an interphase during which the regular chromosomes reproduce but once. Such a mechanism has previously been reported for the flatworm *Polycelis* (7, 75). In the present case in *Crepis,* however, some exceptions to the perfect doubling have occurred and, therefore, more cytological evidence is needed before it can be decided whether the cause of the reduplication is really endomitotic reduplication or possibly directed somatic nondisjunction.

Mechanisms of Numerical Increase in Animals

During the last decade much work on accessory chromosomes has been carried out in various groups of insects and especially among several species of grasshoppers. Nur, for instance, has concentrated on *Calliptamis palaestinensis,* a species collected in Israel, and *Camnula pellucida,* which is native to California. In these species different follicles or cysts in the same testis commonly contain different numbers of acc. chromosomes, suggesting that mitotic nondisjunction occurs in the germ line. Within follicles the chromosome numbers are uniform. In the case of *Calliptamis* (80) about 12% of the males carried accessories and the great majority of these males must have started their development with a single acc. chromosome. Such individuals had some follicles with 0, some with 2 acc. chromosomes, and still others with more than 2 accessories. Simple nondisjunction would have produced a 1:1 ratio of follicles with and without accessories but the observed ratio was approximately 15:1. From this it could be concluded that there is a differential nondisjunction at embryonic mitoses, in which future germ-line cells are segregated from future somatic cells.

In *Camnula pellucida* (81) there is also mitotic instability of such a kind that it leads to an accumulation of acc. chromosomes. Among 780 males, 105 had accessories, and in the testes of these males the number of such chromosomes varied from follicle to follicle with either one of the numbers in the range from 0 to 4. Seventy-two males that had started with 1 acc. chromosome later reached an average number of 1.37 accessories, an increase of 37%. In the grasshopper *Atractomorpha bedeli* (82) about 30% of the males had accessories, and there was strong evidence that in this species also, mitotic instability is associated with an accumulation of accessories. In most grasshopper species for which enough data are available the same kind of situation seems to prevail. Another good example is *Locusta migratoria* (83).

According to recent information concerning *Atractomorpha bedeli* (141), there is a rather strict difference in this species between germ line and somatic cells. The acc. chromosomes are present in the germ line but absent in the somatic line. The difference between germ line and somatic line is explained on the basis of elimination of accessories from the somatic cells. Variation in the number of acc. chromosomes between follicles, which occurs also in this species, was ascribed to mitotic nondisjunction in the germ line prior to differentiation of the follicles.

Nur has also carried out much work with the mealy bug, *Pseudococcus obscurus* (84–86). As in other species with the so-called lecanoid chromosome system, the paternal set of chromosomes of the males becomes heterochromatic in the embryo and remains so throughout development. Meiosis in the group to which this species belongs consists of two divisions during which homologous chromosomes do not pair. The first division is mitosis-like, and the secondary spermatocytes contain a diploid number of chromosomes. During the second division, the euchromatic and heterochromatic sets segregate to opposite poles and, of the four products of meiosis, only two euchromatic derivatives form functional sperm.

The acc. chromosomes in the male are positively heteropycnotic until meiosis but then become negatively heteropycnotic. During the first meiotic division they also divide equationally, but during the second division they usually segregate preferentially with the euchromatic set. Thus, an unreduced number of accessories is usually included in the sperm, and the preferential segregation leads to the creation of an accumulation mechanism.

Another quite different story has been revealed by Hewitt and other workers in *Myrmeleotettis maculatus* (87). Pair matings have shown that there is an accumulation of a large mitotically stable acc. chromosome when it is transmitted through the female. This is presumed to result from preferential segregation of an undivided univalent acc. chromosome at the first division of meiosis in the female to that pole of the spindle from which the egg nucleus will develop. This occurs irrespective of whether the accessories are odd or even in number. In the male, on the contrary, there is loss of acc. chromosomes.

Something similar has recently been reported by Lucov & Nur in the grasshopper *Melanoplus femur-rubrum* (88). This species has deviated from the usual behavior of acc. chromosomes in grasshoppers and has developed an accumulation mechanism on the female side. In crosses of females with 1 B and males with 0 B, 82% of the offspring received the B whereas in the reciprocal cross the corresponding value was 53%. Most probably, the high rate of transmission of the B by the females results from preferential distribution of the B into the secondary oocyte.

In animals other than insects very little is known about mechanisms of numerical increase of acc. chromosomes. Two different cases may be referred to, however. In the flatworm, *Polycelis tenuis* (7), the acc. chromosomes evidently accumulate in the spermatogonia by means of nondisjunction; on the female side they seem to undergo an endomitotic reduplication in the premeiotic resting stage in oogenesis.

In the harvest mouse investigated by Shellhammer (14), mentioned earlier (p. 246), the accessories are small and dot-like. They divide asynchronously and usually late as compared to the larger regular chromosomes. There is evidence that a mechanism of nondisjunction may operate among the accessories and that this mechanism may vary in effectiveness between populations. This would account for the variation in number of accessories between populations.

INFLUENCE ON FERTILITY AND VIGOR

In rye, *Secale cereale,* acc. chromosomes have a very marked negative effect on fertility. Test on seed set as well as pollen fertility have been performed by Müntzing in commercial strains from Scandinavia as well as in primitive strains from Siberia and Korea (28, 89, 90). The following relative values, for instance, were obtained for kernel weight per plant (number of accessories in brackets): 100.0 (0), 85.4 (1), 69.2 (2), 49.0 (3), 30.5 (4), and 17.8 (5–8). The curve goes steeply downwards and plants with more than 6 acc. chromosomes are generally completely sterile.

In the same material, relative values for pollen fertility decreased continuously and significantly from 100.0 to 82.5 as the number of accessories increased from 0 to the class with 5 to 8 accessories. Thus, pollen fertility is less affected than seed set, but the deleterious influence of the acc. chromosomes is obvious.

In the Scandinavian varieties, plant vigor (measured as straw weight) was also negatively correlated with the number of acc. chromosomes, but the depressing effect of the accessories was less pronounced than for fertility, and in the rye strains from Siberia and Korea there was no depressing effect on vigor at all.

In another paper (91) it was reported that heading time in a Japanese variety of rye is proportional to the number of accessories present, and in a British experiment the mean frequency of accessories decreased under high as compared to low density of planting (92). Thus, growth is slower and competition ability reduced in proportion to the number of acc. chromosomes present.

In the grass species *Festuca pratensis,* Bosemark (93) could see no effect on pollen fertility by 1 or 2 acc. chromosomes, but with an increase in the number of accessories there was a corresponding decrease in pollen fertility. Also, seed set appeared to be reduced by high numbers of accessories. The effect of one or two such chromosomes on the vegetative development is in part dependent upon the genotype of the material. In some materials the results indicate a slightly stimulating effect; in other experiments no such effect could be observed.

In general the genetic effect of accessories in *Festuca pratensis,* and probably other species as well, is partly governed by the genotypic constitution of the material and partly by the prevailing environmental conditions. This conclusion was also drawn by Fröst (94) concerning *Centaurea scabiosa.* In this species high numbers of acc. chromosomes reduce vigor as well as pollen fertility, but the effect of a low number of accessories was less decisive. In some cases a stimulating effect on vigor was obtained but in another experiment there was no significant influence. There is no doubt, however, that in most plant species in which this has been recorded the genetic effects of acc. chromosomes are negative with regard to vigor as well as fertility.

The older evidence of this kind has been summarized by Battaglia (5), and in recent papers (54, 70, 95–97) the same trend is obvious. In one case, however, some plants of diploid *Dactylis* with 1 and 3 accessories were reported to have better pollen than plants without acc. chromosomes (95). Also, in *Allium porrum,* individuals with acc. chromosomes were found to germinate prior to those without such chromosomes (98).

In insects with accessory chromosomes and various mechanisms of numerical increase, there is also evidence of deleterious effects on viability and fertility which generally affect males but not females. Thus, for instance, in *Camnula pellucida* males with acc. chromosomes had fewer follicles in their testes and a lower frequency of normal sperm than males without accessories (81). In the mealy bug, *Pseudococcus obscurus,* accessories had little or no effect on females but reduced the fitness of the males (86). By indirect evidence, this fitness in males with 0, 1, 2, 3, and 4 acc. chromosomes was calculated by Nur to be 1.00, 0.64, 0.56, 0.38, and 0.20, respectively. The effect of single accessories was also tested under laboratory conditions (85). In males a single acc. chromosome caused a reduction in the number of sperm in all 15 experiments performed under 4 different temperature regimes. In 12 of the experiments the accessories also increased developmental time in the males, especially at low temperature. In two experiments with females there were no observed effects of the accessories.

In other cases, also in insects, a low number of acc. chromosomes may be favorable but a higher number deleterious. According to White (99, 100) 1 acc. chromosome in *Trimerotropis sparsa* under most conditions increases viability and fertility, whereas 2 or more decrease fitness to some extent.

INFLUENCE ON ECOLOGICAL ADAPTATION

In plants the first large-scale investigations on the possible ecological importance of acc. chromosomes were carried out by Bosemark in various grass species (48) and especially in *Festuca pratensis* (47) and *Phleum phleoides* (51, 101). In the former species there was a significant correlation between soil type and number of accessories, this number being higher in areas with clay soil than in areas with lighter soils. The reason for this correlation, however, is not clear. One possibility is that the more vigorous clay-soil plants are able to stand a higher number of accessories than the weaker plants growing on light soils. A similar conclusion was drawn with regard to *Phleum phleoides* in which acc. chromosomes were found to be most common in populations growing under the conditions most optimal for the species. As Bosemark pointed out, this fact does not support the view first suggested by Darlington (102) that acc. chromosomes increase the variability and thus in the long run the adaptability of the species.

In *Centaurea scabiosa* (family Compositae) the acc. chromosomes are small and often numerous, and from breeding work and other comprehensive investigations with this species carried out by Fröst, it is clear that there is a certain loss of accessories from one generation to the next (103). Since these acc. chromosomes are of frequent occurrence they must have a positive selection value. This value may be higher under some ecological conditions than others and may explain why the frequency of accessories in *Centaurea scabiosa* is significantly higher in certain parts of Europe with a dry continental climate than in western parts with a humid oceanic climate (104).

In rye, *Secale cereale,* the frequency of acc. chromosomes is very much different in various populations, the frequency of plants with accessories ranging from 0 to more than 90% (91, 105–107). In recent years attempts have been made to find ecological factors with an influence on this variation. Thus, Lee in Korea has reported that the frequency of accessories is higher in material growing on acidic soils than on basic soils (108, 109), and in Yugoslavia the frequency of plants with accessories is greater in the arid region of Macedonia than in the humid region of Slovenia (107). Some experimental investigations have also been carried out to elucidate the possible ecological role of acc. chromosomes of rye. Kishikawa worked with clonal plants grown under different temperatures, soil, and moisture conditions and found that the frequency of accessories was lower in progeny populations derived from plants grown under high temperature or dry soil conditions (110). This reduction in the number of accessories was attributed to meiotic elimination, disturbances of the nondisjunction process, and abortion of seeds with many accessories. Cultivation of a rye strain containing accessories under quite different edaphic and climatic conditions may also lead to significant changes in the frequency of acc. chromosomes. Thus, for instance, in a primitive strain of rye from the Transbaikal area in Siberia that was cultivated in southern Sweden for 14 years, the percentage of plants with accessories was gradually reduced from a starting value of 28 to 10% (111). Similar experiences were noted in another investigation with rye material (92).

In the grass species *Dactylis glomerata* spp. judaica, stands occupying different ecological niches differ from one another in regard to frequency of acc. chromosomes (49). This is regarded as indicating different selective advantage of the accessories in different habitats, but it is not known how this balance is achieved. Similarly, in the plant genus *Clarkia* acc. chromosomes are so frequent, and the differences in frequency between different populations so marked, that this suggests the accessories have some adaptive role (112). McClintock, however, in a discussion of the chromosomal constitutions of races of maize from various regions in Latin America and the US, has reached the conclusion that the accessory chromosomes in maize contain no genetic components that alter in any yet recognizable way either the appearance or the physiology of a plant (6).

Though many deeply penetrating investigations of acc. chromosomes in grasshoppers and other insects have been carried out, there is not yet much information about their possible ecological importance. However, Hewitt & John, working with the grasshopper *Podisma pedestris* (113), state that while drive mechanisms (especially in females) may well account for the maintenance of acc. chromosomes in some natural populations, they do not by themselves offer a satisfactory explanation for the wide variation of accessories that can occur in others, and in local areas the levels of frequency of acc. chromosomes fall into clear clinal relationships. Such a situation implies some form of differential selection. Similar conclusions were drawn by Kayano et al from work with *Acrida lata* (114). The frequency of acc. chromosomes was significantly different between populations, which may result from different environmental conditions of the habitats and/or to different genotypes in the populations that affect variously the fitness of individuals with accessories.

INTERACTION BETWEEN A AND B CHROMOSOMES

Chromatin Elimination Induced by B Chromosomes in Maize

Rhoades et al (115–117) have reported that B chromosomes cause elimination of chromatin from knob-bearing A chromosomes at the second microspore mitosis. Two sperm cells are produced, one with an intact and the other with a deficient chromosome, and fertilization of the egg and polar nuclei with dissimilar germ cells results in noncorrespondence in the genetic constitution of embryo and endosperm. The knob has been included almost regularly in the deleted chromatin. All knobbed chromosomes may undergo chromatin loss, but chromosome arms with large knobs are lost more frequently than are those with smaller knobs, while knobless arms are stable. These findings are in harmony with the observations by Longley that among strains of maize in different regions there is a negative correlation between number of heteropycnotic knobs on the A's and presence of B's (118).

Whatever causes B's to undergo nondisjunction affects knobs in a similar way. Since failure of replication at specific sites has been implicated in nondisjunction of B's in rye (32), it was postulated that the two phenomena in maize stem from delayed replication of heterochromatic segments. Thus, it is the failure of the proximal heterochromatic knob of the B and of the more distally placed heterochromatic knobs of A's to replicate normally that is responsible for nondisjunction and loss, respectively. This view is supported by genetic data on the pattern of loss for specific marker genes affecting endosperm color or development.

Evidence has been obtained (119) that the distal euchromatic tip of the B controls nondisjunction of a B^A chromosome having the B centromere. Rhoades et al (117) suggest that the same distal tip of the B suppresses replication of the A chromosome knobs. However, delayed replication of knobs shows a higher threshold of response to the inducer because two or more B's must be present in the microspore before loss is induced in the A chromosomes. Moreover, the frequency of nondisjunction of a B is much greater than is the loss of knobbed A chromosomes, although both occur in the same cell. This does not seem surprising as there is reason to assume that both phenomena are caused by a chemical substance able to affect the A knobs only by diffusion into the cytoplasm. This, apparently, is a phenomenon resembling the interaction between different kinds of B chromosomes in *Festuca pratensis* and rye (40–42). In these cases B's with a changed structure, which therefore had lost the ability to carry out nondisjunction when alone in a microspore, could do so if they were influenced by another, normal B chromosome present in the same pollen grain.

Similar interaction was observed in work with the grass species *Poa alpina*, in a strain with 14 chromosomes in the roots but 16 in the meiosis (142). The extra pair represents B chromosomes that are eliminated from the somatic cells but retained in the germ line. In one exceptional plant the pair of accessories was smaller, about half of the chromosome being eliminated. At the same time the capacity of somatic elimination had been lost, since the two homologues representing this deleted B chromosome were regularly present in the root tips.

In order to test whether somatic elimination or nonelimination would be autonomous for each kind of B, hybrids were produced carrying one normal and one deleted B chromosome (143). Between these chromosomes there was in fact a very pronounced interaction. In 82 of 219 F_1 plants both B's were absent. In the majority of plants, however, B's were found to be present, but their number was variable with differences between root tips as well as between cells in the same root tip. This indicates that the large B (probably an isochromosome) produces a substance that regulates the elimination of this chromosome and also by diffusion affects the small B (probably a telocentric) in the same way. It is also possible that the small B influences the large B in such a way that this chromosome is not always eliminated. In such cases both B's are retained in the same cell.

Chiasma Frequency and Chromosome Pairing

In recent years much work has been performed concerning the influence of B chromosomes on chiasma frequency and chromosome pairing in the A chromosomes. In the majority of cases so far described the presence of B's tends to increase chiasma frequency and genetic recombination in the A chromosomes. In a minority of cases there is a negative correlation or no correlation at all. Reference may be given to Viinikka who has recently discussed the literature in this field (120) and also reported new observations in the plant species *Najas marina*. The results in *Najas* suggest a reduction in chiasma frequency. Thus the B's appear to influence A meiosis through at least two different mechanisms: one increasing the chiasma frequency, the other one causing a decrease. The decrease mechanism in *Najas* may be an interaction between B's and heterochromatic regions in the A chromosomes.

Not only chiasma frequency but also chromosome pairing between different genomes may be influenced by the presence of acc. chromosomes. *Aegilops speltoides* and *mutica* have genotypes that can suppress the activity of *Ph,* the gene that controls meiotic pairing in *Triticum aestivum,* which is situated on the long arm of chromosome 5 B. Hybrids between wheat and these *Aegilops* species show extensive pairing of homoeologous chromosomes of the A, B, and D genomes. However, the presence of acc. chromosomes from the two *Aegilops* species resulted in a marked reduction in the extent of homoeologous pairing, in pronounced contrast to the super-high pairing that occurs in the absence of 5 B (121, 122). Similar effects on homoeologous meiotic pairing have been observed in *Lolium temulentum* X *L. perenne* (123, 124). Presence of acc. chromosomes drastically reduced association of homoeologous chromosomes in both the primary diploid hybrid and in the tetraploid amphidiploid.

Also in the hybrid *Lolium perenne* X *Festuca arundinacea* the presence of an acc. chromosome was associated with a significant reduction in the number of chiasmata per pollen mother cell. In the corresponding amphiploid, individuals without accessories had a higher frequency of univalents and multivalents than naturally occurring polyploids. Incorporation of a pair of acc. chromosomes, however, reduced the number of multivalents and univalents (144).

Effects of New Gene Environments

The effects of acc. chromosomes in new gene environments have been tested in two ways: 1. by transferring the acc. chromosomes from diploid to autotetraploid rye (89) and 2. by transferring accessories from diploid rye to bread wheat, *Triticum aestivum* (125).

In autotetraploid rye all characters measured showed clearly negative correlations with the number of accessories. The acc. chromosomes tend to reduce fertility as well as vigor not only on the diploid level (cf 28, 89, 90), but also on the tetraploid. The most pronounced negative effects on the tetraploid level were observed with regard to kernel weight per plant and percentage of seed set. Straw weight, the number of culms, and plant height as measures of plant vigor, show a significant decline. Pollen fertility showed a slight but steady reduction with increasing B-chromosome numbers.

A detailed comparison, involving various strains of diploid and autotetraploid rye, revealed that it is primarily the absolute number of acc. chromosomes and not the ratio between A and B chromosomes, which is of decisive importance. Only a low degree of buffering effect of the tetraploid chromosome number was indicated. These results strengthen the view (previously based on the absence of cytological homology) that the genes in the A chromosomes are not at all, or only to a slight degree, represented in the acc. chromosome.

Lindström (125) succeeded in transferring acc. chromosomes from diploid rye to bread wheat in the following way: the primary hybrid was chromosome-doubled and the resulting octoploid Triticale type was crossed to bread wheat. The resulting F_1 hybrids ($2n = 49 + B$'s) were back-crossed to the wheat parent. In this way 2 plants with 42 wheat chromosomes and 2 B's from rye were obtained. The offspring of these plants, constituting the so-called "Lindström strain" (L strain), has been studied by Müntzing et al with the following main results (126–128). The B's are retained in the strain without any difficulty in spite of much meiotic elimination and the fact that the strain is just as self-fertilizing as the wheat parent. A striking feature is that the B's in the L strain frequently undergo structural changes resulting in new types of B's. Some of them are isochromosomes, while others must be products of deletion. Newly arisen large iso B chromosomes have exactly the same appearance and tendency to interarm pairing at meiosis as large iso B's in rye (36). In some plants heterobivalents were formed by pairing between large iso B's and standard B's. Though a certain number of bivalents are formed at meiosis in plants with 2 to 6 B's, meiotic pairing of the B's in the L strain is on a average poor and the frequency of B univalents is much higher than in the parental strain of rye. The reason for this difference must be an influence of the wheat chromosomes or the wheat cytoplasm on the B chromosomes.

In progenies of plants with 4 B's a marked variation was obtained, the number of B's ranging from 0 to 10. This variation is caused by meiotic irregularities as well as nondisjunction of B's at the first pollen mitosis. Definite evidence that this mechanism is retained in the L strain (in the pollen as well as in the ovules) was obtained from reciprocal crosses between individuals with 0 and 2 B's.

The B's in the L strain reduce pollen fertility in this strain at least as efficaciously as in rye. There is also a pronounced negative effect on female fertility, but marked or absolute sterility is attained still more rapidly in varieties of rye. The difference is ascribed to a slight buffering effect of the 42 wheat chromosomes. In the L strain, in contrast to rye, a significant positive correlation between B's and vegetative vigor was observed. This also is probably due to a slight buffering effect of the wheat chromosomes. Increase in straw weight may then result from an indirect effect of the sterility induced by the B's. The nutrients not used for seed production may instead favor vegetative development. Thus, in spite of various weak indications of positive functions of the B's in rye, the present data demonstrate that their primary negative effects are predominant in the L strain as well as in rye varieties.

MOLECULAR–GENETIC INFORMATION

Work on acc. chromosomes with molecular-genetic methods has been started and some of these investigations are clearly important for a deeper understanding of the nature, function, and effect of these chromosomes. With regard to plant material most of the work so far has been devoted to the acc. chromosomes of rye (129–133) and maize (134–136). A few points of special interest here may be mentioned.

In rye Kirk & Neil Jones found that the amount of nuclear DNA was directly proportional to the number of accessories (131). According to the same investigation, relative amounts of total nuclear protein and nuclear RNA decreased with increasing numbers of acc. chromosomes but not in a strictly linear fashion. The values were disproportionately low for odd-numbered B classes of plants. Histone protein was found to increase as the number of accessories went up, and in this case the values were disproportionally high for odd-numbered B classes. A negative correlation was found between histone and total nuclear protein and between histone and nuclear RNA amounts. Each B chromosome adds about 4% to the DNA amount attributable to the 14 A chromosomes.

The data gathered seem clear and convincing, but one intriguing aspect of the work is the differential genetic activity probably displayed by the B's in odd- and even-numbered classes. This phenomenon seems to have a counterpart on the organism level. Neil Jones & Rees (137) had observed that in a material of rye, measured by Müntzing (89), straw weight, tiller number, and plant weight indicated a zigzag relationship between phenotypic expression and alternating odd and even numbers of acc. chromosomes. More precisely, odd numbers had lower values of phenotypic expression than the adjoining classes with even numbers.

According to the authors the only feasible way to explain these data would be "an interaction of the kind whereby B chromosomes function differently in complete rather than incomplete pairs." This interaction is supposed to occur in the interphase nuclei. Unfortunately, new data from measurements of rye and of the Lindström strain indicate no clear differences between the odd and even categories (90). On the other hand, the data gathered by Neil Jones & Rees on the cellular level have been amply confirmed by a similar investigation with maize (136), carried out by

Ayonoadu & Rees. In this case the general trend in respect to dry mass, RNA, total protein, and histone protein was complicated by a differential variation between odd- and even-numbered B chromosome plants. Thê dry mass, amount of nuclear RNA, and total protein were disproportionally high in even-numbered B plants, while the reverse was true for histone protein.

Among animals the most detailed molecular-genetic data on acc. chromosomes have been obtained in insects. Of particular interest are the results obtained by Gibson & Hewitt in studies of the grasshopper *Myrmeleotettix maculatus* (138, 139). DNA was isolated from populations in two distinct areas. Two of the populations were from Wales, one of them containing 49% B chromosome individuals and the second 0%. The other two populations were from East Anglia, one with 46% B chromosomes and the other 0%. The isolated DNA was subjected to CsCl density centrifugation and the gradients were then analyzed. The populations without B chromosomes gave only a main A chromosome peak, whereas the populations with B chromosomes had two peaks: one main A peak and one subsidiary B peak. Analyses of base composition revealed that the two areas were significantly different with regard to the base compositions of the A peaks as well as the B peaks. With the isolation of the subsidiary B-chromosome DNA, the way became open to examine the organization of its nucleotide sequences and to see whether it functions to make RNA and protein (138).

New results from this investigation were reported two years later (139). The B-chromosome DNA had been found to contain 28% repeated and 72% unique sequences, and the surprising observation was that the repeated nucleotide sequences of the B-chromosome DNA are specific for each population. Furthermore, there was no sequence homology between the B-chromosome DNA and the main peak (nuclear) DNA. In every case the RNA synthesized from the B-chromosome DNA was found to hybridize to its own source of B-chromosome DNA and not to that from 5 other DNA samples made from B chromosomes. In fact, there was no more hybridization of RNA synthesized from the B-chromosome sequences to other B-chromosome satellite DNA's than to an unrelated source of DNA such as *Paramecium*. Thus the B-chromosome DNA in *Myrmeleotettrix* seems to be specific to an exceedingly high degree.

In contrast, recent work in maize does not indicate any differences of this kind between A and B chromosomes. Chilton & McCarthy (145) found that DNA preparations from 5B and 0B maize seedlings were indistinguishable in their buoyant density distribution in CsCl gradients. The results of several types of analysis of native and homologous and heterologous renatured duplexes of 5B and 0B maize DNA failed to reveal convincing differences ascribable to the presence of foreign DNA sequences in B chromosomes. Thus it was concluded that the DNA of B chromosomes in maize is very closely related to that of A chromosomes.

Evidently, continued work along molecular-genetic lines is needed to elucidate the specificity or nonspecificity of accessory chromosomes and may ultimately lead to a deeper and more general understanding of the nature and function of this kind of chromosome.

Literature Cited

1. Håkansson, A. 1945. Überzählige Chromosomen in einer Rasse von *Godetia nutans* Hiorth. *Bot. Notis.* 1:1–19
2. Müntzing, A. 1945. Cytological studies of extra fragment chromosomes in rye II. Transmission and multiplication of standard fragments and iso-fragments. *Hereditas* 31:457–77
3. McClung, C. E. 1902. The spermatocyte divisions of the Locustidae. *Kans. Univ. Sci. Bull.* 14
4. Randolph, L. F. 1928. Types of supernumerary chromosomes in maize. *Anat. Rec.* 41:102
5. Battaglia, E. 1964. Cytogenetics of B-chromosomes. *Caryologia* 17:245–99
6. McClintock, B. 1960. Chromosome constitutions of Mexican and Guatemalan races of maize. *Carnegie Inst. Wash. Yearb.* 59:461–72
7. Melander, Y. 1950. Accessory chromosomes in animals, especially *Polycelis tenuis*. *Hereditas* 36:19–38
8. Müntzing, A. 1958. Accessory chromosomes. *Trans. Bose Res. Inst. Calcutta* 22:1–15
9. Müntzing, A. 1959. A new category of chromosomes. *Proc. Int. Congr. Genet., 10th* 1:453–67
10. Müntzing, A. 1966. Accessory chromosomes. *Bull. Bot. Soc. Bengal* 20:1–15
11. Müntzing, A. 1954. Cytogenetics of accessory chromosomes (B-chromosomes). *Caryologia* 6:282–301
12. White, M. J. D. 1973. *Animal Cytology and Evolution.* London: Cambridge Univ. Press. 3rd ed. 961 pp.
13. Hayman, D. L., Martin, P. G. 1965. Supernumerary chromosomes in the marsupial *Schoinobates volans* (Kerr.). *Aust. J. Biol. Sci.* 18:1081–82
14. Shellhammer, H. S. 1960. Supernumerary chromosomes of the harvest mouse, *Reithrodontomys megalotis.* *Chromosoma* 27:102–8
15. Hayman, D. L., Martin, P. G., Waller, P. F. 1969. Parallel mosaicism of supernumerary chromosomes and sex chromosomes in *Echymipera kalabu* (Marsupialia). *Chromosoma* 27:371–80
16. Hayata, I. 1973. Chromosomal polymorphism caused by supernumerary chromosomes in the field mouse, *Apodemus giliacus.* *Chromosoma* 42:403–14
17. Yong, H.-S., Dhaliwal, S. S. 1972. Supernumerary (B–) chromosomes in the Malayan house rat, *Rattus rattus diardii* (Rodentia, Muridae). *Chromosoma* 36:256–62

18. Gustavsson, I. 1964. Karyotype of the fox. *Nature* 201:950–51
19. Gustavsson, I., Sundt, C. O. 1966. Chromosome complex of the family Canidae. *Hereditas* 54:249–54
20. Gustavsson, I., Sundt, C. O. 1967. Chromosome elimination in the evolution of the silver fox. *J. Hered.* 58:75–78
21. Vogt, D. W., Arakaki, D. T. 1971. Karyotype of the American red fox (*Vulpes fulva*). *J. Hered.* 62:318–19
22. Buchton, K. E., Cunningham, C. 1971. Variation of the chromosome number in the red fox (*Vulpes vulpes*). *Chromosoma* 33:268–72
23. Low, R.-J., Benirschke, K. 1972. Microchromosomes in the American red fox, *Vulpes fulva.* *Cytologia* 37:1–11
24. Lin, C. C., Johnston, D. H., Ramsden, R. O. 1972. Polymorphism and quinacrine fluorescence karyotypes of red foxes (*Vulpes vulpes*). *Can. J. Genet. Cytol.* 14:573–80
25. Ridler, M. A. C., Berg, J. M., Pendrey, M. J., Saldana, P., Timothy, J. A. D. 1970. Familial occurrence of a small supernumerary metacentric chromosome in phenotypically normal women. *J. Med. Genet.* 7:148–52
26. Müntzing, A., Prakken, R. 1941. Chromosomal aberrations in rye populations. *Hereditas* 27:273–308
27. Levitsky, G. A., Melnikov, A. N., Titova, N. N. 1932. Zytologie der Nachkommenschaft 16-chromosomiger Roggen. (Vorläuf. Mitt.). *Acad. Sci. USSR Bull. Lab. Genet.* 9:90–96
28. Müntzing, A. 1943. Genetical effects of duplicated fragment chromosomes in rye. *Hereditas* 29:91–112
29. Levitsky, G. A. 1931. The morphology of chromosomes. *Bull. Appl. Bot. Genet. Plant Breed.* 27:19–173
30. Hasegawa, N. 1934. A cytological study on 8-chromosome rye. *Cytologia* 6:68–77
31. Håkansson, A. 1948. Behaviour of accessory rye chromosomes in the embryo-sac. *Hereditas* 34:35–59
32. Müntzing, A. 1946. Cytological studies of extra fragment chromosomes in rye III. The mechanism of non-disjunction at the pollen mitosis. *Hereditas* 32:97–119
33. Lima-de-Faria, A. 1948. B chromosomes of rye at pachytene. *Port. Acta Biol. Ser. A* 11:167–74
34. Müntzing, A., Lima-de-Faria, A. 1949. Pachytene analysis of standard frag-

ments and large iso-fragments in rye. *Hereditas* 35:253–68

35. Müntzing, A., Lima-de-Faria, A. 1953. Pairing and transmission of a small accessory iso-chromosome in rye. *Chromosoma* 6:142–48

36. Müntzing, A. 1944. Cytological studies of extra fragment chromosomes in rye I. Iso-fragments produced by misdivision. *Hereditas* 30:231–48

37. Müntzing, A. 1948. Cytological studies of extra fragment chromosomes of rye IV. The position of various fragment types in somatic plates. *Hereditas* 34:161–79

38. Müntzing, A. 1948. Cytological studies of extra fragment chromosomes in rye V. A new fragment type arisen by deletion. *Hereditas* 34:435–42

39. Müntzing, A. 1951. The meiotic pairing of iso-chromosomes in rye. *Port. Acta Biol. Ser. A, R. B. Goldschmidt Vol.:* 831–60

40. Håkansson, A. 1959. Behaviour of different small accessory rye chromosomes at pollen mitosis. *Hereditas* 45:623–31

41. Lima-de-Faria, A. 1962. Genetic interaction in rye expressed at the chromosome phenotype. *Genetics* 47: 1455–62

42. Bosemark, N. O. 1956. On accessory chromosomes in *Festuca pratensis* IV. Cytology and inheritance of small and large accessory chromosomes. *Hereditas* 42:235–60

43. Östergren, G. 1947. Heterochromatic B-chromosomes in *Anthoxanthum*. *Hereditas* 33:261–96

44. Bosemark, N. O. 1950. Accessory chromosomes in *Festuca pratensis* Huds. *Hereditas* 36:366–68

45. Bosemark, N. O. 1954. On accessory chromosomes in *Festuca pratensis* I. Cytological investigations. *Hereditas* 40:346–76

46. Bosemark, N. O. 1954. On accessory chromosomes in *Festuca pratensis* II. Inheritance of the standard type of accessory chromosomes. *Hereditas* 40:425–37

47. Bosemark, N. O. 1956. On accessory chromosomes in *Festuca pratensis* III. Frequency and geographical distribution of plants with accessory chromosomes. *Hereditas* 42:189–210

48. Bosemark, N. O. 1957. Further studies on accessory chromosomes in grasses. *Hereditas* 43:236–97

49. Puteyevsky, E., Zohary, D. 1970. Behaviour and transmission of supernumerary chromosomes in diploid *Dacty-*

lis glomerata. *Chromosoma* 32:135–41

50. Williams, E., Barclay, P. C. 1972. Transmission of B-chromosomes in *Dactylis*. *N. Z. J. Bot.* 10:573–84

51. Bosemark, N. O. 1956. Cytogenetics of accessory chromosomes in *Phleum phleoides*. *Hereditas* 42:443–66

52. Albers, F. 1972. Cytotaxonomie und B-Chromosomen bei *Deschampsia caespitosa (L.)* P. B. und verwandten Arten. *Beitr. Biol. Pflanz.* 48:1–62

53. Baenziger, H., Carr, R. B. 1968. Supernumerary chromosomes in synthetic populations of crested wheatgrass, *Agropyron desertorum*. *Can. J. Genet. Cytol.* 10:813–18

54. Mendelson, D., Zohary, D. 1972. Behaviour and transmission of supernumerary chromosomes in *Aegilops speltoides*. *Heredity* 29:329–39

55. Randolph, L. F. 1941. Genetic characteristics of the B chromosomes in maize. *Genetics* 26:608–31

56. Roman, H. 1947. Mitotic nondisjunction in the case of interchanges involving the B-type chromosome in maize. *Genetics* 32:391–409

57. Roman, H. 1948. Selective fertilization in maize. *Genetics* 33:122

58. Roman, H. 1948. Directed fertilization in maize. *Proc. Nat. Acad. Sci. USA* 34:36–42

59. Roman, H. 1950. Factors affecting mitotic non-disjunction in maize. *Genetics* 35:132

60. Blackwood, M. 1956. The inheritance of B chromosomes in *Zea Mays*. *Heredity* 10:353–66

61. Rutishauser, A. 1956. Genetics of fragment chromosomes in *Trillium grandiflorum*. *Heredity* 10:195–204

62. Kayano, H. 1956. Cytogenetic studies in *Lilium callosum* II. Preferential segregation of a supernumerary chromosome. *Mem. Fac. Sci. Kyushu Univ. Ser. E Biol.* 2:53–60

63. Kayano, H. 1957. Cytogenetic studies in *Lilium callosum* III. Preferential segregation of a supernumerary chromosome in EMCs. *Proc. Jap. Acad.* 33:553–58

64. Fröst, S. 1969. The inheritance of accessory chromosomes in plants, especially in *Ranunculus acris* and *Phleum nodosum*. *Hereditas* 61:317–26

65. Rutishauser, A. 1960. Zur Genetik überzäliger Chromosomen. *Arch. Julius Klaus Stift. Vererbungsforsch. Sozialanthropol. Rassenhyg.* 35:440–58

66. Röthlisberger, E. 1971. Verteilung der B-Chromosomen und Blütenentwick-

264   MÜNTZING

lung bei *Crepis capillaris. Ber. Schweiz. Bot. Ges.* 80:194–224
67. Weng, T. 1962. Intra-individual variation in number of B chromosomes in *Miscanthus Japonicus* Anderss. *Bot. Bull. Acad. Sinica* 3:19–31
68. Jauhar, P. P., Joshi, A. B. 1968. Accessory chromosomes in a new hexaploid species of *Panicum. Caryologia* 21: 105–10
69. Li, N., Jackson, R. C. 1961. Cytology of supernumerary chromosomes in *Haplopappus spinolosus* ssp cotula. *Am. J. Bot.* 48:419–26
70. Pritchard, E. 1968. A cytogenetic study of supernumerary chromosomes in *Haplopappus gracilis. Can. J. Genet. Cytol.* 10:928–36
71. Ehrendorfer, F. 1961. Akzessorische Chromosomen bei *Achillea:* Struktur, Cytologisches Verhalten, zahlenmässige Instabilität und Entstehung (Zur Phylogenie der Gattung *Achillea,* V). *Chromosoma* 11:523–52
72. Fröst, S. 1960. A new mechanism for numerical increase of accessory chromosomes in *Crepis pannonica. Hereditas* 46:497–503
73. Fröst, S. 1962. Numerical increase of accessory chromosomes in *Crepis conyzaefolia. Hereditas* 48:667–76
74. Fröst, S. 1964. Further studies of accessory chromosomes in *Crepis conyzaefolia. Hereditas* 52:237–39
75. Melander, Y. 1949. Cytological studies on Scandinavian flatworms belonging to Tricladida, Paludicola. *Proc. Int. Congr. Genet., 8th, 1948,* pp. 625–26
76. McClintock, B. 1933. The association of non-homologous parts of chromosomes in the mid-prophase of meiosis in *Zea Mays. Z. Zellforsch.* 19:191–237
77. Carlson, W. 1969. Factors affecting preferential fertilization in maize. *Genetics* 62:543–54
78. Carlson, W. 1970. Nondisjunction and isochromosome formation in the B chromosome of maize. *Chromosoma* 30:356–65
79. Carlson, W. 1973. A procedure for localizing genetic factors controlling mitotic nondisjunction in the B chromosome of maize. *Chromosoma* 42:127–36
80. Nur, U. 1963. A mitotically unstable supernumerary chromosome with an accumulation mechanism in a grasshopper. *Chromosoma* 14:407–22
81. Nur, U. 1969. Mitotic instability leading to an accumulation of B-chromosomes in grasshoppers. *Chromosoma* 27:1–19

82. Sannomiya, M., Kayano, H. 1968. Local variation and year-to-year change in frequencies of B-chromosomes in natural populations of some grasshopper species. *Proc. Int. Congr. Genet., 12th* 2:116–17
83. Kayano, H. 1971. Accumulation of B chromosomes in the germ line of *Locusta migratoria. Heredity* 27:119–23
84. Nur, U. 1966. Harmful supernumerary chromosomes in a mealy bug population. *Genetics* 54:1225–38
85. Nur, U. 1966. The effect of supernumerary chromosomes on the development of mealy bugs. *Genetics.* 54:1239–49
86. Nur, U. 1969. Harmful B-chromosomes in a mealy bug: additional evidence. *Chromosoma* 28:280–97
87. Hewitt, G. 1973. Variable transmission rates of a B-chromosome in *Myrmeleotettix maculatus* (Thunb.) (Acrididae:Orthoptera). *Chromosoma* 40: 83–106
88. Lucov, Z., Nur, U. 1973. Accumulation of B-chromosomes by preferential segregation in females of the grasshopper *Melanoplus femur-rubrum. Chromosoma* 42:289–306
89. Müntzing, A. 1963. Effects of accessory chromosomes in diploid and tetraploid rye. *Hereditas* 49:371–426
90. Müntzing, A. 1973. Effects of accessory chromosomes of rye in the gene environment of hexaploid wheat. *Hereditas* 74:41–56
91. Kishikawa, H. 1965. Cytogenetic studies of B chromosomes in rye, *Secale cereale* L., in Japan. *Agr. Bull. Saga Univ. Japan, Oct. 1965,* pp. 1–81
92. Rees, H., Ayonoadu, U. 1973. B chromosome selection in rye. *Theor. Appl. Genet.* 43:162–66
93. Bosemark, N. O. 1957. On accessory chromosomes in *Festuca pratensis* V. Influence of accessory chromosomes on fertility and vegetative development. *Hereditas* 43:211–35
94. Fröst, S. 1958. Studies of the genetical effects of accessory chromosomes in *Centaurea scabiosa. Hereditas* 44: 112–22
95. Williams, E., Barclay, P. C. 1968. The effects of B-chromosomes on vigour and fertility in *Dactylis* hybrids. *N. Z. J. Bot.* 6:405–16
96. Smith, E. B. 1968. Supernumerary chromosomes in *Haplopappus validus* (Rydb.) Cory. *Evolution* 22:748–50
97. Mehra, P. N., Mann, S. K. 1972. Accessory chromosomes in *Pterotheca falconeri* H. F. *Nucleus* 15:123–33

98. Vosa, C. G. 1966. Seed germination and B-chromosomes in the leek (*Allium porrum*). *Chromosomes Today, Vol. 1*, ed. C. D. Darlington, K. R. Lewis, 24–27. Edinburgh: Oliver and Boyd. 274 pp.

99. White, M. J. D. 1951. A cytological survey of wild populations of *Trimerotropis* and *Circotettix* (Orthoptera, Acrididae) II. Racial differentiation in *T. sparsa*. *Genetics* 36:31–53

100. White, M. J. D. 1951. Cytogenetics of Orthopteroid insects. *Advan. Genet.* 4:267–330

101. Bosemark, N. O. 1967. Edaphic factors and the geographical distribution of accessory chromosomes in *Phleum phleoides. Hereditas* 57:239–62

102. Darlington, C. D. 1956. *Chromosome Botany.* London: Allen and Unwin. 186 pp.

103. Fröst, S. 1957. The inheritance of the accessory chromosomes in *Centaurea scabiosa. Hereditas* 43:403–22

104. Fröst, S. 1958. The geographical distribution of accessory chromosomes in *Centaurea scabiosa. Hereditas* 44:75–111

105. Müntzing, A. 1950. Accessory chromosomes in rye populations from Turkey and Afghanistan. *Hereditas* 36:507–9

106. Müntzing, A. 1957. Frequency of accessory chromosomes in rye strains from Iran and Korea. *Hereditas* 43:682–85

107. Zečević, L., Paunovic, D. 1967. B chromosome frequency in Yugoslav rye populations. *Biol. Plant.* 9:205–11

108. Lee, W. J. 1966. On accessory chromosomes in *Secale cereale* III. Relationship between the frequency of accessory chromosomes in rye and soil properties. *Kor. J. Bot.* 9(3–4):1–6

109. Lee, W. J. 1968. The frequency and geographical distribution of rye with accessory chromosomes in Korea. *Proc. Int. Congr. Genet., 12th,* 2:118

110. Kishikawa, H. 1970. Effects of temperature and soil moisture on frequency of accessory chromosomes in rye, *Secale cereale* L. *Jap. J. Breed.* 20:269–74

111. Müntzing, A. 1967. Some main results from investigations of accessory chromosomes. *Hereditas* 57:432–38

112. Mooring, J. S. 1960. A cytogenetic study of *Clarkia unguiculata* II. Supernumerary chromosomes. *Am. J. Bot.* 47:847–54

113. Hewitt, G. M., John, B. 1972. Interpopulation sex chromosome polymorphism in the grasshopper *Podisma pedestris* II. Population parameters. *Chromosoma* 37:23–42

114. Kayano, H., Sannomiya, M., Nakamura, K. 1970. Cytogenetic studies on natural populations of *Acrida lata* I. Local variation in the frequency of B-chromosomes. *Heredity* 25:113–22

115. Rhoades, M. M., Dempsey, E., Ghidoni, A. 1967. Chromosome elimination in maize induced by supernumerary B chromosomes. *Proc. Nat. Acad. Sci. USA* 57:1626–32

116. Rhoades, M. M., Dempsey, E. 1972. On the mechanism of chromatin loss induced by the B chromosome of maize. *Genetics* 71:73–96

117. Rhoades, M. M., Dempsey, E. 1973. Chromatin elimination induced by the B chromosome of maize I. Mechanism of loss and the pattern of endosperm variegation. *J. Hered.* 64:13–18

118. Longley, A. E. 1938. Chromosomes of maize from North American Indians. *J. Agr. Res.* 56:177–95

119. Ward, E. J. 1973. Nondisjunction: Localization of the controlling site in the maize B chromosome. *Genetics* 73:387–91

120. Viinikka, Y. 1973. The occurrence of B chromosomes and their effect on meiosis in *Najas marina. Hereditas* 75:207–12

121. Dover, G. A., Riley, R. 1972. Prevention of pairing of homoeologous meiotic chromosomes of wheat by an activity of supernumerary chromosomes of *Aegilops. Nature* 240:159–61

122. Vardi, A., Dover, G. A. 1972. The effect of B chromosomes on meiotic and premeiotic spindles and chromosome pairing in *Triticum aegilops* hybrids. *Chromosoma* 38:367–85

123. Evans, G. M., Macefield, A. J. 1972. Suppression of homoeologous pairing by B chromosomes in a *Lolium* species hybrid. *Nature New Biol.* 236:110–11

124. Evans, G. M., Macefield, A. J. 1973. The effect of B chromosomes on homoeologous pairing in species hybrids I. *Lolium temulentum X Lolium perenne. Chromosoma* 41:63–73

125. Lindström, J. 1965. Transfer to wheat of accessory chromosomes from rye. *Hereditas* 54:149–55

126. Müntzing, A., Jaworska, H., Carlbom, C. 1969. Studies of meiosis in the Lindström strain of wheat carrying accessory chromosomes of rye. *Hereditas* 61:179–207

127. Müntzing, A. 1970. Chromosomal variation in the Lindström strain of wheat carrying accessory chromosomes of rye. *Hereditas* 66:279–86

128. Müntzing, A. 1973. See Ref. 90

129. Ayonoadu, U. W., Rees, H. 1968. The regulation of mitosis by B-chromosomes in rye. *Exp. Cell Res.* 52:284–90

130. Neil Jones, R., Rees, H. 1968. The influence of B-chromosomes upon the nuclear phenotype in rye. *Chromosoma* 24:158–76

131. Kirk, D., Neil Jones, R. 1970. Nuclear genetic activity in B-chromosome rye, in terms of the quantitative interrelationships between nuclear protein, nuclear RNA and histone. *Chromosoma* 31:241–54

132. John, P. C. L., Neil Jones, R. 1970. Molecular heterogeneity of soluble proteins and histones in relationship to the presence of B-chromosomes in rye. *Exp. Cell Res.* 63:271–76

133. Ayonoadu, U., Rees, H. 1973. DNA synthesis in rye chromosomes. *Heredity* 30:233–40

134. Von Schaik, N., Pitout, M. J. 1966. Base composition of deoxyribonucleic acid from maize with heterochromatic B-chromosomes. *S. Afr. J. Sci.* 62: 53–56

135. Abraham, S., Smith, H. H. 1966. DNA synthesis in the B chromosomes of maize. *J. Hered.* 57:78–80

136. Ayonoadu, U. W., Rees, H. 1971. The effect of B chromosomes on the nuclear phenotype in root meristems of maize. *Heredity* 27:365–83

137. Neil Jones, R., Rees, H. 1969. An anomalous variation due to B chromosomes in rye. *Heredity* 24:265–71

138. Gibson, I., Hewitt, G. 1970. Isolation of DNA from B chromosomes in grasshoppers. *Nature* 225:67–68

139. Gibson, I., Hewitt, G. 1972. Interpopulation variation in the satellite DNA from grasshoppers with B-chromosomes. *Chromosoma* 38:121–38

140. Wigh, K. 1973. Accessory chromosomes in some mosses. *Hereditas* 74:211–24

141. Sannomiya, M. 1973. Cytogenetic studies on natural populations of grasshoppers with special reference to B-chromosomes II. *Atractomorpha bedeli.* *Chromosoma* 44:99–106

142. Müntzing, A., Nygren, A. 1955. A new diploid variety of *Poa alpina* with two accessory chromosomes at meiosis. *Hereditas* 41:405–22

143. Müntzing, A. 1966. Some recent data on accessory chromosomes in *Secale* and *Poa.* See Ref. 98, pp. 7–14

144. Bowman, J. G., Thomas, H. 1973. B chromosomes and chromosome pairing in *Lolium perenne* × *Festuca arundinacea* hybrid. *Nature New Biol.* 245:80–81

145. Chilton, M. D., McCarthy, B. J. 1973. DNA from maize with and without B chromosomes: A comparative study. *Genetics* 74:605–14

SOMATIC CELL GENETICS
OF HIGHER PLANTS[1]

❖3074

R. S. Chaleff [2] *and P. S. Carlson* [3]

Biology Department, Brookhaven National Laboratory, Upton, New York 11973

We must ... be doubly vigilant not only to hold high the banner in the more theoretical branches of our field, but also to hold it even higher in the linking of theory and practice.

H. J. Muller (1936)

INTRODUCTION

Our present understanding of the molecular mechanisms that mediate the expression of genetic information is derived largely from investigations of microorganisms. Genetic studies of these organisms have achieved a highly sophisticated and refined level. In contrast, the extent to which we understand and are able to manipulate the genetic information of higher plants is severely limited. The different degrees of success with which the molecular genetic approach has been met by microbes and higher plants are due largely to the distinct biological organization of these disparate forms. Bacteria and fungi have extended haploid phases and small nutrient reserves which permit the immediate phenotypic expression of genetic variation. The ability to grow large, homogeneous populations with short generation times on defined media makes possible the application of selective screens to enormous numbers of genomes. These organizational features which have nearly restricted the science of molecular genetics to microbes are now becoming available to higher plants. Cells of many plant species may be cultured under defined conditions (1, 2), techniques exist for obtaining haploid cell lines (3–5), and whole plants may be differentiated from cultured cells (6, 7). With the ability to manipulate experimentally higher plants as microorganisms, it should be possible to dissect the functioning of these

[1]Research carried out at Brookhaven National Laboratory under the auspices of the US Atomic Energy Commission and partially supported by US Public Health Service Grant No. GM 18537.

[2]Present address: Department of Applied Genetics, John Innes Institute, Colney Lane, Norwich, England.

[3]Present address: Department of Crop and Soil Science, Michigan State University, East Lansing, Michigan 48823.

267

more complex forms and use this understanding to effect agronomically beneficial changes in those crop species upon which our own existence ultimately depends.

This review considers the various means by which directed modification of the genetic information of higher plant cells may be achieved in vitro.

MUTANT INDUCTION AND SELECTION

The use of tissue culture permits the techniques of microbial genetics to be applied to higher plant cells. Genetic variability may be induced in large, homogeneous populations of plant cells by exposure to chemical or physical mutagens. Conditions that select for defined mutant types may then be imposed upon this population. Dominant and semidominant mutants are obtainable from cultures of diploid cells. If haploid tissue is used, direct selection for recessive mutants becomes possible.

Selection against prototrophs has been used by Carlson to isolate auxotrophic mutants from mutagenized haploid cells of the two fern species *Todea barbara* (L.) and *Osmunda cinnamomea* (L.) (8) and of tobacco, *Nicotiana tabacum* (9). The selective system, which was adapted from a procedure used successfully with mammalian cells (10), depends upon differential incorporation of 5-bromodeoxyuridine (BUdR) into DNA of prototrophic cells dividing actively in a minimal medium. Only auxotrophs, which synthesize little or no DNA during the BUdR pulse, survive subsequent illumination. Because expression of the mutant phenotype requires prior genetic segregation and depletion of nutrient reserves, it is necessary to allow the mutagenized cell population to undergo several divisions in minimal medium before the addition of BUdR. The development of additional selective systems should make possible the recovery of more diverse types of auxotrophic and conditional mutants.

A wide spectrum of vitamin and amino acid requirers was represented among the 33 auxotrophs isolated from *T. barbara.* The technique was modified to select for a specific class of auxotrophs in *O. cinnamomea.* Three uracil requirers of this species were obtained from a population of 20,000 mutagenized spores (8). Six auxotrophs were recovered from *N. tabacum,* among which were mutants requiring amino acids, vitamins, and purines. All mutant cultures grew slowly on a minimal medium. Plants were differentiated from four of the mutant clones, and these also grew slowly without supplementation (9). A sexual analysis of these four regenerated plants demonstrated that three were inherited as single recessive Mendelian factors. One displayed a more complex pattern of inheritance. Spontaneous changes in chromosome number occur in the somatic cells of all six tobacco auxotrophs both in vitro and in vivo (P. S. Carlson, unpublished results).

The recovery of only leaky mutants may be due to a lack of functional diploidization of the *N. tabacum* genome. Haploid cells obtained from this amphiploid species may contain two copies of metabolically essential genes. Thus, a mutational event could not delete any single function, because genetic modification of one locus would still leave a second homologous locus intact (9). Future research must undertake the essential task of isolating a wide spectrum of auxotrophic mutants from true diploid species. These mutants could then be used, as they have been in bacteria and fungi, to elucidate the functioning of the genetic and metabolic machinery of higher plants.

Mutants resistant to antibiotics and nucleic acid base analogs also have been isolated. Resistance to streptomycin has been selected in haploid cultures of *Petunia* (11, 12) and tobacco (13). A diploid plant was derived from one mutant tobacco clone. The results of genetic crosses with this plant suggest that resistance is inherited maternally (13). Resistance to BUdR was also selected in haploid tobacco callus (a proliferating unorganized mass of plant cells). Although plants have been differentiated from this callus, genetic data have not yet been published (14). The isolation of a tobacco cell line resistant to 8-azaguanine has also been reported (15).

Studies of microbial mutants resistant to metabolite analogs have furnished extensive knowledge of genetic regulatory mechanisms (16). Mutants of this type have been isolated in higher plants and are discussed below. Resistant and auxotrophic mutants are important as markers in analyses of genetic linkage, complementation, and recombination. Such mutants also may be exploited in selecting for products of fusion events between cells of different genetic backgrounds. In addition to their utility in answering these basic biological questions, several mutant types have been recovered from cultured plant cells, which suggest that these techniques may contribute to the solution of important nutritional and agricultural problems.

Widholm has isolated cell lines of tobacco and carrot capable of growth in the presence of a normally inhibitory concentration of 5-methyl tryptophan (17, 18). In vitro assays with crude extracts demonstrated that the resistant cell lines produce a species of anthranilate synthetase less sensitive to feedback inhibition by tryptophan and 5-methyl tryptophan than is the wild-type enzyme. The mutant carrot enzyme also has a lower apparent K_m for chorismate and Mg^{2+} than does the wild-type enzyme (18). Because the analog-resistant mutants were isolated from diploid cell lines, they would be expected to be heterozygous and synthesize the wild-type as well as the mutant form of anthranilate synthetase. However, the published data do not indicate whether the mutants contain both enzyme species. Endogenous levels of free tryptophan in resistant lines of tobacco and carrot were 15 and 27 times higher, respectively, than the wild-type levels (17, 18). It would appear that in these mutants, resistance to 5-methyl tryptophan is accomplished by the overproduction of tryptophan, which reduces the effective concentration of the analog. The ability to transport 5-methyl tryptophan into the cell is impaired in another analog-resistant strain of carrot which possesses a wild-type anthranilate synthetase enzyme. This resistance phenotype is transmitted stably from callus to plant and again back to callus (J. M. Widholm, personal communication).

The wildfire disease of tobacco is caused by the bacterial pathogen, *Pseudomonas tabaci* (19). Tobacco cells resistant to wildfire toxin have been recovered among a mutagenized haploid cell population by selecting for growth in the presence of methionine sulfoximine, an analog of both methionine and the wildfire toxin (20). Three diploid plants that were regenerated from these methionine sulfoximine-resistant calluses are less susceptible than the parent plant to the pathogenic effects of bacterial infection. In one mutant plant which contains wild-type levels of free methionine, resistance appears to be genetically complex. The intracellular concentration of free methionine in the other two resistant plants is five times higher than in the wild type. The resistance phenotype of these plants is transmitted in crosses as a single semidominant locus (20). Therefore it appears that, as in the case of

5-methyl tryptophan resistance, resistance to methionine sulfoximine may be achieved by mutationally altering the regulation of amino acid biosynthesis. These results indicate that in higher plants the endogenous concentration of a specific metabolite can be increased by selecting for resistance to a structural analog of that metabolite. It follows, therefore, that by using analogs of amino acids essential to human nutrition, variants of crop species that produce elevated levels of these compounds may be induced and selected in culture. Such research, directed toward improving the nutritional quality of these plants, is now in progress in this laboratory.

The successful recovery of tobacco plants resistant to wildfire disease reveals the potential of selection for toxin resistance as a generalized procedure for generating disease-resistant varieties. Direct selection for resistance to a pathogen also may be possible in culture. Several in vitro systems have been developed for studying host-pathogen interactions under controlled conditions (21). An elegant series of experiments has defined the interaction between a viral pathogen and its host in culture. Infection of protoplasts isolated from mesophyll cells of *N. tabacum* by either the RNA from tobacco mosaic virus (TMV) or the complete virus particle is followed by synchronous multiplication of the virus within the protoplasts (22–24). Viral multiplication involves the synthesis of TMV-specific proteins (25). Research on the crown gall disease has provided a model system for the study of bacterial pathogens (21, 26). Tissue culture has been employed to elucidate the pathology of fungal diseases as well. Ingram & Robertson (27) showed that cultured tissue of susceptible and resistant varieties of *Solanum tuberosum* exhibits the same response as the intact plants to infection with *Phytophthora infestans*. When exposed to fungal spores, resistant *Solanum* cells produce a thermolabile toxic compound which may reflect expression of R genes (hypersensitive resistance) in culture (28). A differential growth rate of *Phytophthora parasitica* var. *nicotianae*, the agent causing black shank disease, has been obtained on calluses derived from resistant and susceptible lines of *N. tabacum* (29). The potential use of these systems for plant breeding is illustrated by the isolation of clones of sugar cane resistant to mosaic virus disease from cultures of a sensitive variety (30). Plants derived from several of these clones proved to be immune under field conditions favorable to mosaic virus spread (R. E. Coleman, cited in reference 31).

Mutants that concentrate metabolites more efficiently may also be selected in culture. Growth of tobacco cells is inhibited by threonine when nitrate is the only available nitrogen source. Heimer & Filner (32) have isolated a threonine-resistant cell line of tobacco in which regulation of nitrate uptake is altered. Although the variant strain was isolated initially as an instrument for investigating the physiological processes involved in nitrate uptake and utilization, other applications are evident. It may be possible to reduce fertilizer requirements by selection for enhanced assimilation of these growth-limiting compounds. Other systems may be developed which would allow in vitro selection for increased tolerance to conditions of stress, such as salinity, cold, and drought.

Several difficulties are encountered in employing cultured plant cells as organisms for genetic studies. The proliferation of plant cells in culture is often accompanied by chromosomal aberrations, changes in ploidy, and loss of totipotency (33–36). Induced changes in chromosome composition have been used to generate polyploid lines of sugar cane in vitro (31). However, in most circumstances, changes of karyotype are undesirable. One possible method for restricting these changes would be to establish an environment in which cells of a given chromosome number would proliferate preferentially. Gupta & Carlson (37) have reported differential growth of cells of various ploidies in the presence of p-fluorophenylalanine. However, these results are not always reproducible.

A means of circumventing the genetic and physiological instability of continuous cell cultures is offered by the technique of anther culture. By culturing detached anthers under the appropriate conditions, immature pollen grains can be induced to develop into haploid plants (4, 5). Mutants may be induced by exposing the anther to a mutagen either when it is in culture (5) or while it is developing on the parent plant. Many of the same selective screens that have been discussed for use with continuous cell cultures could then be applied to the anther cultures. Diploid plants for further studies are easily obtainable from the haploids (5, 38, 39). More recent advances have made possible the culture of isolated pollen grains of several genera in the presence of an extract of cultured developing anthers (40, 41). The identity of the factor active in promoting embryogenesis is being sought (C. Nitsch, personal communication). A completely defined system for plating isolated pollen thus may soon become available. The ability to obtain large, homogeneous populations of haploid cells that may be subjected readily to mutagenesis, selection, and differentiation should provide a powerful tool for genetic modification of higher plants.

The success of the approaches outlined in this section requires the capacity to culture cells of a given species and to regenerate mature plants from these cells. This has not yet been realized for many agriculturally important species. Until the conditions necessary for organogenesis of these species in culture are defined, variants may be isolated by applying selective screens to populations containing homozygous mutant seed. These populations, designated M2, are obtained from the self-fertilization of plants derived from mutagenized seed. By exposing M2 oat seed to a dilute solution of the toxin produced by *Helminthosporium victoriae,* Konzak (42) successfully isolated individuals resistant to Victoria blight, a root-rotting disease. This method has also been used to isolate chlorate-resistant mutants of *Arabidopsis thaliana* (43) and has been suggested as a means for identifying high lysine strains of barley (44).

GENE TRANSFER

Transformation is a common means of transferring genetic information in bacterial systems. This process involves the uptake of exogenously supplied DNA and its subsequent integration into the host genome. Initial studies suggest that this technique may be developed as a method for modifying the genotypes of higher plants. The extensive literature on transformation in higher organisms has been reviewed recently (45). Therefore, we confine this discussion to experiments with somatic cells of higher plants.

The successful transfer of genes from one individual to another via transformation requires at least three processes. First, the DNA must be taken up as an intact macromolecule. Second, it must be able to replicate. This process may not necessarily require integration into the host genome (46). Third, the newly introduced genes must be expressed. Expression demands transcription and translation of the foreign DNA. In addition, any enzyme product must be capable of functioning in the new molecular environment.

Exogenous DNA has been taken up by cells of tobacco (47) and by protoplasts of *Petunia* (48), *Ammi,* soybean, and carrot (49, 50). However, most of the foreign DNA is rapidly degraded. This degradation may be minimized in the future by using recipient cells with low DNase activity or by chemically inhibiting this activity (51). To date, no evidence has been obtained of integration of intact DNA into the genome of cultured plant cells, although the integration (52) and expression (46, 52) of foreign DNA have been reported in plants following treatment of seeds with a DNA solution.

The systems mentioned above are adequate for characterizing the process of DNA uptake. Replication and phenotypic expression may be assayed together by designing a selective system in which cell survival and proliferation are dependent upon the biological activity encoded by the newly introduced gene. In the ideal system, defined genetic information would be introduced into either a cell which never possessed this information initially or a cell from which it has been deleted. The protein product of the transferred gene should be fully characterized so that it can be identified unambiguously by physical and biochemical methods in the transformants. Because several different mechanisms may accomplish antimetabolite resistance, caution is urged in employing such systems. The nature of the resistance must be understood completely, and it must be experimentally demonstrable that the resistance phenotype results from expression of the newly introduced gene. Ultimately, the intracellular localization and mechanism of replication of the transferred DNA must be determined by physical methods such as nucleic acid hybridization and density gradient centrifugation (47, 52). Progress toward the realization of this ideal system has been made in several laboratories.

Recent experiments have employed bacteriophages to effect gene transfer. These viruses may serve to protect the donor DNA from nucleases and facilitate its entry into the plant cell. Following infection of *Hordeum vulgare* protoplasts with the virulent coliphage T3, the synthesis of two phage-specific enzymes is observed. These two enzymes, S-adenosylmethionine-cleaving enzyme (SAMase) and RNA polymerase, are encoded by the early region of the T3 genome and are not normally produced by the plant. Control experiments demonstrated that the appearance of these activities is dependent upon phage infection of intact protoplasts and is not attributable to contamination. Infection of protoplasts with a T3 strain containing an amber mutation in the RNA polymerase structural gene results in significant synthesis of SAMase activity only (53).

Bacterial genes may be carried by specially constructed transducing phages. In addition to the possible protective and transport roles of the phage, incorporation into the smaller phage genome also serves to purify partially the bacterial genes of interest. The transducing phages λ and φ80, containing bacterial genes for galactose and lactose utilization, respectively, have been used to infect callus tissue of *Lycopersicon esculentum* (54). Tomato callus is usually unable to survive when either galactose or lactose is provided as sole carbon source. However, calluses inoculated with λ*pgal⁺* survive and grow slowly on galactose. Growth on galactose apparently requires an intact *gal⁺* operon because tomato calluses inoculated with φ80, φ80 *plac⁺*, or λ*pgal⁻* are unable to utilize this sugar. Experiments on lactose were not as absolute as on galactose. Approximately 10% of uninoculated tomato calluses are able to grow on lactose. Inoculation with φ80 or λ*pgal⁺* does not alter this result. Following inoculation with φ80*plac⁺* alone or together with λ*pgal⁺*, 80% of the treated calluses grow slowly on lactose.

Synthesis of the bacterial β-galactosidase enzyme in the surviving calluses was demonstrated immunologically. Antiserum specific for *E. coli* β-galactosidase affords the same degree of protection against denaturation at 60° C to β-galactosidase activity in extracts of *E. coli* and of an inoculated callus capable of growth on lactose (54, 55).

Johnson, Grierson & Smith (56) have performed related experiments with suspension cultures of sycamore cells. Only infection with λ phage carrying the *lac* genes and not with a wild-type phage enables cells to utilize lactose. These workers have not yet identified the bacterial enzymes in cells growing on lactose.

The phenomenon of gene transfer and subsequent phenotypic expression has been called transgenosis by Doy, Gresshoff & Rolfe (54, 57). This term was assigned deliberately to describe an observation only and to avoid any assumption of molecular mechanisms implicit in bacterial terminology. Certainly the definition of the molecular mechanism underlying transgenosis should be one of the major aims of future research.

The use of bacteriophages as vectors to transport selected bacterial genes into higher plant cells appears promising for both fundamental and applied research. Doy, Gresshoff & Rolfe (57) already have suggested using this technique to introduce bacterial nitrogen fixation genes into higher plants. Other desirable phenotypic changes mentioned earlier may possibly be achieved through transgenosis. However, before these experiments are attempted, many difficulties with the system must be overcome. Among these are achieving stability of the new phenotype (54) and determining whether the bacterial genes can be expressed in the mature plant and transmitted through the gametes.

SOMATIC HYBRIDIZATION

The fusion of protoplasts and the subsequent regeneration of a plant may provide another method for increasing genetic variability. Somatically produced hybrids should bring together genetic material from diverse species and effect genetic combinations not otherwise possible. Protoplasts can be isolated from many different species (58) and stimulated to fuse by a number of techniques (59–64). However, in only a few cases are plants obtainable from the protoplasts (65–68). This regenerative capacity is prerequisite for the construction of hybrid plants. The recognition of a heteroplasmic fusion event is also a major rate-limiting step in this overall procedure. This may be accomplished by regenerating plants nonselectively from the population of fused protoplasts and identifying the hybrids morphologically (69). The use of recessive albino or color mutations may facilitate this approach (62, 70). A more efficient method requires the use of genetic markers to establish preferential growth of hybrid cells in cultures.

To our knowledge, only one interspecific hybrid plant has been produced by fusion of somatic cells (71). The amphiploid hybrid *Nicotiana glauca* X *N. langsdorffii* can also be formed by standard sexual means. The rare interspecific fusion event was selected on a medium in which only hybrid protoplasts are able to divide. Plants derived from the growing calluses were indistinguishable from the sexually produced hybrid by several biological criteria. The recovery and analysis of the hybrid plant depended upon characteristics known to be unique to the sexually formed hybrid. More widely applicable selective procedures must be developed before this method may be generalized. Such selective systems using dominant resistance traits (70; J. Tempe, personal communication) or complementing auxotrophs may be defined.

The technique of protoplast fusion offers the potential for releasing the construction of interspecific hybrids from the constraints imposed by sexual incompatibilities. Agronomically useful blocks of genes may be defined and introduced en masse into other species. This would be especially valuable for transferring polygenic characteristics. Protoplast fusion also introduces the possibility of establishing a parasexual cycle for higher plants (72). The development of such a cycle would permit genetic analyses to be performed in vitro. Genetic complementation patterns could be determined in heterozygotes formed by fusion (73). If chromosome loss is inducible in plant cell hybrids as it is in animal cell hybrids (74), genetic linkage relationships may be determined. Genetic mapping by mitotic recombination, routinely employed in fungal studies (75, 76), may also prove feasible in somatic cell hybrids. The analysis of sterile plants expected from interspecific hybridization would then be possible. The long generation times typical of a sexual cycle may be bypassed. However, at the present time, these proposals are rendered highly speculative by the many experimental difficulties which first must be surmounted.

INCORPORATION OF SUBCELLULAR PARTICLES

A number of studies have demonstrated that protoplasts will take up a wide range of biological and nonbiological material. Besides the uptake of macromolecules previously cited, protoplasts have been observed to engulf viruses (22, 77), chloroplasts (53, 78), and nuclei (79), as well as intact bacteria (80) and polystyrene latex spheres (77). The possible functioning of incorporated chloroplasts has been reported (53). This work utilized albino protoplasts isolated from leaf mesophyll cells of a maternally inherited variegating albino mutant of *N. tabacum*. Whole, variegating plants were regenerated from albino protoplasts which had been exposed to wild-type chloroplasts. Reversion and cross-feeding have been excluded as explanations of the results. However, a definitive interpretation awaits experiments with defined, chloroplast-encoded genetic markers. These markers will serve to identify positively the presence of the introduced chloroplasts in the recipient cell.

HETEROGENEOUS ASSOCIATIONS OF CELLS

Tissue culture provides opportunities for bringing into juxtaposition cells of widely divergent origins. If these cells are derived from different plant species, a chimeral callus, and perhaps a chimeral plant, may be formed. The experimental production of chimeral plants has been accomplished from the organization of chimeral shoots at a graft union (81). This has restricted chimeral production to forms composed of cells from graft-compatible species. The use of chimeral associations of cells in vitro eliminates the requirement for graft compatibility and thus may extend the range of such associations. Several types of chimeral plants between *N. tabacum* and the amphiploid hybrid *N. glauca* X *N. langsdorffii* have been constructed by this method (82). This technique should prove valuable for generating new varieties of vegetatively propagated crops.

The symbiotic association between legumes and bacteria of the genus *Rhizobium* has been reproduced and studied in an in vitro system (83). Similar systems also may be used to create symbioses not found in nature. We have been able to force a symbiotic relationship in culture between the free-living nitrogen-fixing bacterium *Azotobacter vinelandii* and cells of carrot *Daucus carota*. The composite callus is able to proliferate slowly on a defined medium lacking fixed nitrogen (P. S. Carlson and R. S. Chaleff, unpublished results). Although plants have not yet been obtained, it is hoped that a general method will be developed for extending to other crop species the agriculturally important symbiosis which allows many plants to utilize atmospheric nitrogen.

CONCLUSION

We have attempted to outline the current state of research on the somatic cell genetics of higher plants and the directions in which we believe this research should progress. Unfortunately, restrictions on length have forced us to omit mention of many excellent, but ancillary, studies. Hopefully, the experimental approaches described will provide means for achieving the directed modification of plant genotypes. These should serve to mitigate present limitations on genetic and biochemical investigations of higher plants. The potential applicability of these techniques to plant breeding has been recognized widely (31, 57, 69, 84–87). Selection for disease and herbicide resistance and possibly for improved nutritional quality and fertilizer utilization may prove rather straightforward. These developments, however, must await the extension of tissue culture methods to important crop species.

Other questions also remain. For example, it is not known whether phenotypes selected in essentially undifferentiated cells in culture will be expressed in the mature plant or whether expression will occur in the appropriate organ. It may not turn out that cultured cells of a cereal selected for high lysine content by resistance to a lysine analog necessarily will give rise to plants that contain elevated lysine levels in the seed endosperm. Many such changes may even prove deleterious to the plant. Other restrictions also are evident. Many agronomically important characteristics such as increased yield and resistance to lodging cannot yet be recognized in culture.

These qualifications should serve only to illustrate the need for additional experimentation from which the answers to these questions may be derived. It is anticipated that these in vitro methods will provide a significant supplement to current plant breeding practices.

ACKNOWLEDGMENTS

The authors wish to express their gratitude to Drs. M. Berlyn, G. Fink, N. Giles, E. Jaworski, C. Nitsch, D. Parke, J. Polacco, T. Rice, H. Smith, and J. Widholm for their critical review of the manuscript.

Literature Cited

1. Street, H. E., Henshaw, G. G. 1966. In *Cells and Tissues in Culture*, ed. E. N. Willmer, 3:459–532. New York: Academic. 826 pp.
2. Street, H. E., Ed. 1973. *Plant Tissue and Cell Culture*. Berkeley: Univ. Calif. Press. 503 pp.
3. Chase, S. S. 1969. Monoploids and monoploid-derivatives of maize. *Bot. Rev.* 35:117–67
4. Sunderland, N. 1973. See Ref. 2, pp. 205–39
5. Nitsch, J. P. 1972. Haploid plants from pollen. *Z. Pflanzenzuecht.* 67:3–18
6. Street, H. E. 1966. See Ref. 1, pp. 631–89
7. Vasil, I. K., Vasil, V. 1972. Totipotency and embryogenesis in plant cell and tissue cultures. *In Vitro* 8:117–27
8. Carlson, P. S. 1969. Production of auxotrophic mutants in ferns. *Genet. Res.* 14:337–39
9. Carlson, P. S. 1970. Induction and isolation of auxotrophic mutants in somatic cell cultures of *Nicotiana tabacum. Science* 168:487–89
10. Puck, T. T., Kao, F. 1967. Genetics of somatic mammalian cells. V. Treatment with 5-bromodeoxyuridine and visible light for isolation of nutritionally deficient mutants. *Proc. Nat. Acad. Sci. USA* 58:1227–34
11. Binding, H., Binding, K., Straub, J. 1970. Selektion in Gewebekulturen mit haploiden zellen. *Naturwissenschaften* 57:138–39
12. Binding, H. 1972. Selektion in kalluskulturen mit haploiden zellen. *Z. Pflanzenzuecht.* 67:33–38
13. Maliga, P., Sz-Breznovits, A., Márton, L. 1973. Streptomycin-resistant plants from callus culture of haploid tobacco. *Nature New Biol.* 244:29–30
14. Maliga, P., Márton, L., Sz-Breznovits, A. 1973. 5-Bromodeoxyuridine-resistant cell lines from haploid tobacco. *Plant Sci. Lett.* 1:119–21
15. Lescure, A. M. 1973. Selection of markers of resistance to base-analogues in somatic cell cultures of *Nicotiana tabacum. Plant Sci. Lett.* 1:375–83
16. Umbarger, H. E. 1971. Metabolite analogs as genetic and biochemical probes. *Advan. Genet.* 16:119–40
17. Widholm, J. M. 1972. Cultured *Nicotiana tabacum* cells with an altered anthranilate synthetase which is less sensitive to feedback inhibition. *Biochim. Biophys. Acta* 261:52–58
18. Widholm. J. M. 1972. Anthranilate synthetase from 5-methyltryptophan susceptible and resistant cultured *Daucus carota* cells. *Biochim. Biophys. Acta* 279:48–57
19. Braun, A. C. 1955. A study on the mode of action of the wildfire toxin. *Phytopathology* 45:659–64
20. Carlson, P. S. 1973. Methionine-sulfoximine-resistant mutants of tobacco. *Science* 180:1366–68
21. Braun, A. C., Lipetz, J. 1966. See Ref. 1, pp. 691–722
22. Takebe, I., Otsuki, Y. 1969. Infection of tobacco mesophyll protoplasts by tobacco mosaic virus. *Proc. Nat. Acad. Sci. USA* 64:843–48
23. Aoki, S., Takebe, I. 1969. Infection of tobacco mesophyll protoplasts by tobacco mosaic virus ribonucleic acid. *Virology* 39:439–48
24. Otsuki, Y., Takebe, I., Honda, Y., Matsui, C. 1972. Ultrastructure of infection of tobacco mesophyll protoplasts by tobacco mosaic virus. *Virology* 49:188–91
25. Sakai, F., Takebe, I. 1972. A non-coat protein synthesized in tobacco mesophyll protoplasts infected by tobacco mosaic virus. *Mol. Gen. Genet.* 118:93–96
26. Meins, F. Jr. 1972. Stability of the tumor phenotype in crown gall tumors of tobacco. *Progr. Exp. Tumor Res.* 15:93–109

27. Ingram, D. S., Robertson, N. F. 1965. Interaction between *Phytophthora infestans* and tissue cultures of *Solanum tuberosum*. *J. Gen. Microbiol.* 40: 431–37
28. Ingram, D. S. 1967. The expression of R-gene resistance to *Phytophthora infestans* in tissue cultures of *Solanum tuberosum*. *J. Gen. Microbiol.* 49:99–108
29. Helgeson, J. P., Kemp, J. D., Haberlach, G. T., Maxwell, D. P. 1972. A tissue culture system for studying disease resistance: the black shank disease in tobacco callus cultures. *Phytopathology* 62:1439–43
30. Coleman, R. E. 1970. New plants produced from callus tissue culture. *Sugarcane Res. 1970 Rept., ARS, USDA,* p. 38
31. Nickell, L. G., Heinz, D. J. 1973. *Genes, Enzymes and Populations,* ed. A. Srb, 109–28. New York:Plenum. 359 pp.
32. Heimer, Y. M., Filner, P. 1970. Regulation of the nitrate assimilation pathway of cultured tobacco cells. II. Properties of a variant cell line. *Biochim. Biophys. Acta* 215:152–65
33. Partanen, C. R. 1963. Plant tissue culture in relation to developmental cytology. *Int. Rev. Cytol.* 15:215–43
34. Torrey, J. G. 1967. Morphogenesis in relation to chromosomal constitution in long term plant tissue cultures. *Physiol. Plant.* 20:265–75
35. Thomas, E., Street, H. E. 1970. Organogenesis in cell suspension cultures of *Atropa belladonna* L. and *Atropa belladonna* cultivar *lutea* Döll. *Ann. Bot. London* 34:657–69
36. Heinz, D. J., Mee, G. W. P., Nickell, L. G. 1969. Chromosome numbers of some *Saccharum* species hybrids and their cell suspension cultures. *Am. J. Bot.* 56:450–56
37. Gupta, N., Carlson, P. S. 1972. Preferential growth of haploid plant cells *in vitro*. *Nature New Biol.* 239:86
38. Nitsch, J. P., Nitsch, C., Hamon, S. 1969. Production de *Nicotiana* diploides à partir de cals haploides cultivés *in vitro*. *C. R. Acad. Sci.* 269:1275–78
39. Tanaka, M., Nakata, K. 1969. Tobacco plants obtained by anther culture and the experiment to get diploid seeds from haploids. *Jap. J. Genet.* 44:47–54
40. Nitsch, C., Norreel, B. 1973. Effet d'un choc thermique sur le pouvoir embryogéne du pollen de *Datura innoxia* cultivé dans l'anthère ou isolé de l'anthère. *C. R. Acad. Sci.* 276:303–6
41. Debergh, P., Nitsch, C. 1973. Premier résultats sur la culture in vitro de grains de pollen isolés chez la Tomate. *C. R. Acad. Sci.* 276:1281–84
42. Konzak, C. F. 1956. Induction of mutations for disease resistance in cereals. *Brookhaven Symp. Biol.* 9:157–76
43. Oostindiër-Braaksma, F. J., Feenstra, W. J. 1973. Isolation and characterization of chlorate-resistant mutants of *Arabidopsis thaliana*. *Mutat. Res.* 19: 175–85
44. Brock, R. D., Friederich, E. A., Langridge, J. 1973. *Nuclear Techniques for Seed Protein Improvement,* 329–38. Vienna: Int. At. Energy Agency. 422 pp.
45. Hess, D. 1972. Transformationen an höheren Organismen. *Naturwissenschaften* 59:348–55
46. Hess, D. 1973. Transformation versuche an höheren Pflanzen: Untersuchungen zur Realisation des Exosomen-Modells der Transformation bei *Petunia hybrida*. *Z. Pflanzenphysiol.* 68:432–40
47. Bendich, A. J., Filner, P. 1971. Uptake of exogenous DNA by pea seedlings and tobacco cells. *Mutat. Res.* 13:199–214
48. Hoffmann, F., Hess, D. 1973. Die Aufnahme radioaktiv markierter DNS in isolierte Protoplasten von *Petunia hybrida*. *Z. Pflanzenphysiol.* 69:81–83
49. Ohyama, K., Gamborg, O. L., Miller, R. A. 1972. Uptake of exogenous DNA by plant protoplasts. *Can. J. Bot.* 50:2077–80
50. Ohyama, K., Gamborg, O. L., Shyluk, J. P., Miller, R. A. 1973. Studies on transformation: Uptake of exogenous DNA by plant protoplasts. *Colloq. Int. Cent. Nat. Rech. Sci.* 212:423–28
51. Holl, F. B. 1973. Cellular environment and the transfer of genetic information. *Colloq. Int. Cent. Nat. Rech. Sci.* 212:509–16
52. Ledoux, L., Huart, R., Jacobs, M. 1971. *Informative Molecules in Biological Systems,* ed. L. Ledoux, 159–75. Amsterdam: North-Holland. 466 pp.
53. Carlson, P. S. 1973. The use of protoplasts for genetic research. *Proc. Nat. Acad. Sci. USA* 70:598–602
54. Doy, C. H., Gresshoff, P. M., Rolfe, B. G. 1973. Biological and molecular evidence for the transgenosis of genes from bacteria to plant cells. *Proc. Nat. Acad. Sci. USA* 70:723–26
55. Doy, C. H., Gresshoff, P. M., Rolfe, B. G. 1973. Time course of phenotypic expression of *Escherichia coli* gene Z following transgensosis in haploid *Lycopersicon esculentum* cells. *Nature New Biol.* 244:90–91

56. Johnson, C. B., Grierson, D., Smith, H. 1973. Expression of λplac5 DNA in cultured cells of a higher plant. *Nature New Biol.* 244:105–7
57. Doy, C. H., Gresshoff, P. M., Rolfe, B. G. 1973. *The Biochemistry of Gene Expression in Higher Organisms,* ed. J. Pollak, J. W. Lee, 21–37. Artarmon: Aust. and New Zealand Book Co.
58. Cocking, E. C. 1972. Plant cell protoplasts—isolation and development. *Ann. Rev. Plant Physiol.* 23:29–50
59. Power, J. B., Cummins, S. E., Cocking, E. C. 1970. Fusion of isolated plant protoplasts. *Nature* 225:1016–18
60. Eriksson, T. 1971. Isolation and fusion of plant protoplasts. *Colloq. Int. Cent. Nat. Rech. Sci.* 193:297–302
61. Keller, W. A., Harvey, B. L., Kao, K. N., Miller, R. A., Gamborg, O. L. 1973. Determination of the frequency of interspecific protoplast fusion by differential staining. *Colloq. Int. Cent. Nat. Rech. Sci.* 212:455–63
62. Keller, W. A., Melchers, G. 1973. The effect of high pH and calcium on tobacco leaf protoplast fusion. *Z. Naturforsch.* 28c:737–41
63. Withers, L. A. 1973. Plant protoplast fusion: methods and mechanisms. *Colloq. Int. Cent. Nat. Rech. Sci.* 212:517–45
64. Kao, K. N., Michayluk, M. R. 1974. A method of high frequency intergeneric fusion of plant protoplasts. *Planta* 115:355–67
65. Nitsch, J. P., Ohyama, K. 1971. Obtention de plantes à partir de protoplastes haploides cultivés *in vitro. C. R. Acad. Sci.* 273:801–4
66. Takebe, I., Labib, G., Melchers, G. 1971. Regeneration of whole plants from isolated mesophyll protoplasts of tobacco. *Naturwissenschaften* 58:318–20
67. Grambow, H. J., Kao, K. N., Miller, R. A., Gamborg, O. L. 1972. Cell division and plant development from protoplasts of carrot cell suspension cultures. *Planta* 103:348–55
68. Durand, J., Potrykus, I., Donn, G. 1973. Plantes issues de protoplastes de Petunia. *Z. Pflanzenphysiol.* 69:26–34
69. Melchers, G., Labib, G. 1973. Plants from protoplasts. Significance for genetics and breeding. *Colloq. Int. Cent. Nat. Rech. Sci.* 212:367–72
70. Limbourg, B., Prevost, G. 1972. Utilisation de marqueurs génétiques en vue de l'étude de la recombinaison de cellules végétales en culture. *Colloq. Intern. Cent. Nat. Rech. Sci.* 193:241–43

71. Carlson, P. S., Smith, H. H., Dearing, R. D. 1972. Parasexual interspecific plant hybridization. *Proc. Nat. Acad. Sci. USA* 69:2292–94
72. Carlson, P. S. 1973. Towards a parasexual cycle in higher plants. *Colloq. Int. Cent. Nat. Rech. Sci.* 212:497–501
73. Giles, K. L. 1973. Attempts to demonstrate genetic complementation by the technique of protoplast fusion. *Colloq. Int. Cent. Nat. Rech. Sci.* 212:485–95
74. Weiss, M. C., Green, H. 1967. Human-mouse hybrid cell lines containing partial complements of human chromosomes and functioning human genes. *Proc. Nat. Acad. Sci. USA* 58:1104–11
75. Pontecorvo, G., Käfer, E. 1958. Genetic analysis based on mitotic recombination. *Advan. Genet.* 9:71–104
76. Manney, T. R., Mortimer, R. K. 1964. Allelic mapping in yeast by X-ray-induced mitotic reversion. *Science* 143:581–83
77. Cocking, E. C. 1970. Virus uptake, cell wall regeneration, and virus multiplication in isolated plant protoplasts. *Int. Rev. Cytol.* 28:89–124
78. Potrykus, I. 1973. Transplantation of chloroplasts into protoplasts of *Petunia. Z. Pflanzenphysiol.* 70:364–66
79. Potrykus, I., Hoffmann, F. 1973. Transplantation of nuclei into protoplasts of higher plants. *Z. Pflanzenphysiol.* 69:287–89
80. Davey, M. R., Cocking, E. C. 1972. Uptake of bacteria by isolated higher plant protoplasts. *Nature* 239:455–56
81. Jørgensen, C. A., Crane, M. B. 1927. Formation and morphology of *Solanum* chimaeras. *J. Genet.* 18:247–73
82. Carlson, P. S. 1974. *In vitro* production of interspecific chimeral plants. *Proc. Nat. Acad. Sci. USA.* Submitted
83. Holsten, R. D., Hardy, R. W. F. 1972. *Methods in Enzymology,* ed. A. San Pietro, 24:497–504. New York: Academic. 526 pp.
84. Nickell, L. G., Torrey, J. G. 1969. Crop improvement through plant cell and tissue culture. *Science* 166:1068–70
85. Smith, H. H. 1974. Model systems for somatic cell plant genetics. *BioScience.* In press
86. Melchers, G. 1972. Haploid higher plants for plant breeding. *Z. Pflanzenzuecht.* 67:19–32
87. Cocking, E. C. 1973. Plant cell modification: problems and perspectives. *Colloq. Int. Cent. Nat. Rech. Sci.* 212:327–41

FUNGAL GENETICS ❖3075

D. G. Catcheside
Department of Genetics, Research School of Biological Sciences, The Australian National University, Canberra, A.C.T., Australia

In this broad field the topics reviewed are concentrated on two main ones. Recombination continues to occupy a prominent place in fungal genetics, a reflection of the opportunity for tetrad, or octad, analysis ranging from large samples of unselected tetrads heterozygous at many loci in yeast to selected tetrads in *Ascobolus immersus*. There is also the opportunity for analysis of random spores either by selective or unselective procedures. The former permits the survey of very large populations. In consequence, the understanding of the range of genetic control of recombination in eukaryotes is well advanced in fungi. Numerous mutants in *Neurospora crassa, Aspergillus nidulans, Saccharomyces cerevisiae, Schizosaccharomyces pombe,* and *Ustilago maydis* are exceptionally sensitive to ultraviolet light (UV). Several of these mutants are altered in properties of recombination, whether meiotic or mitotic, and of mutation. The results are very complex. Connections between the mechanism of repair from radiation damage and recombination and mutation are evident, and progress toward a general theory has been made.

GENETICS OF RECOMBINATION

Two general types of control of recombination have received extensive study. One is the contribution that differences between the nucleotide pairs at a heterozygous site may make to the probabilities of conversion, the "marker effect." The other is the contribution that other genes concerned with the mechanism and regulation of recombination make to the frequency of the events and to their detailed characteristics.

Marker Effects

It is generally accepted that recombination involves the establishment of a stretch of hybrid DNA (a heteroduplex) in either or both of two interacting chromatids of two paired homologous chromosomes. Correction, so called, of mispaired bases leads to conversion. In several cases, notably in *A. immersus* and *N. crassa,* it has been shown that the frequencies of conversion differ from one site to another, even within the same locus. Sometimes these frequencies are polarized, being higher at

279

one end of a locus and declining toward the other end. This may be expressed as a polarity in recombination, the recombinant asci resulting from preferential conversion at the site nearer to a particular end of the locus. This seems explicable on the ground that this is the end from which the heteroduplex is established; the actual initiation locus may be outside the locus displaying polarity. On the other hand differences in frequency of conversion are apparently not always in such a systematic order (62). A second difference found for different allelic mutants is the direction of gene conversion. In a cross $+ \times m$, four classes of exceptional asci are commonly observed where octads can be analyzed, namely $6+:2m$, $2+:6m$, $5+:3m$, and $3+:5m$. The preferred direction of conversion is defined by the ratio of the frequencies of 6:2 to 2:6 or of 5:3 to 3:5 segregations, the wild types being given first in each ratio. Differences of these kinds, comprising frequency of conversion or recombination, time of conversion, whether meiotic (6:2) and (2:6) or postmeiotic (5:3 and 3:5), and direction of conversion, have been ascribed to the nature of the genetic alterations at the mutant sites.

Rossignol (62) demonstrated the existence among mutants at locus 75 of *A. immersus* of a number of distinct categories in respect of the direction of conversion, which he specified as a dissymmetry coefficient (DC). In this particular case the common exceptions were 6:2 and 2:6. He found five classes: α (DC =11), β (1), γ (0.7), δ (0.04), and ϵ (2), so that the preferred directions were strongly different. The members of each DC class show a regular increase of conversion frequency from the left to the right end of the locus, the slope of increase ranging from a low value for α, through an intermediate one for γ, to a high one for β. Mutants belonging to different classes are randomly mixed. The observations would be accounted for by the formation of hybrid DNA from a fixed starting place and by different probabilities of correction according to the type of mispairing at a heterozygous site.

What molecular differences could define these different responses? Leblon (41, 42) has sought to answer this question by enquiring whether, if the repair of DNA is involved in recombination, excision is provoked by the site of the mismatch and in particular ways. He uses a "conversion spectrum" to classify mutants. This is defined by two criteria observed among exceptional asci in heterozygotes: (*a*) the time of segregation, whether meiotic (6:2 and 2:6) or postmeiotic (5:3 and 3:5) and (*b*) the direction of conversion, whether preferentially wild to mutant (2:6 and (3:5) or mutant to wild (6:2 and 5:3). If the nature of the mutation were responsible for the observed differences, mutants induced by mutagens of a very specific action, producing only one kind of mutation, would have homogeneous conversion spectra, while mutants induced by mutagens of low specificity should have a range of conversion spectra. Different spontaneous mutants, the source of most mutants in earlier work on *A. immersus,* exhibit all possible spectra.

Mutants at two loci (*b1* and *b2*) which result in white, instead of colored, ascospores were induced in stock 28 of *A. immersus* by treating mycelium of one strain of one mating type with mutagen and crossing it to the other mating type. The use of just one strain for producing the mutants is important because consistent differences between different specific mutagens would virtually exclude causation external to the locus itself. The following mutants were obtained: 25 induced by acridine (ICR170), 18 induced by nitrosoguanidine (NG), and 19 induced by ethylmethanesulfonate (EMS). All NG mutants showed frequent postmeiotic segregation and a predominance of conversion to wild type. In contrast, all ICR showed no postmeiotic segregation and a predominance of conversion to mutant. While most EMS mutants behaved like the NG ones, some showed only meiotic conversion with predominance of conversion either to wild type or else to mutant. Leblon (41) concludes that the nature of the mutational differences at a site influences the conversion spectrum.

He has attempted (42) to identify the actual genetic alterations in a sample of five mutants. All revert by back mutation and one, an NG mutant, is suppressed by either of two apparent supersuppressors at other loci. One ICR mutant ($b2$.A38) and one EMS mutant ($b2$.47E) revert by intragenic suppressors and are inferred to be frameshift mutants, the ICR mutant having an extra base or bases and the EMS mutant a deletion. Like all of the ICR mutants they show little or no postmeiotic segregation and show a preponderance of conversion either to the wild allele, in the addition mutant, or the mutant allele, in the deletion mutant. It is inferred that the single strand loops expected to be formed in a heteroduplex are recognized efficiently and excised to give meiotic correction and segregation. It is not evident yet whether a unique nucleotide substitution or a diversity is characteristic of mutants that give postmeiotic segregation.

Leblon & Rossignol (43) have studied the behavior of crosses between wild type and two double mutants at the $b2$ locus of $A.$ $immersus$ obtained by spontaneous reversions by intragenic suppression of an ICR mutant ($b2$.A38) and an EMS mutant ($b2$.A4). Either mutant alone in each double gives colorless ascospores; the double mutants are colored, light brown in one case and pink in the other, contrasting with the deep brown of the wild. In consequence a variety of aberrant types of ascus can be observed. One of the double mutant crosses showed only meiotic segregation for both sites and opposite asymmetries in direction of conversion. The other cross showed postmeiotic segregation for one site but not for the other. Their conclusions are that conversions arise by correction in a heteroduplex, that the correction is not polarized, and that it is initiated by the site, each site acting independently. The majority of the types of ascus showing conversion are explicable by one initial event involving either one chromatid or two nonsister chromatids, on either or both of which a heteroduplex is formed. There is not a polarized process of recognition of mispairing along the heteroduplex nor of polarized excision.

A different type of supposed marker effect has been reported by Gutz (31) in $S.$ $pombe.$ Whereas the $Ascobolus$ cases are strictly local in their effects acting only on the site itself and not causing dependent effects in the rest of the locus, the case in fission yeast has effects that extend to other parts of the locus. The mutant concerned is $ade6$-M26, causing a nutritional requirement for adenine. The mutant increases recombination in the $ade6$ locus by large amounts of up to twenty times. In crosses of $ade6$-M26 to the wild type, about 3–5% of asci show conversion and in these M26 is more often converted to wild type than the homologous normal site is converted to mutant, in the ratio of twelve to one In heterozygotes between M26 and other $ade6$ mutants, the neighboring sites M216, L52, and M375 show conversion, nearly

100% for M216 and about 60% for the more distant L52. Moreover, the direction of conversion at the M26 and L52 sites is commonly opposite to that at the M26 site. Gutz ascribes the effects to the M26 site itself, causing an additional point of preferential breakage within the *ade6* locus. However, there is nothing in the data to exclude a linked factor, analogous to a cog^+ gene (see below) being present outside the *ade6* locus in the M26 stocks. This would account for the observations, especially if the external locus had the effect of producing a heteroduplex in one chromatid only, that carrying M26 and the presumed control gene.

Classes of Mutants That Control Recombination

In *N. crassa,* several genetic factors are known to have quantitative effects on crossing over and conversion. They fall into three classes distinguished by their phenotypic properties: 1. genes (*nuc, uvs, mei*) of general effect yielding recessive mutants that result in the reduction or elimination of recombination as well as sensitivity to ultraviolet light and X rays, 2. genes (*rec*) of local effect, not usually linked to the target regions, in which recessive *rec* genes caused increased recombination, and 3. genes (*cog*) of local effect on the region within which the *cog* genes are located, recessive *cog* genes resulting in decreased recombination. Variant genes of the second and third classes are present in wild populations, which appear to be polymorphic with respect to these genes.

Mutants of the first class are also known in *U. maydis* (3, 32) and *Podospora anserina* (71, 72). Mutants of the second and perhaps third classes are known also in *Schizophyllum commune* (63, 64, 68–70, 83–87) and are considered separately; they probably occur also in *A. immersus* (58, 82). The nature of mutants isolated in *S. cerevisiae* (61) is unknown as yet.

There is indirect evidence in *N. crassa* of a fourth class of genes (*con*) of local action which, like those of the third class, are located within the regions where they cause an effect. It is thought that dominant variants of these genes would cause a local reduction in the frequency of recombination (18). It can be argued that loci of the third class are targets of an endonuclease determined by a gene of the first class, while genes of the fourth class are targets of the products of genes of the second class. The *rec* and *con* systems appear to correspond to the systems proposed by Simchen & Stamberg (70) to be fine controls of recombination.

There is also evidence (16) in *N. crassa* of genes that may modify the main effects due to genes having major local effects, particularly of the *rec* genes, but nothing is known of their genetics and exact function. Their effects are rather small and difficult to handle experimentally.

Recombination Genes of General Effect

Recessive mutants, of genes of general effect, that virtually eliminate recombination would result in high levels of sterility in sexual reproduction. The formation of ascospores in asci and their proper maturation is dependent upon the regularity of meiosis and so on a balanced, haploid set of chromosomes in each of its products, incorporated into ascospores. The regularity of meiosis is dependent on there being at least one chiasma in every chromosome pair, unless there is some other mechanism to keep them associated. The formation of chiasmata is dependent on the occurrence of crossing over. Diagnostic, it would seem, of genes of general effect on recombination is that their recessive mutations cause sterility and that this sterility is manifested by a breakdown at about the stage of meiosis in the relevant cells. However, a few might escape through this block.

Schroeder (66, 67) showed that mutation at two of several loci in *N. crassa* at which mutants sensitive to ultraviolet light occur appear to affect recombination. These loci are *uvs-3*, in IVL near to *cys-10*, and *uvs-5*, in IIIR very close to *vel*. When either mutant is homozygous in a cross, there is complete infertility. Asci do not develop beyond the stage of ascogenous hyphae. Possible effects on mitotic recombination have been examined in strains that are duplicated and heterozygous for mating type, generating them by means of an inversion [In(ILR)H4250] and a transposition [Tp(I→II)39311]. The duplication progeny grows very slowly as long as it is heterozygous for mating type, but release occurs when a mitotic event causes homozygosity or hemizygosity for mating type. With *uvs-3*, but not *uvs-5* or *uvs-4*, this release occurs two days earlier than in controls.

Smith (75) reports a recessive mutant, *mei-1*, which produces abundant ascospores, 90% of which are aborted. Most asci are 0 black: 8 white, with occasional 2:6 and 4:4 types; the absence of 8:0 and 6:2 types is consistent with all meiosis being defective in the first division. Many of the viable ascospores are disomic, often for several chromosomes, indicating that the inviability of ascospores is caused by nondisjunction at meiosis. The disomics rapidly become haploids, and in these there is an absence of recombination wherever a linkage group was marked at several places. Cytologically there appears to be a complete absence of chromosome pairing in crosses homozygous for *mei-1*. Why ascospores can be formed in *mei-1* crosses, but not in *uvs-3* or *uvs-5*, needs careful study.

Recombination Genes of Local Effect in *N. crassa*

Genes of local effect on recombination (10–22, 35–37, 73, 74, 76–78, 89) were first discovered in *N. crassa*. Jessop & Catcheside (35) found genes that controlled the frequency of recombination, presumably by conversion, between alleles at the *his-1* locus in linkage group V. The dominant gene at the *rec-1* locus, later shown to be in the same linkage group about 30 centimorgans distally to *his-1* (20), reduced the frequency of allelic recombination by a factor of about fifteen. No obvious effects on nonallelic recombination in the neighborhood could be detected. However, certain effects upon the neighborhood of *his-1* were evident because the distribution of flanking markers among prototrophic recombinants was strongly affected by the substitution of *rec-1*$^+$ for *rec-1*. The quantitative effects are most simply stated in terms of a standard type of cross $P\,m^1 + D \times p + m^2\,d$, m^1 and m^2 being two allelic mutant sites of difference such that m^1 is proximal, the genes P and p being proximal to the m locus while D and d are distal. Thomas & Catcheside (89) showed that besides a fifteenfold reduction in the frequency of prototrophs, the *rec-1*$^+$ gene caused a change in the ratio of *PD:pd* from 1.5:1 in *rec-1* $\times$ *rec-1* to 0.55:1 in *rec-1*$^+$ $\times$ *rec-1* and *rec-1*$^+$ $\times$ *rec-1*$^+$. The effects can be accounted for if most recombination in the *his-1* locus in the presence of *rec-1* originates distally to *his-1* and if the action of *rec-1*⁺ is to inhibit recombination originating from this position. It was presumed that *rec-1*⁺ acts as a repressor of recombination and that it produces a product that has an affinity for *his-1* or a locus in its neighborhood and so interferes with the initiation of recombination.

An alternative theory was that the effect on allelic recombination was a byproduct of the regulation of transcription. This was disproved decisively (10, 11) in the case of the control by *rec-3*⁺ vs *rec-3* of allelic recombination at the *am-1* locus (see below). The *rec-3* and *rec-3*⁺ genes have no differential effect upon the repressibility of the NADP-specific glutamate dehydrogenase specified by the *am-1*⁺ gene.

Subsequently it was shown that *rec-1*⁺ was highly specific and had no effect on allelic recombination at several other loci (16, 20). Moreover, genes at other loci were found to be effective in controlling recombination at some of these loci, particularly *his-2*, *his-3*, and *am-1*. The first of these to be found was *rec-3*⁺ which has the effect of reducing recombination at the *am-1* locus by factors of 10 to 25 (14). It was observed that *rec-1*⁺ was specific to *his-1*, while *rec-3*⁺ was specific to *am-1*. This implied that there were recognition sites in or adjacent to the target loci and that these were quite distinct in character (14). There would be one type specific for the product of *rec-1*⁺ and a different type specific for the product of *rec-3*⁺.

Smith (73) found a third locus, *rec-2*, genes at which control nonallelic recombination in linkage group IV. It was found through its action in reducing recombination between *pyr-3* and *leu-2*, used as flankers in the study of the fine structure of *his-5*, from 23 to 10%. Subsequently (74) it was found that the whole effect was concentrated in the *pyr-3 his-5* segment, and no other target regions were found in a fairly extensive survey. The loci *rec-1*, *rec-2*, and *rec-3* are all distinct from one another. As mentioned above, some other loci were found to be the targets of *rec* genes. The latter were given letter designations until their relationship with the three loci already known was settled. They were *rec-x* acting on *his-2* (20), *rec-w* (also called *rec-5*) acting on *his-3* (20), and *rec-z* acting on *nit-2* (12). There were indications from similarity of location in the genetic map that these three might respectively be the same as *rec-3*, *rec-2*, and *rec-1*. This has proved to be the case within small limits of experimental probability.

The testing of these indications was particularly important because, to this stage, it appeared that allelic and nonallelic recombination might be controlled by different *rec* genes. The problem lay in showing that two different phenotypes are due to the action of the same gene. The method employed can be illustrated by reference to *rec-3* and *rec-x* both of which were known to be in the left arm of linkage group I between the mating type and *arg-3* loci. Fortunately it turned out that they were between *acr-3* and *arg-3*, which were tightly linked with about 2.5% recombination in the stocks used. Heterozygotes of the constitution A *acr-3*ˢ *rec-x* + *his-2*; + / a *acr-3*ʳ *rec-x*⁺ *arg-3* +; *am-1*² were constructed and selected recombinants a *acr-3*ʳ *arg-3*⁺ *his-2 am-1*² were isolated. Selection of recombinants between the *acr-3* and *arg-3* locus effectively magnifies by about 40 times the chance of separating the *rec-3* and *rec-x* loci, if they are distinct and separable. Each of these progeny was assayed for the level of allelic recombination at the *his-2* and *am-1* loci. It was found

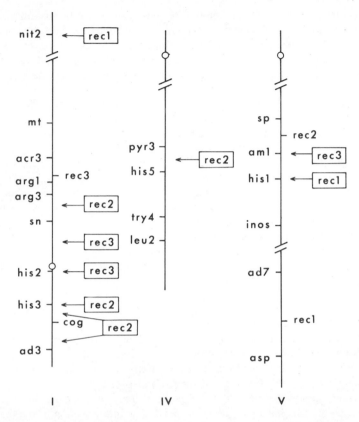

Figure 1   Maps of linkage groups I, IV, and V of *N. crassa,* showing positions of *rec* and *cog* loci and of their sites of action. The symbols on the left of each linkage group are of loci in them; the symbols on the right of linkage groups I and V not in boxes are of *rec* and *cog* loci. The symbols in boxes to which arrows are attached indicate the target regions for the genes at the three *rec* loci. Those pointing at a named locus in a linkage group control allelic recombination at that locus. Those pointing to a segment between two named loci control nonallelic recombination in that segment.

that each individual was either high at both loci or low at both loci. The number of progeny examined in this way (21, 22) is such that with very high probability *rec-3* = *rec-x* (and *rec-3*⁺ = *rec-x*⁺). Similarly *rec-2* = *rec-w* (22) and = *rec-5* and = *rec-4* (36), while *rec-1* = *rec-z* (13). Figure 1 shows the linkage relations of these genes and their targets as known at present.

It is evident that two of the *rec* loci simultaneously affect allelic and nonallelic recombination commonly as parts of the same target area. Thus *rec-3* acts on *his-2* and the adjacent *sn his-2* segment, while *rec-2* acts on *his-3* and the adjacent *his-3 ad-3* segment. There is no detectable effect of *rec-1* on nonallelic recombination near to *his-1* (35), but it appears to have an effect near to *nit-2* (13). The small effects of *rec-3* on nonallelic recombination near to *am-1* can be ascribed to conversion in the *am-1* locus (77). In the case *rec-2*, there is no effect on the frequency of allelic recombination at the *his-5* locus (73) and the *pyr-3* locus cannot be tested due to infertility. The effects of *rec-2* on the distribution of flanking markers among *his-5*⁺ prototrophs can be accounted for by the increase in crossing over in the *pyr-3 his-5* segment (22 *contra* 73).

The *rec-2/his-3* system has disclosed a second class of gene of local effect, namely *cog*. The only locus of this class known definitely is situated between *his-3* and *ad-3* (1). Two alleles occur, their difference being expressed in the absence of *rec-2*⁺. In the presence of *rec-2*⁺, allelic recombination in the *his-3* locus and crossing over between *his-3* and *ad-3* is the same in *cog*⁺ stocks as in *cog* stocks. In *rec-2* X *rec-2* crosses the frequency of allelic recombination in *his-3* is about six times as great in *cog*⁺ crosses as in *cog* X *cog*. Crossing over between *his-3* and *ad-3* is about four times as great in *cog*⁺ as in *cog* X *cog* crosses. Apparently *cog*⁺ is completely dominant, for no difference in frequency is seen between *cog*⁺ X *cog* and *cog*⁺ X *cog*⁺. However, there are other differences. In the *cog*⁺ X *cog* heterozygote the *his-3* site in the *cog*⁺ chromosome undergoes preferential conversion irrespective of whether it is proximal or distal to the *his-3* site in the homologous chromosome. In *cog*⁺ X *cog*⁺ and *cog* X *cog* crosses the distal site is preferentially converted. The behavior suggests that *cog* is the locus at which, after synapsis, the formation of a heteroduplex DNA is initiated perhaps by an endonuclease. There is a strong preference for *cog*⁺ over *cog* in the heterozygote, and perhaps a heteroduplex is formed only on one chromatid of the tetrad. The *cog* locus appears not to be the target of the product of *rec-2*⁺ (or of its allele *rec-2*). The target of the *rec-2*⁺ product must be yet another locus in the *his-3 ad-3* region. This is the *con* locus suggested in Figure 3*c* of Angel, Austin & Catcheside (1).

The evidence for this third class of genes of local effect on recombination is indirect, no variants having as yet been found. The interpretation of the function of *cog*⁺ is reinforced by the behavior of heterozygotes in which one *his-3* mutant, TM429, is due to an interchange with one break in the *his-3* locus itself (1, 19). This mutant has *cog*⁺ in the distal segment of linkage group I which is translocated to linkage group VII (*fide* Dr. D. D. Perkins). In heterozygotes in which *cog*⁺ is associated only with TM429 no differential effect of *rec-2* and *rec-2*⁺ is observed for recombination proximal to the site of TM429. If *cog*⁺ is associated both with TM429 and its structurally normal homologue, the usual large differential effects of *rec-2* vs *rec-2*⁺ are observed. The action leading to recombination is confined to the chromosome, or chomosomes, that carry *cog*⁺.

The action of the products of *rec*⁺ genes at several different targets implies the existence of a class of *recognition* or *control* genes (14). There should be a species of *control* gene (*con*) corresponding to each of *rec-1, rec-2,* and *rec-3*. For example, there should be a *con* locus for *rec-3* products near to *am-1* and also near to *his-2*. There is now evidence that the *con* genes in these positions are not exactly alike (18). It has been found that there are three alleles (*rec-3, rec-3*⁺, and *rec-3*ᴸ) at the *rec-3* locus that differ in their action on allelic recombination at the *am-1* and *his-2* loci. The *rec-3*ᴸ gene is in the Lindegren A wild strain and its laboratory derivatives 1A and 25a. The relative effects of *rec-3, rec-3*^{L} , and *rec-3* on allelic recombination are 1:8:25 at the *am-1* locus and 8:8:48 at the *his-2* locus. The *con* gene near to *his-2* evidently does not differentiate between the products of *rec-3*$^+$ and *rec-3*^{L} respectively and must therefore be different in some minor respect from the *con* gene near to *am-1*.

The failure of the *con* gene near to *his-2* to differentiate between *rec-3*$^+$ and *rec-3*^{L} raises some doubts about the interpretation of negative tests for the possible action of *rec-1, rec-2,* and *rec-3* on a number of loci. Some 15 loci have been tested for the action of these *rec* genes on allelic recombination; 10 appear insensitive. One could argue that if recombination everywhere is controlled then there must be further, as yet undiscovered, *rec* loci, but the data are too limited to make any sensible estimate of how many.

Two other points deserve comment. In *rec-2* X *rec-2* crosses, the crossing over between *pyr-3* and *his-5* is about 23.3%, compared with 1.84% in *rec-2*$^+$ X *rec-2* crosses. The absence of *rec-2*$^+$ from a cross adds some 21.5 centimorgans to the "length" of the *pyr-3 his-5* segment, with the basic value of 1.8. This is a large effect on a "small" region. Each meiotic cell in which a crossing over, presumably reciprocal, occurs between *pyr-3* and *his-5* would produce an equal number of recombinant and nonrecombinant spores. Therefore the observed increase means the occurrence of an event leading to crossing over between *pyr-3* and *his-5* in 47% of meiotic cells in *rec-2* X *rec-2,* compared with 3.7% in *rec-2*$^+$ X *rec-2.* Work in recent years has tended to show that allelic (conversion) and nonallelic (crossing over) recombinations are different manifestations of one common mechanism. Moreover, half of the tetrads in which conversion has occurred show correlated crossing over between flanking markers, while the other half do not (28). We may infer that a conversion is initiated or occurs between *pyr-3* and *his-5* in 94% of meiotic cells in *rec-2* X *rec-2.* This is very high compared with 26% in the *his-3 ad-3* and 12% in the *arg sn* segments. Collectively *rec-2*$^+$ removes an average of about half a chiasma in respect of these three segments.

The other matter is that crossing over in *N. crassa* appears to occur in a succession of segments. Thus in the region from *arg-3* to *ad-3* in the linkage group I, spanning the centromere, the segments appear to be as follows:

$$\text{arg-3} \xrightarrow{rec-2} \text{sn} \xrightarrow{rec-3} \text{his-2} \xrightarrow{rec-u} \text{his-3} \xrightarrow{rec-2} \text{ad-3}$$

The genetic length of this segment is 8.1 centimorgans if all dominant *rec*$^+$ genes are present and 39.4 centimorgans if all are absent.

Many years ago Fincham (27) demonstrated a significant difference between the linkage maps of the mating type chromosomes (linkage group I) in *N. crassa* and *N. sitophila.* The latter showed much higher frequencies of crossing over near the centromere than did *N. crassa,* the differences apparently not extending to more distal parts of this linkage group. Newcombe & Threlkeld (49) have studied this difference further and have shown that a genetic factor (or factors) in the neighborhood of the centromere of *N. sitophila* acts as a dominant enhancer of crossing over frequencies on both sides of the centromere. There is some analogy evident with *cog* genes in that greater recombination is dominant to lesser. This will be an interesting system to analyze fully.

It is possible that recombination in eukaryotes is not initiated by the action of endonuclease at defined loci, but by a failure to complete the previous chromosome replication, so leaving short gaps at specified places. Hotta & Stern (33) have demonstrated in *Lillium longiflorum* that some synthesis of DNA, amounting to about 0.3% of the total DNA, occurs at zygotene of meiosis and represents delayed replication. There is evidence that the delayed synthesis occurs in regions of special character. If these gaps are the places from which recombination is initiated, the action of the products of the dominant repressors (*rec*$^+$) of recombination in local regions might be to seal the gaps or otherwise to inhibit the formation of hybrid DNA. To do so specifically there would need to be specific recognition sites adjacent to the gaps. After the conditions necessary for the establishment of hybrid DNA had passed, the repressor would presumably detach and allow a DNA polymerase and ligase to fill in the gap. The *cog* and *cog*$^+$ genes might be the recognition sites for the products of *rec* and *rec*$^+$, with *cog*$^+$ being less receptive to the *rec* product than is *cog*. In a *cog*$^+$ *cog* heterozygote which is also *rec rec* an opportunity for forming a heteroduplex might be afforded to the *cog*$^+$ strand more frequently than to the *cog* strand.

Recombination Control in Schizophyllum commune

Simchen & Stamberg (70) have proposed a distinction between fine and coarse controls of recombination. Extensive studies (39, 40, 63, 64, 68–70, 83–87) of the systems of genes controlling mating type in *S. commune* have been adduced in support. Close analogies to the system described in *N. crassa* are evident. Genes of coarse control (equivalent to genes of general effect) are those that have extreme effects on recombination ("all or none") throughout the entire genome, with only rare variants in nature. they occur in prokaryotes and eukaryotes and presumably control steps (synapsis, breakage, repair, and separation) which must operate sequentially. Genes of fine control (equivalent to genes of local effect) are those of small effect on small, highly specific regions of the genome, with variants occurring commonly in natural populations, but only of eukaryotes, the regional specificity being imposed on at least one step of the coarse control. The implication is that there are two kinds of genes of fine control, one producing the controlling materials and the other constituting the recognition sites situated in the regions in which control is exercised. The prediction is that these occur and are effective in virtually all eukaryotes.

In *S. commune* two unlinked systems *A* and *B* (often referred to as two mating type loci) control different parts of the process of dikaryon formation. In fact each of *A* and *B* consists of linked genetic loci, of different function, designated α and β, and yet other genes can occur between the α and β factors. Complete success in mating between two strains depends upon their being different at *A*α, *A*β, *B*α, and *B*β loci. Identity at some loci permits the formation of a dikaryon to proceed in part. Studies of control of recombination have involved the of α and β factors in both the A and B systems.

Simchen (68) selected for high (14%) and low (4%) recombination between $A\alpha$ and $A\beta$, the difference being due to genes of major effect occurring at a postulated *rec* locus linked to the A system and to genes of minor effect at other loci. Low frequency of recombination was dominant to high, a property that could be related to the breeding behavior. Namely, close linkage between α and β allows a high number of specific A factors to be maintained in a population as well as a high potential for outbreeding, while the potential for inbreeding is kept low and near its minimum.

Stamberg (83) showed that recombination in the A and B systems was under separate control. The methods of study can be illustrated from this particular paper. Strains 14 (mating type A4 B4) and 699 (mating type A41 B41) were crossed and progeny selected with the mating types of the two parents, eight of each. Each of the progeny was then backcrossed with a compatible parent. The recombination between the α and β loci in each of the A and B systems was then measured in a sample grown from random basidiospores, the sample size ranging from 84 to 196. The data were analyzed by χ^2 tests. Among the progeny after meiosis at 32°C there was considerable heterogeneity in A system recombination, but not in the B system. After meiosis at 23°C there was heterogeneity in both of the systems in respect of the same progeny as were tested at 32°C, but the variations in recombination frequency for each system were uncorrelated. Variations due to temperature were also uncorrelated. A deficiency of this study and of others by Simchen, Stamberg and their colleagues is a lack of replication of the tests on progeny of crosses. Undoubtedly the experiments are tedious and time consuming, but too complex a system seems to be founded on rather limited evidence.

For instance, Stamberg & Koltin (86) present acceptable evidence for a genetic difference between strains due to a factor located within the recombining region itself. They argue, on the basis of the analysis of modest samples of a limited number of progeny, that the segregation among the progeny indicates that the factor consists of a number of sites, with additive effects some positive, others negative. This together with the dominance relationships (high frequencies dominant to low frequencies), suggested to them that these sites may be recognition sites that comprise part of the fine control of recombination. The evidence that the progeny can really be divided into more than two classes is not convincing. It is clear that there are genetic factors of the *rec* kind in *Schizophyllum*. Reduction of recombination in unlinked or loosely linked target segments is due to the action of dominant variants. There are also recognition sites within the target segments, but whether they are of the *cog* or *con* type or both is uncertain.

GENETICS OF SENSITIVITY TO RADIATIONS

The agents that may kill or alter a cell through its genetic apparatus include ionizing radiations (X rays, γ rays, and radioactive decay), ultraviolet light, heat, and chemicals that can interact with DNA. They do so by inducing damage, most of which is normally repaired or modified in the cell in significant ways. The metabolic response of an organism to altered DNA is highly significant, and the responses are dependent on genes. A single mutational step in one of these genes can result in any one or more of the following changes: increased sensitivity to one or more types of radiation, defectiveness in the removal of damage to DNA, inability to participate in genetic recombination, alteration in the probability of spontaneous or induced mutation. Evidently, there are, in normal organisms, repair mechanisms that restore damaged DNA to its original state and so permit the organism to recover from the effect of irradiation or other treatment. As with other biological systems, a comparative study of mutations that alter the capacity to repair DNA or to carry out mutation allows analysis of the system itself. The genetic apparatus and its function is under its own control. Much of the understanding of the mechanisms is based on studies of bacteria, but fungi and other organisms have made their contributions.

The alterations to DNA include breaks in the phosphodiester chains, coupling of pyrimidines to form dimers, gain or loss of nucleotides in chains, and change in the chemical structure of bases. Breaks in single strands may be caused by decay of radioactive elements, such as ^{32}P incorporated into DNA, and by ionizing radiations. In the case of ^{32}P, one lethal event occurs for every radioactive decay event in the single-stranded DNA of the phage ϕX174; there is evidently no significant way of rejoining the separated strands. In double-stranded DNA the efficiency is much smaller, more than 90% of the breaks being rejoined. Damage due to X rays and other ionizing radiations is of three kinds in normal DNA (double-stranded), namely single strand breaks which are reparable, double strand breaks which are probably always lethal, and damage to bases. The breakages are caused either directly by ionization in the target molecule or indirectly by a free radical produced from water and capable of being quenched by sulfhydryl compounds or by histidine. Damage to bases appears to result from a second class of free radical, requiring oxygen for expression and causing a chemical change in pyrimidine bases.

Ultraviolet light acts chiefly by causing the formation of covalently linked dimers between neighboring pyrimidines in DNA. At least 50% of such dimers induced in *Escherichia coli* are thymine-thymine pairs, perhaps 40% occur between thymine and cytosine, and fewer than 10% are cytosine-cytosine pairs. At biologically significant doses of ultraviolet light, the dimers are exclusively between pyrimidines in the same strand. Dimers may be formed between pyrimidines in sister strands, but only at very high doses, well beyond that necessary for quite high degrees of lethality.

Chemicals that have effects, causing lethality or other genetic changes at a higher rate than occurs spontaneously, may act by alkylating bases, substituting for bases, deaminating bases, or intercalating in DNA. Aspects of these events are considered only in so far as they appear to relate to the main topic.

Commonly, organisms with an active recovery process from damage caused by ultraviolet light display curves, in plots of the logarithm of the proportion surviving against dosage, with a shoulder at relatively low doses before extensive inactivation occurs. After the shoulder, the survival declines exponentially in proportion to increasing dosage. If irradiated organisms are exposed to strong light for short times, the survival rate is considerably improved compared to those kept under similar physiological conditions in the dark for the same times, before placing them in conditions for growth. This photoreactivation is due to an enzyme that combines with DNA having pyrimidine dimers and that splits these dimers using visible light as its source of energy. The photoreactivation enzyme has no affinity for DNA after the dimers have been split. It splits the bonds holding the pyrimidine rings together, but has no effect upon the phosphodiester backbone of the DNA molecule.

Other recovery will occur in the dark and is dependent upon the action of a series of enzymes. It begins with an endonuclease, sensitive to pyrimidine dimers, that breaks a phosphodiester bond in the affected DNA chain at or near a dimer. The damaged bases together with a contiguous portion of the same strand of DNA are removed and a polymerase, identical with or similar to the Kornberg DNA polymerase, restores the lost portion by synthesis using the undamaged complementary strand as a template. A final step in the patching process would require the action of a ligase to join the free ends of the DNA chains by a phosphodiester bond.

The replication of DNA is impeded by the presence of pyrimidine dimers. A dimer distorts the phosphodiester backbone locally and stops or slows the DNA polymerase in its movement along the template that it is replicating. Sometimes, it appears, the polymerase can skip over the dimer to give a new chromosome with a gap in one chain opposite to the dimer in the parent strand, the other new chromosome being complete and without any gap. Of course, this description of completeness versus a gap refers only to the local situation. Any parental chromosome that had several dimers would give daughters with gaps at different complementary places. The surviving dimers may be eliminated later. This post replication repair may involve exchanges with a sister chromosome (or chromatid) having gaps at different places so that between them they generate a complete chromosome. The gap opposite to a dimer may be filled by a patching process which, because it lacks the correct guidance of the sister strand, is prone to error compared with what should be in this part of a chromosome.

Thus there are two general kinds of dark repair, one depending on excision of the lesion, the other on post replication repair by patching or by recombination. The former is known only for lesions due to ultraviolet light or to agents having similar effects. Several chemicals (acriflavine, caffeine, and actinomycin for example) inhibit repair. Repair is often enhanced by incubating the irradiated cells in buffer in the dark for fairly long periods. The full amount of possible recovery in yeast requires several days. This liquid-holding recovery is inhibited by 2, 4-dinitrophenol, cyanide, or azide indicating that metabolic processes are involved. Caffeine and acriflavin also inhibit this recovery when present in the buffer.

Neurospora crassa

The following loci, mutation at which leads to greater sensitivity to ultraviolet light, have been found: *uvs-1* (23), *uvs-2* on IVR some 5 map units distal to *cys-4* (81), *uvs-3* on IVL near to *cys-10* (66), *uvs-5* also on IIIR less than 1 map unit from *vel* (66), *uvs-6* (reported by A. L. Schroeder, F. J. de Serres, and M. Schupbach), and *upr-1* on IL about 1.6 map units proximal to the mating type locus (90–92). A summary of the main properties of mutants at these loci is given in Table 1. All appear to be recessive in that the enhanced sensitivity is not shown in heterokaryons with wild type. Although *upr-1* shows diminished photoreactivation, it has a func-

Table 1   Differences from wild type in responses of mutants of *N. crassa* selected for sensitivity to ultraviolet light[a]

	US	UE	XR	HN	NM	PR	DR	F	REC
uvs-1	0	+		+		N	0	-	N
uvs-2	0	+				N	0	N	N
uvs-3	0	+	+	+	+	-	0	0	0
uvs-4	-	+		-	N	N	N	N	N
uvs-5		+	+?				0	0	0?
uvs-6	0	+	+						
upr-1		+				0/-		N	N

[a]The column headings are: US, presence of shoulder with low doses of UV light; UE, the exponential rate found usually with higher doses of UV light; XR, sensitivity to X rays; HN, sensitivity to nitrite or to nitrosoguanidine; NM, sensitivity to nitrogen mustard; PR, photoreactivation; DR, dark repair as judged by action of acriflavin; F, fertility in homozygote; REC, recombination (allelic and/or nonallelic) at meiosis in homozygote. The symbols signify: N, same reaction as wild type; -, less than wild type; +, more than wild type; 0, totally lost.

tional PR enzyme (90), and it is suggested that *upr-1* may control a nuclease (or nucleases) that competes with the PR enzyme for substrate.

Clearly there are several classes of mutant. Three are sensitive also to X rays, and two of these are known to affect meiosis adversely to the degree that it cannot be determined whether any allelic or nonallelic recombination occurs. There is a causal relation between many defects leading to X-ray sensitivity and to meiotic breakdown. As regards spontaneous mutability, *uvs-2, uvs-5,* and *upr-1* are similar to wild type, *uvs-1* and *uvs-4* are less mutable, while *uvs-3* is about ten times more mutable (25). As regards mutation induced by ultraviolet light it is necessary to use a wide range of UV doses, due to differences in survival curves and also to differences in dose-effect curves for the induction of mutation. However, it appears that *uvs-3* and *uvs-4* have markedly reduced induced mutation rates, *uvs-2* has a markedly increased rate, while *uvs-1, uvs-5,* and *upr-1* are similar to or slightly less mutable than is the wild type. The differences from wild type in respect to spontaneous and induced mutation are not correlated in a way to suggest that identical mechanisms are involved.

Saccharomyces cerevisiae

A very large number of studies have been made of radiation-sensitive mutants in yeast (2, 4–9, 24, 26, 29, 30, 34, 38, 45, 47, 50–57, 79, 80, 88, 96, 97) and related studies have been concerned with the genetics of induced mutation (44–46) and induced allelic recombination (conversion) at mitosis (59, 60). The latter classes of mutant have been sought on the assumption that not all mutants that affect these processes need affect sensitivity to ultraviolet light.

The number of different loci so far identified is large. Cox & Parry (24), in an effort to exhaust the different loci, found 22 and estimated that there were probably 8 to

15 further loci for which they had not recovered mutants. A more recent paper (7) records that there are mutations at 23 loci giving sensitivity to ultraviolet light and at 7 other loci giving sensitivity to X rays. Because a number of laboratories collected mutants independently, separate nomenclatures developed. These were collated at Chalk River in September 1970 during the International Yeast Genetics Conference. There it was agreed to use the locus symbols *rad-1* to *rad-49* for those at which mutations primarily affected sensitivity to ultraviolet light and the symbols (*rad-50* onwards for loci mutations at which primarily affect sensitivity to ionizing radiations. A table of synonyms for *rad-1* to *rad-22* was published by Game & Cox (29). Table 2 attempts to summarize the loci known and their principal properties, as may be culled from the literature. However, these summary statements must be treated with caution, because alleles show significant differences.

Table 2 Properties of yeast mutants that are sensitive to radiations[1]

Locus (Interlab)	Original Locus name	Ref. No.	UV	XR	H	MMS	HN	NM	PUVT	SPOR	MUT	REC
rad-1	*uvs1*	48	S	N	N	N	N	N	4	N	N	N
rad-2	*uvs2*	48	VS	N		S-	S	N	4	N	N/-	N
rad-3	*uvs4*	79	VS	N		S+	VS		4	N	N	N
rad-4	*uvs1*	79	S	N		S-	S		2	N	N	N
rad-5	*uvs10*	79	S	N	N	S	S-		1	N	-	N
rad-6	*uvs6*	24	S+	S		VS	VS		1	0		
rad-7	*uvs7*	24	S	N		N	S-		1	N		
rad-8	*uvs8*	24	S	S-		S-	S		1	N		
rad-9	*uvs9*	24	S	S		S+	S+		1	N		
rad-10	*uvs14*	79	S	N		S	S-		1	N		
rad-11	*uvs11*	24	S-	N		S	N		1	N		
rad-12	*uvs18*	24	S	N		S	S+		1	N		
rad-13	*uvs3*	24	S	N		S	S+			N		
rad-14	*uvs11*	79	S	N		S	N		1	N		N
rad-15	*uvs15*	24	S	S-		N	S-		3	N		
rad-16	*uvs16*	24	S	N		S-	N		2	N		
rad-17	*uvs17*	24	S	N		S	S+		1	N		
rad-18	{*uxs1* / *uvs18*}	56 / 79	S	S	N?	S		S	1		+	
rad-19	*uvs19*	24	S-	N		N	S-		1		-	
rad-20	*uvs20*	24	S	N		S+	VS		1		-	
rad-21	*uvs21*	24	S-	N		VS	S+		1		-	
rad-22	*uvs22*	24	S	N		S+	S+		1	N	-	
	uvs5	24	S-	S+		S+	S		3		-	
rad-50	*xs1*	48	S	S	S	S		S				
rad-51	*xs1*	56	N	S	S	S				0	+	
rad-52	*xs2*	56	S-	S	S	S					N	
rad-53	*xs3*	56	N	S-	S	S				0	+	

[1]The Interlab locus symbols are given in column 1, followed by the locus symbol assigned first by the reference in column 3; columns 4 to 9 give relative sensitivities to ultraviolet light (UV), X- or γ rays (XR), heat (H), methylmethanesulfonate (MMS), nitrous acid (HN), and nitrogen mustard (NM). PUVT, response to post UV irradiation treatment of holding in the dark in buffer solution and then irradiating with visible light, the figures 1 to 4 being the groups described in the text; SPOR, sporulation of homozygotes; MUT, spontaneous mutability; REC, spontaneous allelic recombination at mitosis. The symbols mean: N, same reaction as wild type; S-, S, S+, VS, increasing degrees of sensitivity; 0, none; -, reduced; +, increased. Gaps in the columns indicate an absence of information.

The mutants sensitive to ultraviolet light in general lack a shoulder to their survival curves so that at a given relatively low dosage there is less survival than is shown by the wild type treated with the same dosage. Most of the sensitive mutants show an exponential decline in survival with increasing dosage, the slope being greater (e.g. *rad-2, rad-3*), about equal to (e.g. *rad-5, rad-8, rad-13*) or even less (e.g. *rad-19*) than the slope finally shown by the wild strain. In several cases the shape of the survival curve is complex. As examples, *rad-6* shows a relatively resistant tail, *rad-20* a slope increasing with dosage, and *rad-21* a plateau at 20% survival persisting over a dose range of about 600 ergs. These and other variants cannot be explained readily.

The dose-modifying factor (DMF) is frequently used to compare reactions; it is the factor of increase of dosage required to obtain equal survival or equal response between two strains or two conditions of irradiation. It is of course troublesome to use if two survival curves to be compared are markedly different in shape. Moreover, it appears that different mutant genes at the same locus differ in their responses to exposure to ultraviolet light, to other treatments, and to treatments used after irradiation. The extent to which this is due to the alleles themselves rather than to other genetic differences between the strains is unknown. There is a strong tendency to ascribe all differences in properties to the gene mutation known rather than to what may be unknown. Hence the precise differences recorded as the result of mutation at each locus must be viewed with caution.

Nevertheless, different alleles show correlations in their properties. Thus, eleven different alleles at the *rad-3* locus compared (94) in haploids and diploids show a range of sensitivity to ultraviolet light that is almost exactly paralleled by their sensitivity to nitrous acid. They also show variation in their response to post irradiation treatments (95). Positive liquid holding recovery is shown by those diploids with alleles conferring the highest levels of UV resistance, indicating that liquid holding recovery requires the activity of the excision repair pathway for expression. The different *rad-3* alleles show a range in the degree of deficiency of the excision repair, so that there is effectively a negative correlation between the degree of deficiency and capacity to recover in the dark when incubated in buffer.

As with other fungi and bacteria, some of the mutants sensitive to ultraviolet light are also more sensitive than the wild strains to ionizing radiations, and many of them are also more sensitive to inactivation by various chemicals. A few are insensitive to methylmethanesulfonate (MMS) (*rad-1*, -7, -15, -19) and others no more sensitive to nitrous acid (*rad-14*, -16), while *rad-1* and *rad-11* are less sensitive to nitrous acid than is the wild type (97). There are marked correlations between sensitivity to UV and/or ionizing radiations (XR) and sensitivity to various chemicals, especially mono- and bifunctional alkylating agents such as MMS and nitrogen mustard (NM) respectively. Whereas the wild type is resistant to all four, mutants in yeast fall into three classes: I, sensitive to UV and NM (e.g. *rad-1*, -2); II, sensitive to XR and MMS (e.g. *rad-50*); III sensitive to all four (e.g. *rad-18*). Some double mutants are much more sensitive to irradiation than are the single mutants (7). Combinations that do not increase sensitivity over that of the more sensitive parent can be regarded as conferring sensitivity by defects in different parts of the same mechanisms. Thus

Game & Cox (30) showed that double combinations of rad-1, rad-2, rad-3, and rad-4, which belong to class I, have this behavior. Their wild alleles mediate excision repair, and it is presumed that the genes are concerned with consecutive steps in a linear path. Double mutants of classes I + I and I + II are usually no more sensitive to any of the four agents than is the more sensitive singly mutant parent. However, the sensitivity to UV and to NM may be slightly increased ("modified") as in the combinations rad-1 rad-50 and rad-2 rad-50 (7). The double I + II is supersensitive to UV and to NM, whereas the double II + III is supersensitive to XR and to MMS. Modifications of response are also shown in that the I + III doubles rad-1 rad-18 and rad-2 rad-18 are relatively more sensitive to MMS than is expected on simple additivity. The II + III double rad-18 rad-50 is also more sensitive to UV and to NM, but not supersensitive. Mutants in class I are defective in excision repair, and the large number of them indicates several steps in this function or that several enzymes are composed of more than one unit. The mutants in the other classes are defective in either of two other pathways of dark repair. One, that in II, is capable of repairing X-ray and MMS damage, while that in III competes with the other two for lesions in DNA. The very large number of genes in yeast concerned with sensitivity to radiations can be arranged into these three classes, with some 17 loci in group I, at least 4 loci in group II, and 4 in group III. Among the members of each group there are certainly phenotypic differences, but their meanings are obscure in relation to attachment to finer functional differences.

Subjection to the post irradiation treatments of holding the irradiated cells in buffer solution in the dark (dark holding) for several days (rather than the few hours that suffices with bacteria) and then exposing the cells to strong visible light has marked differential effects (53). The rad-1 to rad-22 mutants, all of which are photoreactivable, are divisible into four classes (Table 2) by the following responses: 1. survival increases during dark holding, but exposure to light afterwards has no further effect, this being the same response as the wild type; 2. survival increases during dark holding and exposure to light causes a further increase in survival; 3. survival decreases during dark holding, exposure to light having no further effect; 4. survival decreases during dark holding, after which exposure to light increases survival. These observations and others show that more than one kind of lesion could be caused by irradiation with ultraviolet light and that there is more than one pathway of repair even in the one starting with excision.

Mutants that are sensitive to ultraviolet light may show apparent changes in properties of induced mutation and of induced allelic recombination. However, exact studies have been made of only a few mutants. Precise comparisons are required of the survival curves and of the dose-action curves; inferences cannot be made reliably from one or a few observations. Induced mutations are thought to arise as a result of enzymic processes acting on damaged DNA as a substrate. While the molecular mechanisms are still obscure, the evidence from bacteria suggests that mutations are produced during repair, after replication, of the gaps left through replication leaping lesions (pyrimidine dimers) induced by ultraviolet light. Gaps caused by X rays could also be fed into such a system. The gaps are filled somewhat randomly. The post replication repair enzymes in bacteria are determined by rec^+

and *exr*⁺ (or *lex*⁺) genes. Mutation of these gives strains which are sensitive to UV and to X rays, are defective in recombination to varying degrees, and show reduced or no mutability induced by UV; they still possess excision repair.

The study in fungi of mutation induced by UV has assumed that it is related to the dark repair of damage to DNA and that mutants defective in such repair will be defective in mutability. Reports of decreases and of increases in UV-sensitive strains of the frequency of UV-induced mutation are abundant. However, Lemontt (44) has tested the correlation by selecting strains of yeast that are defective in induced mutation, so avoiding the prior condition that all such mutants be sensitive to UV. Although the selection was based on one gene, the *arg4–17* allele, the selected mutants act on other genes, e.g. *lys1-1* and *arg4–6*. Three loci were found, *rev1*, *rev2*, and *rev3*. All confer moderate sensitivity to UV and slight sensitivity to X rays; *rev2* is allelic to *rad-5* mutants. The general relation to UV sensitivity seems vindicated. Lemontt suggests that *rev1–1* and *rev3–1* may have a general nonspecific action in reducing induced reversion, while *rev2–1* might specifically block reversion only of ochre-suppressible alleles like *arg4–17* for which it was selected. Double mutants with *rad-9* (= *uvs9*) are considerably more sensitive to UV than is *rad-9* alone (45). Doubly and trebly mutant *rev* strains are severely reduced in UV-induced mutability but are no more sensitive to killing than is *rev3*, which is twice as sensitive as *rev1* or *rev2*. Lemontt suggests that *rad9*⁺ repairs primary UV damage through the excision pathway, *rev*⁺ genes acting on damage not repaired in this way. It is suggested that *rev1*⁺ and *rev2*⁺ act through different parallel pathways of repair that generate a common intermediate acted upon by *rev3*⁺ to produce repaired DNA containing mutations induced by ultraviolet light. Recombination at meiosis in *rev* homozygotes occurs (46) at control frequencies between *leu1* and *trp5* and between *arg4–6* and *arg4–17*. The frequency of mitotic recombination induced by UV or by X rays, measured for the centromere to *ade2* segment and for *arg4–6/arg4–17* increased more sharply with radiation dose in *rev/rev* diploids than in +/+ ones. Mitotic recombination, although induced by UV damage, is not correlated with UV mutagenesis.

The possible relation of radiation sensitivity to spontaneous mutability has been examined most exactly by von Borstel, Cain & Steinberg (6) using a thousand compartment fluctuation test. The method used a medium limited for lysine and tester stocks carrying the nonsense mutant *lys1–1*, or a medium limited for uracil and tester stocks carrying the frameshift mutant *ura4–11*. Markers at several other loci were also usually present. Mutation rates for lysine independence were measured and the proportion determined of mutants at the *lys1* locus itself as compared with mutants to supersuppressors of class I type, active also on nonsense auxotrophic mutants at the *ade2*, *arg4*, *his5*, and *trp5* loci. Seven genes, at five loci, conferring radiation sensitivity were compared with the wild type. Mutation of *lys1–1* itself to prototrophy was increased by *rad-51*, while mutation rates to class I supersuppressors were increased by *rad18*, *rad51*, and *rad53*. Mutation of the frameshift mutant *ura4–14* is also raised by *rad-18* and *rad-51*, the only ones tested. One allele at the *rad-2* locus appears to reduce the mutation rate for the suppressors, but neither *rad-52* allele affects mutation rate. There is a clear indication of a connection between some of the functions in the repair of damage due to ionizing radiations and the minimizing of spontaneous mutation.

Rodarte-Ramón & Mortimer (59, 60) have examined the genetic control of recombination induced by radiation. Mutants were isolated from cells disomic for chromosome VIII and marked at the *arg4* locus, namely *arg4-2* +/+ *arg4-17*. Colonies were grown up on solid complete medium from cells in which mutations had been induced. They were then replicated to solid synthetic medium lacking arginine and irradiated with X rays (2.5 krad) to induce reversions, about 20 per plate. Colonies without induced reversions were saved as potentially deficient in recombination and tested further. Ten mutants were found and seven of these were studied in detail. Complementation and other allelism tests show a minimum of four loci: *rec1, rec2, rec3, rec4*. Two mutants are sensitive to X rays (one of these is *rec2* and the other is not located), one is sensitive to X rays and to ultraviolet light (no locus designation), and the other four are all insensitive to UV and to X rays. Besides the lack of induction of mitotic recombination by X rays, all except *rad2* show a similar lack of induction by UV. There are differences between these mutants and the wild type in respect to meiotic recombination (60). In homozygotes of *rec2* and *rec3*, meiosis is abortive. In *rec1* spontaneous and induced allelic recombination at *arg4* at meiosis is about equal to the control, but in *rec4* it is depressed. Nonallelic recombination appears to be normal in *rec1* and *rec4*. It appears that some, but not all steps in mitotic and meiotic recombination may be in common. This is shown by the correlated effects in *rec2, rec3,* and *rec4*.

Literature Cited

1. Angel, T., Austin, B., Catcheside, D. G. 1970. Regulation of recombination at the *his-3* locus in *Neurospora crassa. Aust. J. Biol. Sci.* 23:1229–40
2. Averbeck, D., Laskowski, W., Eckardt, F., Lehmann-Brauns, E. 1970. Four radiation sensitive mutants of *Saccharomyces. Mol. Gen. Genet.* 107:117–27
3. Badman, R. 1972. Deoxyribonuclease-deficient mutants of *Ustilago maydis* with altered recombination frequencies. *Genet. Res.* 20:213–29
4. Bandas, E. L., Bekker, M. L., Luchkina, L. A., Tkatchenko, V. P., Zakharov, I. A. 1973. Temperature sensitive radiosensitive mutants of the yeast *Saccharomyces paradoxus. Mol. Gen. Genet.* 126:153–64
5. Bandas, E. L., Zakharov, I. A. 1972. Conditional lethal radiation sensitive mutants of yeast. I. Isolation, genetic and radiobiological study. *Genetika* 8:101–8
6. von Borstel, R. C., Cain, K. T., Steinberg, C. M. 1971. Inheritance of spontaneous mutability in yeast. *Genetics* 69:17–27

7. Brendel, M., Haynes, R. H. 1973. Interactions among genes controlling sensitivity to radiation and alkylation in yeast. *Mol. Gen. Genet.* 125:197–216
8. Brendel, M., Khan, N. A., Haynes, R. H. 1970. Common steps in the repair of alkylation and radiation damage in yeast. *Mol. Gen. Genet.* 106:289–95
9. Brown, A. M., Kilbey, B. J. 1970. Hyper-UV-sensitive yeast. I. Isolation and properties of two such mutants. *Mol. Gen. Genet.* 108:258–65
10. Catcheside, D.E.A. 1968. The mechanism of genetic regulation of recombination and gene expression in *Neurospora crassa.* In *Replication and Recombination of Genetic Material,* ed. W. J. Peacock, R. D. Brock, 227–28. Canberra: Aust. Acad. Sci.
11. Catcheside, D.E.A. 1968. Regulation of the *am-1* locus in Neurospora: evidence of independent control of allelic recombination and gene expression. *Genetics* 59:443–52
12. Catcheside, D.E.A. 1970. Control of recombination within the *nitrate-2* locus of *Neurospora crassa:* an unlinked dominant gene which reduces proto-

troph yields. *Aust. J. Biol. Sci.* 23: 855–65

13. Catcheside, D.E.A. 1974. Unpublished
14. Catcheside, D. G. 1966. A second gene controlling allelic recombination in *Neurospora crassa. Aust. J. Biol. Sci.* 19:1039–46
15. Catcheside, D. G. 1966. Behaviour of flanking markers in allelic crosses. *Aust. J. Biol. Sci.* 19:1047–59
16. Catcheside, D. G. 1968. The control of genetic recombination in *Neurospora crassa.* See Ref. 10, pp. 216–26
17. Catcheside, D. G. 1971. Regulation of crossing over in *Neurospora crassa. Genet. Lect.* 2:7–18. Corvallis: Oregon State Univ. Press
18. Catcheside, D. G. 1974. The occurrence in wild strains of *Neurospora crassa* of genes controlling genetic recombination. *Aust. J. Biol. Sci.* 27:In press
19. Catcheside, D. G., Angel, T. 1974. A *histidine-3* mutant, in *Neurospora crassa,* due to an interchange. *Aust. J. Biol. Sci.* 27:219–29
20. Catcheside, D. G., Austin, B. 1969. The control of allelic recombination at *histidine* loci in *Neurospora crassa. Am. J. Bot.* 56:685–90
21. Catcheside, D. G., Austin, B. 1971. Common regulation of recombination at the *amination-1* and *histidine-2* loci in *Neurospora crassa. Aust. J. Biol. Sci.* 24:107–15
22. Catcheside, D. G., Corcoran, D. 1973. Control of non-allelic recombination in *Neurospora crassa. Aust. J. Biol. Sci.* 26:1337–53
23. Chang, L. T., Tuveson, R. W. 1967. Ultraviolet-sensitive mutants in *Neurospora crassa. Genetics* 56:801–10
24. Cox, B. S., Parry, J. M. 1968. The isolation, genetics and survival characteristics of ultraviolet light-sensitive mutants in yeast. *Mutat. Res.* 6:37–55
25. De Serres, F. J. 1971. Mutability of ultraviolet-sensitive strains of *Neurospora crassa. Genetics* 68: Suppl., s14
26. Evans, M. E., Parry, J. M. 1972. The cross sensitivity to radiations, chemical mutagens and heat treatment of X-ray sensitive mutants of yeast. *Mol. Gen. Genet.* 118:261–71
27. Fincham, J. R. S. 1951. A comparative study of the mating type chromosomes of two species of *Neurospora. J. Genet.* 50:221–29
28. Fogel, S., Mortimer, R. K. 1971. Recombination in yeast. *Ann. Rev. Genet.* 5:219–36

29. Game, J. C., Cox, B. S. 1971. Allelism tests of mutants affecting sensitivity to radiation in yeast and a proposed nomenclature. *Mutat. Res.* 12:328–31
30. Game, J. C., Cox, B. S. 1972. Epistatic interactions between four *rad* loci in yeast. *Mutat. Res.* 16:353–62
31. Gutz, H. 1971. Site specific induction of gene conversion in *Schizosaccharomyces pombe. Genetics* 69:317–37
32. Holliday, R. 1967. Altered recombination frequencies in radiation sensitive strains of *Ustilago. Mutat. Res.* 4: 275–88
33. Hotta, Y., Stern, H. 1971. Analysis of DNA synthesis during meiotic prophase in *Lilium. J. Mol. Biol.* 55:337–55
34. Hunnable, E. G., Cox B. S. 1971. The genetic control of dark recombination in yeast. *Mutat. Res.* 13:297–309
35. Jessop, A. P., Catcheside, D. G. 1965. Interallelic recombination at the *his-1* locus in *Neurospora crassa* and its genetic control. *Heredity* 20:237–56
36. Jha, K. K. 1967. Genetic control of allelic recombination at the *histidine-3* locus of *Neurospora crassa. Genetics* 57:865–73
37. Jha, K. K. 1969. Genetic factors affecting allelic recombination at the *histidine-3* locus of *Neurospora crassa. Mol. Gen. Genet.* 105:30–37
38. Khan, M. A., Brendel, M., Haynes, R. H. 1970. Supersensitive double mutants in yeast. *Mol. Gen. Genet.* 107:376–78
39. Koltin, Y. 1970. Studies on mutations disruptive to nuclear migration in *Schizophyllum commune. Mol. Gen. Genet.* 106:155–61
40. Koltin, Y., Stamberg, J., Simchen, G. 1971. Three tests for shared allelic specificities in the B incompatibility factor of *Schizophyllum commune. Genetica* 42:313–18
41. Leblon, G. 1972. Mechanism of gene conversion in *Ascobolus immersus.* I. Existence of a correlation between the origin of mutants induced by different mutagens and their conversion spectrum. *Mol. Gen. Genet.* 115:36–48
42. Leblon, G. 1972. Mechanism of gene conversion in *Ascobolus immersus.* II. The relationship between the genetic alterations in *b1* or *b2* mutants and their conversion spectrum. *Mol. Gen. Genet.* 116:322–35
43. Leblon, G., Rossignol, J. L. 1973. Mechanism of gene conversion in *Ascobolus immersus.* III. The interaction

of heteroalleles in the conversion process. *Mol. Gen. Genet.* 122:165–82

44. Lemontt, J. F. 1971. Mutants of yeast defective in ultraviolet light-induced mutation. *Genetics* 68:21–33

45. Lemontt, J. F. 1971. Pathways of ultraviolet mutability in *Saccharomyces cerevisiae*. I. Some properties of double mutants *uvs9* and *rev. Mutat. Res.* 13:311–17

46. Lemontt, J. F. 1971. Some properties of double mutants *uvs9* and *rev.* II. The effect of *rev* genes on recombination. *Mutat. Res.* 13:311–17

47. Mori, S., Nakai, S. 1972. Induction and repair of gene conversion in UV-sensitive mutants of yeast. *Mol. Gen. Genet.* 117:187–96

48. Nakai, S., Matsumoto, S. 1967. Two types of radiation-sensitive mutant in yeast. *Mutat. Res.* 4:129–36

49. Newcombe, K. D., Threlkeld, S. F. H. 1972. Interspecific crosses and crossing over in *Neurospora*. *Genet. Res.* 19: 115–19

50. Parry, J. M. 1969. Comparison of the effects of UV and ethylmethylsulphonate upon the frequency of mitotic recombination in yeast. *Mol. Gen. Genet.* 106:66–72

51. Parry, J. M. 1971. The genetic effects of liquid holding recovery in UV light sensitive mutants of yeast. *Mol. Gen. Genet.* 111:51–60

52. Parry, J. M. 1972. A quantitative analysis of "negative liquid holding" in some UV sensitive mutants of yeast. *Mol. Gen. Genet.* 118:33–43

53. Parry, J. M., Parry, E. M. 1969. The effects of UV-light post-treatments on the survival characteristics of 21 UV-sensitive mutants of *Saccharomyces cerevisiae*. *Mutat. Res.* 8:545–56

54. Parry, J. M., Parry, E. M. 1972. The genetic implications of ultraviolet light exposure and liquid holding post-treatment in the yeast *Saccharomyces cerevisiae*. *Genet. Res.* 19:1–16

55. Parry, E. M., Parry, J. M., Waters, E. 1972. Genetic and physiological analysis of UV-sensitive mutants of *Saccharomyces cerevisiae*. *Mutat. Res.* 15:135–46

56. Resnick, M. A. 1969. Induction of mutations in *Saccharomyces cerevisiae* by ultraviolet light. *Mutat. Res.* 7:315–32

57. Resnick, M. A. 1969. Genetic control of radiation sensitivity in *Saccharomyces cerevisiae*. *Genetics* 62:519–31

58. Rizet, G., Rossignol, J.-L., Lefort, C. 1969. Sur la variété et la spécificité des spectres d'anomalies de ségrégation chez *Ascobolus immersus*. *C. R. Acad. Sci. Paris* 269:1427–30

59. Rodarte-Ramón, U. S. 1972. Radiation-induced recombination in *Saccharomyces:* The genetic control of recombination in mitosis and meiosis. *Rad. Res.* 49:148–54

60. Rodarte-Ramón, U.S., Mortimer, R. K. 1972. Radiation-induced recombination in *Saccharomyces:* Isolation and genetic study of recombination deficient mutants. *Rad. Res.* 49:133–47

61. Roth, R., Fogel, S. 1971. A system selective for yeast mutants deficient in meiotic recombination. *Mol. Gen. Genet.* 112:295–305

62. Rossignol, J.-L. 1969. Existence of homogeneous categories of mutants exhibiting various conversion patterns in gene *75* of *Ascobolus immersus*. *Genetics* 63:795–805

63. Schaap, T. 1971. Recognition sites and coordinate control of recombination in *Schizophyllum*. *Genetica* 42:219–30

64. Schaap, T., Simchen, G. 1971. Genetic control of recombination affecting mating factors in a population of Schizophyllum, and its relation to inbreeding. *Genetics* 68:67–75

65. Schroeder, A. L. 1968. Ultraviolet sensitive mutants in *Neurospora crassa*. *Genetics* 60:233 (Abstr.)

66. Schroeder, A. L. 1970. UV-sensitive mutants of Neurospora. I. Genetic basis and effect on recombination. *Mol. Gen. Genet.* 107:291–304

67. Schroeder, A. L. 1970. UV-sensitive mutants of Neurospora. II. Radiation studies. *Mol. Gen. Genet.* 107:305–20

68. Simchen, G. 1967. Genetic control of recombination and the incompatibility system of *Schizophyllum commune*. *Genet. Res.* 9:195–210

69. Simchen, G., Connolly, V. 1968. Changes in recombination frequency following inbreeding in Schizophyllum. *Genetics* 58:319–26

70. Simchen, G., Stamberg, J. 1969. Fine and coarse controls of genetic recombination. *Nature* 222: 329–32

71. Simonet, J. M. 1973. Mutations affecting meiosis in *Podospora anserina*. II. Effect of *mei2* mutants on recombination. *Mol. Gen. Genet.* 123:263–81

72. Simonet, J. M., Zickler, D. L. 1972. Mutations affecting meiosis in *Podospora anserina*. I. Cytological studies. *Chromosoma* 37:327–51

73. Smith, B. R. 1966. Genetic controls of recombination. I. The *recombination-2*

gene of *Neurospora crassa. Heredity* 21:481–98

74. Smith, B. R. 1968. A genetic control of recombination in *Neurospora crassa. Heredity* 23:162–63

75. Smith, D. A. 1973. A mutant affecting meiosis in *Neurospora. Genetics* 74: Suppl., s259

76. Smyth, D. R. 1971. Effect of *rec-3* on polarity of recombination the *amination-1* locus of *Neurospora crassa. Aust. J. Biol. Sci.* 24:97–106

77. Smyth, D. R. 1973. Action of *rec-3* on recombination near the *amination-1* locus of *Neurospora crassa. Aust. J. Biol. Sci.* 26:439–44

78. Smyth, D. R. 1973. A new map of the *amination-1* locus of *Neurospora crassa,* and the effect of the *recombination-3* gene. *Aust. J. Biol. Sci.* 26:1355–70

79. Snow, R. 1967. Mutants of yeast sensitive to ultraviolet light. *J. Bacteriol.* 94:571–75

80. Snow, R. 1968. Recombination in UV-sensitive strains in *Saccharomyces cerevisiae. Mutat. Res.* 6:409–18

81. Stadler, D. R., Smith, D. A. 1968. A new mutation in *Neurospora* for sensitivity to ultraviolet. *Can. J. Genet. Cytol.* 10:916–19

82. Stadler, D. R., Towe, A. M., Rossignol, J.-L. 1970. Intragenic recombination of ascospore color mutants in Ascobolus and its relationship to the segregation of outside markers. *Genetics* 66:429–47

83. Stamberg, J. 1968. Two independent gene systems controlling recombination in *Schizophyllum commune. Mol. Gen. Genet.* 102:221–28

84. Stamberg, J. 1969. Genetic control of recombination in *Schizophyllum commune:* separation of the controlled and controlling loci. *Heredity* 24:306–9

85. Stamberg, J., Koltin, Y. 1971. Selectively recombining B incompatibility factors of *Schizophyllum commune. Mol. Gen. Genet.* 113:157–65

86. Stamberg, J., Koltin, Y. 1973. Genetic control of recombination in *Schizophyllum commune:* evidence for a new type of regulatory site. *Genet. Res.* 22: 101–11

87. Stamberg, J., Simchen, G. 1970. Specific effects of temperature on recombination in *Schizophyllum commune. Heredity* 25:41–52

88. Suslova, N. G., Zakharov, I. A. 1970. The gene controlled radiation sensitivity of yeast. VII. Identification of the genes for the X-ray sensitivity. *Genetika* 6:158–63

89. Thomas, P. L., Catcheside, D. G. 1969. Genetic control of flanking marker behaviour in an allelic cross of *Neurospora crassa. Can. J. Genet. Cytol.* 11:558–66

90. Tuveson, R. W. 1972. Genetic and enzymatic analysis of a gene controlling UV sensitivity in *Neurospora crassa. Mutat. Res.* 15:411–24

91. Tuveson, R. W. 1972. Comparison of two transformation systems for the assay of the *Neurospora* photoreactivating enzyme. *Genet. Res.* 20:9–18

92. Tuveson, R. W., Mangan, J. 1970. An ultraviolet-sensitive mutant of *Neurospora* defective for photoreactivation. *Mutat. Res.* 9:455–66

93. Unrau, P., Wheatcroft, R., Cox, B. S. 1971. The excision of pyrimidine dimers from DNA of UV irradiated yeast. *Mol. Gen. Genet.* 113:359–62

94. Waters, R., Parry, J. M. 1973. The response to chemical mutagens of the individual haploid and homoallelic diploid UV-sensitive mutants of the *rad3* locus of *Saccharomyces cerevisiae. Mol. Gen. Genet.* 124:135–43h

95. Waters, R., Parry, J. M. 1973. A comparative study of the effects of UV irradiation upon diploid cultures of yeast defective at the *rad3* locus. *Mol. Gen. Genet.* 124:145–56h

96. Zakharov. I. A., Kozina, T. N., Federova, I. V. 1970. Effects de mutation vers la sensibilité du rayonnement ultraviolet chez la levure. *Mutat. Res.* 9:31–39

97. Zimmermann, F. K. 1968. Sensitivity to methylmethanesulphonate and nitrous acid of ultraviolet light-sensitive mutants in *Saccharomyces cerevisiae. Mol. Gen. Genet.* 102:247–56

GENETICS OF DNA TUMOR VIRUSES

♦3076

Walter Eckhart

The Salk Institute, San Diego, California 92112

Tumor viruses are able to cause cancer in animals and transform the growth properties of infected cells in culture. Genetic analysis of tumor viruses is aimed at identifying the viral genes and understanding how these genes affect cell growth regulation. Polyoma virus and Simian Virus 40 (SV40) are the smallest DNA tumor viruses, having genomes approximately 3.4×10^6 mol wt. This review describes what is known about the genetic properties of polyoma and SV40. Because of their small size and ease of propagation, polyoma and SV40 have been studied more thoroughly than other DNA tumor viruses, such as human adenoviruses ($20–25 \times 10^6$ mol wt) and herpesviruses (100×10^6 mol wt). Genetic analysis of adenoviruses (1) and herpesviruses (2) is also underway, but will not be described in detail here. Much of the basic information about these viruses is contained in a book published recently by Cold Spring Harbor (3).

INFECTION AND TRANSFORMATION BY POLYOMA AND SV40

The genetic information of polyoma and SV40 is expressed in an orderly way which depends upon the type of cell that has been infected. Productive or lytic infection of mouse cells by polyoma, or monkey cells by SV40, results in the synthesis of infectious progeny virus and death of the host cell. Abortive infection of hamster or rat cells by polyoma, or mouse cells by SV40, results in the expression of only part of the viral genetic information and does not lead to the synthesis of infectious progeny virus or death of the host cell. A fraction of the surviving cell population is transformed and exhibits growth properties different from those of the normal cell population. Viral DNA is permanently associated with the transformed cells, and a fraction of the viral genetic information is transcribed into stable mRNA.

Early during SV40 productive infection, virus-specific mRNA corresponding to approximately one third of the viral genome is transcribed (4–6). Cellular RNA synthesis continues during this time, and the virus-specific RNA constitutes less than 0.01% of the total RNA of the cell. After viral DNA replication begins, the

301

pattern of virus-specific mRNA changes. Sequences corresponding to the entire viral genome now are transcribed, and the proportion of virus-specific RNA in the cell increases to 3–6% of the newly synthesized RNA at late times after infection (7). During the early phase of SV40 productive infection, stable mRNA is transcribed from one strand of the viral DNA [designated the E or (–) strand], whereas during the late phase of infection additional sequences of stable mRNA are transcribed from the opposite strand [designated the L or (+) strand] (8–10). Aloni (11) has suggested that primary transcription occurs symmetrically and that stable transcripts are produced by preferential breakdown of one of the symmetrical products. It seems likely that regulation of viral transcription is important in determining the pattern of viral gene expression in the infected cell. Two outstanding questions arise from these observations: how the regulation of "early" and "late" viral mRNA synthesis is accomplished and how the switch occurs from early to late patterns of transcription, which occurs coincidentally with viral DNA replication. Among the obvious candidates to explain the features of transcriptional regulation are alterations in the viral DNA template and alterations in the polymerases responsible for transcription. Little is known about the state of either the viral template or the polymerases involved in transcription.

As a consequence of early viral gene expression during polyoma and SV40 infection, a number of changes occur in the host cell. Cellular DNA synthesis is induced, and the activity of a number of enzymes involved in DNA synthesis increases. Several new antigens appear in the infected cell, including the nuclear T antigen and the surface-associated transplantation antigen (TSTA). The cell surface changes so that infected cells become more agglutinable by wheat germ agglutinin or Concanavalin A. The transport of certain sugars is increased in infected cells (12). Because the amount of viral information expressed early during infection is so small (corresponding to approximately 100,000 daltons of protein) the early viral gene products must have pleiotropic effects, producing a number of secondary changes in the infected cell.

The DNA of polyoma and SV40 is a double-stranded, closed circular supercoiled molecule. The features of replication of the viral DNA have been analyzed extensively, both in vivo and in vitro, and at least one of the early viral gene products is involved directly in viral DNA replication (see below).

Purified virions of polyoma and SV40 contain a number of distinct proteins (for review see 13). The major protein has a molecular weight of 46,000–47,000 and is most likely coded for by the virus. In polyoma, two other virion proteins of molecular weights 35,000 and 23,000 may also be virus coded. These two proteins may be derived from a common precursor (14). Three small basic proteins of molecular weight 12,000–14,000 appear to be cellular histones associated with the viral DNA (15). None of the virion proteins is absolutely required for infection, because purified viral DNA is infectious, but some of the virion proteins may be multifunctional and act as control elements during viral replication.

Transformed cells display a number of properties that distinguish them from normal cells. The properties used most often for defining a cell as transformed are altered morphology, saturation density, nutritional requirements (including serum), agglutinability by plant lectins, ability to grow in agar or methyl cellulose, and ability to produce tumors. The properties of virus-transformed cells are probably determined largely by cellular genes and selection during growth in culture. A number of lines of evidence suggest that viral gene expression is also able to influence the properties of transformed cells. If we knew more about viral gene regulation we could infer a great deal about cellular gene regulation. Therefore, it is important to understand the nature of the viral genetic information expressed in the transformed cell, how the expression of this information is controlled, and how viral functions interact with cellular functions.

The number of genes of polyoma and SV40 is very small, perhaps as few as four or five. Several kinds of mutants of polyoma and SV40 have been selected and used to investigate viral gene functions.

TEMPERATURE-SENSITIVE MUTANTS OF POLYOMA AND SV40

Temperature-sensitive mutants of polyoma were isolated first by Fried (16) and M. Vogt (unpublished results) on the basis of differential growth at low and high temperatures (32 and 39°C). Subsequently Eckhart (17) and diMayorca (18, 19) described the isolation and characterization of larger numbers of mutants. For reasons that are not clear it seemed to be difficult at first to isolate temperature-sensitive mutants of SV40, but recently several groups have reported the isolation of SV40 mutants, and characterization of these mutants is proceeding rapidly (20–27).

Analysis of the temperature-sensitive mutants of polyoma and SV40 by functional tests and complementation shows that they fall into at least four classes (17–19, 24, 26, 28–31). Two of the classes are late mutants, affecting steps that occur after viral DNA replication in the lytic cycle (17–19, 24–26, 28, 30). All of these mutants are able to transform cells at the restrictive temperature, indicating that the gene functions in which they are defective are not required for transformation. Some of the late mutants are more heat labile than are the wild-type when heated to 64–68°C, suggesting that virion proteins are altered in these mutants (17, 24).

The other two classes of temperature-sensitive mutants of polyoma and SV40 affect early functions. One class of mutants is defective in induction of cellular DNA synthesis and in viral DNA synthesis at the restrictive temperature (26, 29, 31, 32). This class includes the ts3 mutant of polyoma and the ts*101 mutant of SV40. The other class of early mutants (the tsA class) is defective in viral DNA synthesis at the restrictive temperature, but induces cellular DNA synthesis normally (17, 33–36). The functional defects of various early and late mutants are discussed in more detail below.

HOST RANGE MUTANTS

Benjamin (37) has used a different selective technique to isolate mutants that might be defective in viral gene functions expressed in transformed cells. He has selected polyoma host range mutants that are able to grow in polyoma-transformed mouse

3T3 cells, which can be superinfected by polyoma, but not in untransformed cells. The rationale for this selection is that if viral gene functions are expressed in the wild-type transformed cell, a mutant defective in these functions might be able to grow in the transformed cell but not in its untransformed counterpart. In agreement with this prediction, a number of host range mutants have been isolated, all of which are defective in lytic growth in 3T3 cells and in transformation of hamster or rat cells (37). These mutants induce the synthesis of cellular DNA in the nonpermissive 3T3 host cell but do not cause the cell surface alteration detectable by enhanced agglutination of infected cells by wheat germ agglutinin or Concanavalin A (38).

RESTRICTION ENDONUCLEASES AND GENOME MAPPING

Although it has been possible to demonstrate recombination between pairs of temperature-sensitive mutants of polyoma (39) and SV40 (27), the recombination frequency is low, and it is doubtful that recombination will be a reliable tool in defining the structure of the viral genome. In recent years, restriction endonucleases have emerged as new and powerful tools to provide the information that might otherwise have been provided by conventional genetic analysis. Danna & Nathans (40) were the first to employ restriction endonucleases to map the physical position of replication and transcription regions on the SV40 chromosome. Restriction endonucleases introduce double-strand cuts into DNA at particular nucleotide sequences. Danna & Nathans (40) used the restriction endonuclease of *Hemophilus influenzae* to produce 11 fragments of the SV40 DNA molecule. By combining the endonucleases of *H. influenzae* and *H. parainfluenzae* with the RI endonuclease of *Escherichia coli*, Danna, Sack & Nathans (41) arranged the fragments in a circular array corresponding to their position in the circular SV40 DNA molecule. By comparing the specific activity of pulse-labeled fragments from replicating SV40 DNA molecules, Nathans & Danna (42) were able to locate the site of initiation of viral DNA replication at or near fragment C and infer that replication proceeds bidirectionally from this origin. The arrangement of the *H. influenzae* fragments on the SV40 chromosome is shown in Figure 1.

Fareed, Garon & Salzman (43) used the *E. coli* RI restriction endonuclease to analyze the origin and direction of SV40 DNA replication. The RI enzyme cleaves SV40 at a unique site (44, 45), generating fragments with cohesive ends (46). Because SV40 replication proceeds on a parental template whose strands remain in a closed circular configuration (47, 48), it was possible for Fareed, Garon & Salzman (43) to map the origin and direction of replication relative to the RI cleavage site. This was done by isolating molecules that had replicated to different extents on the basis of their density in ethidium bromide-containing cesium chloride gradients. When these molecules were cleaved with the RI enzyme and examined by sedimentation analysis and electron microscopy, the replicated regions could be visualized and their distance from the ends of the linear molecule generated by the RI enzyme measured. The results of this experiment showed that SV40 DNA replication begins at a site 33% of the genome length from the RI cleavage site and proceeds bidirectionally to terminate at a position approximately 180° from the origin.

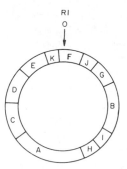

Figure 1   A physical map of the SV40 genome. The fragments produced by cleavage with a restriction endonuclease from *H. influenzae* are arranged in a circular map, using the cleavage site of the *E. coli* RI restriction endonuclease as the origin (41).

Crawford, Syrett & Wilde (49) have performed a similar analysis on polyoma DNA. They concluded that replication begins at a site approximately 30% of the genome length from the RI cleavage site and proceeds bidirectionally from this origin.

DELETION MUTANTS

Another class of mutants potentially important for defining the functions of the genomes of polyoma and SV40 are deletion mutants. Defective viral genomes, carrying deletions of portions of the viral DNA, arise during the growth of polyoma and SV40, and are enriched in virus stocks passed at high multiplicities of infection. These defective particles can replicate by virtue of functions provided by the nondefective particles of the population during mixed infection at high multiplicity.

Deletion mutants can be selected in a number of ways. Particles carrying less than a full complement of viral DNA can be isolated because of their lighter density in cesium chloride equilibrium density gradients or because of their slower sedimentation in velocity gradients (50, 51). Deletions at specific sites can be isolated by cleaving a population of viral DNA molecules with restriction enzymes and selecting molecules that have lost the restriction enzyme site and are thereby rendered resistant to cleavage (52). The deletion mutants can be propagated by mixed infection with nondefective wild-type virus or by mixed infection with temperature-sensitive mutants (52). The position of a deletion in the viral DNA can be located by examining heteroduplexes of mutant and wild-type viral DNA in the electron microscope. The availability of deletion mutants located in various positions of the viral DNA should permit correlations to be made between mutant defects and specific functional defects.

HYBRID VIRUSES

A partial functional map of the early region of the SV40 genome has been constructed from studies of the adeno-SV40 hybrid viruses, which contain various portions of the SV40 genome inserted into the adenovirus genome. Five hybrid viruses, designated Ad2$^+$ND$_1$, Ad2$^+$ ND$_2$, Ad2$^+$ND$_3$, Ad2$^+$ND$_4$, and Ad2$^+$ND$_5$, have been studied. These viruses differ in their biological properties (53). Ad2$^+$ND$_1$ induces the synthesis of the SV40 U-antigen, but not T-antigen or tumor-specific transplantation antigen (TSTA). Ad2$^+$ND$_2$ induces U- and TSTA, but not T-antigen. Ad2$^+$ND$_4$ induces U-, TSTA, and T-antigens. Ad2$^+$ND$_3$ and Ad2$^+$ND$_5$ do not induce serologically detectable SV40 antigens. The hybrids contain different amounts of SV40 DNA (54) and induce the synthesis of different species of SV40-specific mRNA (55, 56). The amount of SV40-specific RNA synthesized after infection by each of the hybrids is proportional to the amount of SV40 DNA contained in the hybrid (56). The RNA synthesized after infection by any hybrid is also synthesized after infection by all hybrids having longer SV40 segments (56).

The structure of the SV40 regions in the adeno-SV40 hybrid viruses has been established by heteroduplex mapping (57, 58). The locations of the SV40 regions of the hybrids on the SV40 chromosome have been mapped (58). These results are illustrated in Figure 2, using the RI endonuclease cleavage site as a reference point. The SV40 DNA segment contained in Ad2$^+$ND$_4$ is colinear with the region between 0.11 and 0.59 SV40 DNA map units (58). The SV40-specific RNA synthesized after infection by Ad2$^+$ND$_4$ contains all of the species made during the early phase of infection by SV40 (56). The region between 0.11 and 0.59 SV40 DNA map units may also contain regions that are transcribed only late in SV40 infection (58).

The control of transcription after Ad2$^+$ND$_4$ infection is still unclear: at least some of the SV40-specific RNA appears to be linked to Ad2 RNA (59), and studies of the interferon sensitivity of SV40 T-antigen after Ad2$^+$ND$_4$ infection suggest that transcription of the information for SV40 T-antigen synthesis is initiated in a region

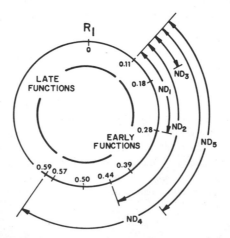

Figure 2   Adenovirus-SV40 hybrid viruses. The portions of the SV40 genome contained in each of the nondefective adenovirus-SV40 hybrid viruses are indicated by arrows on the SV40 genome map (58).

of the Ad2 DNA (59). Experiments utilizing separated strands of the *H. influenzae* cleavage products of SV40 DNA and RNA from cells infected with the various adeno-SV40 viruses suggest that transcription of the SV40 region of the hybrid virus Ad2$^+$ND$_4$ begins on adeno DNA and proceeds on the (–) strand of SV40 DNA past the termination point for early SV40 RNA and into fragment G (Figure 1), thereby transcribing both early RNA sequences and "antilate" sequences (60).

The topographical location and direction of synthesis of SV40 early and late RNA made during SV40 lytic infection have been deduced by the use of restriction enzymes. Westphal (61) made the original observation that *E. coli* RNA polymerase transcribes the closed circular form of SV40 DNA in vitro in an asymmetric manner, generating RNA (cRNA) complementary to only one strand of the SV40 DNA. This cRNA can be used to separate the strands of SV40 DNA. The strand of SV40 DNA that is complementary to cRNA is designated the (–) strand; the opposite strand is designated the (+) strand. Khoury & Martin (62) used separated strands of the 11 fragments of SV40 DNA produced by the *H. influenzae* restriction enzyme to determine which strand of each fragment anneals to SV40 early and late RNA. The results showed that early RNA is transcribed from the (–) strand of fragments A, H, I, and B, whereas late RNA is transcribed predominantly from the (+) strands of fragments A, C, D, E, K, F, J, G, and B (see Figure 1). Each of these sets of fragments is continuous on the physical map. The direction of transcription was inferred by measuring the reassociation kinetics of separated strands of SV40 linear DNA molecules (produced by cleavage with the RI enzyme) that had been digested with *E. coli* exonuclease III. Reassociation of these "half molecules" in the presence of denatured *H. influenzae* fragments led to the conclusion that transcription proceeds in a counterclockwise direction from fragment A to fragment B on the (–) strand, and in a clockwise direction on the (+) strand. This result is illustrated in Figure 3. Sambrook et al (63) reached similar conclusions by using fragments generated by the *H. parainfluenzae* restriction enzyme, HpaI, and an independent method of inferring the direction of transcription.

From these results, it is clear that a good deal is known about the physical structure of the SV40 genome and that this is due in large part to the great impetus provided by the restriction enzymes and the adeno-SV40 hybrid viruses. Studies with polyoma have proceeded more slowly because of the lack of appropriate restriction enzymes, but recently Griffin, Fried & Cowie (64) have derived a physical map of polyoma DNA, and it can be expected that information about the pattern of transcription of polyoma DNA will be available soon.

THE *tsA* MUTANTS OF POLYOMA AND SV40

Both polyoma and SV40 have an early gene whose function is required for viral DNA synthesis. This has been designated the *tsA* gene for SV40 (65) and the *ts-a* gene for polyoma. (The designation *tsA* gene will be used here for both viruses.) Temperature-sensitive mutants in this gene are defective in cell transformation at the restrictive temperature (17, 18, 36, 66, 67). Cells transformed by these mutants at the permissive temperature retain their transformed characteristics when grown

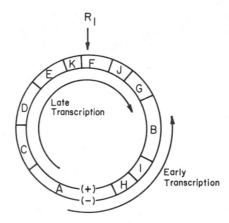

Figure 3   Transcriptional patterns of SV40. The direction and strand orientation of SV40 transcription during lytic infection are indicated by arrows on the SV40 genome map.

at the restrictive temperature (17, 18, 36, 66, 67), indicating that the function of the *tsA* gene is required for the initiation or stabilization of transformation, but not for maintenance of the transformed phenotype.

Tegtmeyer (36) examined the participation of the *tsA* gene function in SV40 DNA replication by temperature shift experiments in which cells infected with *tsA* mutants were shifted from the permissive to the restrictive temperature after viral DNA synthesis had begun. He found that a round of viral DNA synthesis initiated at the permissive temperature proceeded to completion at the restrictive temperature, but that initiation of new rounds of replication was blocked at the restrictive temperature. In similar experiments, Francke & Eckhart (35) found that completion of rounds of polyoma DNA synthesis at the restrictive temperature, after initiation at the permissive temperature, proceeds at the same rate in *tsA* mutant-infected cells as in wild-type–infected cells, but that initiation of new rounds of replication is blocked in the mutant-infected cells at the restrictive temperature.

These results suggest that the product of the *tsA* gene of polyoma and SV40 is required to initiate each round of viral DNA replication. The *tsA* mutants of polyoma and SV40 are the only mutants whose function appears to be involved directly in viral DNA replication; therefore, most of the replicative functions must be carried out by cellular enzymes. The *tsA* gene product may promote a site-specific association of the viral DNA with the cellular replicative apparatus or may function by introducing a nick in an early replicating form of the viral DNA to permit replication to proceed by allowing unwinding of the strands of the parental template molecule. These alternatives have not yet been tested.

Oxman, Takemoto & Eckhart (68) reported that the polyoma *tsA* mutants do not synthesize T-antigen detectable by immunofluorescence during lytic infection at the restrictive temperature. More recently, Robb et al (69) reported that the *tsA* mutants of SV40 produced T-antigen in reduced numbers of cells during lytic infection at the restrictive temperature (30% compared to the permissive temperature) and nearly undetectable levels of T-antigen during abortive infection at the restrictive temperature (0.1% compared to the permissive temperature). In addition, the *tsA* mutants of SV40 were defective in U-antigen production during both lytic and abortive infections at the restrictive temperature (69). These results could be interpreted to mean that the *tsA* gene is the structural gene for these antigens or that the product of the *tsA* gene is required for synthesis or detection of the antigens in the infected cell. Viral DNA synthesis is not required for the production of T- or U-antigens, so the mutant defect is not a consequence of the lack of viral DNA replication at the restrictive temperature.

A different line of experiments suggests that the *tsA* gene function of polyoma may be involved in integration and excision of viral DNA from cellular DNA. Permissive mouse 3T3 cells can be transformed by the polyoma *ts-a* mutant by infecting the cells for several days at the permissive temperature and then shifting the infected cells to the restrictive temperature (70). These transformed *ts-a*–3T3 cells retain the transformed phenotype when propagated at the restrictive temperature and produce little or no virus. When these cells are shifted to the permissive temperature, virus replication begins and a variable proportion of the cell population is killed, releasing infectious mutant virus. An unusual feature of this system is the presence of a high proportion of oligomeric viral DNA molecules in the intracellular replicating pool (71). Originally, two hypotheses were advanced to explain the appearance of oligomeric molecules: the molecules could be generated by excision of viral DNA molecules integrated in tandem in the cellular genome or by the presence of plasmids of viral DNA of various sizes (71). According to the first hypothesis, excision would result from expression of the *tsA* gene function after the shift to the permissive temperature. The state of the polyoma genome in the *ts-a*–3T3 cells (whether integrated into the cellular chromosome or free as a plasmid) has not been defined; therefore, although it is attractive to speculate that the *tsA* gene product is involved in excision, the nature of the process of virus induction after shift of *ts-a*–3T3 cells to the permissive temperature is still unknown.

Infection of hamster BHK cells by *tsA* mutants of polyoma at the restrictive temperature results in abortive transformation (72, 73). The infected cells divide a few times in agar or methylcellulose but do not become permanently transformed as they do at the permissive temperature. This result suggests again that *tsA* mutants are defective in a function needed to fix or stabilize the transformed state, and integration of the viral DNA into the cellular genome has been regarded as a likely candidate for this fixation step. BHK cells transformed by the *tsA* mutant of polyoma and propagated at the restrictive temperature will yield virus when fused to mouse cells at the permissive temperature (74). Viral DNA replication can be induced by shifting such transformants from the restrictive to the permissive temperature without fusion to mouse cells (75), indicating that some BHK-21 cells are partially permissive for polyoma replication. As in the case of the *ts-a*–3T3 cells, which yield virus after being shifted to the permissive temperature, the role of the *tsA* function in excision remains a matter of speculation.

Cuzin et al (76) have described the properties of an endonuclease associated with purified polyoma virions. The endonuclease is tightly bound to the virus particle, as the specific activity increases with the degree of purity of the virus. The enzyme has a high affinity for polyoma DNA, compared to mouse DNA, and induces a small number of breaks in the viral DNA molecule. The enzyme was not detected in two separate preparations of *tsA* mutant virions, suggesting that the enzyme might be the product of the *tsA* gene and that the mutant protein might be inactivated during virus purification. The virion endonuclease is not needed for viral infectivity, because purified viral DNA is infectious, and *tsA* virions are infectious at the permissive temperature. An endonuclease involved in viral DNA replication and integration of viral DNA into cellular DNA could account for the properties of the *tsA* gene product, but whether the virion endonuclease represents such an activity is still an open question.

THE *ts3* MUTANT OF POLYOMA

The *ts3* mutant of polyoma is blocked early after lytic infection at the restrictive temperature. Induction of cellular DNA synthesis is blocked in resting mouse BALB/3T3 cells at the restrictive temperature (32). Viral DNA synthesis and the cell surface alterations detectable by enhanced agglutinability with wheat germ agglutinin or Concanavalin A are also blocked (77). Temperature sensitivity of mutant growth is different in different cell lines, being greatest in BALB/3T3 cells and less in mouse embryo cells (29). The mutant does not show complementation in mixed infection with other temperature-sensitive mutants of polyoma (29). After infection with viral DNA, mutant growth is not temperature-sensitive (29). Stimulation of resting infected cells with serum reduces the temperature sensitivity of mutant growth, but only to a limited extent (29).

Cells transformed by the *ts3* mutant at the permissive temperature exhibit temperature-dependent expression of some aspects of the transformed phenotype. When propagated at the restrictive temperature, *ts3*-transformed BHK cells show agglutinability and density-dependent inhibition of DNA synthesis characteristic of normal cells, whereas they show properties characteristic of transformed cells when propagated at the permissive temperature (77). Chinese hamster embryo cells transformed by the *ts3* mutant exhibit temperature-dependent nutritional requirements (D. Paul, unpublished observations). BALB/3T3 cells transformed by *ts3* show temperature-dependent morphology, saturation density serum dependence, and responses to a growth factor derived from the pituitary (W. Eckhart, unpublished observations; P. Rudland and W. Seifert, unpublished observations). The temperature-dependent expression of cellular growth properties by *ts3*-transformed cells could result from a requirement for a functional viral gene product for expression of transformed properties or because the *ts3* mutant transforms only a special class of cells that exhibit temperature-dependent growth properties because of cellular mutations. The second alternative is less likely than the first because the frequency of transformation of BHK cells by *ts3* at the permissive temperature is approximately the same as by wild-type polyoma.

The properties of the *ts3* mutant during lytic infection could be explained in either of two ways. The mutant could contain an altered virion protein, which might prevent expression of early viral functions at the restrictive temperature. Alternatively, the mutant could carry a temperature-sensitive mutation in a viral gene required for early regulatory changes in the infected cell, such as induction of cellular DNA synthesis and cell surface alteration. It is likely that there is an early gene function of polyoma required for induction of cellular DNA synthesis because the *tsA* mutants induce cellular DNA synthesis normally at the restrictive temperature, and interferon treatment or inactivation of wild-type polyoma by various agents blocks induction of cellular DNA synthesis (78–80). The lack of normal complementation by *ts3* and lack of temperature sensitivity of growth after viral DNA infection are consistent with an alteration in a virion protein, but could be explained in other ways (such as by a *cis*-acting gene product and by multiplicity-dependent leakiness following viral DNA infection).

THE *ts*101 MUTANT OF SV40

The *ts*101 mutant of SV40 has many properties similar to the *ts3* mutant of polyoma. During lytic infection at the restrictive temperature, induction of cellular DNA synthesis, viral DNA synthesis, and appearance of early antigens are blocked (31). Growth of the mutant is not temperature-sensitive following infection with viral DNA, and the mutant does not show complementation in mixed infection with other temperature-sensitive mutants of SV40 (31).

Transformation of mouse cells by the *ts*101 mutant occurs at both permissive and restrictive temperatures, but the frequency of transformation is 10–20-fold lower than wild-type SV40 (81). Virus can be rescued from the *ts*101-transformed cells by fusion with permissive monkey cells, and the rescued virus is temperature-sensitive in lytic growth. Expression of T-antigen is temperature-dependent in *ts*101-transformed 3T3 cells; wild-type transformed cells usually show more than 90% T-antigen positive nuclei at both the permissive and the restrictive temperatures by immunofluorescence, whereas *ts*101-transformed cells show 85–95% positive nuclei at the permissive temperature, but only 10–15% positive nuclei at the restrictive temperature (82).

Chou & Martin (26) have described a class of early mutants of SV40 (Class D), which includes the *ts*101 mutant. These mutants are defective in viral DNA synthesis at the restrictive temperature (26) and can be distinguished from the *tsA* mutants on the basis of complementation. The Class D mutants do not show complementation with other temperature-sensitive SV40 mutants under the usual conditions of growth, but exhibit "delayed complementation," giving increased yields in mixed infection at the restrictive temperature after prolonged incubation (26). The basis of this delayed complementation by Class D mutants is not understood.

THE LATE MUTANTS OF POLYOMA AND SV40

The late mutants of polyoma fall into two (or possibly three) classes on the basis of complementation analysis (18, 19, 28). The mutants synthesize viral DNA normally at the restrictive temperature, but do not make infectious progeny (17–19, 28).

The mutants transform normally at the restrictive temperature, indicating that the functions in which they are defective are not required for transformation (17–19, 28).

Analysis of nine SV40 late mutants by Tegtmeyer & Ozer (30) showed that the mutants could be divided into two classes: one class made empty capsids at the restrictive temperature; the other did not. The mutants synthesized viral DNA at the restrictive temperature, but did not produce infectious progeny. All nine mutants fell into one class on the basis of complementation. Kimura & Dulbecco (24) reported that four late mutants of SV40 could be placed in two complementation groups. All four of the late mutants were more heat labile than wild-type SV40 (24).

Robb et al (69) made an analysis of the antigens present in cells infected at the restrictive temperature by the SV40 late mutants of Tegtmeyer (30) and Ishikawa & Aizawa (25). They used sera directed against early antigens (U and T), against intact virions (V), and against detergent-disrupted virus (C). Some of the late mutants synthesize all antigens normally at the restrictive temperature. Other mutants synthesize T-antigen, but little or no U- and V-antigen. The latter mutants synthesize C-antigen at the restrictive temperature, but the antigen is located in the cytoplasm and the nucleolus rather than in the nucleus (69, 83, 84). Therefore, some of the SV40 late mutants may be defective in processing virion proteins in the infected cell at the restrictive temperature (83, 84).

An extensive complementation analysis of 76 SV40 temperature-sensitive mutants has been carried out by Chou & Martin (26). The 55 late mutants analyzed in this survey fall into three classes: two classes (B and C) complement each other and the early mutants; a third class (BC) complements the early mutants, but does not complement either of the other two classes of late mutants (26). One explanation for the BC class would be that the B and C cistrons correspond to structural proteins of SV40 and that some mutations in either of these cistrons would not complement late mutants in the other cistron because small amounts of mutant protein would render the virions produced by complementation labile (26).

Another explanation for the BC class could be that the B and C cistrons are normally translated into a single polypeptide, which subsequently is cleaved into two separate polypeptides. A mutation affecting the cleavage might be unable to complement mutants in either of the two cistrons.

VIRAL GENES AND CELL TRANSFORMATION

The genetic analysis of polyoma and SV40 suggests that these viruses have at least two early functions and two late functions. From the results of Chou & Martin (26) these four functions may be the only ones required for the lytic growth of SV40. Several lines of evidence suggest that viral gene functions may influence at least some of the properties of cells transformed by polyoma and SV40. Viral DNA is found in transformed cells, and most or all of the early region of the viral genome is transcribed into stable mRNA (10, 85). Late viral mRNA and late viral proteins are absent in most or all transformed cells. The strongest evidence for the direct involvement of a viral gene function in transformation comes from the temperature-dependent properties of polyoma *ts3*-transformed cells (77) and SV40 *ts*101*-transformed cells (82).

Because the stable viral mRNA in transformed cells corresponds to the early region of the viral genome, the postulated "transforming function" must be an early gene function of the virus. The small genomes of polyoma and SV40 seem to have evolved an early viral gene that exerts a regulatory effect in the infected cell, responsible for inducing the synthesis of cellular DNA and stimulating resting cells to provide the factors required for efficient virus replication. The *tsA* gene function is not required for the induction of cellular DNA synthesis (33, 34) and is not required to maintain the properties of transformed cells (17, 18, 66, 67). Therefore, a very likely candidate for a "transforming function" of polyoma and SV40 is the early gene required for the initiation of cellular DNA synthesis.

How could an early viral regulatory gene function act? It is particularly interesting in this connection to consider the mechanism of action of the SV40 helper function for adenovirus replication in monkey cells. When adenovirus infects monkey cells, all species of mRNA are synthesized, but some late proteins are made in sharply reduced amounts, and virus particle assembly does not occur (86, 87). This defect can be overcome by coinfection with SV40 or by transformation with SV40 (88, 89). The SV40 helper function appears to be due to expression of an early gene of SV40 because the helper function is intact in the nondefective adeno-SV40 hybrid viruses. The *tsA* gene of SV40 is not responsible for helper activity because the *tsA* mutants of SV40 are not defective in helper activity at the restrictive temperature (J. Robb, personal communication). Therefore, a study of the SV40 helper function may give clues to the mechanism of action of an early SV40 regulatory gene.

Recently Hashimoto et al (90) have studied the nature of the defect in capsid protein synthesis after infection of monkey cells with adenovirus type 2. They found that the defect is unlikely to be in the elongation or release of nascent polypeptides, but rather appears to reside in the polyribosomes. By analyzing the late adenovirus mRNA, they found that certain species of adeno-mRNA appear to be missing from the polyribosomes of infected monkey cells. Their results suggest that the inability of monkey cells to support the growth of adenovirus derives from an inability of the ribosomes to associate with certain species of late adeno-mRNA, and that the SV40 helper function may act by promoting the association of these viral mRNA species with ribosomes.

The results of Hashimoto et al (90) suggest a possible mechanism by which an early gene product of SV40 could exert a regulatory effect in the infected cell by influencing the association of specific mRNA species with ribosomes, thereby controlling the synthesis of specific proteins. It will be important to determine whether such a mechanism operates in other systems, for example, in cells showing temperature-dependent properties after transformation with temperature-sensitive mutants.

VIRUSES AND CELL GROWTH CONTROL

It is now ten years since the first mutant of polyoma virus was isolated. During that time the techniques of molecular biology have provided a great deal of insight into the structure and function of the genomes of polyoma and SV40. Some of the developments have been surprisingly rapid, for example, the mapping of the genome by restriction enzymes and heteroduplex analysis. Others, such as the relationship between viral gene functions and cell growth control, have been frustratingly slow. It seems obvious that cell growth control is very complex and that viral gene functions are only one factor that may contribute to determining the properties of transformed cells. A complete understanding of cell growth control will require a great deal of information not yet in sight. The properties of a cell are determined by interactions at many different levels among many cellular gene products. The importance of the DNA tumor viruses as a tool for understanding cell growth control lies in what we are able to learn from them about the nature of these cellular regulatory interactions.

ACKNOWLEDGMENTS

I am grateful to Drs. M. Fried, R. Martin, J. Mertz, and J. Robb for providing data prior to publication.

Literature Cited

1. Williams, J. F., Ustacelebi, S. 1971. Temperature restricted mutants of human adenovirus type 5. In *Ciba Found. Symp. Strategy Viral Genome*, ed. G. E. W. Wolstenholme, M. O'Connor, 275–94. Edinburgh & London: Churchill Livingstone

2. Brown, S. H., Ritchie, D. A., Subak-Sharpe, J. H. 1973. Genetic studies with herpes simplex virus type 1. The isolation of temperature-sensitive mutants, their arrangement into complementation groups, and recombination analysis leading to a linkage map. *J. Gen. Virol.* 18:329–46

3. Tooze, J. Ed. 1973. *The Molecular Biology of Tumor Viruses,* Cold Spring Harbor, NY: Cold Spring Harbor Lab. 743 pp.

4. Oda, K., Dulbecco, R. 1968. Regulation of transcription of the SV40 DNA in productively infected and in transformed cells. *Proc. Nat. Acad. Sci. USA* 60:525–32

5. Aloni, Y., Winocour, E., Sachs, L. 1968. Characterization of the Simian Virus 40-specific RNA in virus-yielding and transformed cells. *J. Mol. Biol.* 31:415–29

6. Sauer, G., Kidwai, J. R. 1968. The transcription of the SV40 genome in productively infected and in transformed cells. *Proc. Nat. Acad. Sci. USA* 61:1256–63

7. Tonegawa, S., Walter, G., Bernardini, A., Dulbecco, R. 1970. Transcription of the SV40 genome in transformed cells and during lytic infection. *Cold Spring Harbor Symp. Quant. Biol.* 35:823–31

8. Lindstrom, D. M., Dulbecco, R. 1972. Strand orientation of SV40 transcription in productively infected cells. *Proc. Nat. Acad. Sci. USA* 69:1517–20

9. Khoury, G., Byrne, J. C., Martin, M. A. 1972. Patterns of Simian Virus 40 transcription after acute infection of permissive and nonpermissive cells. *Proc. Nat. Acad. Sci. USA* 69:1925–28

10. Sambrook, J., Sharp, P. A., Keller, W. 1972. Transcription of SV40. I. Separation of the strands of SV40 and hybridization of the separated strands to RNA extracted from lytically infected and transformed cells. *J. Mol. Biol.* 70: 57–73

11. Aloni, Y. 1972. Extensive symmetrical transcription of SV40 DNA in virus-yielding cells. *Proc. Nat. Acad. Sci. USA* 69:2404–9

12. Eckhart, W., Weber, M. 1974. Uptake of 2-deoxyglucose by BALB/3T3 cells: changes after polyoma infection. *Virology.* In press

13. Crawford, L. V. 1973. Proteins of polyoma virus and SV40. *Brit. Med. Bull.* 29:253–58

14. Friedmann, T. 1974. Novel genetic economy of polyoma virus: capsid proteins are cleavage products of the same

viral gene. *Proc. Nat. Acad. Sci. USA* 71:257–59
15. Frearson, P., Crawford, L. V. 1972. Polyoma virus basic proteins. *J. Gen. Virol.* 14:141–55
16. Fried, M. 1965. Isolation of temperature-sensitive mutants of polyoma virus. *Virology* 25:669–71
17. Eckhart, W. 1969. Complementation and transformation by temperature-sensitive mutants of polyoma virus. *Virology* 38:120–25
18. DiMayorca, G., Callender, J., Marin, G., Giordano, R. 1969. Temperature sensitive mutants of polyoma virus. *Virology* 38:126–33
19. DiMayorca, G., Callender, J. 1970. Significance of temperature-sensitive mutants in viral oncology. *Progr. Med. Virol.* 12:284–301
20. Tegtmeyer, P., Dohan, C., Reznikoff, C. 1970. Inactivating and mutagenic effects of nitrosoguanidine on Simian Virus 40. *Proc. Nat. Acad. Sci. USA* 66:745–52
21. Robb, J. A., Martin, R. 1970. Genetic analysis of Simian Virus 40. I. Description of microtitration and replicating techniques for virus. *Virology* 41:751–60
22. Kit, S., Tokuno, S., Nakajima, K., Trkula, D., Dubbs, D. R. 1970. Temperature-sensitive Simian Virus 40 mutant defective in late function. *J. Virol.* 6:286–94
23. Takemoto, K. K., Martin, M. A. 1970. SV40 thermosensitive mutant: synthesis of viral DNA and virus-induced proteins at nonpermissive temperature. *Virology* 42:938–45
24. Kimura, G., Dulbecco, R. 1972. Isolation and characterization of temperature-sensitive mutants of Simian Virus 40. *Virology* 49:394–403
25. Ishikawa, A., Aizawa, T. 1973. Characterization of temperature-sensitive mutants of SV40. *J. Gen. Virol.* 21:1–11
26. Chou, J. Y., Martin, R. G. 1974. Complementation analysis of SV40 mutants. *J. Virol.* 13:1011–19
27. Dubbs, D. R., Rachmeler, M., Kit, S. 1974. Recombination between temperature-sensitive mutants of Simian Virus 40. *Virology* 57:161–74
28. Eckhart, W. 1974. Properties of temperature-sensitive mutants of polyoma virus. *Cold Spring Harbor Symp. Quant. Biol.* In press
29. Eckhart, W., Dulbecco, R. 1974. Properties of the ts3 mutant of polyoma

virus during lytic infection. *Virology.* In press
30. Tegtmeyer, P., Ozer, H. L. 1971. Temperature-sensitive mutants of Simian Virus 40: infection of permissive cells. *J. Virol.* 8:516–24
31. Robb, J. A., Martin, R. G. 1972. Genetic analysis of Simian Virus 40. III. Characterization of a temperature-sensitive mutant blocked at an early stage of productive infection in monkey cells. *J. Virol.* 9:956–68
32. Dulbecco, R., Eckhart, W. 1970. Temperature-dependent properties of cells transformed by a thermosensitive mutant of polyoma virus. *Proc. Nat. Acad. Sci. USA* 67:1775–81
33. Fried, M. 1970. Characterization of a temperature-sensitive mutant of polyoma virus. *Virology* 40:605–17
34. Eckhart, W. 1971. Induced cellular DNA synthesis by early and late temperature-sensitive mutants of polyoma virus. *Proc. Roy. Soc. London B.* 177:59–63
35. Francke, B., Eckhart, W. 1973. Polyoma gene function required for viral DNA synthesis. *Virology* 55:127–35
36. Tegtmeyer, P. 1972. Simian Virus 40 deoxyribonucleic acid synthesis: the viral replicon. *J. Virol.* 10:591–98
37. Benjamin, T. L. 1970. Host range mutants of polyoma virus. *Proc. Nat. Acad. Sci. USA* 67:394–99
38. Benjamin, T. L., Burger, M. M. 1970. Absence of a cell membrane alteration function in nontransforming mutants of polyoma virus. *Proc. Nat. Acad. Sci. USA* 67:929–34
39. Ishikawa, A., DiMayorca, G. 1971. Recombination between two temperature-sensitive mutants of polyoma virus. *Lepetit Colloq. Biol. Med.* 2:294–99
40. Danna, K., Nathans, D. 1971. Specific cleavage of Simian Virus 40 DNA by restriction endonuclease from *Hemophilus influenzae. Proc. Nat. Acad. Sci. USA* 68:2913–17
41. Danna, K., Sack, G., Nathans, D. 1973. Studies of Simian Virus 40 DNA. VII. A cleavage map of the SV40 genome. *J. Mol. Biol.* 78:363–76
42. Nathans, D., Danna, K. 1972. Specific origin in SV40 DNA replication. *Nature New Biol.* 236:200–2
43. Fareed, G. C., Garon, C. F., Salzman, N. P. 1972. Origin and direction of Simian Virus 40 deoxyribonucleic acid replication. *J. Virol.* 10:484–91

44. Morrow, J., Berg, P. 1972. Cleavage of Simian Virus 40 DNA at a unique site by a bacterial restriction enzyme. *Proc. Nat. Acad. Sci. USA* 69:3365–69
45. Mulder, C., Delius, H. 1972. Specificity of the break produced by restriction endonuclease RI in Simian Virus 40 DNA as revealed by partial denaturation mapping. *Proc. Nat. Acad. Sci. USA* 69:3215–19
46. Mertz, J., Davis, R. 1972. Cleavage of DNA by RI restriction endonuclease generates cohesive ends. *Proc. Nat. Acad. Sci. USA* 69:3370–74
47. Jaenisch, R., Mayer, A., Levine, A. 1971. Replicating SV40 molecules containing closed circular template DNA strands. *Nature New Biol.* 233:72–76
48. Sebring, E. D., Kelly, T. J., Thoren, M. M., Salzman, N. P. 1971. Structure of replicating SV40 DNA molecules. *J. Virol.* 8:479–90
49. Crawford, L. V., Syrett, C., Wilde, A. 1973. The replication of polyoma DNA. *J. Gen. Virol.* 21:515–22
50. Fried, M. 1974. Isolation and partial characterization of different defective DNA molecules derived from polyoma virus. *J. Virol.* 13:939–46
51. Brockman, R. W., Nathans, D. 1974. The isolation of Simian Virus 40 variants with specifically altered genomes. *Proc. Nat. Acad. Sci. USA* 71:942–46
52. Mertz, J. E., Berg, P. 1974. Defective Simian Virus 40 genomes: isolation and growth of individual clones. *Virology.* In press
53. Lewis, A. M. et al 1973. Studies of nondefective adenovirus 2-Simian Virus 40 hybrid viruses. V. Isolation of additional hybrids which differ in their Simian Virus 40-specific biological properties. *J. Virol.* 11:655–64
54. Henry, P. H. et al 1973. Studies of nondefective adenovirus 2-Simian Virus 40 hybrid viruses. VI. Characterization of the DNA from five nondefective hybrid viruses. *J. Virol.* 11:665–71
55. Oxman, M. N. et al 1971. Studies of nondefective adenovirus 2-Simian Virus 40 hybrid viruses. IV. Characterization of the Simian Virus 40 ribonucleic acid species induced by wild type Simian Virus 40 and by the nondefective hybrid virus Ad2$^+$ND$_1$. *J. Virol.* 8:215–24
56. Levine, A. S., Levin, M. J., Oxman, M. N., Lewis, A. M. 1973. Studies of nondefective adenovirus 2-Simian Virus 40 hybrid viruses. VII. Characterization of the Simian Virus 40 RNA species induced by five nondefective hybrid viruses. *J. Virol.* 11:672–81
57. Kelly, T. J., Lewis, A. M. 1973. Use of nondefective adenovirus-Simian Virus 40 hybrids for mapping the Simian Virus 40 genome. *J. Virol.* 12:643–52
58. Morrow, J. F., Berg, P., Kelly, T. J., Lewis, A. M. 1973. Mapping of Simian Virus 40 early functions on the viral chromosome. *J. Virol.* 12:653–58
59. Oxman, M. N., Levin, M. J., Lewis, A. M. 1974. Control of Simian Virus 40 gene expression in adenovirus-Simian Virus 40 hybrid viruses. Synthesis of hybrid adenovirus 2-Simian Virus 40 RNA molecules in cells infected with a nondefective adenovirus 2-Simian Virus 40 hybrid virus. *J. Virol.* 13: 322–30
60. Khoury, G., Lewis, A. M., Oxman, M. N., Levine, A. S. 1973. Strand orientation of SV40 transcription in cells infected by nondefective adenovirus 2-SV40 hybrid viruses. *Nature New Biol.* 246:202–5
61. Westphal, H. 1970. SV40 DNA strand selection by *E. coli* RNA polymerase. *J. Mol. Biol.* 50:407–20
62. Khoury, G., Martin, M. A. 1973. A map of Simian Virus 40 transcription sites expressed in productively infected cells. *J. Mol. Biol.* 78:377–89
63. Sambrook, J., Sugden, B., Keller, W., Sharp, P. H. 1973. Transcription of Simian Virus 40. III. Orientation of RNA synthesis and mapping of early and late species of viral RNA extracted from lytically infected cells. *Proc. Nat. Acad. Sci. USA* 70:3711–15
64. Griffin, B. E., Fried, M., Cowie, A. 1974. Polyoma DNA—a physical map. *Proc. Nat. Acad. Sci. USA* 71:2077–81
65. Robb, J. A., Tegtmeyer, P., Martin, R. G., Kit, S. 1972. Proposal for a uniform nomenclature for Simian Virus 40 mutants. *J. Virol.* 9:562–63
66. Fried, M. 1965. Cell transforming ability of a temperature-sensitive mutant of polyoma virus. *Proc. Nat. Acad. Sci. USA* 53:486–91
67. Kimura, G., Dulbecco, R. 1973. A temperature-sensitive mutant of Simian Virus 40 affecting transforming ability. *Virology* 52:529–34
68. Oxman, M. N., Takemoto, K. K., Eckhart, W. 1972. Polyoma T antigen synthesis by temperature-sensitive mutants of polyoma virus. *Virology* 49:675–82
69. Robb, J. A., Tegtmeyer, P., Ishikawa, A., Ozer, H. L. 1974. Antigenic phenotypes and complementation groups

of temperature-sensitive mutants of Simian Virus 40. *J. Virol.* 13:662–65

70. Vogt, M. 1970. Induction of virus multiplication in 3T3 cells transformed by a thermosensitive mutant of polyoma virus: I. Isolation and characterization of ts-a–3T3 cells. *J. Mol. Biol.* 47:307–16

71. Cuzin, F., Vogt, M., Dieckmann, M., Berg, P. 1970. Induction of virus multiplication in 3T3 cells transformed by a thermosensitive mutant of polyoma virus. II. Formation of oligomeric polyoma DNA molecules. *J. Mol. Biol.* 47:317–33

72. Stoker, M. 1968. Abortive transformation by polyoma virus. *Nature* 228:234–38

73. Stoker, M., Dulbecco, R. 1969. Abortive transformation by the ts-a mutant of polyoma virus. *Nature* 223:397–98

74. Summers, J. W., Vogt, M. 1971. Recovery of virus from polyoma-transformed BHK-21. *Lepetit Colloq. Biol. Med.* 2:306–12

75. Folk, W. R. 1973. Induction of virus synthesis in polyoma-transformed BHK-21 cells. *J. Virol.* 11:424–31

76. Cuzin, F., Rouget, P., Blangy, D. 1973. Endonuclease activity of purified polyoma virions. *Lepetit Colloq. Biol. Med.* 4:188–201

77. Eckhart, W., Dulbecco, R., Burger, M. M. 1971. Temperature-dependent surface changes in cells infected or transformed by a thermosensitive mutant of polyoma virus. *Proc. Nat. Acad. Sci. USA* 68:283–86

78. Dulbecco, R., Johnson, T. 1970. Interferon sensitivity of the enhanced incorporation of thymidine into cellular DNA induced by polyoma virus. *Virology* 42:368–74

79. Gershon, D., Hansen, P., Sachs, L., Winocour, E. 1965. The induction of cellular DNA synthesis by Simian Virus 40 in contact-inhibited and X-irradiated cells. *Proc. Nat. Acad. Sci. USA* 54:1584–92

80. Basilico, C., Marin, G., DiMayorca, G. 1966. Requirement for integrity of the viral genome for the induction of host DNA synthesis by polyoma virus. *Proc. Nat. Acad. Sci. USA* 56:208–15

81. Robb, J. A., Smith, H. S., Scher, C. D. 1972. Genetic analysis of Simian Virus 40. IV. Inhibited transformation of BALB/3T3 cells by a temperature-sensitive early mutant. *J. Virol.* 9:969–72

82. Robb, J. A. 1973. Simian Virus 40—Host cell interactions. I. Temperature-sensitive regulation of SV40 T antigen in 3T3 mouse cells transformed by the ts*101 temperature-sensitive early mutant of SV40. *J. Virol.* 12:1187–90

83. Tegtmeyer, P., Robb, J. A., Widmer, C., Ozer, H. L. 1974. Characterization of Simian Virus 40 mutant ts B11. *J. Virol.* In press

84. Widmer, C., Robb, J. A. 1974. SV40-host cell interactions. II. Temperature-sensitive cytoplasmic and nucleolar accumulation of virion protein by a capsid assembly defective Simian Virus 40 mutant, ts B11. *J. Virol.* In press

85. Khoury, G., Byrne, J. C., Takemoto, K. K., Martin, M. A. 1973. Patterns of Simian Virus 40 deoxyribonucleic acid transcription. II. In transformed cells. *J. Virol.* 11:54–60

86. Fox, R., Baum, S. 1972. Synthesis of viral ribonucleic acid during restricted adenovirus infection. *J. Virol.* 10:220–27

87. Lucas, J. J., Ginsberg, H. S. 1972. Transcription and transport of virus-specific ribonucleic acids in Africa Green monkey kidney cells abortively infected with type 2 adenovirus. *J. Virol.* 10:1109–17

88. Shiroki, K., Shimojo, H. 1969. Replication of human adenovirus in green monkey kidney cells transformed by adeno 7 - SV40 hybrid virus. *Jap. J. Microbiol.* 13:125–28

89. Rapp, F. T., Trulock, S. C. 1970. Susceptibility to superinfection of Simian cells transformed by SV40. *Virology* 40:961–70

90. Hashimoto, K., Nakajima, K., Oda, K., Shimojo, H. 1973. Complementation of translational defect for growth of human adenovirus type 2 in Simian cells by a Simian Virus 40-induced factor. *J. Mol. Biol.* 81:207–23

FRAMESHIFT MUTATIONS ❖3077

John R. Roth
Department of Molecular Biology, University of California, Berkeley, California 94720

INTRODUCTION

Since the original descriptions of frameshift mutations in T4 (1, 2), a large body of literature has accumulated characterizing such mutations and the mutagens which induce them. A model suggested by Streisinger (3) accounts for some aspects of the data. In spite of these advances, many questions remain unanswered as to the mechanism by which frameshift mutations occur and the mode of action of frameshift mutagens. Particularly puzzling are differences in effects of mutagens on various organisms and the apparent specificities of particular mutagens for certain sites in the DNA. Although informational suppressors of frameshift mutations have been described, their exact mode of action is not clear. Many of these questions have been clearly described and critically discussed by Drake in his book *The Molecular Basis of Mutation* (4). The present review covers selected topics in this area, most of which have developed since Drake's review.

BACKGROUND

Frameshift Mutations Defined

Alterations of DNA which result in addition or removal of one or several base pairs can have drastic consequences for the protein whose structure is encoded in that DNA. Since the code is read in blocks of three bases from a fixed starting point, addition or removal of bases from the DNA (and thus from the mRNA) causes the translation process to lose the proper frame of reference. Although the sequence of bases in the message is correct beyond the point of the alteration, the translation process now chooses its blocks of three improperly; this results in a run of contiguous, improper amino acids. Usually this series of improper missense insertions is cut short when a nonsense codon (termination signal) is encountered in the course of reading a normal message in the wrong frame of reference. Mutations which cause such a shift in the frame of reference are known as frameshift mutations.

   Both addition and removal of base pairs from the DNA can have these consequences so long as the number of bases added or deleted is not a multiple of three. If an integral number of code words (each a block of three bases) is added or

319

removed, no shift in reading frame occurs. Removal of one base pair is equivalent to the addition of two bases in terms of its effect on the translational frame of reference. Both mutations cause the reading frame to shift one base ahead (a −1 frameshift). Conversely, a single base addition, like removal of two bases, causes the reading frame to shift one base backwards (a +1 frameshift).

Determining the Mutational Alteration

The exact DNA changes involved in many frameshift mutations have been inferred from sequence studies of revertant proteins. This method takes advantage of the fact that frameshift mutations of opposite sign (i.e. a +1 and a −1 frameshift) are mutually corrective. When both mutations are present in a gene, the first (say the −1 type) causes the frame to shift ahead one base. This out-of-frame reading continues until translation reaches the site of the second frameshift (the +1 type); here the frame shifts one base backwards into the proper reading frame. Thus, only the region between the two mutations is read improperly. If two mutations are to be mutually corrective, no nonsense codons can be encountered in the region between the two mutant sites. Therefore, usually the two mutations must be close together for correction to be possible. Mutually corrective pairs of mutations like these are frequently encountered among revertants of frameshift mutations. A functional (though not normal) protein is produced if a second frameshift mutation (of appropriate sign of correction) occurs near the site of the original lesion. These revertant proteins have a normal amino acid sequence except for the amino acids encoded by the region between the mutant sites. This part of the revertant protein is found to have a short stretch of contiguous, improper amino acids. By comparing the improper amino sequence of the revertant protein to the corresponding region of the wild-type protein, with the aid of a table of code words, one can frequently infer the base sequence of the messenger RNA and thus the nature of the two frameshift mutations involved. All mutant DNA sites discussed here were determined in this way.

An example involving T4 lysozyme mutants (3, 5) is presented in Figure 1. Here the mutually corrective mutations eJ17 and eJ44 are seen to cause respectively a two-base and a one-base addition to the message. Thus, the double mutant has three additional bases, and consequently the protein has one additional amino acid residue.

A MODEL FOR FRAMESHIFT MUTAGENESIS

A model mechanism for frameshift mutation was suggested by Streisinger and co-workers (3) based on their work with T4 lysozyme mutants. This model (diagrammed in Figure 2) proposes that frameshift mutations are generated at gaps in the DNA (see line *b*). These gaps might be produced during DNA repair, recombination, or replication. Such structures have the capability of opening base pairs, slipping, and re-forming pairs improperly (see line *c*). If ends are mispaired at the time the DNA is resynthesized and repaired, additions or deletions of bases in one strand can occur (see line *d*). In the example presented, the mispairing occurred

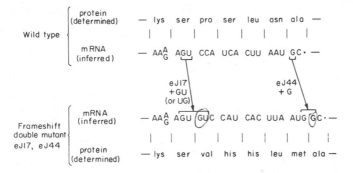

Figure 1   Determination of the nature of frameshift mutations. Redrawn after Streisinger et al (3). The amino acid sequence of a portion of wild-type T4 lysozyme is presented at the top. The amino acid sequence found in a revertant of a lysozyme frameshift mutant is shown at the bottom. The message sequences inferred are presented in the middle two lines. Comparing these message sequences makes it possible to determine the probable nature of the two mutational events. Due to degeneracy of the code, it is not always possible to infer a unique message sequence (i.e. the lysine residue above may be inserted in response to an AAA or an AAG codon).

following backward slippage and involved bases that were already replicated. This results in addition of bases. Slippage in the opposite direction would cause a deficiency of bases in the same strand. The example presented in Figure 2 is a mechanism by which the two-base addition mutation, eJ17 (Figure 1), might have occurred according to this model. The mutation adds a G/C and a T/A pair to the DNA; this results in addition of G and U in the message.

The mistaken pairing required by this model is hard to predict. Frequently many pairing arrangements are possible which account equally well for the same mutation. The model does predict, however, that base additions will be identical to adjacent bases. In Figure 2 it can be seen that the –CA– residues added to the bottom strand (line *d*) are adjacent to identical –CA– residues already present.

Thus the model predicts that frameshift mutations should frequently occur by addition or removal of single bases from monotonous runs and by addition and removal of two bases from regions of alternating doublets (i.e. CACACA in one strand). Both kinds of sequences provide good opportunities for slipping and mispairing. In considering the data that have accumulated since this model was originally presented, we will see impressive examples which fit these predictions.

The role of frameshift mutagens in this model is more difficult to assign. These mutagens are generally planar, heterocyclic compounds, known or presumed to intercalate into the DNA double helix and stack with DNA base pairs (6–8). This intercalation stabilizes the helix, increasing the melting temperature of DNA appreciably and changing the helix dimensions. Streisinger suggested that intercalation of proflavin or similar heterocyclics into mispaired regions of the DNA might increase the half-life of structures arising by slippage of one strand. This would increase the probability that the mispaired configuration would exist at the moment that synthesis and rejoining occurred to permanently assure the addition or removal of bases from one strand. Several problems with this aspect of the model will be discussed later (cf Mechanism of Frameshift Mutagenesis).

PROFLAVIN-INDUCED MUTATIONS OF T4 LYSOZYME

A summary of proflavin-induced mutational changes in T4 lysozyme is presented in Figure 3. With one exception all of the proflavin-induced mutations can easily be accounted for by the Streisinger model. That is, the bases added or deleted are identical to adjacent bases. This is especially clear for the loss of A from a run of As (eJ42) and for the two-base additions, which all seem to be duplications of a preexisting two-base sequence. The five-base addition (eJ13) may also be a duplication of a preexisting sequence. Mutation eJ44 represents duplication of a single G residue. Three additional proflavin mutations (not presented) affect runs of the same base. Mutations JD11 and J335 are additions of A to runs of five A residues. Mutant J320 is a deletion of a U from a run including four or five U residues. (14; Joyce Owen, personal communication). All of the above mutations can be explained by "strand slippage." The exceptional mutation (eJ28), which is more complicated, has four bases added which do not represent a contiguous duplication of preexisting material. However, the sequence GAUG (indicated in the figure) does recur three bases away. Schemes can be devised which explain this addition in terms of the Streisinger model (10). Other frameshift revertants have been analyzed but are not

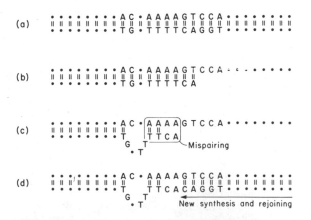

Figure 2   The Streisinger model for frameshift mutation. This figure depicts the sequence of events which might have led to the frameshift mutation eJ17 (see Figure 1), according to the Streisinger model. Line *a* is the original DNA molecule. Line *b* is the same molecule after breakage and removal of one strand. Line *c* depicts the proposed "slipped" pairing. Finally line *d* shows that resynthesis of the removed material and rejoining to the mispaired strand can result in addition of two bases to that strand. Subsequent replication would lead to a daughter chromosome with two added base pairs (C/G and A/T). Transcription of the mutant DNA can yield message with the added GU doublet seen in Figure 1. Data are from Streisinger et al (3).

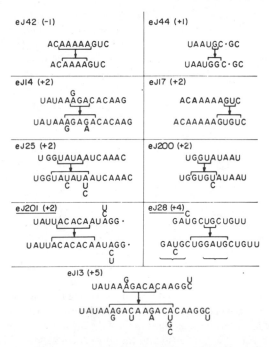

Figure 3 Proflavin-induced mutations of T4 lysozyme. Above are the base changes inferred for a series of lysozyme frameshift mutants of phage T4. All mutations were induced by proflavin, and the changes are presented as they would affect messenger RNA. In each case, changes were inferred from amino acid analysis of revertant (double mutant) proteins. Only one mutational event is present in each panel above. The data are derived from the following sources: eJ42 and eJ44 (9); eJ14 and eJ13 (10); eJ17 (5); eJ25 and eJ200 (11); eJ201 (12); eJ28 (13).

presented due to uncertainties in their base changes or in deciding which of the two events was induced by proflavin (see 15–17).

Two features of the proflavin-induced mutations seem noteworthy. First is the high incidence of multibase changes. Of the 12 proflavin-induced mutations discussed, 8 involve more than single-base additions or deletions. The second feature of the proflavin mutations is that most are base additions. Of the 12 mutations, 10 are additions. At least one other mutation (not presented) is known to be due to base addition. This tendency toward additions is not predicted by the Streisinger model and may reflect an unsuspected complexity of the mutational process.

FRAMESHIFT MUTATIONS IN BACTERIA

The ICR Compounds

Proflavin, which is highly mutagenic for T4, proved to be a very poor mutagen for bacteria. (See, however, section on proflavin mutagenesis of bacteria.) The first studies of frameshift mutations in bacteria employed a series of acridine-like compounds known as ICR compounds (in honor of the Institute for Cancer Research where they were synthesized by Creech and his associates) (18). The ICR compounds are like proflavin in being planar heterocyclic compounds, but in addition they carry a polyamine side chain. The most highly mutagenic ICR compounds have an alkylating side chain and so are probably capable of covalently linking the acridine moiety to the bacterial DNA. In most of the work described here, ICR-191 was used.

Several sorts of evidence suggest that ICR compounds cause frameshift mutations (18). Most mutations caused by ICR are not induced to revert by base analogues, but are induced to revert by ICR. Most known base substitution mutations are *not* induced to revert by ICR (however, see below). ICR-induced mutations show polarity effects, which is expected of frameshifts due to the nonsense codons encountered in the out of phase translation. Martin (19) and Newton (20) have shown that reversion of ICR-induced mutations frequently occurs by mutations at a second site within the mutant gene. When separated, these secondary mutations also show the properties typical for frameshift mutations. Thus, as in the case for T4, two frameshift mutations compensate each other so as to provide a functional gene product.

The most unambiguous demonstration of frameshift mutations is obtained through analysis of protein produced by genes carrying two mutually compensatory frameshift mutations. These studies have been done on bacteria by two groups. Yanofsky and co-workers have used mutants of the tryptophan synthetase *trpA* gene of *Escherichia coli* (21–23); Yourno and his co-workers have studied the histidinol dehydrogenase gene (*hisD*) of *Salmonella typhimurium* (24–29). Figure 4 presents the ICR-induced mutations whose base changes have been inferred from these studies.

These mutations differ strikingly from the proflavin-induced mutations of T4 lysozyme. First, all mutations seem to have affected runs of G/C pairs in the DNA; in every case, several such pairs were already present in the target DNA. Each ICR-induced mutation results from addition or removal of a single base pair from the preexisting run. The fact that five of the six mutations presented are additions is probably not significant since Yanofsky's ICR-induced mutations were obtained through reversion and thus depend on the parent frameshift mutation. Yourno's sample is biased in favor of +1 types. After finding that *hisD3018* was suppressible, Yourno chose other suppressible types for sequencing; only +1 mutations are suppressible. (See section on informational suppression of frameshift mutations.)

It may be argued that the alkylating side chain of ICR-191 accounts for its specificity for G/C rich regions. This is reasonable since alkylating agents generally show a preference for the N-7 position of guanine. However, one of the mutations described (*hisD3068*) was induced by the nonalkylating mutagen ICR-364OH. Data of Oeschger & Hartman (30), discussed below, also suggest a similarity in the specificity of ICR-191 and ICR-364OH.

Mutations induced by ICR-191 and ICR-364OH have been classified as to map position, reversion properties, and polarity effects by Oeschger & Hartman (30) and Newton (20). Results of these studies are in essential agreement. Mutants induced

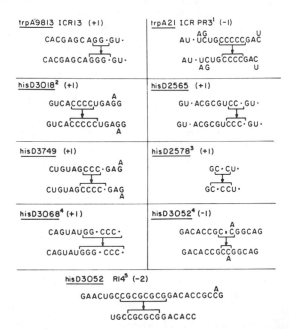

Figure 4   Mutagen-induced frameshifts in bacteria. Above are mutational changes inferred for a series of frameshift mutants of *E. coli* and *S. typhimurium*. Changes are presented as they would appear in the messenger RNA. Each panel presents the change inferred for *one* mutation. The *trp* mutations (top two panels) are reversion events induced in the tryptophan synthetase gene of *E. coli* K12 by ICR-191 (21, 22). All other mutations affect the histidinol dehydrogenase gene of *S. typhimurium*. Except for the reversion event (*hisD3052R14*, bottom) all *his* mutations occurred as forward mutations. Unless otherwise noted, all events were induced by ICR-191.

Data presented above are derived from the following sources: *trpA9813* ICR13 (21); *trpA21* ICR PR3 (22); *hisD3018* (24); *hisD2565* (27); *hisD3068* (26); *hisD3749* (25); *hisD2578* (28); *hisD3052*; and *hisD3052* R14 (29).

Footnotes

[1]This change is one of two mutations: one was induced by ICR-191, the other occurred spontaneously. The event depicted was chosen as the probable ICR-induced mutation because of its similarity to other ICR-induced events.

[2]Reversion of *hisD3018* to a wild-type sequence can also be induced by ICR (24).

[3]Available data do not exclude the possibility that this mutation might represent addition of two C residues (C•C⟶C•CCC).

[4]This mutation was induced by ICR364OH.

[5]This mutation is a reversion event induced by nitrosofluorene.

by ICR-191 and ICR-364OH fall into four classes (I–IV in Table 1). The class IV mutations are induced to revert by both ICR–191 and NG. These mutations are probably all additions of single G/C pairs to runs of G/C pairs in the DNA. This is based on sequencing data of Yourno and co-workers (see Figure 4) and substantiated by the fact that almost all mutations of this type prove to be suppressible by frameshift suppressors which read four-base codons CCC• or GGG• (see section on suppressors). One exceptional class IV mutant which is *not* suppressed (*hisD2578* in Figure 4) is probably also due to addition of a G/C pair to a run of G/C pairs in the DNA (28).

The specificity of ICR-191 for G/C runs, shown by these one-base addition mutations, suggests that the class III mutations (induced to revert only by ICR-191) might involve frameshifts in such runs. Since class III mutations are not suppressible, they probably do not represent additions to G/C runs (see section on suppression). Thus it seems likely that class III contains –1 frameshifts in G/C runs or possibly at nonsuppressible sites near G/C runs.

Class II contains mutants which revert spontaneously but are not induced to revert by ICR or nitrosoguanidine. This class might contain frameshift mutations which occurred in a variety of sequences other than G/C runs and therefore are not revertible by ICR compounds. Some may represent loss of G/C pairs from a run of G/C pairs such that the mutant site no longer has a sufficient G/C run to be revertible by ICR.

The stable mutations of class I are probably more extensive deletions. One such mutation (*hisD3050*) is a deletion of the entire histidine operon (30).

Both of the above studies have revealed "hot spots" or preferred sites for ICR mutagenesis. Of the 31 *lac* mutants studied by Newton, 6 mapped at one single site, 4 at another site, and 3 at a third site. In the *hisC* gene of Salmonella, 17 mutants induced by ICR-364OH were investigated; 11 mutations seemed to affect a single site and 3 additional mutants were very closely linked to that site. Similarly a "hot spot" containing 4 ICR-191-induced mutations was identified in the *hisD* gene. One of the *lac* "hot spots" and both of the *his* "hot spots" included mutants of both the class III and class IV types (see Table 1). That is, some of the mutations at a single site were NG mutable, others were not. This might be accounted for if, for example, both one and two-base deletions were occurring in the same long run of G/C pairs.

In general, the ICR compounds seem to be specifically frameshift mutagens. They are unable to induce reversion of most missense and amber mutations. However, several lines of evidence suggest that ICR-191 may cause base substitution mutations with low frequency at specific sites in the DNA. This was noted by Whitfield et al (31) who found several exceptional missense and nonsense mutations which were induced to revert by ICR compounds. By analyzing revertant proteins, Berger et al (22) demonstrated such an ICR-induced substitution event in the *trpA* gene of *E. coli.* While it is difficult to quarrel with these data of Berger et al, there are reasons to suspect that ICR-induced revertants of other base substitution mutants may in fact be the result of base addition or deletion. Several missense (base substitution) mutations affecting residue 210 of tryptophan synthetase seem to be correctable by addition of a single base pair ahead of the mutant site in the gene (23). The exact nature of these revertant proteins has not been established.

Table 1 Classification of ICR-induced mutations in bacteria[a]

Mutational Class	Reversion induced by			Distribution of his mutations induced by		Distribution of lac mutations induced by ICR-191[b]	Distribution of spontaneous his mutations	Probable nature of mutation
	Spont.	ICR	NG	ICR-191 (alkylating)	ICR-3640H (nonalkylating)			
I	−	−	−	6	5	2	19	multibase deletion
II	+	−	−	7	4	0	11	frameshift
III	+	+	−	8	12	19	9	(i) −1 frameshift in runs of G/C pairs (ii) other sorts of frameshifts
IV	+	+	±	27	9	10	0	+1 frameshift in runs of G/C pairs
V	−	−	+	0	2[c]	0	44	base substitution
Totals				48	32	31	83	

[a] A summary is presented above of mutant types induced by ICR-191 in the histidine operon of *S. typhimurium* (30) and the lactose operon of *E. coli* K12 (20). For comparison, the distribution of spontaneous histidine mutants is presented (33).

[b] The sample of ICR-induced *lac* mutations is biased in that leaky *lac* mutations were not classified (20).

[c] These two mutations may well be spontaneous, since this type is not generally found among ICR-induced mutants and since ICR-3640H is relatively weak mutagen.

Most of the ICR-induced revertants of nonsense mutations in the histidine operon are due to creation of nonsense suppressors. It is possible that such suppressors may arise by base addition or deletion mutations in the suppressor gene (32). Thus, while ICR compounds induce frameshift mutations preferentially, there are good reasons to think that base substitution mutations are induced in particular regions and with low frequency. Care must be exercised in drawing conclusions from data on induced reversion of base substitution mutations, since some reversion events may prove to be base additions or deletions.

Alkylating Agents as Frameshift Mutagens

The observation that +1 frameshifts in runs of G/C pairs are induced to revert by DES and NG suggests that such mutagens can induce −1 frameshift mutations. In one instance, a NG-induced reversion occurred by deletion of a G/C pair from such a run (24). Almost no spontaneous frameshifts are induced to revert by NG (30, 33). This suggests that both spontaneous and NG-induced frameshift mutations may frequently involve one-base deletions in G/C runs. One example of an NG-induced +1 frameshift base has been reported (29). These events may be a consequence of depurination which has been shown to occur following alkylation of guanine residues (reviewed in 34). It should be pointed out that both NG and DES are rather weak agents for inducing reversion of frameshift mutations. The revertants are detected under strong selective conditions. It is not surprising that NG-induced frameshifts have not been encountered in forward mutagenesis experiments.

Spontaneous Frameshift Mutations

With very few exceptions, the spontaneous frameshift events which have been analyzed in detail occurred as highly selected events causing reversion of proflavin- or ICR-induced frameshift mutants. This strong selection may give a very biased view of what sort of frameshift events usually occur spontaneously. Reversion events must occur rather close to the site of the original mutation; thus one detects only mutations in particular small regions of the DNA, regions in which spontaneous frameshift mutations may rarely occur. Thus, many of the frameshift mutations carried by highly selected revertants are likely to be atypical of frameshift mutations in general. This may explain why some of the spontaneous frameshift reversion events in T4, bacteria, and yeast are hard to rationalize in terms of the Streisinger model, while forward mutational events induced by mutagens generally fit with the predictions of that model.

Mutations which have extremely high reversion frequencies may well represent exceptional frameshift mutations which create or reside in a region extremely susceptible to spontaneous mutations. Therefore, these reversion events are more likely to represent typical spontaneous frameshift events. Several of these have been investigated. In the T4 lysozyme gene, a set of 130 proflavin-induced mutations were checked for spontaneous reversion frequency; 24% of the mutants showed extremely high reversion frequencies. Twenty-two of these "unstable" mutants have been mapped and found to reside at five sites in the lysozyme gene. The two mutant sites showing highest spontaneous reversion frequency (4×10^{-5}) both resulted from addition of an A/T pair to a run of five A/T pairs (runs of A in the message).

Another site (reversion frequency 5×10^{-6}) results from removal of a U residue from five Us or possibly from UAUUU (as seen in the message). (14; Joyce Owen, personal communication).

A set of spontaneous lysozyme mutants has been selected directly (not by reversion of existing frameshifts). Sixteen percent of these spontaneous mutations map at the C-terminal end of the lysozyme gene in a region known to carry a run of five A/T base pairs. These mutations are not induced to revert by proflavin or 2-aminopurine, but are probably frameshifts since an oversized lysozyme is produced. The failure to be induced to revert by proflavin might be explained if these mutations were deletions of a substantial portion of the base pair run (Joyce Owen, personal communication).

A positive selection has been devised for isolation of *his* mutations with a polarity effect (32). This positive selection method gives a spectrum of mutants comparable to that recovered following standard penicillin selection techniques but lacking missense mutations which constitute about 34% of the mutants isolated by standard techniques. Roughly 50% of the mutants recovered by the positive selection method are frameshift mutations. Thus, frameshifts represent an appreciable fraction of detectable spontaneous mutations. This is in substantial agreement with data for *r*II region of phage T4 where frameshifts are a large proportion of detected *r*II mutants [review by Drake (4)].

Very few of the spontaneous frameshifts in bacteria are of the suppressible, NG-revertible type induced by ICR-191 (i.e. addition of a G/C pair to a run of G/C pairs). This suggests that if spontaneous mutations occur in G/C runs, then base removal must be the rule. Alternatively, the spontaneous mutations may affect a wider variety of sequences.

Several "hot spots" for spontaneous frameshift mutations have been observed in the histidine operon. Of 68 spontaneous frameshift mutants mapped by Fink, Klopotowski & Ames (32), 26 affected a single site in the *hisG* gene: four mutations were found at a single site in the *hisC* gene. Another "hot spot" for spontaneous forward mutagenesis has been identified and characterized by Yourno, Ino & Kohno (35). At least five independent occurrences of mutations at this site have been observed. In addition, one ICR-induced mutation maps at the same site. One of the spontaneous mutations (*hisD2550*) has been studied by sequence analysis of revertant proteins. A likely interpretation of these results is that five bases have been deleted from a sequence containing a nearly perfect repeat of that sequence. That is, the sequence GUCC•GU(C)(UC) may have been converted to GU(CUC). This interpretation would suggest that repeated longer sequences, as well as monotonous runs of base pairs, are prone to spontaneous mispairing, as suggested by the Streisinger model. The ICR-induced mutation at this site seems to be a C addition to the CC• region of the wild-type sequence above. Thus, in this region prone to five base deletions, ICR-191 still seems to show its proclivity for single base additions in G/C runs.

Proflavin Mutagenesis of Bacteria

The strong mutagenicity of proflavin for T4 prompted attempts to use this mutagen in bacteria. The early attempts were generally unsuccessful although several papers report detection of a weak mutagenic effect in particular strains (cf 4). The suggestion was made that recombination might be required for proflavin mutagenesis. Since T4 shows much more recombination than *E. coli,* this might account for the difference in mutagenic effect of proflavin in these two systems (7, 8).

In pursuing this possibility, Sesnowitz-Horn & Adelberg (36–38) demonstrated that proflavin is mutagenic for *E. coli* if cells are exposed during the course of a conjugational cross. The results were puzzling, however, since recombination did not need to occur within the region being mutagenized. Furthermore, many (23%) of the mutants isolated were apparently base substitution types, not frameshift mutations.

Recent work on proflavin mutagenesis in *Salmonella* indicates that proflavin mutagenizes bacteria in an indirect way (39). Proflavin is only mutagenic if three conditions are met. First, cells must be made permeable. This can be done by EDTA treatment, by use of mutants which are more permeable, or by P22 infection. There is well-documented precedent for phage-induced proflavin permeability in the case of T2 phage (40). High concentrations of proflavin can also overwhelm the permeability barrier, but cell killing increases under these conditions (39, 41). The second requirement for mutagenesis is a functional *rec* system. A similar requirement is seen for uv mutagenesis (42). This suggests that proflavin mutagenizes by causing DNA damage repairable by the error-prone *rec* repair system. This interpretation is supported by the types of mutations induced by proflavin; 40% of the proflavin-induced mutants isolated are base substitution types, 40% seem to be some sort of frameshift mutations, and 20% are stable mutants, many of which are demonstrable multisite deletions. Mutants induced by uv show roughly the same frequency distribution (33, 43). This would be expected if error-prone repair of DNA damage is causing the mutations in both cases. The third requirement for proflavin mutagenesis is that genes to be mutagenized must be derepressed or transcribed at a high frequency. This may give a clue as to how proflavin might cause repairable DNA damage. Perhaps transcription of DNA containing intercalated proflavin leads to nuclease-sensitive structures. Proflavin has been shown to block transcription in some systems (44, 45).

Informational Suppressors of Frameshift Mutations

The first report of an external suppressor of a frameshift mutation was that by Ryasaty & Atkins (46). They found a frameshift mutation in the tryptophan operon of *Salmonella* which was suppressed weakly by an unlinked UGA suppressor. Subsequent work (47; J. Atkins and J. Ryce, personal communication) has revealed that this suppressor is identical to the recessive UGA suppressor, *supK* (48). Apparently these suppressor mutations lead to miscoding which can correct both UGA and frameshift mutations. The mode of action of this suppressor is not known, but its recessive nature and unpublished data of R. Reeves (personal communication) suggest that the suppressor causes a defect in tRNA modification.

Frameshift-specific suppressors were first reported by Yourno and co-workers (49, 50). In the course of selecting revertants of a (+1) frameshift mutant (see Figure 4; *hisD3018*), they observed revertants which maintained the original frameshift mutation and carried, in addition, an unlinked suppressor mutation. Subsequent work by Yourno and his co-workers and by others has revealed the probable mode of action of these suppressors.

Six distinct suppressor loci have been identified (51, 52). They fall into two general groups. Suppressors *sufA, B,* and *C* all show cross-suppression of one series of *his* frameshift mutations. Suppressors *sufD, E,* and *F* show cross suppression of a second series of *his* frameshift mutations. Thus, there are two types of suppressible frameshift mutations, each type having its own suppressors. No cross suppression of the two groups has been observed (51).

The nature of the site of action of *sufA, B,* and *C* has been determined for three different suppressible *his* mutants: *hisD3018* (CCCU), *hisD2565* (CCC•) and *hisD3749* (CCC•) (24, 25, 27). Thus, all of these suppressors seem to read a four-base codon, CCC•. One *sufB* suppressor has been shown to insert proline in response to this four-base code word and thus restore the proper reading frame (50). Since the wild-type message had a normal proline codon (CCC) at this position, the suppressor permits synthesis of a wild-type protein from the mutant message.

These findings made it seem likely that the suppressors *sufA, B,* and *C* might affect proline tRNA. This has been confirmed for *sufA* and *B* (53). These suppressor mutations cause an alteration in different proline tRNAs. The exact nature of these changes is not known.

The site of action of suppressors *sufD, E,* and *F* has been determined for only one mutant. This mutant has the sequence GGG• at the mutant site (26).

Two of these suppressors, *sufD* and *F,* affect glycine tRNA (53). One of these (*sufD*) was analyzed in detail by Riddle & Carbon (54). They find that the suppressor mutation causes addition of a C residue to the anticodon region of a glycine tRNA which normally has CCC as its anticodon; this tRNA normally reads the glycine codon GGG. Apparently the mutant tRNA can use its larger, four-base anticodon to recognize a four-base code word (GGG•) and thereby restore translation to the proper reading frame. As in the case of the proline-inserting suppressors, these suppressors would be expected to permit formation of a wild-type protein. This has not been demonstrated for *sufD, E,* or *F.*

All characterized frameshift-specific suppressors seem to act at repeats of C or G in the message. The apparent specificity may be due to the sorts of mutations used in selecting suppressors. Most of the available frameshift mutations in bacteria were induced by ICR-191. Since this mutagen is specific for runs of G/C pairs (see above), a highly biased sample of frameshift mutations has been used in seeking suppressors. Perhaps as different types of frameshift mutations are checked, a wider variety of frameshift suppressor types will be encountered. Already three new suppressors have been identified in revertants of proflavin-induced frameshift mutations (T. Kohno, unpublished results). Atkins & Ryce have found suppressors of a second site revertant of their original frameshift mutation. That is, they have identified two mutually compensatory frameshift mutations and have found that both mutations are suppressible by external suppressors. Since one of these mutations is probably a −1 type, suppressors of −1 frameshifts may be possible (47).

The exact mode of action of frameshift suppressors is not clear, even though one suppressor tRNA has been shown to result from addition of a base to its anticodon

(54). Several alternatives still exist. (*a*) The suppressor tRNA may read a specific four-base codon; in this case suppressors may yet be found which recognize codons that are not monotonous runs (CCC• or GGG•). (*b*) The suppressor tRNA may recognize its normal three-base codon but cause ribosomes to advance four bases (instead of three) along the message. (*c*) The suppressor tRNA may recognize a three-base codon but have difficulty determining where to select those three from a monotonous run of four bases (CCC• or GGG•) where two positions permit good hydrogen bonding possibilities.

Streptomycin-induced miscoding has frequently been used in classifying mutant types (cf 31). Generally it has been assumed that this miscoding can cause phenotypic suppression of missense and nonsense mutants; one criterion for identifying frameshift mutations is their failure to show this streptomycin-induced phenotypic suppression. Most of these tests have been done using growth response on solid medium as the criterion for streptomycin-induced suppression. The results are internally consistent and seem to justify the underlying assumption. However, more sensitive tests, using β-galactosidase assays, reveal that frameshifts *are* subject to some streptomycin-induced suppression or ribosomal ambiguity (55). Apparently the magnitude of this suppression is very low for frameshift mutations. The most efficiently suppressed frameshift gained only 0.05% of wild-type activity. Presumably this level of correction is not detected when growth is used as a criterion for suppression.

FRAMESHIFT MUTATIONS IN YEAST

Frameshift mutations affecting iso-1-cytochrome *c* have been intensively investigated by Stewart & Sherman and their co-workers (56–58). Their work has centered on the N-terminal 10% of this cytochrome (the first 44 translated bases of the cytochrome message). This region of the protein seems to tolerate a variety of amino acid sequences and still permit a functional product. Therefore few restrictions are placed on the sort of frameshift revertants which are detected. More frameshift revertant proteins have been characterized in this region of yeast cytochrome *c* than in any phage or bacterial systems.

The first frameshift mutation characterized in this system is *cyc1-183* (an addition of A to a run of four As as seen in the message). This mutation is probably identical to *cyc1-134* which is at the same site. The nature of these mutations is diagrammed in Figure 5. Many revertants have been selected and the amino acid sequence of the affected part of iso-1-cytochrome *c* determined. Both single-base and multiple-base mutations were identified which compensate for the original (+1) mutation and restore the reading frame. Single-base deletions are summarized in Figure 5. These were induced by a variety of agents. It is apparent that base deletions have been detected at most of the possible sites. However, the frameshift types which have been observed repeatedly are at sites containing several repeated bases. Thus, even though the target for frameshift mutation is small in selecting these revertants, mutations seem to occur preferentially at sites which show properties predicted by the Streisinger model.

In order to check the distribution of +1 frameshift mutations, a starting –1 frameshift mutation *cyc1-239* was used. (This mutation was derived from a revertant of +1 mutant *cyc1-183*). A series of revertants of *cyc1-239* were selected and their proteins analyzed. The base changes deduced are presented in Figure 6. In the case of these +1 reversion events, base addition can occur at a variety of sites. However, the more common events agree with the predictions of the Streisinger model in that they represent base additions to a sequence of repeated bases (as seen in the message).

Among the revertants of +1 mutations *cyc1-183* and *134* are several deletions and duplications which are consistent with the Streisinger model. Three reversion events were found to involve tandem duplications of preexisting sequences. One duplicated sequence was only two bases long; however, the other two duplications involved 11 bases and 14 bases. Two deletion mutations were found which remove material separating two regions of very similar or identical sequence. These similar regions would provide the base pairing required if these events occurred by the Streisinger model.

One sort of mutational event has been observed in these yeast mutants which has not been reported for phage or bacteria. These are frameshift mutations that seem to result from *removal* of a block of base pairs and *replacement* of that block by a different block of base pairs (the number of bases added is one less than the number removed). The added base sequence is not a duplication of adjacent sequences. No model has been suggested to account for these events, yet they are relatively com-

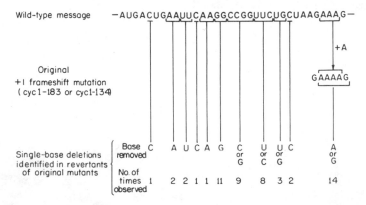

Figure 5 Single-base deletions in iso-1-cytochrome *c*. A series of revertants of the +1 frameshift mutations (*cyc1-183* or *cyc1-134*) were selected. The revertant proteins were analyzed and used to infer the nature of the genetic alteration that restored proper reading frame. The nature of mutation *cyc1-183* is presented in the middle (*right*) of the figure. (Mutation *cyc1-134* is similar but may have GAAA or AAAG at the mutant site.) Various single-base deletion events which restore proper reading frame are presented toward the bottom. Vertical lines and brackets indicate the position at which base deletion has occurred. These mutations were induced by a variety of means and are pooled here. All data are those of Stewart & Sherman (57).

mon. Sixteen have been identified, and this number represents 15% of the characterized frameshift reversion events.

Frameshift mutagen ICR-170 has been used to induce 30 iso-1-cytochrome *c* mutants. Unfortunately, none of these mutations map in the N-terminal region that is amenable to sequencing. (It may be significant that this region of the gene does not include a run of G/C pairs which is the preferred site of action of this mutagen in bacteria). Twenty-eight of the 30 ICR-170-induced mutants produce no detectable iso-1-cytochrome *c*. This would be expected if their mutations are frameshifts. Another feature of the ICR-induced mutants is their distribution into "hot spots." Of the 30 ICR-induced mutants, 9 were found at a single site, 3 at another, and 2 at a third site. This tendency is also seen for proflavin-induced mutations of T4 and ICR-191-induced mutations in bacteria. Three of the ICR-induced cytochrome mutants map at a site occupied by 15 other mutations induced by a variety of mutagens. Only the three ICR-induced mutations at this site prevent production of a detectable gene product (58).

These findings are consistent with induction of frameshift mutations by ICR-170. However, one of the 30 ICR-induced cytochrome mutations is suppressible by ochre-specific (UAA) suppressors. Two additional mutations are apparently missense since they lead to production of a detectable but abnormal protein. These mutations are probably base substitutions, but deletion or addition of $3n$ bases could also lead to strains with these properties.

In the *his4* gene cluster of yeast, a series of ICR-induced mutants has been isolated by G. Fink and his associates. Roughly 70% of these mutations show strong

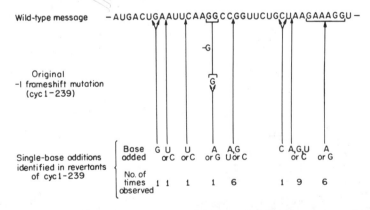

Figure 6 Single-base additions in iso-1-cytochrome *c*. A series of revertants of −1 frameshift mutant (*cyc1-239*) were selected. The revertant proteins were analyzed and used to infer the nature of the genetic alteration that restored the proper reading frame. The original mutational (*cyc1-239*) change is diagrammed in the center. The various reversion events that serve to restore phase are diagrammed toward the bottom. Vertical arrows indicate the position in the messenger sequence where the indicated bases are added. As seen at the bottom, several reversion events have been detected more than once. The mutations included were induced by a variety of means and are pooled for presentation. All data presented are those of Stewart & Sherman (57).

polarity effects but are not suppressible by nonsense suppressors and thus are probably not nonsense mutations. The polarity effects are characteristic for nonsense mutations but would be expected for frameshift mutations due to the nonsense codons encountered by out-of-phase translation. These ICR-induced mutations are induced to revert by ICR and by uv (G. Fink, personal communication).

ICR compounds have been found mutagenic in other fungal systems as well as yeast (59–62). It is likely that many ICR-induced mutations will prove to be frameshifts; however, since *some* ICR mutations seem to be base substitutions, revertibility of a mutation by an ICR compound should not be accepted as sufficient evidence for classification of that mutation as a frameshift.

A FRAMESHIFT MUTATION AFFECTING HUMAN HEMOGLOBIN

A recently analyzed human α-chain hemoglobin variant, Hemoglobin Wayne (HbW1) is probably the only clearly demonstrated example of a frameshift mutation in humans (63). This variant is particularly interesting when its sequence is compared to that of another variant Hemoglobin Constant Spring (HbCS) (64). Using these two sequences it is possible to infer the base sequences that may serve as termination signals in human cells. The amino acid sequences and inferred base sequences are seen in Figure 7. Normal hemoglobin α chain (HbA) is presented in the center of Figure 7. Only the four amino acids at the carboxyl end of the protein are depicted.

Hemoglobin Wayne is longer than normal hemoglobin and seems to be caused by a –1 frameshift mutation; this results in extended miscoding and a failure to read the termination signal in the proper reading frame. (See bottom of Figure 7.) Based on the amino acid sequence, one can infer that one residue, possibly A, is missing from a sequence • A A A/G in the message. Thus, the HbW mutation may well result from removal of an A/T pair from a run of as many as four A/T pairs. However, other possibilities, such as removal of G from the lysine codon or the undetermined base from the preceding serine codon (UC•), are not excluded.

Hemoglobin Constant Spring has an abnormally long α chain due to addition of 31 extra residues beyond the normal sequence. (Seven of these extra amino acids are shown at the top of Figure 7.) It seems likely that a base substitution mutation has affected the termination codon and converted it to a glutamine codon (CAA). The out-of-frame reading seen in Hemoglobin Wayne supports this interpretation and requires the last two bases of the termination codon to be A. Thus, the single codon which terminates the normal HbA chain is probably UAA (ochre). The other possibilities, GAA and AAA, are not excluded by these data, but seem unlikely, since these code words have been assigned to particular amino acids (Glu and Lys).

The out-of-phase reading generated by the Wayne mutation proceeds to add only eight improper residues. The codon following the last amino acid of Hemoglobin Wayne can be inferred by looking at amino acids found in the corresponding position of Hemoglobin Constant Spring. Thus, one can infer that the codon which may terminate Hemoglobin Wayne is U•G. It seems likely that the codon UAG (amber) may serve as the termination signal. The above conclusions are based on the assumption that neither normal HbA nor the Wayne Hemoglobin has been shortened by proteolysis at the C-terminal end.

Several additional human hemoglobin variants may have arisen by a mechanism similar to the Streisinger model. These variants lack one or more amino acid residues and must result from deletion of a multiple of three bases. For this reason, they are not frameshift mutations. In each case the missing sequence is bracketed by a repeated sequence of bases. A number of these cases are reviewed by Fitch (65), who suggests that this mutation might have arisen by unequal crossing-over. These mutations can also be explained according to the Streisinger model for frameshift mutagenesis. One variant in particular, Hemoglobin Freiburg (66), is more easily accounted for by the model for frameshift mutagenesis than by recombination. *Adjacent* to the site of the deleted bases is the sequence •GG•GG•G. This sequence could provide the needed pairing if one strand slipped three bases relative to the other.

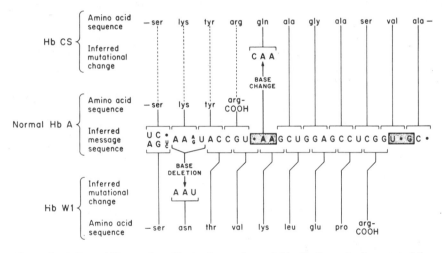

Figure 7 A frameshift mutation affecting human hemoglobin. In the center is presented the last four residues of the normal hemoglobin α chain (HbA). Two variant hemoglobins, HbCS and HbW1, are presented at the top and bottom. Both of these variant hemoglobins are *longer* than the normal HbA. Variant HbCS has 31 residues in *addition* to the normal sequence (only seven of those residues are presented). Variant hemoglobin HbW1 differs from normal by three changed amino acids and five added residues. From these amino acid sequences one can infer the normal message sequence and the nature of the mutational events which lead to HbCS and HbW1. Possible termination signals are •AA (probably UAA) for normal HbA and U•G (probably UAG) for HbW1. These codons are indicated in shaded rectangles. Hemoglobin Wayne is found in two forms. HbW1 (presented above) and HbW2 which has Asp in place of Asn. It is presumed that HbW2 is formed from HbW1 by enzymatic deamidation. Data on HbCS are from Clegg et al (64); data on HbW are from Seid-Akhavan et al (63).

SCREENING SYSTEMS FOR DETECTION OF MUTAGENS

A variety of systems have been used in detection of mutagens. We will concentrate here on only one system, since it has been used predominantly for detection of frameshift mutagens and because evidence for the nature of the mutations involved is well established.

Over the past several years, Ames and his co-workers have developed a sensitive bacterial test for use in detecting mutagenic compounds (67–70). Their system detects both base substitution mutagens and frameshift mutagens but has been used most extensively for screening intercalating compounds which are most likely to be frameshift mutagens (see section on frameshift mutagens and carcinogenesis).

The principle of the test is to assemble a series of bacterial strains carrying well-characterized mutations whose reversion can be observed by a positive selection technique. Ames uses histidine auxotrophs of *Salmonella typhimurium*. Strains have been selected whose *his* mutations are extremely sensitive to induced reversion by particular classes of known mutagens. Many of these mutations have been characterized by Yourno and his co-workers so the nature of the mutant site is firmly established (24, 25, 29). Potential mutagens are placed on a series of petri dishes, each seeded with a lawn of a different *his⁻* tester strain. By observing which of the tester strains are induced to revert (to His+) one can determine which compounds are mutagenic and the general nature of the mutations induced. These tester strains have been made more sensitive to mutagenesis by adding to each strain a mutation (*rfb*) which increases permeability and another mutation (*uvr*) which destroys the excision repair systems (71).

Many compounds are mutagenic only after modification (activation) by enzyme systems present in mammalian tissues (reviewed in 72). The bacterial test system can detect mutagenicity of these compounds if a rat liver microsomal preparation is added to the test plate. Enzymes present in this preparation can generate mutagens which then affect the bacterial cells (73, 74).

A similar test system has been used by Hartman and his collaborators to screen antischistosomal drugs for mutagenicity (75, 76). These studies revealed the frameshift-inducing mutagenicity of hycanthone, a drug that has been used extensively in treatment of schistosomiasis. This test system also has permitted the screening of related compounds with activity against schistosomes in search of drug structures active against schistosomes but not mutagenic (77).

It is likely that this sort of bacterial test system will not detect all mutagens active in higher organisms; conversely, the system may reveal some mutagens which are uniquely mutagenic for bacterial systems. Nevertheless, since all these systems use DNA, they must be subject to common problems in faithful replication and repair of their genetic material. Therefore it is likely that a compound mutagenic for one system will also prove mutagenic for others.

An additional problem inherent in this system is that it will miss mutagens which act at DNA sites not represented in the series of tester mutations used. As a wider variety of mutational types are added to the series of tester strains, this objection will weaken. However, in spite of some possible objections, the bacterial system provides a fast, easy, low-cost method of screening many compounds. Thus it is a very powerful method for detecting potentially dangerous compounds.

This test system not only provides a practical way of screening drugs and food additives, but it has also revealed two points which seem to be of basic biological interest. The first point is the strong correlation observed between carcinogenicity and ability to act as frameshift mutagens (see next section). The second observation is the apparent sequence specificity of frameshift mutagens. The ICR compounds show a strong preference for runs of G/C pairs; nitrosoflorene and nitroquinoline-N-oxide show high mutagenicity at a site that contains alternating G/C and C/G pairs. This will be discussed later (see section on mechanisms of frameshift mutagenesis).

FRAMESHIFT MUTAGENS AND CARCINOGENESIS

The possibility that chemical carcinogens might act by inducing mutations has been considered at various times, but has been discarded by most cancer workers. The primary evidence against a mutational basis of carcinogenesis was that no correlation was previously seen between mutagenicity and carcinogenicity when lists of compounds were tested for these properties. Another argument against a mutational theory was the high frequency of transformation observed when cultured cells were treated with carcinogens; the transformation frequencies were thought to be much too high to be accounted for by the mutation rates estimated for animal cells.

Several new findings suggest that the mutational theory of chemical carcinogenesis should be reconsidered. The new lines of evidence center around three points:
1. Many carcinogens are metabolized by cellular enzyme systems to active forms.
2. Many carcinogens have proven to be frameshift mutagens.
3. Frameshift mutagens have effects on DNA that are highly nonrandom with respect to DNA's base sequence and physiological state.

Some chemical carcinogens have been shown to be metabolized by microsomal oxidases to yield reactive epoxides. These epoxides can bind covalently to DNA, RNA, and protein, and are more active than the parent compound in transforming cells in tissue culture (reviewed in 72 and 78). These epoxides are strongly mutagenic, both in mammalian cells and in bacteria where they are found to act as frameshift mutagens (70, 71, 73). The ICR compounds that act as frameshift mutagens in bacteria are of the order of 100–1000 times more mutagenic if they carry an alkylating side chain and can be covalently linked to DNA. Similarly the reactive epoxides of carcinogens can also form covalent bonds to DNA and become more mutagenic. If many carcinogenic compounds must be metabolized to an active form, then any test for mutagenicity of that carcinogen must include the activated form of the carcinogen. Earlier tests did not take this into consideration. Now that tests can be set up which use the reactive derivatives of the carcinogen, the correlation between carcinogenicity and mutagenicity is greatly strengthened.

The observation that many carcinogens are active as frameshift mutagens dictates that the special features of frameshift mutations should be kept in mind when evaluating the mutagenicity of these compounds. The frameshift mutagens studied in bacteria and phage show strong preference for particular sites in DNA; these sites are probably either long monotonous sequences or tandemly repeated base sequences. This means that particular genes might be strongly mutagenized by a particular frameshift mutagen while other genes would be insensitive. This effect is seen graphically in the case of nitrosoflorene, which is highly active in inducing reversion of the *Salmonella* mutation, *hisD3052*. This is an easily detected mutagenic effect on a very short sequence of base pairs. In spite of its high mutagenicity at this site, nitrosofluorene cannot be shown to induce forward mutations in a situation where the target is at least one entire gene [azetidine carboxylic acid resistance (70)]. Thus, the target sequences affected by this mutagen may not be represented in some genes. A related problem arises in that different, frameshift mutagens seem to show specificity for different sorts of sequences in DNA (reviewed in earlier sections). Thus it is conceivable that a highly mutagenic compound could escape detection in any particular experimental situation.

Detection of the mutagenic activity of proflavin in bacteria depends on derepression or active transcription of the genes being mutagenized. If this effect proves to be general for other frameshift mutagens, it may further increase the specificity of these compounds for particular regions of the genome.

An additional example of nonrandom effects of intercalating agents is seen in the case of compounds which induce aberrations of mitochondrial DNA in yeast. Growth of yeast in ethidium bromide causes gross rearrangements and eventually loss of mitochondrial DNA with extremely high frequency (79). These effects may be due to inhibition of replication or to particular structural features of these DNA molecules. In any event, here is a mutagenic event, mediated by an intercalating agent, that occurs at a frequency approaching 100% of the treated cells. It is clearly a nonrandom effect on a particular type of DNA.

Earlier tests of mutagenicity of chemical carcinogens thus failed to take into account that activation may be required for biological activity. These earlier tests also assumed, incorrectly, that mutagenesis is random. That is, they assumed that the presence or absence of mutagenicity, measured in one test system, applied equally to the entire genome. These problems make it more difficult to demonstrate meaningful correlations between mutagenicity and carcinogenicity. These problems also provide food for speculation. The effects of frameshift mutagens on a cell may well be strongly influenced by the physiological state of that cell. The metabolism of mutagens to active forms could vary with physiological state. The depression of particular genes might well influence which genes are most subject to the action of those mutagens. Thus the effect of mutagenic carcinogens on cells might be strongly dependent on cell type, preconditions of cell growth, and the presence or absence of viral infection. It seems reasonable that, in the light of recent findings, mutagenesis should be reconsidered as a mode of action of chemical carcinogens.

THE MECHANISM OF FRAMESHIFT MUTAGENESIS

The Streisinger model for frameshift mutagenesis (outlined in a previous section), accounts for many aspects of how these mutations occur. It suggests the primary event is single-strand slippage at a point in the DNA when the linear continuity of one strand has been interrupted. The model is strongly supported by the base sequences at the site where mutations occur. The model does not satisfactorily explain the different effects of frameshift mutagens in T4 and bacteria. It also does not account for the role of frameshift mutagens in causing these mutations. These aspects of the mutagenic mechanism remain unclear.

When the difference was seen between the action of proflavin on T4 and bacteria, it was suggested that the higher recombination frequencies of T4 might explain this (7, 8). Recombination would be expected to involve intermediate DNA structures such as those required for the Streisinger model. This suggestion has not been supported by subsequent data. Drake (80) did not find the expected correlation between occurrence of frameshift mutations and recombination for outside markers. Sesnowitz-Horn & Adelberg (36–38) did observe proflavin mutagenesis during conjugational crosses; however, recombination was not required at the site where mutations arose. Evidence that recombination is not required for frameshift mutagenesis is the observation of Newton and co-workers that both ICR-191 and 9-aminoacridine induce frameshift mutations at normal frequencies in $recA^-$ and $recB^-$ cells of *E. coli* (81). This suggests that recombination is not essential to the action of these mutagens. Although proflavin was found nonmutagenic in rec^- strains of *Salmonella,* it seems likely that proflavin is not primarily a frameshift mutagen in that system (see earlier section).

Recombination need not be invoked to provide single-strand interruptions. Replication of DNA involves synthesis and later joining of short fragments on an uninterrupted template. These "Okazaki" fragments should provide an ample supply of the sort of structures needed for the Streisinger model. Several lines of evidence are consistent with use of these fragments for frameshift mutagenesis. Sarabhai & Lamfrom (82) observed an enhancement of proflavin mutagenesis in ligase-deficient T4 phage. Failure to rejoin fragments promptly might allow time for strand slippage or for repeated cycles of synthesis and degradation before fragments were finally joined. Newton and co-workers (81) found that both ICR-191 (which has an alkylating side chain) and 9-aminoacridine (which is nonalkylating) seem to mutagenize at the replication point of the chromosome. Thus it seems unlikely that these frameshift mutagens act on recombining DNA. In all probability the high recombination frequency observed in T4 does not account for the action of proflavin in this system. More likely, the mutability of T4 is due to T4's mode of replication or the nature of its replication enzymes. Another possible explanation may lie in the vastly different base compositions of T4 and *E. coli* DNA. T4 DNA has a G/C content of only 34%; *E. coli* has 50% G/C pairs in its DNA. It is conceivable that the abundance of A/T pairs in phage T4 permits more unpairing and slipping at the ends of interrupted strands of DNA. Alternatively the high A/T content may favor the sort of proflavin–DNA interaction which can lead to mutations.

A fully satisfactory model for the occurrence of frameshift mutations must account for the action of the compounds that act as frameshift mutagens. Common to these mutagens is their heterocyclic ring structure and their ability to bind to DNA by intercalating into the base pair stack. Considerable evidence has accumulated for intercalation and for the resultant stabilization of the double helix to thermal denaturation (6–8). It has been suggested that this stabilizing influence is the basis of mutagenic activity; according to the Streisinger model, intercalated acridines (or similar compounds) would stabilize the imperfect pairing that results when one strand "slips" relative to its complementary strand (3). This stabilization might maintain the mispaired configuration long enough to permit the resynthesis and ligase action needed to seal the strand interruption. This would permanently establish addition or deletion of one or more bases. While this model is attractive, two rather serious problems exist which do not yet fit well with the model.

The first problem is the poor correlation between intercalation ability and mutagenicity. The model, as outlined above, would predict that any intercalating compound should stabilize the helix and promote mutagenesis. However, many compounds intercalate strongly and have not been found mutagenic [reviewed and discussed by Drake (4)]. This suggests that intercalation per se might not be sufficient for mutagenesis. If intercalation is required at all for mutagenesis, perhaps only particular types of interactions between DNA bases and intercalated compounds can lead to mutagenic events.

The second problem is the apparent tendency for particular dyes to cause mutations in particular base sequences. The ICR compounds and hycanthone seem to prefer monotonous runs of G/C base pairs (see earlier sections). In contrast, nitrosoflorene and many of the carcinogens tested by Ames and co-workers show a preference of the alternating sequence of G/C and C/G pairs present at the site of mutant *hisD3052*. Even closely related mutagens such as ICR-191 (with alkylating side chain) and ICR-364OH (nonalkylating) show slightly different preferred sites of mutation in the *his* operon, although both mutagens seem to act at runs of G/C pairs. Hopefully an understanding of interactions between intercalating dyes and DNA may reveal a basis for this specificity.

At present detailed information is not available on the variety of possible interactions between intercalating compounds and particular base sequences in DNA. Several general classes of sites have been identified. Binding seems to occur to the outside of the helix in addition to the stronger binding by intercalation (83, 84). At the level of intercalation, several sorts of interactions have been seen (85–89). Both proflavin and quinacrine have been found to interact with poly dA·poly dT to yield enhanced fluorescence. When the same dyes interact with poly dG·poly dC, fluorescence is strongly quenched. Enhanced fluorescence seems to depend on the A:T base pair; quenching is caused by the G residues alone. Thus the intercalation of proflavin and quinacrine can be shown to have different properties depending on the base composition of the DNA. A different situation is seen with acridine orange; this dye shows enhanced fluorescence on binding to DNA, regardless of the base composition of DNA used. These studies show that the nature of the interaction between DNA and an intercalating dye depends both on the base composition of the DNA and on the particular dye. None of these results suggests which sort of interaction is required for mutagenesis, or indeed if intercalation is required at all. However, it seems likely that intercalation will be involved, and these results suggest that understanding the nature and consequences of different modes of binding may be essential to an explanation of the activity and specificity of these mutagens.

A novel model has recently been suggested by Schreiber & Daune to account for the different mutagenicity of various intercalating dyes (94). The model is based on a possible correlation between dyes mutagenic for phage T4 and those whose fluorescence is quenched by binding to guanine residues. The dye, acridine orange, which shows no quenching by binding to G/C-rich DNA, is a very poor mutagen for T4. It is suggested that the interaction that results in quenching might be the interaction critical to mutagenesis. In pursuing this possibility, the authors suggest a set of rules governing induction of base addition and deletion. According to these rules, all frameshift mutations are initiated at G residues to which a dye molecule is bound. The amino acid sequences seen for revertants of frameshift mutations by Streisinger and others can be reinterpreted in the light of these rules. When this is done, some of the base changes inferred are different from those presented in Figure 3. This model is attractive in that it is consistent with the frequent occurrence of bacterial frameshift mutations in G/C-rich sequences. The association of G/C pairs with frameshifts in T4 might be obscured by the high A/T content of T4 DNA. Further work should reveal how well this interesting model can account for the variety of observations made on frameshift mutations in various organisms.

A persistent possibility should be kept in mind in considering the action of frameshift mutagens. It is conceivable that intercalation is not involved in mutagenesis. Dyes may act as analogs of base pairs and thereby cause errors in replication. Alternatively, the hydrophobic nature of the dyes causes them to bind to proteins and to membranes; either sort of interaction could conceivably lead to infidelity in DNA replication. There is ample evidence that the fidelity of replication is affected by changes in the enzymes involved (90–93). In spite of these possibilities, intercalation into DNA remains the most attractive means by which frameshift mutagens might exert their effects.

ACKNOWLEDGMENTS

The author's research is supported by a Public Health Service grant GM18633.

I would like to thank the following individuals who provided helpful suggestions and access to unpublished material: Bruce Ames, John Atkins, Gerald Fink, Phil Hartman, Joyce Owen, Don Rucknagel, Shala Ryce, Fred Sherman, Bea Singer, John Stewart, and Joe Yourno.

Literature Cited

1. Brenner, S., Barnett, L., Crick, F. H. C., Orgel, A. 1961. The theory of mutagenesis. *J. Mol. Biol.* 3:121–24
2. Crick, F. H. C., Barnett, L., Brenner, S., Watts-Tobin, R. J. 1961. Triplet nature of code. *Nature* 192:1227–32
3. Streisinger, G. et al 1966. Frameshift mutations and the genetic code. *Cold Spring Harbor Symp. Quant. Biol.* 31:77–84
→4. Drake, J. W. 1970. *Molecular Basis of Mutation.* San Francisco: Holden-Day
5. Okada, Y. et al 1966. A frameshift mutation involving addition of two base pairs in the lysozyme region of phage T4. *Proc. Nat. Acad. Sci. USA* 56:1692–98
6. Lerman, L. S. 1961. Structural considerations in the interaction of DNA and acridines. *J. Mol. Biol.* 3:18–30
7. Lerman, L. S. 1963. The structure of the DNA-acridine complex. *Proc. Nat. Acad. Sci. USA* 49:94–102
8. Lerman, L. S. 1964. Acridine mutagens and DNA structure. *J. Cell. Comp. Physiol.* 64:Suppl. 1, 1–18
9. Terzaghi, E. et al 1966. Change in a sequence of amino acids in phage T4 lysozyme by acridine-induced mutations. *Proc. Nat. Acad. Sci. USA* 56:500–7
10. Okada, Y., Amagase, S., Tsugita, A. 1970. Frameshift mutation in the lysozyme gene of bacteriophage T4: Demonstration of the insertion of five bases, and a summary of in vivo codons and lysozyme activities. *J. Mol. Biol.* 54:219–46
11. Tsugita, A. et al 1969. Frameshift mutations resulting in the changes of the same amino acid residue (140) in T4 bacteriophage lysozyme and the in vivo codons for trp tyr met val and ile. *J. Mol. Biol.* 41:349–64
12. Inouye, M., Akaboshi, E., Tsugita, A., Streisinger, G., Okada, Y. 1967. A frameshift mutation resulting in the deletion of two base pairs in the lysozyme gene of bacteriophage T4. *J. Mol. Biol.* 30:39–47
13. Imada, M., Inouye, M., Eda, M., Tsugita, A. 1970. Frameshift mutation in the lysozyme gene of bacteriophage T4: Demonstration of the insertion of four bases and the preferential occurrence of base addition in acridine mutagenesis. *J. Mol. Biol.* 54:199–217
14. Okada, Y. et al 1972. Molecular basis of a mutational hot spot in the lysozyme gene of bacteriophage T4. *Nature* 236:338–41
15. Okada, Y. et al 1968. Frameshift mutations near the beginning of the lysozyme genes of bacteriophage T4. *Science* 162:807–8
16. Lorena, M. J., Inouye, M., Tsugita, A. 1968. Studies on the lysozyme from bacteriophage T4 eJD7 eJD4 carrying two frameshift mutations. *Mol. Gen. Genet.* 102:69–78
17. Tsugita, A. 1971. Phage lysozyme and other lytic enzymes. In *The Enzymes,* ed. P. D. Boyer, 5:343–411. New York: Academic
18. Ames, B. N., Whitfield, H. J. Jr. 1966. Frameshift mutagenesis in *Salmonella. Cold Spring Harbor Symp. Quant. Biol.* 31:221–25
19. Martin, R. G. 1967. Frameshift mutants in the histidine operon of *Salmonella typhimurium. J. Mol. Biol.* 26:311–28
20. Newton, A. 1970. Isolation and characterization of frameshift mutations in the *lac* operon. *J. Mol. Biol.* 49:589–601
21. Brammer, W. J., Berger, H., Yanofsky, C. 1967. Altered amino acid sequences produced by reversion of frameshift mutants of tryptophan synthetase A gene of *E. coli. Proc. Nat. Acad. Sci. USA* 58:1499–1506
22. Berger, H., Brammer, W. J., Yanofsky, C. 1968. Spontaneous and ICR-191-A-induced frameshift mutations in the A gene of *Escherichia coli* tryptophan synthetase. *J. Bacteriol.* 96:1672–79
23. Berger, H., Brammer, W. J., Yanofsky, C. 1968. Analysis of amino acid replacements resulting from frameshift and missense mutations in the tryptophan synthetase A gene of *E. coli. J. Mol. Biol.* 34:219–38
24. Yourno, J., Heath, S. 1969. Nature of the *hisD3018* frameshift mutation in *Salmonella typhimurium. J. Bacteriol.* 100:460–68
25. Yourno, J. 1971. Similarity of cross-suppressible frameshifts in *Salmonella typhimurium. J. Mol. Biol.* 62:223–31
26. Yourno, J. 1972. Externally suppressible +1 "glycine" frameshift: Possible quadruplet isomers for glycine and proline. *Nature New Biol.* 239:219–21
27. Yourno, J., Kohno, T. 1972. Externally suppressible proline quadruplet CCC•. *Science* 175:650–52
28. Isono, K., Yourno, J. 1974. Non-sup-

pressible addition frameshift in Salmonella. *J. Mol. Biol.* 82:355–60

29. Isono, K., Yourno, J. 1974. Chemical carcinogens as frameshift mutagens: Salmonella DNA sequence sensitive to polycyclic carcinogens. *Proc. Nat. Acad. Sci. USA* 71:1612–17

30. Oeschger, N., Hartman, P. E. 1970. ICR-induced frameshift mutations in the histidine operon of *Salmonella. J. Bacteriol.* 101:490–504

31. Whitfield, H. Jr., Martin, R., Ames, B. 1966. Classification of aminotransferase (C gene) mutants in the histidine operon. *J. Mol. Biol.* 21:335–55

32. Fink, G. R., Klopotowski, R., Ames, B. N. 1967. Histidine regulatory mutants in *Salmonella typhimurium.* IV. A positive selection for polar histidine mutants. *J. Mol. Biol.* 30:81–95

33. Hartman, P. E., Hartman, Z., Stahl, R. C. 1971. Classification and mapping of spontaneous and induced mutations in the histidine operon of *Salmonella typhimurium. Advan. Genet.* 16:1–34

34. Singer, B. The chemical effects of nucleic acid alkylation and their relation to mutagenesis and carcinogenesis. *Progr. Nucl. Acid. Res.* In press

35. Yourno, J., Ino, I., Kohno, T. 1971. A "hot spot" for spontaneous frameshift mutations in the histidinal dehydrogenase gene of *Salmonella typhimurium. J. Mol. Biol.* 62:233–40

36. Sesnowitz-Horn, S., Adelberg, E. A. 1968. Proflavin treatment of *E. coli:* Generation of frameshift mutagens. *Cold Spring Harbor Symp. Quant. Biol.* 33:393–402

37. Sesnowitz-Horn, S., Adelberg, E. A. 1969. Proflavin-induced mutations in the L-arabinose operon of *E. coli.* I. Production and genetic analysis of such mutations. *J. Mol. Biol.* 46:1–16

38. Sesnowitz-Horn, S., Adelberg, E. A. 1969. Proflavin-induced mutations in the L-arabinose operon of *E. coli.* II. Enzyme analysis of the mutants. *J. Mol. Biol.* 46:17–23

39. Kohno, T., Roth, J. 1974. *J. Mol. Biol.* In press 89:17-32

40. Silver, S. 1967. Acridine sensitivity of bacteriophage T2: A virus gene affecting cell permeability. *J. Mol. Biol.* 29:191–202

41. Zampieri, A., Greenberg, J. 1965. Mutagenesis by acridine orange and proflavin in *E. coli. Mutat. Res.* 2: 552–56

42. Witkin, E. 1969. The mutability toward ultraviolet light of recombination-defi-

cient strains of *E. coli. Mutat. Res.* 8: 9–14

43. Drake, J. W. 1963. Properties of ultraviolet-induced rII mutants of bacteriophage T4. *J. Mol. Biol.* 6:268–83

44. Conde, F., Del Campo, F., Ramirez, J. 1971. Cyclic adenosine-3'5-monophosphate and the inhibition of ribonucleic acid synthesis by proflavin. *FEBS Lett.* 16:150–58

45. Sankaran, L., Pogell, B. 1973. Differential inhibition of catabolite sensitive enzymes by intercalating dyes. *Nature New Biol.* 245:257–60

46. Ryasaty, S., Atkins, J. 1968. External suppression of a frameshift mutant in *Salmonella. J. Mol. Biol.* 34:541–57

47. Atkins, J., Ryce, S. 1974. UGA and non-triplet suppression of the genetic code. Submitted to *Nature*

48. Reeves, R., Roth, J. R. 1971. A recessive UGA suppressor. *J. Mol. Biol.* 56:523–33

49. Yourno, J., Barr, D., Tanemura, S. 1969. Externally suppressible frameshift mutant of *Salmonella typhimurium. J. Bacteriol.* 100:453–59

50. Yourno, J., Tanemura, S. 1970. Restriction of in-phase translation by an unlinked suppressor of a frameshift mutation in *Salmonella typhimurium. Nature* 225:422–26

51. Riddle, D., Roth, J. 1970. Suppressors of frameshift mutations in *Salmonella typhimurium. J. Mol. Biol.* 54:131–44

52. Riddle, D., Roth, J. 1972. Frameshift suppressors. II. Genetic mapping and dominance studies. *J. Mol. Biol.* 66: 483–93

53. Riddle, D., Roth, J. 1972. Frameshift suppressors. III. Effects of suppressor mutations on transfer RNA. *J. Mol. Biol.* 66:495–506

54. Riddle, D., Carbon, J. 1973. A nucleotide addition in the anti-codon of a glycine tRNA. *Nature New Biol.* 242: 230–37

55. Atkins, J., Elseviers, D., Gorini, L. 1972. Low activity of β-galactosidase in frameshift mutants of *E. coli. Proc. Nat. Acad. Sci. USA* 69:1192–95

56. Sherman, F., Stewart, J. W. 1973. Mutations at the end of the iso-1-cytochrome *c* gene of yeast. In *The Biochemistry of Gene Expression in Higher Organisms,* ed. J. K. Pollak, J. W. Lee, pp. 56–86. Sydney: Australian and New Zealand Book Co.

57. Stewart, J. W., Sherman, F. 1974. Yeast frameshift mutations identified by sequence changes in iso-1-cytochrome *c.*

In *Molecular and Environmental Aspects of Mutagenesis,* ed. L. Prakash, F. Sherman, C. W. Lawrence, H. W. Tuber, 102–27. Springfield, Illinois: Thomas

58. Sherman, F., Stewart, J. W., Jackson, M., Gilmore, R., Parker, J. 1974. Mutants of yeast defective in iso-1-cytochrome *c. Genetics* 77:255–84

59. Malling, H. W. 1967. The mutagenicity of the acridine mustard (ICR-170) and the structurally related compounds in *Neurospora. Mutat. Res.* 4:265–74

60. Magni, G. E., Puglisi, P. P. 1966. Mutagenesis of super-suppressors in yeast. *Cold Spring Harbor Symp. Quant. Biol.* 31:699–704

61. Magni, G. E., Von Borstel, R. C., Sora, S. 1964. Mutagenic action during meiosis and anti-mutagenic action during mitosis of 5-amino acridine in yeast. *Mutat. Res.* 1:227–30

62. Leblon, G. 1972. Mechanism of gene conversion in *Ascobolus immersus.* II. The relationship between the genetic alterations and the conversion spectrum. *Mol. Gen. Genet.* 116:322–35

63. Seid-Akhavan, M., Winter, W., Abramson, R., Rucknagel, D. 1972. Hemoglobin Wayne: A frameshift variant occurring in two distinct forms. *Blood* 40:927 (Abstr.)

64. Clegg, J. B., Weatherall, D. J., Milner, P. F. 1971. Haemoglobin Constant Spring—a chain termination mutant? *Nature* 234:337–40

65. Fitch, W. M. 1973. Aspects of molecular evolution. *Ann. Rev. Genet.* 7:343–80 (See Table 1)

66. Jones, R. T., Brimhall, B., Huisman, T. H. J., Kleihauer, E., Betke, K. 1966. Hemoglobin Freiburg: Abnormal hemoglobin due to deletion of a single amino acid. *Science* 154:1024–27

67. Ames, B. N. 1971. The detection of chemical mutagens with enteric bacteria. In *Chemical Mutagens: Principles and Methods for their Detection,* ed. A. Hollaender, 1: 267–82. New York: Plenum

68. Ames, B. N. 1972. A bacterial system for detecting mutagens and carcinogens. In *Mutagenic Effects of Environmental Contaminants,* ed. E. Sutton, M. Harris, pp. 57–66. New York: Academic

69. Ames, B. N., Sims, P., Grover, P. L. 1972. Epoxides of polycyclic hydrocarbons are frameshift mutagens. *Science* 176:47–49

70. Ames, B. N., Gurney, E. G., Miller, J. A., Bartsch, H. 1972. Carcinogens as frameshift mutagens: Metabolites and derivatives of acetylaminofluorene and other aromatic amine carcinogens. *Proc. Nat. Acad. Sci. USA* 69:3128–32

71. Ames, B. N., Lee, F. D., Durston, W. E. 1973. An improved bacterial test system for the detection and classification of mutagens and carcinogens. *Proc. Nat. Acad. Sci. USA* 70:782–86

72. Miller, E. C., Miller, J. A. 1971. In *Chemical Mutagenesis,* pp. 83–120. New York: Plenum

73. Ames, B. N., Durston, W. E., Yamasaki, E., Lee, F. D. 1973. Carcinogens are mutagens: A simple test system combining liver homogenates for activation and bacteria for detection. *Proc. Nat. Acad. Sci. USA* 70:2281–85

74. Durston, W. E., Ames, B. N. 1974. A simple method for the detection of mutagens in urine: Studies with the carcinogen 2-acetylamino fluorene. *Proc. Nat. Acad. Sci. USA* 71:737–41

75. Rogers, S., Bueding, E. 1971. Hycanthone resistance: Development in *Schistosoma mansoni. Science* 172:1057–58

76. Hartman, P. E., Levine, K., Hartman, Z., Berger, H. 1971. Hycanthone: A frameshift mutagen. *Science* 172:1058–69

77. Straus, D., Hartman, P. E., Hartman, Z. 1973. Actions of mutagens on *Salmonella:* Molecular mutagenesis. In *Molecular and Environmental Aspects of Mutagenesis. Proc. 6th Ann. Rochester Int. Conf. Environ. Toxicity*

78. Heidelberger, C. 1973. Current trends in chemical carcinogenesis. *Fed. Proc.* 32:2154–61

79. Goldring, E., Grossman, L., Krupnik, D., Cryer, D., Marmur, J. 1970. The petite mutation in yeast. *J. Mol. Biol.* 52:323–35

80. Drake, J. W. 1967. The length of the homologous pairing region for genetic recombination in bacteriophage T4. *Proc. Nat. Acad. Sci. USA* 58:962–66

81. Newton, A., Masys, D., Leonardi, E., Wygal, D. 1972. Association of induced frameshift mutagenesis and DNA replication in *Escherichia coli. Nature New Biol.* 236:19–22

82. Sarabhai, A., Lamfrom, H. 1969. Mechanism of proflavin mutagenesis. *Proc. Nat. Acad. Sci.* 63:1196–97

83. Schmechel, D. E. V., Crothers, D. M. 1971. Kinetic and hydrodynamic studies of the complex of proflavin with poly A· poly U. *Biopolymers* 10:465–80

84. Li, H. J., Crothers, D. M. 1969. Relaxation studies of the proflavin-DNA complex: The kinetics of an intercalation reaction. *J. Mol. Biol.* 39:461–77
85. Löber, G., Achtert, G. 1969. On the complex formation of acridine dyes with DNA. VII. Dependence of the binding on dye structure. *Biopolymers* 8:595–608
86. Thomas, J. C., Weill, G., Daune, M. 1969. Fluorescence of proflavine-DNA complexes: Heterogeneity of binding sites. *Biopolymers* 8:647–59
87. Bidet, R., Chambron, J., Weill, G. 1971. Hétérogénéité des sites de fixation de la proflavine sur le DNA. *Biopolymers* 10:225–42
88. Weisblum, B., de Haseth, P. L. 1972. Quinacrine, a chromosome stain specific for deoxyadenylate-deoxythymidylate-rich regions of DNA. *Proc. Nat. Acad. Sci. USA* 69:629–32
89. Tubbs, R. K., Ditmars, W. E. Jr., Van Winkle, Q. 1964. Heterogeneity of the interaction of DNA with acriflavin. *J.*

Mol. Biol. 9:545–57
90. Allen, E., Albrecht, I., Drake, J. W. 1970. Properties of bacteriophage T4 mutants defective in DNA polymerase. *Genetics* 65:187–200
91. Hall, Z. W., Lehman, I. R. 1968. An in vitro transversion by a mutationally altered T4-induced DNA polymerase. *J. Mol. Biol.* 36:321–33
92. Speyer, J. F. 1969. Base analogues and DNA polymerase mutagenesis. *Fed. Proc.* 28:348
93. Muzyczka, N., Poland, R., Bessman, M. 1972. Studies on the biochemical basis of spontaneous mutation. I. A comparison of the DNA polymerases of mutator, anti-mutator and wild type bacteriophage T4. *J. Biol. Chem.* 247:7116–22
94. Schreiber, J. P., Daune, M. P. 1974. Fluorescence of complexes of acridine dye with synthetic polydeoxyribonucleotides: A physical model of frameshift mutagenesis. *J. Mol. Biol.* 83:487–501

GENETIC ANALYSIS OF ❖3078
THE CHLOROPLAST AND
MITOCHONDRIAL GENOMES

Nicholas W. Gillham
Department of Zoology, Duke University, Durham, North Carolina 27706

INTRODUCTION

An understanding of the rules governing the inheritance of chloroplast and mito-chondrial genes has been slow in coming despite the isolation of presumptive plastid mutants as long ago as 1909 (1, 2) and mitochondrial mutants over two decades ago (3, 4). Several reasons can be cited. First, until the discovery of unique species of chloroplast DNA in 1963 (5, 6) and mitochondrial DNA (mitDNA) at about the same time (7) there was no physical evidence for hereditary carriers in chloroplasts and mitochondria. Hence, specific mutations have been convincingly associated with organelle DNA only recently. Second, the paucity of well-defined, phenotypi-cally distinguishable organelle gene markers has posed a serious problem in all but a few organisms. Third, even in those organisms where suitable markers are avail-able unique experimental problems arise because one cannot directly score the phenotype of the organelle carrying a specific mutation but only the phenotype of the cell within which the organelle resides. Furthermore, it is difficult to control the input and overflow of organelle genomes in a cross. One is tempted to make an analogy to a phage cross in which neither the average multiplicity of infection, average burst size, nor frequency of phenotypically different progeny phages can be measured directly.

Despite the aforementioned obstacles to the study of organelle genetics, experi-ments with the yeast *Saccharomyces cerevisiae* and the green alga *Chlamydomonas reinhardtii* are beginning to yield important insights into the genetics of mito-chondria and chloroplasts respectively. This review is devoted to a discussion of these experiments. The specific topics to be considered include the evidence that genes are localized in the chloroplast of *Chlamydomonas* and the mitochondrion of *Saccharomyces,* segregation and recombination of organelle genes, genetic mapping and the formal models underlying the chloroplast and mitochondrial genetic maps, and finally a comparison of the organelle genetic systems of *Chlamydomonas* and *Saccharomyces.*

347

Within the past decade, an enormous amount has been learned about the structure, size, and coding capacity of chloroplast and mitDNA, the unique protein synthesizing systems of chloroplasts and mitochondria, and which component proteins of chloroplasts and mitochondria are synthesized in the organelles themselves or in the cytoplasm. Molecular hybridization experiments have revealed that chloroplast and mitochondrial ribosomal RNA are transcripts of the DNA in the respective organelles and that many of the unique mitochondrial transfer RNAs are also transcripts of mitDNA. Since most of this material is not covered in this review the interested reader is referred to several recent books (8–11), symposia (12–16), and review articles (17–40). This review is, in a sense, a lineal descendant of the excellent article on extrachromosomal inheritance by Preer published in this series in 1971 (41). The author also wishes to note the recent and interesting experiments with mitochondrial mutations to antibiotic resistance in *Paramecium* (42–47). This system holds excellent promise but is not discussed in this review because it has not yet shed much light on the problems of recombination and mapping of organelle genes.

MITOCHONDRIAL GENETICS OF *SACCHAROMYCES*

The Petite Mutation and Mitochondrial DNA

S. cerevisiae is a facultative anaerobe. When the organism is supplied with glucose as a carbon source, it ferments the glucose to ethanol, which can then be respired. Respiration-deficient ("petite") mutations arise at a high frequency (0.1 to 10%) in cells growing fermentatively. The properties of these mutations including their induction preferentially by a wide variety of mutagens, notably polynuclear heterocyclic aromatic dyes such as euflavin and ethidium bromide, their pleiotropic effects on the respiratory chain, and their genetics have been reviewed exhaustively elsewhere (9–11, 26, 27).

The vast majority of spontaneous and induced petite mutations exhibit non-Mendelian inheritance and are termed *vegetative* petites (48). A minority exhibit typical Mendelian inheritance and are called *segregational* petites (48, 49). Over twenty such Mendelian genes are known (50–58). Vegetative petites are nonreverting, pleiotropic mutations whose respiration deficiency can be traced to their inability to elaborate a functional electron transport chain including cytochromes a + a_3, b, and c_1, and the attendant system of ATP generation (for review see 9, 27). As a result these conditional lethal mutants grow only on fermentable carbon sources such as glucose and not on carbon sources such as glycerol which must be respired.

Vegetative petites fall into two subclasses called *neutral* and *suppressive.* The neutral petite phenotype is not transmitted in crosses to wild type either among the vegetative diploid progeny of the resulting zygotes or the haploid meiotic progeny of these zygotes (4). Suppressive petites transmit the petite phenotype to a fraction of the vegetative diploid progeny in crosses (59). The level of suppressiveness is characteristic of a given petite strain as shown by cloning experiments (60) but is frequently altered as the result of cytoplasmic mutation (60, 61). Suppressiveness varies from 1% or less petite diploid progeny in some strains to over 99% petite progeny in other strains. Suppressive petite diploids do not sporulate, but the zygotes formed between wild-type and suppressive petite strains do. Ephrussi et al (59) found that in crosses between a highly suppressive petite strain and wild type, the zygotes almost always yielded 0 wild type: 4 petite ascospores. Heterokaryon tests of a suppressive petite and wild type have confirmed the cytoplasmic transmission of the petite phenotype (63).

A cytoplasmic factor was originally postulated to explain the transformation of a wild-type cell to a vegetative petite (64) and was later given the name rho (ρ) (50). A wild-type cell is ρ^+ and a petite ρ^-. Suppressiveness was initially explained in terms of a suppressive factor which prevented the perpetuation of the normal factor by destroying it, interfering with its replication, or preventing its distribution during cell division (59). Later Ephrussi et al (62) suggested that the suppressive factor interfered with the replication of the "normal factor" (ρ^+) as the result of a "mutual repression." The mechanism underlying suppressiveness has since been the subject of much speculation and will be discussed later in this article (pp. 355–56).

Two lines of evidence relate the ρ factor and hence the vegetative petite mutation to mitDNA. First, the mutagen ethidium bromide, which induces petite mutations with 100% efficiency (65), causes the degradation of mitDNA after prolonged exposure of cells (66). The petites arising as the result of such exposure are neutral petites which, while they contain petite mitochondrial structures, lack mitDNA (67, 72). Second, suppressive petites, whether they arise spontaneously or are induced by mutagens, contain mitDNA that is often grossly altered in base composition with respect to wild-type mitDNA (68–74). These results can be interpreted as meaning that the ρ factor and mitDNA are one and the same. Neutral petites lacking mitDNA are designated ρ^0 and suppressive petites containing mutant mitDNA are designated ρ^- (26, 75).

Antibiotic Resistance and the Mitochondrial Genome

Antibiotics blocking mitochondrial protein synthesis (e.g. chloramphenicol and erythromycin) or uncoupling respiration from phosphorylation (e.g. oligomycin and rutamycin) are selectively lethal to *S. cerevisiae* when the organism is grown on a nonfermentable carbon source (76). Mutants resistant to some of these antibiotics were reported in 1967 (76) and in 1968 Linnane et al (78) and Thomas & Wilkie (77) showed that certain erythromycin-resistant mutants were non-Mendelian in inheritance.

When a mitochondrial antibiotic-resistant mutant (ant^R) is crossed to a wild-type strain (ant^S), the zygotes segregate ant^R and ant^S progeny mitotically. Sporulation of ant^S diploids from such a cross yields only ant^S progeny (asci: 4 ant^S:0 ant^R) whereas sporulation of an ant^R diploid yields only ant^R progeny (asci: 0 ant^S:4 ant^R) (77–79). The *ant* phenotypes are expressed only when yeast cells are grown on a nonfermentable carbon source. Since petite mutants cannot themselves grow on such a carbon source, the method used to determine the *ant* condition in petites involves a cross to a respiration-sufficient tester stock and an examination of the diploid progeny of the cross. For example, suppose an ant^R stock is treated with ethidium bromide and the induced petites are isolated and crossed to a respiration-sufficient ant^S tester. Theoretically, the cross can yield one of four results for any given petite. If the diploid progeny are all ρ^+ ant^S, the genotype of the petite parent must have been ρ^0 ant^0. That is, both ρ and the ant marker have been lost in the induced petite. If the diploid progeny include a mixture of petite and respiration-sufficient progeny and the respiration-sufficient progeny are all ant^S, the genotype of the petite parent must have been ρ^- ant^0. That is, the mutation from ρ^+ to ρ^- was accompanied by the loss of the antR marker. If the diploids include petite progeny and respiration-sufficient progeny, some of which are ant^R and some ant^S, the petite parent must have been ρ^- ant^R. That is, the ant^R allele was retained in the petite parent although it could not, by definition, be expressed. Finally, if the diploids are all respiration-sufficient but include ant^R and ant^S diploid progeny, the petite parent must have been ρ^0 ant^R. That is, the petite parent had retained the ant^R allele but lost the ρ factor. Isolation of a petite of this type would mean that the ant marker did not lie in the ρ factor, and hence mitDNA, since the ant^R allele had been retained while the ρ factor had been lost.

Thomas & Wilkie (77) reported that conversion of an erythromycin-resistant (E^R) mutation of the ρ^- phenotype resulted in loss of the ability of the petite to transmit the E^R allele. They therefore argued that the E^R determinant was located in the mitochondrion. Linnane et al (78) found a case in which a petite derived from an erythromycin-sensitive strain (E^S) when crossed to a ρ^+ E^R strain yielded both ρ^+ E^R and ρ^+ E^S diploid progeny. They suggested that a yeast cell carries two separate cytoplasmic factors: one for ρ, the other for the E^R gene. More extensive experiments of this type by Gingold et al (80) revealed that petites frequently retained the ant alleles that they carried. These authors concluded that either ρ^+ and the E^R allele are on different cytoplasmic determinants or that ρ^+ and E^R are on the same piece of mitDNA and recombined with high frequency.

Recent experiments on the retention of ant markers in ethidium bromide-induced petites provide strong evidence that the ant genes are linked to ρ and hence within mitDNA (75, 81). Ethidium bromide inhibits both replication and transcription of mitDNA in $S.$ $cerevisiae$ but has little or no effect on other cellular processes as long as an adequate energy supply is available. In the presence of ethidium bromide, degradation of mitDNA in growing cells also occurs and leads to a population of molecules decreasing in size as a function of exposure. When ethidium bromide is removed, mitDNA synthesis again commences and some of the surviving mitDNA fragments can then serve as templates for the production of mutant progeny mitDNA. After about one cell generation, the rate of synthesis of the mutant mitDNA approaches that of wild type.

Nagley & Linnane (75) treated growing cells of $S.$ $cerevisiae$ carrying the E^R allele for 4 hr with ethidium bromide and then transferred them to medium lacking ethidium bromide for eight generations. They isolated ten petite clones, among which five lacked mitDNA and behaved genotypically as ρ^0 E^0 in crosses. These findings, although limited, suggest that loss of mitDNA is accompanied by the loss of both factors as would be expected if both are localized in mitDNA.

Similar evidence is provided by the experiments of Deutsch et al (81) in which the kinetics of loss of ρ^+ and the alleles for erythromycin resistance (E^R), chloramphenicol resistance (C^R), and oligomycin resistance (O^R) were measured on nongrowing cells as a function of time of treatment with ethidium bromide. The conversion of respiration-sufficient cells to the petite phenotype proceeded at a faster rate than loss of the *ant* alleles. All ρ^+ cells also carried all three of the antibiotic-resistance alleles. While C^R and E^R were usually lost or retained as a pair (i.e. $C^R E^R + C^0 E^0 \gg C^R E^0 + C^0 E^R$), this pair of markers and the O^R marker represented two different but not independent groups of genes. The conclusion that C^R and E^R are linked to each other but not to O^R is consistent with mapping data to be discussed later (pp. 359–63). Deutsch et al also found that the kinetics of petite induction and marker loss with uv light were very similar to those seen with ethidium bromide.

Application of target theory to the marker survival curves indicated that the C^R and E^R targets were each about two thirds the size of ρ^+, and O^R was one half the size of ρ^+. The estimates of target number for each marker including ρ^+ was about two. This number, as well as numbers between two and twenty reported previously (65, 82–86), is small compared to the number of mitDNA molecules per cell which is between 50 and 100 (87).

Another line of evidence supporting localization of the *ant* markers in mitDNA comes from the experiments of Michaelis et al (88). These authors crossed a $C^0 E^R$ petite in which the mitDNA had a buoyant density of 1.688 g/cc with two different $C^R E^0$ petites, one having mitDNA of buoyant density 1.682 g/cc and the other 1.683 g/cc. The resulting diploid subclones were selected using appropriate nuclear, auxotrophic markers and then crossed to a haploid ρ^+ $C^S E^S$ tester using yet other selective markers. The ρ^+ triploid progeny were screened to see if the ρ^- diploid parent had been of genotype $C^0 E^R$, $C^R E^0$, $C^R E^R$, or $C^0 E^0$. The first two genotypes are parental; the third must be recombinant since it carries a resistance marker from each parent; while the fourth could either be recombinant or have arisen because the ρ^- mitDNA of one parent or the other had undergone a spontaneous change leading to the loss of the resistance marker it originally carried. The buoyant densities of the mitDNA of 23 diploid subclones whose genotype was determined in the aforementioned fashion were then compared. The buoyant densities of ρ^- mitDNA extracted from the $C^R E^R$ recombinants were intermediate between those of diploid cells having the parental genotypes. The authors concluded that recombination of the *ant* markers was accompanied by formation of a recombinant species of mitDNA of intermediate buoyant density.

The ρ^- Mutation

Nagley & Linnane (75) introduced the symbol ρ^0 to denote absence of the ρ factor and mitDNA in certain petites. All such petites are neutral and have a suppressiveness of 0%. All other petites contain mitDNA, usually drastically modified in base composition (68–74), and are designated ρ^- (75). Such petites may range in suppressivity from near 0% to 100% (89). The following discussion considers the genetic structure of ρ^- mitDNA; retention of genes in ρ^- mitDNA (i.e. the genotypic stability of ρ^- mitDNA); and the mechanism of suppression.

The buoyant density of mitDNA in ρ^- petites ranges from 1.672 g/cc in the "low density" petites, some of which have an A + T content as high as 96% (69, 74, 90), to 1.689 g/cc in other petites (88). Wild-type ρ^+ mitDNA has a buoyant density of 1.683–1.684 g/cc and an A + T content of 83% (69, 74). Studies with low density petites suggest that the mitDNA of such cells consists in large part of dA and dT tracts arranged in both alternating and nonalternating sequences (68, 69, 74). Borst and his colleagues (90–92) showed in one such petite (95% A + T) that 0.3% of the mutant mitDNA is complementary to wild-type mitDNA based on renaturation studies, that the mutant mitDNA does not contain within its strands self-complementary sequences of appreciable length, and that the mutant mitDNA is composed of a perfectly repeating sequence of 300 or fewer nucleotides. These data suggest that the mutant mitDNA represents a reasonably faithful copy containing some replication errors (<4% mismatching) of a sequence present in wild-type mitDNA. This sequence would be some 300 nucleotides long and amplified many times.

These experiments on the molecular structure of ρ^- mitDNA predict that ρ^- mitDNA should contain certain of the genes present in ρ^+ mitDNA while having lost others. Indeed, the kinetic experiments of Deutsch et al (81) on the retention of mitochondrial markers following ethidium bromide treatment show this to be the case (pp. 350–51). However, the molecular experiments also predict that those nucleotide sequences retained in ρ^- mitDNA are amplified.

In principle, two sorts of methods, one genetic and the other molecular, are available for the detection of amplification or deletion of specific genes in ρ^- mitDNA. In the genetic method deletion of a gene, as already discussed (pp. 349–50), is detected by crossing a ρ^- petite (which was derived, for example, from a ρ^+ ant^R strain) to a ρ^+ ant^S tester and then seeing whether any ρ^+ ant^R diploid progeny are to be found. If so, the petite must have been ρ^- ant^R, but if not, the petite must have been ρ^- ant^0. That is, there has been a *functional* deletion of the ant^R gene in the ρ^- ant^0 petite. Amplification of specific genes in ρ^- mitDNA can be detected by following the kinetics of marker inactivation during ethidium bromide treatment and comparing these kinetics to the kinetics of inactivation of the same marker in ρ^+ mitDNA. If a specific marker has been amplified in a ρ^- petite, its rate of inactivation will be slower than in ρ^+ mitDNA where, presumably, only one copy of each gene is to be found in each mitDNA molecule.

The molecular method of detecting gene deletion and amplification in ρ^- mitDNA takes advantage of the fact that mitochondrial ribosomal RNA (rRNA) and certain transfer RNAs (tRNA) are known to be gene products of mitDNA in yeast from molecular hybridization experiments (93–95). If the sequence for a specific tRNA is deleted in the ρ^- mutant, there will be no significant hybridization of this tRNA to denatured mitDNA from the mutant. On the other hand, if the tRNA genes have been amplified in the ρ^- mutant, a given amount of ρ^- mitDNA will hybridize more tRNA than a comparable amount of ρ^+ mitDNA. A combination of these methods provides a powerful way for mapping the mitochondrial genome. For example, if deletion of a specific tRNA gene in the molecular sense is usually accompanied by deletion of a specific *ant* marker in the genetic sense, it is likely that the two genes lie close together. Coordinate amplification of both genes in other petites would further support the notion of tight linkage. Coupling of these methods with the physical methods of Attardi and his colleagues (96, 97) for mapping mitochondrial rRNA and 4S RNA genes in the HeLa cell mitochondrial genome could provide precise physical locations for the known mitochondrial genes in yeast.

Amplification and deletion of specific mitochondrial tRNA genes in ρ^- mitDNA was reported by Cohen et al in 1972 (98). These authors found that mitDNA of a spontaneous ρ^- mutant hybridized mitochondrial valyl tRNA from the parent ρ^+ strain very poorly while hybridizing twice as much mitochondrial leucyl tRNA as the ρ^+ strain. The authors then showed that not only leucyl tRNA, but also valyl tRNA could be isolated from the ρ^- mutant which would hybridize with ρ^+ mitDNA. The results for leucyl tRNA were not unexpected and showed that the ρ^- mitDNA was capable of transcribing a normal leucyl tRNA molecule, but the presence of mitochondrial valyl tRNA in the ρ^- strain was remarkable in view of the fact that valyl tRNA hybridized so poorly with ρ^- mitDNA. Subcloning experiments showed that the ρ^- mutant was heterogeneous with respect to retention of the valyl tRNA cistrons. When leucyl tRNA was hybridized to mitDNA from the subclones, it was found in every case that leucyl tRNA was hybridized to an extent the same as or greater than the corresponding amount of ρ^+ mitDNA. These results, in toto, suggest that the valyl tRNA cistron is deleted in certain subclones of the ρ^- mutant and retained in others, while the leucyl tRNA cistron is retained in all and amplified in some.

Nagley et al (99), in a study of the effects of ethidium bromide on ρ^- mitDNA, have now provided the complementary genetic evidence for amplification. These authors grew ρ^+ cells and cells of several different ρ^- petites in the presence of varying concentrations of ethidium bromide for eight generations and then analyzed the cells for mitDNA and retention of the E^R marker. They found that the amount of mitDNA and retention of the E^R marker decreased in a concentration-dependent manner in all clones. However, in the ρ^- clones both the decrease in mitDNA and loss of the E^R gene proceeded at slower rate than in ρ^+ mitDNA. This apparent resistance to ethidium bromide in ρ^- was interpreted to be the result of the amplification of E^R genes in ρ^-. The authors found this was true of two ρ^- clones tested when compared to ρ^+. Most striking was the observation that although the kinetics of elimination of ρ^- mitDNA was the same for both ρ^- clones, the ratio of E^R genes to mitDNA retained was much greater for one clone than the other. These results are interpreted in terms of greater amplification of the E^R genes in one of the ρ^- mutants.

Recently, Rabinowitz, Slonimski and their colleagues (100) have published the most comprehensive study to date on gene deletions and repetitions in ρ^- mitDNA. The paper examines the genetic and molecular properties of a series of particularly stable ρ^- mutants whose genotypes are defined with respect to the C^R and E^R markers. The authors discuss the evidence from genetic experiments for retention, loss, and amplification of specific markers in ρ^- mitDNA showing as did Nagley et al (99) that retention of the C^R and E^R markers by ρ^- mutants in the presence of ethidium bromide is greater than it is in ρ^+ strains. They also point out a strong correlation between retention of mitochondrial genes and kinetic complexity of mitDNA. With respect to specific genotypes the kinetic complexity of mitDNA increases in the sequence

$$\rho^- \ C^0E^0 \ < \ \rho^- C^0 E^R \ < \ \rho^- C^R E^0 \ < \ \rho^- C^R E^R \ < \ \rho^+ C^R E^R.$$

These results suggest that genes become deleted in both the molecular and genetic senses in ρ^- mitDNA and this is accompanied by a reduction in kinetic complexity of mitDNA.

Having dealt with the evidence for deletion of specific sequences in ρ^- mitDNA, the authors come to the main focus of the paper, which is how the retained sequences are organized in ρ^- mitDNA. Two models are considered. First, repetition of the retained sequences, either intermolecular or intramolecular, leads to an increase in the number of copies of the retained sequences in the cell. In the case of intermolecular repetition, an increased number of very short molecules should be seen, while in the case of intramolecular repetition, large molecules containing linear repeats of the retained segment will be found. Second, a new sequence of a unique type (e.g. an A + T polymer) or new sequences of different types are added to the retained sequence. In this model, ρ^- mitDNA would be composed of two types of sequences: one identical in sequence and number of copies to a part of ρ^+ mitDNA and the other a nonsense sequence of low complexity devoid of genetic information.

The experimental results support the amplification model and can be summarized as follows. First, $\rho^- \ C^0 E^R$ mitDNAs are denser than ρ^+ mitDNA and $\rho^- \ C^R E^0$ mitDNAs are less dense. Hence, the mitDNA segment carrying the E^R gene is richer in G + C pairs than the segment carrying the C^R gene, and the differences in buoyant densities of the two ρ^- mitDNAs can be explained in terms of amplification of G + C-rich segments in $\rho^- \ C^0 E^R$ petites and A + T segments in $\rho^- \ C^R E^0$ petites. This interpretation is supported by the results of differential melting experiments with mitDNAs of the various genotypes. These experiments show that while $\rho^+ \ C^R E^R$ mitDNA and $\rho^- C^R E^R$ mitDNA have relatively similar melting profiles, $\rho^- \ C^R E^0$ mitDNA has a high proportion of low temperature melting regions (i.e. is A + T-rich), and $\rho^- C^0 E^R$ mitDNA lacks these regions but has a number of high temperature melting regions (i.e. is G + C-rich). Second, there is considerable sequence homology between ρ^- mitDNAs of the same genotype but a low degree of homology between genotypically different ρ^- mitDNAs, indicating that similar DNA sequences are retained in genotypically similar ρ^- mutants and different sequences in genotypically different ρ^- mutants. Third, specific tRNA genes are amplified in certain ρ^- mitDNAs. Similarly, one or both mitochondrial rRNA cistrons appear to be amplified in all $\rho^- \ C^R E^R$ mitDNAs and in some $\rho^- \ C^0 E^R$ mitDNAs, while these regions are retained at a very low level if at all in $\rho^- \ C^0 E^0$ mitDNA. Both the tRNA and rRNA cistrons appear to be expressed in the petites retaining them.

Evidence for repeating units in ρ^- mitDNA was also obtained from ultrastructural studies. A $\rho^- \ C^0 E^0$ clone (F13) was found to synthesize four species of mitochondrial RNA with molecular weights estimated at 0.20, 0.38, 0.58, and 0.81 $\times$ 10^6

respectively. Ultrastructural studies on partially denatured mitDNA of F13 revealed tandemly repeating units each comprised of a denatured (G + C-poor) and an undenatured (G + C-rich) segment; the two totaling 0.18 μ in length, close to the length of the smallest molecular weight RNA described above. Finally, it is known that repeated DNA sequences can form a series of open circular structures whose contour length is a multiple of the repeating unit (101). This appears to be true of the ρ^- mitDNAs and in the case of F13 the contour length of randomly collected circles falls into a polymeric series 0.15 μ in length. The authors conclude that the basic features of the ρ^- mutation are the loss of long segments of mitDNA coupled with a periodic intramolecular repetition of the remaining ρ^+ segments. Any segment can be lost or retained, and when the repeated segment contains a gene, the gene is amplified.

The stability of ρ^- clones with respect to allele retention and suppressiveness has been studied by Saunders et al (89) and Nagley & Linnane (75). Saunders et al (89) showed that spontaneous ρ^- E^R mutants varied widely (5–50%) in suppressiveness. They also examined subclones of a single spontaneous ρ^- mutant for changes in suppressiveness and retention of the E^R allele. They concluded that those ρ^- subclones retaining the E^R allele tended to be similar in suppressiveness to the parent clone whereas those subclones that had mutated to E^0 were either more or less suppressive than the original clone. Nagley & Linnane (75) extended these experiments on the stability of spontaneous mutants to those induced by ethidium bromide. These authors treated ρ^+ E^R cells with a mild dose of ethidium bromide (10 μg/ml) for 4 hr, washed part of the culture free of ethidium bromide, and allowed growth of those cells to continue for eight generations. They then compared those cells treated with ethidium bromide for 4 hr with those that had grown for eight generations in the absence of ethidium bromide for amounts of mitDNA, and suppressiveness and retention of the E^R allele.

Superficially, the cultures appeared quite similar in all respects, but clones from the cells grown in the absence of ethidium bromide proved very heterogeneous. About half the clones were neutral petites lacking mitDNA. Several clones contained mitDNA and were E^0 but in each case the level of suppressiveness was different. In one suppressive petite the E^R marker had been retained to the level of 20%. When the suppressive petite clones were subcloned, all contained a fraction of neutral petites and fell into two classes of suppressiveness. In one class the subclones were similarly suppressive; the second class, in contrast, had not stabilized, and there was a wide variation in the suppressiveness of the subclones.

The authors explained the instability of the ethidium bromide-induced ρ^- mutants in terms of a heterogeneity of mitDNA molecules present in the mutant cells. Segregation of mitDNA molecules unable to replicate themselves into buds produces ρ^0 petites. Molecules capable of replicating themselves produce ρ^- petites, but these molecules form a heterogeneous collection in terms of retention of the E^R allele and suppressiveness. From among these cells some are produced that contain relatively homogeneous collections of ρ^- mitDNA molecules and these cells produce clones of stable petites.

What is the nature of suppressiveness? Recent conjecture has focused principally on two models: 1. preferential replication of the suppressive ρ^- mitDNA and 2. the spreading of errors contained in ρ^- mitDNA through recombination with ρ^+ mitDNA. The first hypothesis, proposed by Slonimski in 1968 (102), attracted considerable attention immediately (70, 103–106). It was supposed that the suppressive petite mitDNA competed with wild-type mitDNA in a replicative sense and that the mitDNA of a highly suppressive petite was at a strong competitive advantage to ρ^+ mitDNA. The molecular analog was the demonstration by Mills et al (107) that in an in vitro system virus Qβ RNA could be selected for faster and faster replication rate and that the final product of this selection was an RNA that had lost 83% of its original genome and had a very high affinity for the Qβ replicase.

Rank and his co-workers in a series of genetic experiments (103–106) attempted to test the "Darwinian" selection model. The first paper (103) dealt with the segregation of ρ^+ and ρ^- cells by colonies of cells in the "premutational" state in which both wild type and progeny are segregated (60–62). Both diploid colonies in the premutational state and haploid colonies produced by ascospores were studied. On the basis of these experiments it was concluded that a cell in the premutational state arises originally by a spontaneous event in a ρ^+ cell which makes the cell heterozygous for ρ^+ and ρ^- factors. Petite cells would be segregated because of the competitive replicative advantage of ρ^- mitDNA over ρ^+ mitDNA, but opposed to this *intracellular* selection would be *intercellular* selection for those cells that had received ρ^+ mitDNA and were therefore wild type.

In another paper Rank (104) postulated, again by analogy to the Qβ selection experiment, that the most highly suppressive ρ^- mitDNA would contain the least genetic information. Hence, ρ^- E^R clones should give rise to E^0 clones because the latter mitDNA should have a replicative advantage. His results showed that this was indeed the case, but, as we have already seen, results of this type can be explained by the segregation of ρ^0 petites lacking mitDNA (75). In the next paper Rank (105) presented the results of crosses between ρ^- strains of high and low suppressiveness. The diploid progeny of these crosses were selected and, through the use of appropriate nuclear selective markers, mated to ρ^+ diploids. The level of suppressiveness of the parental ρ^- diploids could then be ascertained. It was found that these diploids were highly suppressive, as expected from the competitive model. Most recently Rank & Bech-Hansen (106) have shown that nuclear mutations to chloramphenicol resistance are retained in ρ^- cells while mitochondrial mutations tend to be lost. They argue that this result is consistent with Darwinian selection in cells that are already resistant by virtue of nuclear mutations.

The hypothesis that suppressiveness results from a high frequency of recombination between ρ^- mitDNA and ρ^+ mitDNA was postulated by Coen et al (79). According to this hypothesis errors present in ρ^- mitDNA will be propagated into ρ^+ mitDNA through successive rounds of recombination resulting in conversion of ρ^+ mitDNA to the ρ^- phenotype. Michaelis et al (72) pointed out that their experiments showed no correlation between molecular weight of ρ^- mitDNA and degree of suppressiveness. Such a correlation would be expected on the Darwinian selection model but is not a requirement of the recombination model. In another paper discussed earlier (p. 351) Michaelis et al (88) suggested that the level of suppressiveness in ρ^- $C^R E^R$ recombinants emanating from a cross of ρ^- $C^R E^0$ and ρ^- $C^0 E^R$

parents could be correlated with recombination. Despite these observations it appears that the mechanism underlying suppressiveness is still not understood clearly.

Recombination of Mitochondrial Genes in Yeast

Physical evidence for recombination of mitochondrial genes in yeast has been discussed and we turn now to the formal aspects of mitochondrial gene recombination. The principal focus of the discussion is a group of papers by Slonimski and his colleagues (79, 88, 108–111) which develop a general theory of mitochondrial gene recombination in *S. cerevisiae* documented by extensive genetic data. A number of other papers (112–124) are also discussed in relation to the general theory.

1. METHODOLOGY AND TERMINOLOGY   Three methods have been used to study segregation and recombination of mitochondrial genes in yeast: pedigree analysis of diploid buds produced by individual zygotes and their progeny (79, 113, 119), analysis of the genotypic composition of the diploid progeny of individual zygote clones after many generations of vegetative growth (79, 112), and analysis of the genotypic composition of the diploid progeny of a population of zygotes irrespective of lineage after some 20 mitotic divisions (79, 108–111, 116–118, 120–124). The latter method, referred to as the *standard cross* (79), has yielded most of the important data on recombination and linkage of mitochondrial genes. Since the most recent version of the theory of mitochondrial gene recombination likens the process to phage recombination (111), it is worth noting that zygote clone analysis and the standard cross have their analogs in the single burst experiment and phage cross respectively. A zygote clone, then, reveals the genotypic composition of the mitochondrial genomes derived from a single zygote, whereas in the standard cross the population of mitochondrial genomes is analyzed irrespective of the zygote of origin in the same way that progeny phage are analyzed in a phage cross irrespective of the bacterium of origin.

With reference to the results of standard crosses, Coen et al (79) introduced the following terminology, to which new words have been added (109, 111) as more facts about mitochondrial gene recombination have been deduced.

Anisochromosomal isomitochondrial (79): Crosses in which the chromosomal genomes are different but the mitochondrial genomes are the same.

Asymmetry (109, 111): A situation in which the frequency of a pair of alleles or a pair of reciprocal recombinants is unequal among the diploid progeny of a cross due to the influence of a nuclear gene.

Bias (109, 111): This term is applied when the frequency of a pair of alleles or a pair of reciprocal recombinants is unequal among the diploid progeny of a cross for unknown reasons. The definition of this term should be compared to those for asymmetry and polarity.

Isochromosomal anisomitochondrial (79): Crosses in which the chromosomal genomes of two crosses are the same, but the mitochondrial genomes are different.

Polarity (79): The term polarity is applied when the frequency of a pair of alleles or a pair of reciprocal recombinants is unequal among the diploid progeny of a cross because of the influence of a genetic determinant known to be localized in the mitochondrial genome.

Transmission (79): In a cross of two pure haploid clones differing by a pair of mitochondrial alleles (e.g. C^R vs C^S), the transmission measures the frequency of diploid cells possessing one of the alleles. The allele is highly transmitted if the fraction of cells carrying the allele is high.

The definition of transmission and its relationship to polarity has caused confusion probably related to the rapid accumulation of new and sometimes seemingly contradictory observations by different groups of workers (116, 120). The terms *asymmetry* and *bias* have been introduced by Avner et al (109) to distinguish effects on the ratio of reciprocal recombinants or the transmission of allelic pairs which cannot be clearly attributed to polarity. Dujon et al (111) point out that "all crosses which show a strong bias in the frequency of reciprocal recombinants display, thereby, a bias in transmission. The converse is not however true. There exist crosses which show a bias in transmission without showing any bias in the frequency of recombinants." Numerous examples of the latter situation, presumably attributable to asymmetry or bias rather than polarity, are to be found in the papers of Rank & Bech-Hansen (120) and Howell et al (116).

2. TIMING OF RECOMBINATIONAL EVENTS   Most mitochondrial gene recombination apparently takes place in the zygote and, to a lesser extent, in the first buds of the zygote, following which segregation of genetically pure lines usually proceeds rapidly (79, 113, 119). Indirect evidence that most mitochondrial gene recombination took place in the zygote was provided by Coen et al (79). These authors examined the clonal distribution of $C^S E^S$ recombinants in a cross between $C^R E^S$ and $C^S E^R$ parents. They argued that the proportions of $C^S E^S$ recombinants in each zygote clone could be arranged in the series $(\frac{1}{2})^n$ where $n = 1, 2, 3, 4$, etc and that a simple model accounting for this distribution would be one in which successive zygote buds had equal probabilities of producing a pure $C^S E^S$ clone. Thus, if the first bud yielded a pure $C^S E^S$ clone, 50% of the cells in the zygote clone would have the genotype $C^S E^S$; if the second bud did so, then 25% of the cells would be $C^S E^S$; and so forth.

Pedigree analysis (113, 119) has subsequently shown that although the model of Coen et al is probably too simplistic in its assumption that buds pure for mitochondrial genotype are usually produced by the zygote, it is correct in assigning the majority of recombinational events to the zygote. These pedigree studies show that most buds contain a mixture of different mitochondrial genotypes which segregate rapidly during subsequent mitotic cell divisions. Smith et al (125), searching for physical evidence in support of mitochondrial genome recombination in the zygote, have argued, based on electron microscopic observations, that the mitochondria dedifferentiate in the young zygote and redifferentiate in older zygotes. They suggest that the disorganized condition of the mitochondria in young zygotes may correspond both to the period of genome recombination and to the time when a heterogeneous complement of mitochondrial genomes is transmitted to the buds.

Despite the fact that most zygotic progeny appear to segregate their mitochondrial genomes rapidly, some progeny cells may remain heterozygous in mitochondrial genotype for many cell generations as shown by Rank & Bech-Hansen (120).

These authors crossed $C^R E^S$ and $C^S E^R$ stocks, allowed the resulting zygotes to form colonies under nonselective conditions, isolated a number of daughter cell colonies from each zygote clone under nonselective conditions, and tested each daughter clone on selective media to see if these clones were heteroplasmic for parental or recombinant mitochondrial genomes or both. About 6% of the daughter clones contained a mixture of mitochondrial genotypes. These results show that the heteroplasmic state is maintained in a few cells for many generations.

3. THE POLAR GENES AND ω The polar genes are R_I, R_{II}, and R_{III} (109). Mutations at the R_I locus confer resistance to chloramphenicol (C^R) while those at the R_{II} and R_{III} loci usually cause resistance to erythromycin and spiramycin together but occasionally to one antibiotic or the other alone (108, 115). Unfortunately, the erythromycin-resistant mutants at both loci are designated E^R and the spiramycin-resistant mutants S^R by the Slonimski group. Since these symbols are generally used in the literature rather than R_{II} and R_{III}, it is frequently necessary to check allele numbers to be certain which locus is being discussed. For example, E^R_{354} is an R_{II} mutant but E^R_{221} is an R_{III} mutant. The three loci R_I, R_{II}, and R_{III} are so designated because each is believed to specify a mitochondrial ribosomal function (126).

Mitochondrial mutations resistant to chloramphenicol, erythromycin, and spiramycin have been reported by a number of workers, but the genetic homology of the mutations of one group to those of another is usually not clear. For example, Trembath et al (115) have obtained spiramycin-resistant mutants that fall into two and possibly three loci. One locus is defined by the mutations eryl-r, ery2-r, and spi2-r. The first two mutants are cross-resistant to erythromycin and spiramycin but the latter is resistant only to spiramycin. The mutants spi3-r and spi4-r are resistant only to spiramycin and fall into a second locus or possibly two tightly linked loci. It seems likely that the two or possibly three loci for erythromycin and spiramycin resistance defined by Trembath et al (115) will prove to be homologous with the R_{II} and R_{III} loci.

In the original paper on polarity and transmission of mitochondrial genes in S. cerevisiae Coen et al (79) showed that independently arising C^R and E^R mutations isolated in two mating type α strains were highly transmitted in monofactorial crosses whereas similar mutations isolated in an a mating type strain showed low or moderate transmission. For example, if the C^R allele was in an α mating type strain, the polarity of transmission was $C^R > C^s$, but if it was in the a mating type strain, the polarity of transmission was $C^R < C^S$. When bifactorial crosses were made between α mating type $C^R E^S$ parents and a mating type $C^S E^R$ parents it was found that the polarity of recombination was $C^S E^S / C^R E^R < 1.0$. In the reciprocal cross the polarity of recombination was $C^S E^S / C^R E^R > 1.0$. The authors were careful to point out that the relationship of polarity to cellular mating type was probably more apparent than real, but other workers, at the time, felt that the relationship was not fortuitous (127).

Coen et al also found that while the polarity of transmission of independently isolated C^R mutants was always high in α mating-type strains, it was quite variable in the a mating-type strains. They pointed out that these differences in polarity of transmission could be explained by assuming either that the C^R mutations were not allelic or by a phenomenon having to do with polarity itself. The authors presented evidence suggesting that the C^R mutations were indeed allelic and hypothesized that variations in the polarity of transmission could be explained if a polarity point (O) was postulated which varied in its position relative to the C^R gene in independently isolated C^R mutants. They used a vectorial representation to describe the polarity of transmission in which each mutant was placed at a distance from O reflecting its polarity of transmission. Those C^R alleles placed closest to O showed the highest transmission and those furthest away showed the lowest. Four polarity groups were distinguished using these criteria and were found to correspond to four polarity groups distinguished on the basis of polarity of recombination (i.e. the ratio of $C^S E^S / C^R E^R$ recombinants). The relationship between polarity of transmission and recombination could then be described by a model in which the a mating-type strain had a polarity point on its mitochondrial genome and was analogous to a donor in a bacterial cross with the α mating-type mitochondrial genome being a recipient.

The bacterial sexuality model was extended in the second paper by Bolotin et al (108) using specific alleles at the R_I, R_{II}, and R_{III} loci. These authors showed that polarity could not be attributed to cellular sex and that certain crosses showed polarity whereas others did not. These findings were explained in terms of a mitochondrial sex factor, ω, which had two states designated ω^+ and ω^-. In heterosexual crosses ($\omega^+ \times \omega^-$) polarity was seen, but in homosexual crosses ($\omega^+ \times \omega^+$; $\omega^- \times \omega^-$) there was no polarity.

Using the convention of mitochondrial sex, it was found that in heterosexual crosses of the type

$$\omega^- \ C^R E^S \times \omega^+ \ C^S E^R \text{ and } \omega^- \ C^S E^R \times \omega^+ \ C^R E^S$$

the majority recombinant carried the C allele from the ω^+ parent and the E allele from the ω^- parent with the converse being true of the minority parent. The convention used to represent this observation is to write the majority recombinant in a heterosexual cross as $C_\omega^+ E_\omega^-$ or $C^+ E^-$ and the minority recombinant as $C_\omega^- E_\omega^+$ or $C^- E^+$. The polarity of transmission is then indicated by the ratios C^+/C^-, E^+/E^-, etc and the polarity of recombination by the ratio $C^+ E^-/C^- E^+$ etc.

Bolotin et al (108) extended their experiments, which had been done with a C^R mutant in the R_I locus and an E^R mutation in the R_{III} locus, to an R_{II} mutant resistant to spiramycin (S^R). They proposed that the mutants were linked in the order $C^R–S^R–E^R$ for two reasons. First, in homosexual crosses the sum of recombinant types for the outside genes (i.e. R_I and R_{III}) was much lower than the sum of the nonrecombinants. Second, in heterosexual crosses the polarities of recombination were

$$C^+ S^-/C^- S^+ > C^+ E^-/C^- E^+$$

indicating the order R_I, R_{II}, R_{III}.

The experiments reported in the paper also showed that mitochondrial sex must be transmitted according to the following rules. In $\omega^- \times \omega^-$ homosexual crosses all diploid progeny, whether parental or recombinant, were ω^-. Similarly in the homosexual cross $\omega^+ \times \omega^+$ all diploid progeny were ω^+. In heterosexual crosses, cells of parental genotype retained their original sex but recombinants were ω^+. In addition it was found that uv irradiation of the ω^+ parent decreased the frequency of diploid cells in a heterosexual cross carrying the ω^+ parental genotype while increasing the frequency of recombinant cells and cells of the ω^- parental genotype. Transmission of the R_I and R_{III} alleles by the ω^+ parent was also affected with the R_{III} allele being more uv sensitive than the R_I allele. No effect of uv was noted on the ω^- parent.

These results were used by the authors in support of a bacterial model of mitochondrial heredity which supposed that injection of the ω^+ mitochondrial genome into ω^- mitochondria in heterosexual crosses occurred in the order $\omega^+–R_I–R_{II}–R_{III}$. This model seemed to explain why the majority recombinants were $R_I^+ R_{III}^-$, $R_I^+ R_{II}^-$ etc; why uv irradiation only affected transmission of ω^+ genetic markers; and why uv had a greater effect on the R_{III} allele than the R_I allele. As the authors pointed out, the latter observation could be explained in terms of the bacterial analog by assuming that R_{III} was further from ω than R_I as already suggested from the polarity of recombination. However, the analogy could not be strictly drawn since $\omega^- \times \omega^-$ homosexual crosses are possible and $F^- \times F^-$ crosses in bacteria are not.

Interestingly, there is an important inconsistency between the models of Coen et al (79) and Bolotin et al (108). The first model hypothesized that the polarity point O was located in a strain shown in the second paper to be ω^- (i.e. 55R5-3C). The model of Coen et al supposed that the ω^- parent was the donor and that the polarity point is localized at different points in different mutants. In the model of Bolotin et al (108) the ω^+ parent is the donor. This point may seem academic since the newest model of mitochondrial heredity (pp. 363–64) discards the bacterial analog entirely but it is brought up here because the relationship between ω, O, and the different polarity groups has yet to be explained satisfactorily.

4. THE NONPOLAR GENES   Two recent papers by Avner et al (109) on the mapping of the mitochondrial genes for oligomycin resistance (O_I^R, O_{II}^R) and by Wolf et al (110) on the mitochondrial gene for paromomycin resistance (P^R) are the most exhaustive studies yet published on mitochondrial gene recombination. The biochemical basis of resistance to oligomycin and paromomycin is discussed in other papers (114, 128–130). Suffice it to say that the oligomycin-resistant mutants probably fall in mitochondrial genes coding for membrane proteins associated with the mitochondrial ATPase complex and the mutations to paromomycin resistance cause protein synthesis in isolated mitochondria to become resistant to the drug.

The papers by Avner et al and Wolf et al serve, in part, as the departure point for the model of mitochondrial gene recombination discussed in the next section. Mitochondrial sex was established in their experiments by crossing the new strains which had the genotype $C^S E^S$ with respect to the R_I and R_{III} loci to $C^R E^R$ strains of known mitochondrial sex. The sex of the new strain was then ascertained from the polarity of recombination. For example, if the recombinant ratio $C^R E^S/C^S E^R \cong 1$ in a cross of the new strain with a ω^+ tester but is $\ll 1$ in a cross with a ω^- tester, the sex of the unknown must have been ω^+. In the course of sexing the new strains Avner et al discovered a probable case of asymmetry caused by a nuclear gene or genes in a new strain D6. It was found that while the mitochondrial sex of D6 could be established as ω^+ from the polarity of recombination in heterosexual crosses, one of the parental genotypes ($C^R E^R$) was highly transmitted in homosexual crosses. Analysis of a tetrad from a cross of the ω^+ D6 strain to a ω^- strain showed the expected nonsegregation of mitochondrial sex (see Bolotin et al 108) but also showed a 2:2 segregation for the tendency to transmit the parental $C^R E^R$ genotype in homosexual crosses.

The existence of two loci for oligomycin resistance was established by allele testing and by the use of genetically marked ρ^- mutants derived from a ρ^+ O^S strain.

Multifactorial heterosexual and homosexual crosses established that the O_I, O_{II}, and P loci are polar only with respect to the R_I, R_{II}, and R_{III} genes and that they are genetically unlinked to each other or to the R_I–R_{III} segment. In heterosexual crosses, recombinant ratios such as

$$C^+O^-/C^-O^+ \text{ or } C^+P^-/C^-P^+ \text{ were } > 1$$

but the ratio $O^+P^-/O^-P^+ = 1$. In homosexual crosses 20–25% recombinants were found when O_I^R, O_{II}^R, or P^R mutants were crossed to each other or to mutants in the R_I–R_{III} segment while recombination in the R_I–R_{III} region is only about 10%. Since the upper limit for recombination of genetically unlinked mitochondrial genes is probably 25% rather than 50% (see p. 363), it appears that R_I–R_{III} form one linkage group and O_I, O_{II}, and P are unlinked to each other or to the R_I–R_{III} linkage group.

Despite the apparent absence of formal genetic linkage between O_I, O_{II}, and R_I–R_{III} in recombination analysis, other experimental data support the notion that these genes are linked and the order of the genes is either ω–R_I–R_{II}–R_{III}–O_I–O_{II} or O_{II}–ω–R_I–R_{II}–R_{III}–O_I. One piece of evidence comes from a ρ^- mutant that was used to distinguish between the O_I and O_{II} loci by deletion mapping. The complete genotype of this petite is

$$\rho^- R_I^O R_{III}^R O_I^R O_{II}^O.$$

If one assumes that the retained sequences in a ρ^- mutant are adjacent, then R_{III} and O_I must be adjacent and O_{II} cannot fall between them. Furthermore, the order cannot be ω–R_I–O_{II}–R_{III}–O_I since R_I and R_{III} behave as if they are linked in homosexual crosses whereas O_{II} does not. Neither can the order be ω–O_{II}–R_I–R_{II}–R_{III}–O_I because this order would yield the $O_{II}^+ R_I^-$ recombinant as the majority recombinant in heterosexual crosses whereas this is the minority recombinant. Also, uv has a differential effect on transmission of the R_I, R_{III}, O_I, and O_{II} markers by ω^+ strains in heterosexual crosses with the least affected marker being R_I followed by R_{III}, O_I, and O_{II} in that order. The difference in the slopes of inactivation of marker transmission for O_I and O_{II} is slight but the authors suggest it is compatible with the order $\omega-O_I-O_{II}$ rather than $\omega-O_{II}-O_I$. However, this difference is of dubious significance since Dujon et al (111) point out in a later paper that in homosexual crosses the output of markers at different loci is equally affected by uv and the same is true of markers located in the nonpolar region in heterosexual crosses. So far the P locus does not seem to have been mapped by methods similar to those used for O_I and O_{II}.

Finally, the experimental data of Wolf et al make it possible to eliminate the bacterial conjugation model of polarity proposed by Bolotin et al (108). In the conjugation model the degree of polarity seen between a pair of markers in a heterosexual cross depends on the distance of each marker from ω. Suppose the order is $\omega-R_I-R_{III}-O_I$ and those recombinants carrying the ω^+ O_I marker in a heterosexual cross are examined for genotype with respect to the R_I and R_{III} genes. A sequential transfer hypothesis predicts that this class of recombinants should not be polar with respect to the R_I and R_{III} genes, as, by definition, the most distal marker (O_I) has been selected. However, polarity does exist and the polarity values for $C-E$, $C-O$, and $C-P$, which ought to differ on a sequential transfer hypothesis, are similar.

5. A PHAGE MODEL OF MITOCHONDRIAL GENE RECOMBINATION   The discovery of nonpolar mitochondrial genes, the finding that recombinants for supposedly "distal" nonpolar genes in heterosexual crosses are still polar, and extensive data from homosexual crosses have made the bacterial conjugation model of recombination unappealing. Dujon et al (111) have proposed a new model composed of two parts.

The first part of the model supposes that a cross of mitochondrial genomes is analogous to a phage cross. It is inspired by the population-genetics approach of Visconti & Delbrück (131) to the phage cross. Three assumptions are involved in the model. First, there is a *panmictic pool* of mitDNA molecules in a yeast zygote. Pairings occur between genetically homozygous mitDNA molecules and between heterozygous molecules with a probability related to the number of molecules of each type and not to genetic constitution. The genetically detectable events result from matings between molecules of different genotype. The observation that the upper limit of recombination between unlinked mitochondrial genes is 20–25% rather than 50% is largely responsible for the notion of a panmictic pool.

Second, the genetic composition of the panmictic pool is determined by the proportion of molecules derived from one parent (*input fraction*) relative to the total number of molecules. The input fraction cannot be determined experimentally but must be deduced after the fact from the fraction of diploid progeny of a cross carrying a given allele (*output fraction*). It appears that the output is a good estimate of the input since there is a *covariance of transmission* of markers derived from one parent in homosexual crosses and for nonpolar markers in heterosexual crosses. The authors give an example from a homosexual cross in which the output fractions for the mating-type a parent were 0.48 for the C allele, 0.46 for the O allele, and 0.49 for the P allele. Results such as these also indicate linkage of these genes on the same mitDNA molecule even though they are unlinked in terms of percent recombination. Third, the panmictic pool of molecules can undergo several random in time *mating rounds.* There are several pairing events in the line of ancestry of an average mitDNA molecule, therefore, before a cell line becomes genetically homozygous for a given species of mitDNA. The genetic composition of the panmictic pool evolves as a function of the number of rounds of mating.

The second part of the model deals with the mechanism of recombination of mitDNA molecules. It assumes that mitDNA recombination is fundamentally nonreciprocal. One elementary act produces one type of recombinant, another elementary act produces another recombinant, and so forth. In the case of polar genes it is assumed that in a heterosexual cross an obligatory recombinational event takes place every time a ω^+ molecule pairs with a ω^- molecule. The process is initiated at the ω^+ locus, and the probability of "conversion" of an allele carried by the ω^- parent to that carried by the ω^+ parent decreases as one proceeds from ω^+ to R_{III}. Thus all recombinants are ω^+ and the polarity with respect to the $R_I - R_{III}$ segment is $R_I^+ > R_{II}^+ > R_{III}^+$. The scheme generates one ω^+ parental genome and one ω^+ recombinant for each mating round.

The model postulates that the number of mating rounds is similar in homosexual and heterosexual crosses and for both polar and nonpolar regions in heterosexual crosses. Given this assumption the act of recombination must also be nonreciprocal for nonpolar genes in heterosexual crosses and for both polar and nonpolar genes in homosexual crosses because reciprocal recombination would generate two rather than one recombinant per elementary act. If reciprocal recombination occurred for both nonpolar genes and polar genes in homosexual crosses, one would have to postulate additional rounds of mating for polar genes in a heterosexual cross.

The model of Dujon et al makes certain testable quantitative predictions. First, as mentioned already, the output of every allele should be proportional to its input and therefore the output of all alleles at different loci on the same mitDNA molecule should be identical. This proves to be the case. The model also predicts that the output of alleles in nonpolar regions is an adequate estimate of the input fraction of mitDNA molecules and therefore of polar markers in a heterosexual cross. This also appears to be true. Second, the frequency of recombination between markers should vary from cross to cross as a function of input. In a homosexual cross the frequency of recombinants should be a parabolic function of the input with the maximum frequency of recombinants predicted for an equal input of the two parents. In heterosexual crosses involving polar markers, the model predicts that the frequency of recombinants should be maximal for an input biased in favor of the ω^- parent. These predictions are fulfilled. Third, the model predicts the frequency of every type of mitDNA molecule, parental or recombinant, after any number of mating rounds when the input fraction and probabilities of exchange are known. The authors show by example that the model quantitatively predicts the actual data.

The model of Dujon et al is revolutionary in its approach to the recombination of organelle genomes in terms of the population genetics of a phage cross. As we shall see, the same kind of model can be applied to recombination of putative chloroplast genomes in *Chlamydomonas* (pp. 377–78).

CHLOROPLAST GENETICS OF CHLAMYDOMONAS

Non-Mendelian Inheritance and the Chloroplast Genome

Since 1954 when Sager (132) first reported that mutations resistant to high levels of streptomycin exhibited non-Mendelian inheritance, an array of new mutations, most resistant to other antibacterial antibiotics, has been isolated showing the same pattern of inheritance (9, 133–135). In the literature these mutations have been referred to with a variety of terms including, "cytoplasmic," "chloroplast," "non-Mendelian," "non-chromosomal", and "uniparental." There is now sufficient evidence that these mutations are in the chloroplast to refer to them as chloroplast mutations. This evidence, summarized in several sections which follow, consists of experiments suggesting that chloroplast DNA shows the same pattern of inheritance as putative chloroplast mutants, that chloroplast mutants can be induced selectively at the time of chloroplast DNA replication, that mitochondrial mutations show a pattern of inheritance distinct from that seen for chloroplast mutations, and that chloroplast mutations to antibiotic resistance alter chloroplast ribosomal proteins.

Inheritance of Chloroplast Mutations

In crosses, the mating-type plus (mt^+) or *maternal* parent transmits the chloroplast marker(s) it carries to all four meiotic products in 90% or more of all zygotes (*maternal zygotes*). Chloroplast genes from the mating-type minus (mt^-) or *paternal* parent are transmitted in 10% or less of the zygotes (*exceptional zygotes*) (9, 133). Two kinds of exceptional zygotes occur. *Biparental zygotes* transmit chloroplast genes from both parents to the meiotic products. In the progeny of such zygotes chloroplast genes, unlike Mendelian genes, continue to segregate during the post-meiotic, mitotic divisions as well as during the initial meiotic divisions (9, 133). *Paternal zygotes*, the rarest class of all, transmit only chloroplast genes carried by the mt^- parent. The terms maternal and paternal are used advisedly since, in zygote formation in *Chlamydomonas*, gametes of opposite mating type fuse completely with the chloroplasts as well as the nuclei fusing (136, 137).

Inheritance of Chloroplast DNA

Four species of DNA are known to occur in vegetative cells of *C. reinhardtii* (138). The major species, α, is assumed to be nuclear DNA since it constitutes 85% of the total DNA of the cell. This DNA has a buoyant density of 1.723 g/cc; a kinetic complexity of 4.5×10^{10} daltons; and, in a haploid cell, is composed mostly of unique sequences (139). The β DNA, localized in the chloroplast (140), has a buoyant density of 1.695 g/cc (138) and is composed of slow and fast renaturing fractions (139). The slow renaturing fraction has kinetic complexity of 1.9 to 2×10^8 daltons and is repeated some 25 times in a gamete and 50 times in a vegetative cell (139, 141). It is not known whether this DNA is organized in separate pieces or into repeating sequences in larger molecules. The fast renaturing fraction comprises 10% of the chloroplast DNA and has a kinetic complexity of $1–10 \times 10^6$ daltons (139). γ DNA, which hybridizes specifically with cytoplasmic ribosomal RNA (142), has a buoyant density of 1.715 g/cc and comprises 1% of the total DNA in a vegetative cell (138). Mitochondrial DNA, very recently isolated from *Chlamydomonas* (143), amounts to 0.5–1.0% of the cellular DNA and has a buoyant density of 1.706 g/cc. This DNA was originally reported to renature at the same rate as T4 (143) implying a kinetic complexity around 2×10^8 daltons, but more recent work shows the kinetic complexity to be about 1.6×10^7 daltons (144) making it similar in complexity to vertebrate mitDNA rather than the more complex mitDNA of fungi and higher plants (17, 145). It can be estimated that there are about 40 copies of mitDNA per cell (144). From the kinetic complexity, nuclear, chloroplast, and mitDNA can be estimated respectively to contain 99.5, 0.4, 0.04% of the total genetic information of the cell. A fifth species of DNA, called M (maturation)-band DNA, was reported to be synthesized in large amounts early in zygote maturation by Sueoka & Chiang (138). This DNA has the same buoyant density as γ DNA. M-band DNA has not been reported by Sager & Lane (146) in their studies of DNA replication during zygote maturation.

Because of the uniparental pattern of inheritance of chloroplast genes, one might expect only the chloroplast DNA of the mt^+ parent to be transmitted in a cross since the great majority of zygotes are maternal. Chiang (147) was the first to ascertain whether this was the case. He concluded that α, β, and γ DNA from both parents was conserved in the zygote from experiments in which the DNA of one parent was labeled with ³H-adenine and the other with ¹⁴C-adenine. A cold adenine chase during gametogenesis and zygote maturation was used to minimize problems which might arise from DNA turnover or a pool of radioactive precursors. Since unlabeled M-band was synthesized in the zygote, Chiang assumed the chase was effective. Later Chiang (148) combined radioactive and density labeling of the DNA in the same experiment. The two parents were labeled with different radioactive isotopes as before but in addition one was labeled with ¹⁴N and the other with ¹⁵N. At 30 hr after mating, heavy and light species of α DNA were observed (α DNA does not replicate until late in zygote maturation), but only a single density species of β DNA was seen and it contained radioactivity from both parents. Once again, chasing with cold precursors was used to eliminate problems with pools of radioactive precursors and turnover of radioactive DNA. Chiang interpreted his results as indicating extensive recombination between chloroplast DNA from both parents shortly after zygote formation.

Sager & Lane (146) report on the basis of density labeling experiments that within 6 hr after mating only the chloroplast DNA of the mt^+ parent remains. In these experiments optical density profiles were followed directly and radioactive isotopic tags were unnecessary since M-band synthesis did not occur. If the cross mt^+ $(^{14}N) \times mt^-(^{15}N)$ was made, the chloroplast DNA was light, but if the reciprocal cross was made, it was heavy. These results were interpreted to mean that chloroplast DNA from the mt^+ but not the mt^- parent was transmitted in crosses and that chloroplast DNA from the mt^- parent was destroyed. A minor complication in these experiments was the density shifting of the chloroplast DNA of the zygote to the light side of where it was expected. Sager & Lane suggested that this shift could be accounted for by methylation.

The seeming conflict between the results of Sager and Chiang may be caused by a pool problem. Siersma & Chiang (149) found that approximately 90% of the chloroplast and cytoplasmic ribosomes were degraded during gametogenesis. Much of the radioactivity released by ribosome degradation in the gametes was used in gametic DNA synthesis or was to be found in the cellular pool. The labeled compound used in these experiments, adenine, was the same one used by Chiang in his experiments on the inheritance of chloroplast DNA. If the pool of precursors used for the synthesis of chloroplast DNA in the zygote is effectively isolated from the rest of the cell, the apparent chloroplast DNA conservation seen by Chiang can be accounted for. At the time of chloroplast fusion in the zygote this pool would include both ^{14}C and ^{3}H precursors derived by degradation of chloroplast ribosomes in both parents. A pool problem of this type was postulated by Reich & Luck (150) to account for the fact that mitDNA in *Neurospora* did not appear to replicate semiconservatively.

Obviously, pool problems of this kind may also influence the experiments of Sager and Lane. If there were preferential degradation of chloroplast ribosomes in their mt^+ gametes, the nitrogen used for chloroplast DNA synthesis in the zygote would be derived largely from the mt^+ parent. However, such preferential ribosome degradation seems unlikely in view of Chiang's original results on the conservation of radioactive label from both parents in chloroplast DNA.

The finding that thymidine is preferentially incorporated into chloroplast DNA in *Chlamydomonas* (151) should provide the escape from the dilemma of pool problems. Experiments with radioactive thymidine should unequivocally prove or disprove the uniparental inheritance of chloroplast DNA in *Chlamydomonas*. Since recent genetic experiments suggest a biparental mode of inheritance for mitochondrial genes in *C. reinhardtii* (pp. 368–69) it seems likely that mitDNA will prove to be transmitted uniparentally.

Selective Induction of Mutations in Chloroplast DNA

Lee & Jones (152) have used selective mutagenesis of replicating DNA as a means of localizing chloroplast genes in *Chlamydomonas*. They mutagenized synchronously growing cells of the alga with N-methyl-N'-nitro-N-nitrosoguanidine (MNNG) which is an effective mutagen for both Mendelian and chloroplast genes in *C. reinhardtii* (153). In *E. coli* MNNG selectively mutates replicating DNA (154). As an assay Lee & Jones measured the frequency of Mendelian (*sr-1*) and chloroplast (*sr-2*) mutations to streptomycin resistance. These mutations are easily distinguished on the basis of resistance level (132).

MNNG mutagenesis during nuclear DNA replication increased the frequency of *sr-1* mutants 15–30-fold with no increase in *sr-2* mutants, but mutagenesis at the time of chloroplast DNA replication yielded only a 1.5–1.6-fold increase in the frequency of *sr-2* mutants. This was accompanied by a similar increase in the level of Mendelian *sr-1* mutants. Careful repetition revealed that this increase was highly significant statistically for both mutant classes.

The explanation for the increase in *sr-1* mutants at the time of chloroplast DNA replication appears to be that between 2 and 3% of the cells are out of synchrony and replicating nuclear DNA while the rest are replicating chloroplast DNA. Since *sr-1* mutants increase 15–30-fold following MNNG treatment during the normal period of nuclear DNA replication, the increase seen as a result of asynchrony at the time of chloroplast DNA replication is not surprising.

The reason for the small increase of *sr-2* mutants at the time of chloroplast DNA replication is unknown but might have to do with problems of expression since there are many copies of chloroplast DNA in each cell. Although the Lee and Jones experiments indicate that γ DNA is not involved in the increase in *sr-2* mutants since this DNA replicates at a later time than chloroplast DNA, they do not eliminate mitDNA since the time of replication of this DNA is unknown.

Mitochondrial Mutations in Chlamydomonas

C. reinhardtii can be grown photosynthetically with CO_2 as the sole carbon source (*phototrophic growth*); in the dark with acetate as the carbon source (*heterotrophic growth*); or in the light with both acetate and CO_2 as carbon sources (*mixotrophic growth*). Recently, Alexander et al (155) have found that acriflavin preferentially inhibits heterotrophic growth in *Chlamydomonas* and induces a class of lethal mutations termed "minutes" because they divide 8–9 times to form a small colony before dying. These mutants are induced with 100% efficiency no matter what growth condition is used. Alexander et al argue that both heterotrophic growth inhibition and minute induction are the result of preferential interaction of acriflavin with mitDNA.

The inhibition of heterotrophic growth of cells by acriflavin is assumed to result from inhibition of mitDNA transcription by the dye (cf 27). Support for this idea comes from several observations. First, acetate assimilation is not impaired by the dye. Second, acriflavin-treated cells show clear ultrastructural changes in the mitochondria but the chloroplast appears normal. Third, cyanide-sensitive respiration is reduced 50% after one generation of heterotrophic growth in the presence of acriflavin. Cytochrome oxidase is the terminal cytochrome system in cyanide-sensitive respiration, and the synthesis of cytochrome oxidase in yeast (156) and *Neurospora* (157) is dependent on mitochondrial protein synthesis and therefore on mitDNA since mitDNA codes for some of the components of the mitochondrial protein synthesizing system (cf 27). Other support for the preferential interaction of acriflavin with mitDNA in *Chlamydomonas* comes from the experiments of Stegeman & Hoober (158). These authors report that acriflavin and ethidium bromide preferentially inhibit synthesis of polypeptides believed to be of mitochondrial origin.

Despite the eventual lethality of minute mutations, they can be crossed since a cell in which a minute mutation occurs divides several times. In reciprocal crosses of minute mutants to wild type virtually all tetrads segregate either 4 wild type:0 minute or 4 minute:0 wild type. The latter tetrads are the most frequent up to about five days following zygote formation in both crosses but after that time a repair process is evident particularly when mt^+ minute mutants are crossed to wild-type mt^- cells. Repair is seen as the conversion of a tetrad segregating 4 minute:0 wild type to one segregating 4 wild type:0 minute progeny. These results show the pattern of inheritance of the minute phenotype to be distinct from that seen for chloroplast genes.

Alexander et al (155) argue that they are inducing mitochondrial mutations equivalent to highly suppressive yeast petites with acriflavin but that these mutations are lethal in *Chlamydomonas* after several cell divisions. Since the minute mutation is almost always transmitted to all four progeny in reciprocal crosses, the authors argue the mitochondrial genomes derived from both parents cannot remain separated in the zygote. Instead, the mitochondrial genomes from both parents must form a pool in the zygote and the mutant phenotype is then transmitted to all four progeny because it is highly suppressive to wild-type mitDNA (see pp. 354–56).

Flechtner & Sager (159) reported a reversible inhibition of chloroplast DNA replication by ethidium bromide under phototrophic conditions. Under heterotrophic conditions both nuclear and chloroplast DNA replication were inhibited by ethidium bromide. Alexander et al (155) point out that ethidium bromide induces minute mutations at concentrations lower than those used by Flechtner & Sager (159). Alexander et al argue that the effects seen with acriflavin and ethidium bromide are probably highly concentration dependent with the mitochondrion being more sensitive than the chloroplast. They also note that the differential effects of ethidium bromide on DNA synthesis seen by Flechtner and Sager under phototrophic and heterotrophic conditions can be explained if the dye, by blocking mitochondrial transcription, prevents cells from generating the energy required for DNA replication under heterotrophic conditions. This energy would be provided under phototrophic conditions by the chloroplast. Interestingly, Nass & Ben-Shaul (160) report a preferential effect of ethidium bromide on the mitochondria of *Euglena*.

The Biogenesis of Organelle Ribosomes and the Phenomenon of Antibiotic Resistance

There is considerable direct evidence that the biogenesis of chloroplast ribosomes in *Chlamydomonas* is under the control of both chloroplast and nuclear genes. Chloroplast ribosomes from wild-type cells are sensitive to antibiotics such as chloramphenicol, erythromycin, streptomycin, spectinomycin, and neamine as shown by in vitro studies of radioactive antibiotic binding and polyuridylic acid-directed phenylalanine incorporation in the presence of antibiotic (161–167). Chloroplast mutations resistant to these antibiotics have resistant chloroplast ribosomes by one or both of these criteria (161–166). Specific ribosomal protein changes similar to those seen in bacterial ribosomes have been demonstrated in three resistant chloroplast mutants (161, 163, 164).

Mendelian mutations resistant to antibacterial antibiotics in *C. reinhardtii* are also known. Mets & Bogorad (135) have shown that mutations in either of two nuclear genes cause chloroplast ribosomes to lose their ability to bind radioactive erythromycin. A mutation in one of these genes alters a single protein of the large chloroplast ribosomal subunit (163). In addition nonallelic nuclear mutants are known having gross deficiencies in their ability to make normal amounts of chloroplast ribosomes (167–173, 175, 178). Although these mutants show a characteristic syndrome of defects in photosynthesis, including inability to make the CO_2-fixing enzyme ribulose diphosphate carboxylase (RuDPCase) in normal amounts, they will grow when acetate is supplied as a carbon source. Hence, chloroplast protein synthesis appears to be dispensable as long as acetate is supplied to the organism. This notion is also supported by long-term growth of cells on acetate in the presence of rifampicin, a specific inhibitor of chloroplast DNA transcription (174–176). Cells grown in this fashion cannot form chloroplast ribosomes (175, 176) since chloroplast RNA is a transcript of chloroplast DNA and are phenocopies of chloroplast ribosome-deficient mutants (177).

It now appears that both chloroplast and nuclear genes may also be involved in the biogenesis of mitochondrial ribosomes. Surzycki & Gillham (134) proposed a classification for antibiotic-resistant mutants in *Chlamydomonas* which supposed that both the chloroplast and mitochondrion were sensitive to antibacterial antibiotics. Their hypothesis was prompted by the observation that chloroplast protein synthesis is dispensable when acetate is supplied as a carbon source, but all antibiotics tested, with the possible exception of spectinomycin (175), inhibit growth at similar concentrations whether CO_2 or acetate is the carbon source. Surzycki & Gillham supposed that antibiotic-resistant mutants might fall into one of three phenotypic classes. Class I mutations would have a resistant chloroplast and a sensitive mitochondrion. Such mutations could grow photosynthetically in the presence of antibiotic but not in the dark with acetate as the carbon source. It was imagined that photophosphorylation might provide the necessary energy for growth in the light. Class II mutants, with a resistant mitochondrion and a sensitive chloroplast, would grow in the presence of antibiotic when acetate was supplied as a carbon source but could not grow phototrophically in the presence of antibiotic. Class III mutants were those in which both chloroplast and mitochondrion were resistant to antibiotics. Surzycki & Gillham (134) pointed out that mutations reducing cellular permeability to antibiotic would also have the class III phenotype.

Most mutants so far isolated appear to belong to class III and are not cell permeability mutants since they have antibiotic-resistant chloroplast ribosomes (161–166). Class I mutants have not been found, but two nuclear mutations resistant to spectinomycin and belonging to class II are known (134, 162). These results suggest that while chloroplast protein synthesis is dispensable when acetate is provided as a carbon source, mitochondrial protein synthesis may be indispensable under all conditions. Hence, class I mutants would die when plated on antibiotic and so would never be found, but class II mutants could be isolated on antibiotic plates as long as acetate was present in the medium while class III mutants could be isolated under any growth condition.

Since class III mutants appear to make chloroplast ribosomes antibiotic resistant, one might suppose that the mechanism of mitochondrial resistance would involve a resistant protein shared between chloroplast and mitochondrial ribosomes. Recently, Boynton et al (161) have proposed just such an hypothesis in explanation of their experiments with a chloroplast mutation to spectinomycin resistance (*spr-1-27*). In vitro the chloroplast ribosomes of wild type bind radioactive spectinomycin but those of the mutant do not (162). Examination of the kinetics of spectinomycin binding indicate that the spectinomycin binding site, which resides in the small ribosomal subunit (161), is actually lost in the mutant chloroplast ribosomes. This is supported by the finding that a specific small ribosomal protein is missing in gels. The missing protein corresponds in electrophoretic mobility to the spectinomycin (S5) protein of *E. coli* (179). In vivo, however, it is clear that chloroplast protein synthesis must be as sensitive to spectinomycin in the mutant as it is in wild type. Only if acetate is provided as a carbon source can the mutant grow in high concentrations of spectinomycin. Under these conditions the typical syndrome of defects associated with loss of chloroplast protein synthesis occurs including loss of the ability to synthesize RuDPCase. On the other hand, respiration remains normal in the mutant and mitochondrial ultrastructure does not change. Boynton et al (161) propose that *spr-1-27* alters a ribosomal protein shared by both chloroplast and mitochondrial ribosomes.

Because of differences in structure of the two classes of ribosomes, the protein alteration is expressed in vivo in terms of resistance in the mitochondrion and sensitivity in the chloroplast. During ribosome isolation it is supposed that the protein dissociates from the mutant ribosomes because, although its physical alteration does not confer antibiotic resistance to the chloroplast ribosome, it does cause the protein to be associated with the chloroplast ribosome in a less stable fashion. Dissociation of the mutant protein from chloroplast ribosomes during isolation leads to the loss of the spectinomycin binding site and so resistance is expressed in vitro in the binding assay.

In summary, the data so far accumulated show unequivocally that the biogenesis of chloroplast ribosomes is under the control of nuclear chloroplast genes. They also suggest that these genes may contribute to the biogenesis of mitochondrial ribosomes in *Chlamydomonas*.

The Mechanism of Uniparental Inheritance

Conceivably the uniparental pattern of inheritance of chloroplast genes in *Chlamydomonas* could be explained in the same way as polarity in yeast. Such an hypothesis would predict that the mt^+ strains studied thus far were also ω^+ and the mt^- strains ω^-. In her original paper on non-Mendelian inheritance in *Chlamydomonas,* Sager (132) presented evidence that this was not the case for the *sr-2* mutation. She showed, on the basis of backcrosses, that the inheritance of the *sr-2* mutation was strictly related to cellular mating type. This point has been documented in considerable detail by Chu Der & Chiang (180) for three different chloroplast markers.

Recently, Sager & Ramanis (181) have proposed a mechanism for uniparental inheritance of chloroplast genomes in *Chlamydomonas* akin to host-induced modification and restriction in bacteria (182). This model was set forth several years ago when Sager & Ramanis (183) found that uv irradiation of the mt^+ parent prior to mating markedly increased the frequency of exceptional zygotes in a cross while irradiation of the mt^- parent had no such effect. The exceptional zygotes included both biparental and paternal zygotes with the latter class gradually replacing the former as the uv dose was increased. When zygotes were treated with photoreactivating light shortly after mating, the yield of biparental zygotes as a fraction of the total increased at the expense of paternal zygotes.

Sager & Ramanis (183), postulated two uv effects, both photoreactivable: the conversion of maternal zygotes to biparental zygotes, and the conversion of biparental zygotes to paternal zygotes. It was supposed that the first process might involve an enzyme made in the mt^+ parent and required for the loss of paternal chloroplast genomes. The synthesis of this enzyme would be blocked by uv irradiation of the mt^+ parent prior to mating by a mechanism analogous to uv blockage of repressor synthesis during induction of lysogenic bacteriophage such as λ. The second process would involve the direct interference of uv with the replication of maternal genomes causing the formation of paternal zygotes. Sager & Lane (146) subsequently presented evidence that only the chloroplast DNA from the maternal parent was transmitted in crosses but that this DNA is density-shifted to the light side of where it would be expected (see pp. 366–67). They postulated that the density shift could be accounted for by methylation of the mt^+ chloroplast DNA by a modification enzyme with the paternal chloroplast DNA being degraded by a restriction enzyme.

The most recent paper by Sager & Ramanis (181), based on experiments with inhibitors and a new mutant called mat, makes the model even more explicit. They find that treatment of gametes or cells undergoing gametogenesis with certain inhibitors increases the frequency of exceptional zygotes. However, this increase depends on which mating type is treated with a specific inhibitor. For example, treatment of the mt^+ parent with ethidium bromide or ethionine increases the frequency of exceptional zygotes while erythromycin, spiramycin, and cycloheximide cause a similar increase when the mt^- parent is treated. Cycloheximide treatment of the mt^- but not the mt^+ parent in addition causes a large decrease in zygote viability with the proportion of exceptional zygotes conspicuously enriched among the surviving zygotes. The mat mutation, isolated in an mt^- stock, seems to have the same sort of effect as cycloheximide.

Based on these and the earlier results Sager and Ramanis proposed that the chloroplast of mt^- parent contains an inactive restriction enzyme while that of the mt^+ parent includes an inactive modification enzyme. In addition mt^+ cells have in their cytoplasm two regulatory substances called G_1 and G_2. In the newly formed zygote prior to chloroplast fusion G_1 activates the modification enzyme and G_2 the restriction enzyme. By the time of chloroplast fusion the modification enzyme has methylated the maternal chloroplast DNA accounting for the density shifting seen 6 hr after zygote formation, while the restriction enzyme has destroyed the paternal chloroplast DNA so only maternal chloroplast DNA is observed in the young zygote. Obviously, neither of these enzymes could be active in vegetative cells or the chloroplast DNA of the mt^+ parent would have the buoyant density seen in the young zygote and the mt^- parent would have no chloroplast DNA. The effect of uv would be to prevent the release of G_2 and thus the activation of the restriction enzyme. The inhibitor results do not add a great deal to this picture except to show that processes initiated in both parents may be important in determining the frequency of exceptional zygotes. Neither is it clear why the *mat* gene causes such low survival with the selection of exceptional zygotes.

Sager (personal communication) has found that the original *mat* gene, now designated *mat-1*, is tightly linked to the mt^- allele and a new gene, *mat-2*, is tightly linked to the mt^+ allele. It is proposed that the *mat-1* gene is the structural gene for the restriction enzyme and that the *mat-1* mutation results in an altered enzyme with decreased recognition for paternal chloroplast DNA and an increased recognition for maternal chloroplast DNA. The *mat-2* mutation causes a decrease in the frequency of both spontaneous and uv-induced exceptional zygotes. It is proposed that the *mat-2* gene codes for G_2 and that the *mat-2* mutation either makes a mutant G_2 more effective in activating the restriction enzyme than wild-type G_2 or else makes a greater total quantity of G_2 than wild type. The tight linkage of *mat-1* and *mat-2* to mating type is not the first such instance of tight linkage. Complementary auxotrophic mutations requiring thiamine, nicotinimide, and acetate are also tightly linked to the mating-type alleles and attempts to obtain recombinants in this region have proven fruitless (133). It may be that crossover suppression operates in the mating-type region to ensure, among other things, that the genes responsible for maternal transmission of chloroplast DNA are retained in the proper configuration.

Gillham et al (187) have proposed a different mechanism for uniparental inheritance. They hypothesize that maternal and paternal chloroplast genomes can be likened to imcompatible plasmids in bacteria (199) and there is a competition for a fixed number of attachment sites in the zygote between maternal and paternal chloroplast genomes. Such attachment sites are preferentially occupied by maternal genomes and the "unattached" paternal DNA is then either destroyed or diluted out because it cannot replicate. The observation that following uv paternal zygotes are photoreactivated to biparental zygotes at a faster rate than biparental zygotes are photoreactivated to maternal zygotes is explained by a single uv effect. Conversion of a paternal zygote to a biparental zygote requires the activation of only a single nonfunctional maternal genome, but conversion of a biparental zygote to a maternal zygote requires conversion of all nonfunctional maternal genomes to an active state. This model is inspired by that of Uzzo & Lyman for *Euglena* (200) in which it was hypothesized that only half of the DNA molecules in a plastid are replicative copies and that the replicative copies are tightly membrane bound.

Segregation and Recombination of Chloroplast Genes in Chlamydomonas

Most of the published data have been reviewed recently (9, 133, 184–186), notably in Sager's monograph (9), and a formal model of chloroplast genome segregation and recombination has been proposed by Sager. Recently, a new model has been proposed by Gillham et al (187) which retains some of the elements of Sager's model but rejects the notion that a vegetative cell is diploid for its chloroplast genome as proposed by Sager. The model of Gillham et al has similarities to that of Dujon et al (111) for the recombination of mitochondrial genomes in yeast. To see why a new model has been proposed it is first important to review the methodology that has been used in *Chlamydomonas* to see how different experimental approaches have led to conflicting results.

1.METHODOLOGY AND TERMINOLOGY   In contrast to *Saccharomyces,* where the inheritance of mitochondrial genomes has been studied in the vegetative diploid progeny of a cross, most experiments on the inheritance of chloroplast genes in *Chlamydomonas* have been done on the haploid progeny of biparental sexual zygotes despite the fact that vegetative diploids can be used for this purpose too (133). A method comparable to the standard cross in *Saccharomyces* (p. 357) has not been used in *Chlamydomonas.* Instead the progeny of individual biparental zygotes have been examined either by pedigree or zygote clone analysis. Sager has used pedigree analysis as a principal tool in studying the inheritance of chloroplast genes in *Chlamydomonas.* In Sager's method the progeny of a biparental zygote are replated after one postmeiotic mitotic division (octospore stage) and the two daughter cells produced by each octospore are separated and allowed to form colonies (188). The meiotic product from which each pair of octospores was derived is then determined by classifying each of the 16 octospore daughter colonies for three pairs of unlinked Mendelian genes which, because of recombination, usually mark the progeny of each meiotic product in a tetrad in a distinctive fashion.

The principal advantage of Sager's pedigree method is that one can accurately discern the patterns of segregation and recombination of chloroplast genes at meiosis and during the first two postmeiotic mitotic doublings. The principal disadvantage is that it is necessary to use some selective technique to enrich for biparental zygotes in the population. In their earlier experiments Sager & Ramanis (189, 190) obtained spontaneous biparental zygotes through selection *for* a specific chloroplast allele carried by the mt^- parent and *against* an allele carried by the mt^+ parent. For example, in a cross of an acetate-requiring (ac_1) streptomycin-dependent (sd) mt^+ strain with a mt^- streptomycin-resistant (sr-2) strain carrying a different acetate mutant (ac_2), biparental zygotes were recovered by plating the zygotes on an acetate-containing medium lacking streptomycin. Although this method eliminates maternal zygotes from the population since they are streptomycin dependent, it may also be selective for that class of biparental zygotes with the most favorable ratio of sr-2 to sd alleles for survival in the absence of streptomycin. This in turn could bias the results of experiments designed to determine the normal pattern of segregation of chloroplast genes.

The marker selection method was abandoned by Sager & Ramanis (188) when they found that ultraviolet irradiation of the mt^+ parent greatly increased the frequency of biparental zygotes. However, the use of uv to increase the frequency of biparental zygotes may also bias the results of segregation studies since, once again, one may be selecting against chloroplast genomes from the mt^+ parent (see p. 373). Since uv light induces the production of biparental zygotes with high efficiency, it is relatively easy to study the rate of segregation of different chloroplast genes among their progeny either directly in pedigrees or by a modification of the pedigree method (9, 186, 191). In the modified method the progeny of biparental zygotes are individually distributed to tubes, sampled, and classified by plating at successive doublings. In this way it is possible to determine the fraction of cells heterozygous for a pair of alleles at each doubling. The disappearance of heterozygous cells from the population appears to be characteristic for each marker and forms the basis of one of the principal methods used by Sager and Ramanis to map chloroplast genes in *Chlamydomonas* (see pp. 380–81).

Zygote clone analysis rather than pedigree analysis has been used by Gillham (187, 192, 193) to study the inheritance of chloroplast genes in *Chlamydomonas*. The method has the advantage of allowing one to work with large populations of biparental zygotes including those that have not been obtained by selection against the chloroplast genomes of the maternal parent. In the zygote clone method a cross is made between stocks carrying different chloroplast markers to antibiotic resistance. The zygotes are plated on nonselective medium and allowed to form colonies which are replica-plated to nonselective medium and to differentially selective media which select for the resistance markers carried by each parent. Zygote colonies containing cells capable of growing on antibiotics to which both maternal and paternal parents were resistant are biparental zygote colonies. These colonies are picked from the nonselective replica, the cells replated on nonselective medium, and the colonies derived from these cells are replica-plated to a set of diagnostic selective media which allow determination of the frequency of each chloroplast genotype in the zygote clone. This method has the disadvantages that it does not permit the determination of segregation pattern and that the zygote clone is sampled for genotype frequencies many cell generations after meiosis. Hence, intercellular selection for or against certain genotypes may introduce artifacts in zygote clone analysis.

As in the case of yeast mitochondrial genetics, it will be useful to introduce several terms here.

Allelic ratio: This term refers to the ratio of a pair of chloroplast alleles among the progeny of a biparental zygote as determined by pedigree (189) or zygote clone (187) analysis.

Polarity (188): This term is used in a sense totally different than it is in yeast. In *Chlamydomonas* polarity refers to the segregation of different pairs of chloroplast alleles in pedigrees with respect to one another. Thus a polarity of segregation for different pairs of alleles is seen such that certain pairs appear to segregate more rapidly than others.

Segregation type (191): In theory three segregation patterns can be distinguished for pairs of chloroplast alleles and all three are seen in pedigrees. *Type I* segregation produces two daughter cells heterozygous for a pair of alleles. In *type II* segregation one daughter cell is heterozygous for the allele pair and the other is homozygous for one allele or the other of the pair. In *type III* segregation one daughter is homozygous for one of the alleles (e.g. a) and the other daughter is homozygous for the second allele (e.g. a^+).

2. SEGREGATION PATTERNS AND ALLELIC RATIO   Sager (189–191) has reported that cells heterozygous for a pair of chloroplast alleles on the average segregate daughters homozygous for each member of an allelic pair in a 1:1 ratio. This is true whether the heterozygous cells are derived from uv-induced biparental zygotes or from spontaneous biparental zygotes obtained by selection for a chloroplast marker carried by the mt^- parent. Since type III segregations, by definition, segregate alleles in a 1:1 ratio, these observations lead to the conclusion that for any

pair of alleles type II segregations must yield daughters homozygous for each member of the allelic pair with equal frequency. That is, if we consider an allelic pair a^+/a, approximately half of all type II segregations will yield a daughter homozygous for a^+ and the other half a daughter homozygous for a. However, Sager (9) does remark that temporary deviations from this 1:1 ratio can be seen in the early postmeiotic mitotic divisions as the result of a distortion in the frequency of type II events.

The 1:1 allelic ratio is the keystone of the model proposed by Sager & Ramanis (191) which assumes that a vegetative cell is diploid for its chloroplast genome. On the basis of the 1:1 allelic ratio Sager and Ramanis reject models in which the number of copies of chloroplast genomes per vegetative cell is random and variable as well as multicopy models in which the number of genomes per cell is fixed. They say that in such models deviations from a 1:1 allelic ratio would be expected. That is, type II segregations for a pair of alleles would not be expected for each allele with equal frequency. Originally, Sager & Ramanis (191) proposed that type II and III segregations could be accounted for by haploidization. In type II segregation one daughter cell would have been haploid for its chloroplast genome and the other diploid while in a type III segregation both daughter cells would have haploidized for their chloroplast genomes. Later Sager & Ramanis (188) rejected the haploidization hypothesis since they found in recombination analysis that certain markers became homozygous while others were still heterozygous. Sager & Ramanis (9, 188) have hypothesized that type III events resulted from reciprocal recombination by a mechanism akin to mitotic recombination and type II events from nonreciprocal recombination by a mechanism akin to gene conversion.

In analyzing the progeny of spontaneous, unselected zygote clones, Gillham (133, 187, 194) has reported deviations from a 1:1 allelic ratio such that in most zygote clones an excess of cells, sometimes very great, carry chloroplast alleles derived from the mt^+ parent. A similar deviation has been seen among the progeny of hybrid meiotic products obtained by tetrad analysis of spontaneous biparental zygotes and among heterozygous cells that have undergone several postmeiotic mitotic divisions (133, 194). These experiments suggest that the excess of cells carrying chloroplast markers from the mt^+ parent seen in most spontaneous biparental zygote clones can be accounted for in terms of an excess of type II segregation for markers carried by the mt^+ parent.

Gillham's results showing deviations from a 1:1 allelic ratio for unselected biparental zygotes are inconsistent with a two-copy model and argue in favor of a multicopy model in which the majority of copies are derived from the mt^+ parent. Gillham et al (187) considered two possible explanations for the difference between their results with unselected biparental zygotes and those of Sager with selected biparental zygotes and uv-induced biparental zygotes. First, spontaneous, unselected biparental zygotes might have abnormally high numbers of chloroplast genomes contributed by the maternal parent and uv-induced biparental zygotes only one copy of the chloroplast genome from each parent. If this were the case, an allele ratio skewed in favor of maternal markers would be expected in the unirradiated biparental zygote clones but a 1:1 allelic ratio would be expected among uv-induced biparental zygote clones. Second, the skewed allelic ratio seen in spontaneous, unselected biparental zygote clones is normal. The uv effectively reduces the number of chloroplast genomes contributed by the maternal parent with a concomitant increase in the number of paternal genomes transmitted and, hence, an increase in the frequency of exceptional zygotes (see p. 373). This hypothesis predicts that the skew in allelic ratio will disappear gradually with uv dose until a 1:1 allelic ratio is approximated.

To distinguish these possibilities Gillham et al (187) made reciprocal crosses between a stock carrying chloroplast markers for streptomycin and erythromycin resistance and one carrying a chloroplast marker for spectinomycin resistance. In each cross, aliquots of the mt^+ parent were subjected to increasing doses of uv prior to mating. The cells were then mated and zygotes given a dose of photoreactivating light in accordance with the methods of Sager & Ramanis (188). The frequency of biparental, paternal, and maternal zygotes was determined at each time point. Biparental zygote clones were replated and the allelic ratio determined for each pair of alleles. As expected, there was a dramatic increase in the frequency of biparental zygotes at low uv doses and a slower increase in the frequency of paternal zygotes as the dose was increased. Without uv or with low uv doses the allelic ratios in most biparental zygote clones were skewed in favor of maternal markers and this skew decreased with increasing uv dose. The experiments also showed that at high uv doses the mean allelic ratio began to approach 1:1 but the distribution was not normal around this mean. The authors observed a high cotransmission frequency of alleles from the same parent in zygote clones. They also found that without uv, zygote clones containing recombinants tended to contain only one of a reciprocal pair. This recombinant carried two maternal and one paternal markers. With uv this tendency disappeared and recombinants with two paternal and one maternal allele were equal in frequency to the reciprocal type.

Gillham et al (187) interpreted their results in terms of the multi-copy plasmid model described earlier. It was postulated that in most zygotes only maternal genomes are transmitted and these attach preferentially to available attachment sites. In most spontaneous biparental zygotes one or usually a few paternal genomes are transmitted in comparison to the number of maternal genomes. The uv treatment of the mt^+ parent reduces the number of maternal genomes making more attachment sites available for paternal chloroplast genomes. This not only increases the number of biparental zygotes but also serves to equalize the ratio of paternal and maternal genomes and hence the allelic ratio. Eventually all maternal genomes are inactivated by uv and a paternal zygote results. The fact that the majority of recombinants among the progeny of unirradiated biparental zygotes carry two maternal and one paternal marker is explained by the ability of chloroplast genomes to undergo successive rounds of mating and recombination. Since the genomic ratio is skewed in favor of maternal genomes in spontaneous biparental zygotes, one would expect recombinants produced by successive rounds of mating to carry two maternal and one paternal marker. As uv equalizes the genomic input, the frequencies of reciprocal recombinants become more nearly equal. The model also predicts the types of segregation pattens seen by Sager and is in better agreement with the physical data suggesting 26 copies of the chloroplast genome per gamete (139, 141).

There are striking similarities between the model of Gillham et al (187) for the chloroplast genome and that of Dujon et al (111) for the mitochondrial genome. Both models assume a variable input of genomes from each parent which can be determined from the output by measuring allelic frequencies; both models predict successive rounds of mating between organelle genomes; and both models assume that the Visconti-Delbrück theory for recombination in phage can be applied to the recombination of organelle genomes.

Evidence that the number of copies of chloroplast genes per vegetative cell is greater than two also comes from the experiments of Schimmer & Arnold (195–198), with a maternally inherited streptomycin-dependent mutant (sd-3-18) which yielded streptomycin-sensitive (ss) revertants at a frequency of 10^{-6} to 10^{-7}. The ss revertants in turn segregated secondary sd mutants, which are never observed spontaneously in wild-type strains, at frequencies up to 10^{-2} with the frequency declining with the number of doublings in the absence of streptomycin. The initial frequencies of sd mutants segregated varied depending on the sensitive revertant tested. The secondary sd mutants were then shown to segregate sensitive cells with frequencies as high as 10^{-2}. The frequency of sensitive cells per secondary sd mutant varied from clone to clone over three orders of magnitude, declining with time to a level comparable to the reversion frequency of sd. From these secondary ss revertants tertiary sd mutants could be obtained at frequencies of 10^{-2} to 10^{-3}.

Schimmer & Arnold argued that their results could best be explained by a multiple-copy model in which the ratio of sd to ss alleles varied. For example, if one isolates two ss revertants with different ratios of sd to ss alleles, they will segregate secondary sd mutants at different rates. Similarly, the secondary sd mutants will segregate ss cells at distinctive rates for the same reason. These experiments argue strongly against a two-copy model not only for these reasons but on the basis of dominance considerations. The ss revertants are heterozygous for dependence but phenotypically sensitive whereas the secondary sd mutants are phenotypically dependent but heterozygous for sensitivity. This implies dominance of sensitivity to dependence in the first case and the converse situation in the second. Such results are paradoxical on a two-copy model but are easily explained in a multiple-copy model since the phenotype expressed will then depend on the ratio of sd to ss alleles.

Schimmer & Arnold argued that the sd allele must be mitochondrial in location since it was present in multiple copies, while Sager (191) had proposed a two-copy model for the chloroplast genome. This argument is of dubious validity in view of the evidence presented by Gillham et al (187) for a multiple-copy chloroplast genome. Sager (9) also believes the results of Schimmer & Arnold can be explained in terms of the chloroplast genome but for a different reason. She argues that Schimmer & Arnold were studying what she has termed persistent cytoplasmic heterozygotes. Persistent heterozygotes arise at a low frequency among the progeny of biparental zygotes and appear to carry chloroplast genomes from both parents without segregating these genomes at an appreciable rate during vegetative growth.

In a cross the two genomes usually segregate in a 2:2 ratio if carried by the mt^+ parent but are not transmitted (4:0 segregation) when carried by the mt^- parent. Sager argues that the low rates of segregation of secondary sd mutants by ss revertants as well as segregation of ss cells by secondary sd mutants could be explained if Schimmer and Arnold had been studying persistent heterozygotes. However, persistent heterozygotes cannot explain the dominance relationships discussed above unless one assumes that persistent heterozygotes contain multiple copies of the chloroplast genome.

3. MAPPING CHLOROPLAST GENES   Three lines of evidence suggest that the chloroplast genes studied so far in *Chlamydomonas* form a single linkage group. First, all available experimental data suggest that in biparental zygotes all markers from the mt^- parent are transmitted *en bloc* (180, 187, 192, 193). Second, among biparental zygote progeny there is usually a strong correlation in the output frequency of all paternal markers and similarly maternal markers over a wide range of allelic ratios (187). Third, as discussed below, the experiments of Sager & Ramanis (186, 188) show that there is a polarity of segregation of different chloroplast genes among the progeny of biparental zygotes. All of these observations are most easily interpreted by assuming that chloroplast genes fall in a single linkage group.

In 1970 Sager & Ramanis (188) published the first map of chloroplast genes in *Chlamydomonas*. The data published in that paper have been reviewed several times (9, 184–186), but very few new experimental data have been forthcoming to substantiate that map as well as the recent claim by Sager (9) that the chloroplast linkage group is circular. We review mapping methods here and then discuss these in relation to Sager's map.

1. *Recombination analysis in octospore daughters:* In this method the progeny of biparental zygotes are analyzed at the octospore daughter stage and the progeny classified as parental (P), recombinant (R), or heterozygous (H) for different chloroplast markers. Sager & Ramanis (185, 188) have computed recombination frequencies for marker pairs both by ignoring the heterozygotes ($R_A = R/P + R$ × 100) or including them in the denominator ($R_B = R/P + R + H$ × 100). Computation using the R_A method gives recombination frequencies severalfold higher than the R_B method (185). Sager & Ramanis (188) have chosen to use the R_B method for mapping. Since heterozygotes are included in the denominator in the R_B method, not only is the recombination frequency minimized, but also it is necessary to normalize map distances from different crosses to account for differing numbers of heterozygotes. For example, suppose the recombination frequency (R_A) between two genes is 25%. If 80% of the cells analyzed are still heterozygous, R_B equals 5%, while if 60% of the cells are heterozygous, R_B equals 10%. Therefore, the map interval in the second case will be twice as large as in the first and comparison between crosses will require normalization.

The normalization method used by Sager and Ramanis is to make one cross the reference cross and to establish a normalization factor with respect to common intervals between the reference cross and other crosses. This normalization factor is then applied to all map intervals in the cross being compared to the reference cross. For example, if we suppose that in the reference cross R_B = 10% for the map interval used for normalization and R_B = 5% for the same interval in the second cross, all map intervals in the second cross would be multiplied by two. The normalization procedure seems to work reasonably well for the few intervals measured in common between the standard cross and three other crosses reported by Sager & Ramanis (188).

2. *Recombination analysis in zygote clones:* This method, described earlier, has been used by Gillham (187, 192, 193) and involves measuring the frequency of each genotype among the progeny of a biparental zygote following many vegetative cell generations. Since marker segregation by this time should be virtually complete, the method used for measuring recombination frequency is essentially similar to the R_A method used by Sager & Ramanis (185) for octospore daughter cells. That is, $R = R/P + R \times 100$. This method has the advantage that many biparental zygote clones are readily dealt with, and zygotes with similar allelic ratios can be grouped and recombination frequencies measured for pairs of markers as a function of allelic ratio.

3. *Frequency of Type III segregations:* Sager & Ramanis (188) noted that type III segregations, which they postulate result from a mechanism akin to mitotic recombination, showed a polarity that could be related to gene order by recombination analysis. An acetate marker (ac_2) showed the lowest type III segregation frequency at the octospore daughter stage and other markers showed progressively higher frequencies. The markers, with the exception of streptomycin resistance (sm_2) and erythromycin resistance (ery), could be ordered by the polarity of type III segregations with respect to a hypothetical attachment point (ap) assumed to function in a manner similar to a mitotic centromere. Sagar & Ramanis (188) postulated that ac_2 was closest to ap and the other markers were progressively further away, increasing the probability of reciprocal recombination leading to type III segregation as distance increased.

To compare the polarity map with the recombination map, further normalizations were required. In the first two crosses where ac_2 and sm_2 were segregating, the difference between type III segregations for the two markers was normalized to 20, the interval by which these markers appear separated by recombination analysis (i.e. normalization factor = 20/% type III for sm_2 − % type III for ac_2). On this basis it was calculated that ac_2 lay 15.6 map units from ap by multiplying the frequency of type III segregations for ac_2 in each cross by the normalization factor. In two other crosses where ac_1 rather than ac_2 was used the data were normalized by assuming that ac_2 is 15.6 map units from ap and, from recombination analysis, that ac_1 is 2.7 map units from ac_2. Therefore, the distance from ap to ac_1 would be approximately 18 map units. In these crosses normalization was achieved by adding eight to the percentage of type III segregations for each marker since ac_1 prior to normalization yielded about 10% type III segregations. These normalizations appear to yield map distances not greatly different from those obtained by recombination analysis.

There are, however, some important anomalies. One of the five crosses could not be normalized at all due to the high frequency of type III segregations and would have given a map order inconsistent with the other four crosses. In another cross the normalization factor could not be applied to the *ery* marker as it would have given an inconsistent order for this marker. Interestingly, the map placement of *ery* in this paper appears to be incorrect in that it is placed too close to *ap*. In later publications Sager & Ramanis (9, 186) report it is the furthest marker from *ap*.

Recently, mapping using the type III segregation method has been modified based on the observation that segregation and recombination proceed at a constant rate per doubling during growth of zoospore clones and that type II events occur with a constant frequency (9, 186). This means that the disappearance of heterozygotes from the population should be proportional to the frequency of type III segregations. Since the frequency of type III segregations varies from marker to marker one should obtain a characteristic slope for rate of disappearance of heterozygotes for each marker depending on the frequency of type III segregations. The slope of each curve can then be used to compute gene distance with respect to *ap* using zoospore clones growing synchronously in liquid medium.

4. *Cosegregation mapping:* In this method (185) the frequency with which two markers become homozygous together, irrespective of whether they are in a recombinant or parental configuration, is used as the basis of mapping. The cosegregation frequency (R_c) is assumed to be inversely proportional to the map distance since closely linked markers should cosegregate more frequently than distant markers (i.e. $R_c = P + R + H/P + R$). This method seems to assume that if two markers are far apart it is more likely that one or the other will be heterozygous than if they are close together. This assumption might not be unreasonable if it were not for the fact that markers close to *ap* yield a higher frequency of heterozygotes than those further away at the octospore daughter stage when cosegregation frequencies are apparently analyzed (9, 186). It should also be noted that R_c will have a value of one or greater instead of being a fraction less than one, which requires multiplication by 100 to yield a whole number as in the case of R_A and R_B.

Having considered the methods used by Sager & Ramanis (9, 185, 186, 188) to construct a map of the chloroplast linkage group in *Chlamydomonas*, let us see to what extent the published data support the contention that these genes lie in a single, circular linkage group. The map constructed using the characteristic segregation rate for each marker seems the most reliable since it depends strictly on the polarity of segregation of different markers in heterozygotes. However, an assumption implicit in this method, at least when the disappearance of heterozygotes from the population is followed, is that the frequency of type II segregations is constant for every marker. If this is not the case, as the data of Gillham and his colleagues (187) indicate, it might affect the slopes of the marker curves from experiment to experiment but should not affect ordering of markers, assuming that type II segregations also occur by a mechanism akin to mitotic recombination rather than gene conversion (see Gillham et al 187).

Maps obtained by recombination analysis and cosegregation frequencies, on the other hand, seem more suspect both in terms of absolute distances and, in some cases, map order, because heterozygotes are included in the calculations and these vary from marker to marker and possibly from cross to cross as well. In the first place, intervals measured by recombination analysis using the R_B method will depend upon the frequency of heterozygotes (H) for each marker at the octospore daughter stage. H is clearly characteristic for each marker at the octospore daughter stage as shown by the fact that the disappearance of heterozygotes from the population is marker dependent. In effect R_B will be lower for markers close to ap and higher for markers further away because the value of H is greater in the first instance.

This effect of H on map intervals can be seen by comparing data published by Sager & Ramanis (185) using the formulas $R_A = R/P + R$ and $R_B = R/P + R + H$. The ratios of R_A/R_B for the intervals ac_2-ery, ac_2-sm_2, and ery-sm_2 are respectively 4.06, 5.57, and 2.70. That is, R_B begins to approach R_A in the case of the ery-sm_2 interval for which the heterozygote frequency is lowest (70%) but increases as the heterozygote frequency increases ($H = 75\%$ for ac_2-ery and 82% for ac_2-sm_2). The effect is to diminish the apparent recombination frequency between ac_2 and the other two markers while there is a relative increase in the frequency of recombination between ery and sm_2. These effects are readily apparent in the review by Sager & Ramanis (185) where distances computed by the R_B method are not additive whereas they are nearly additive when computed by the R_A method. Furthermore, if, as stated by Sager & Ramanis (186), recombination and segregation are constant per doubling the R_A method should be a much more accurate way of determining recombination frequencies since heterozygotes do not enter into the picture and normalization should not be required.

Cosegregation mapping seems to be of questionable validity. The equation $R_C = P + R + H/P + R$ must by definition yield a number of 1 or greater. If $P + R + H$ is set equal to one and H is varied with respect to $P + R$, a hyperbolic curve is generated which will be the same for any pair of markers no matter how close or far apart they may be. At high heterozygote frequencies, markers will appear farther apart while at low heterozygote frequencies they will appear close together. Therefore, at the octospore daughter stage pairs of markers close to ap will appear relatively farther apart than markers more distant from ap because the proportion of heterozygotes will be greatest for the markers close to ap. The reason for apparent agreement between mapping using the R_B and R_C methods in the review published by Sager & Ramanis (185) can be described as follows. For any percent recombination (R_A) a linear curve can be calculated for R_B by varying the proportion of heterozygotes in the population. The value of R_B will be lowest when H is high and will approach R_A at low values of H.

Sager & Ramanis (185) used whole numbers to construct their cosegregation map and converted the fractions of recombinants obtained by the R_B method to a whole number by multiplying by 100. If one plots R_B and R_C as whole numbers on the ordinate and the fraction of heterozygotes on the abcissa, one obtains the hyperbolic curve described previously for R_C and a series of linear curves for R_B, each corresponding to a different percentage of recombination (R_A). These curves intersect with the curve for R_C at high heterozygote levels but diverge markedly as the fraction of heterozygotes in the population declines. Therefore, one obtains spurious correlations between R_B and R_C at high heterozygote levels. In fact the apparent correlations seen by Sager & Ramanis (185) between R_B and R_C can be predicted exactly from these curves.

It seems from the foregoing discussion that recombination analysis using the R_A method which eliminates heterozygotes from consideration and polarity of segregation ought to be the most reliable methods. To evaluate the recent claim that chloroplast genes in *Chlamydomonas* form a single, circular linkage group (9) it will be necessary to have available a detailed discussion of the methods used as well as considerably more data.

CONCLUDING REMARKS

A satisfactory comparison between the mitochondrial genetic system of yeast and that of the chloroplast of *Chlamydomonas* is difficult at present because of the very disparate methods used to generate the data necessary for model building. In *Chlamydomonas* pedigree analysis has been the principal tool whereas in *Saccharomyces* the analysis of the vegetative progeny of a cross *en masse* after many cell generations has been the principal genetic method. Thus genetic results derived from individual clones have been paramount in *Chlamydomonas* whereas results obtained from populations have provided most of the data in yeast. The recent experimental results of Gillham et al (187) and Dujon et al (111) certainly suggest similarities between the two systems. At the same time, though, there are important differences. Polar genes *sensu Saccharomyces* have not been found in *Chlamydomonas*. Cellular mating type, or more probably genes associated with the mating-type locus, control the transmission of chloroplast genomes in *Chlamydomonas* but not the transmission of mitochondrial genomes in yeast. An interesting question in the case of yeast is whether pedigree analysis will reveal a polarity of marker segregation in the sense that it is seen in *Chlamydomonas*. It remains for the future to determine how closely similar or different these two genetic systems are.

ACKNOWLEDGMENTS

It is a particular pleasure to acknowledge the intellectual contributions of my colleague Dr. John Boynton to this review as well as several students and postdoctorals in our laboratory, notably Dr. Robert Lee, G. M. W. Adams, and Mrs. Karen Van Winkleswift. Papers in press and other insights were kindly provided by Drs. K.-S. Chiang, B. Dujon, P. P. Slonimski, and R. Sager. This work was supported by grants NSF-GB 22769 and NIH-GM 19427 to J. Boynton and N. W. Gillham and by an NIH RCDA Award GM 70437 to N. W. Gillham.

Literature Cited

1. Baur, E. 1909. Das Wesen und die Er-blichkeitsverhaltnisse der "Varietates albomarginatae hort" von *Pelargonium zonale. Z. Vererbungs.* 1:330–51
2. Correns, C. 1909. Vererbungsversuche mit blass (gelb) grunen und bluntblattrigen sippen bei *Mirabilis, Urtica,* und *Lunaria. Z. Vererbungs.* 1:291–329
3. Ephrussi, B., Hottinguer, H., Chimenes, A.-M. 1949. Action de l'acriflavine sur les levures. I. La mutation "petite colonie." *Ann. Inst. Pasteur Paris* 76:351–67
4. Ephrussi, B., Hottinguer, H., Tavlitzki, J. 1949. Action de l'acriflavine sur les levures. II. Étude génétique du mutant "petite colonie." *Ann. Inst. Pasteur Paris* 76:419–50
5. Chun, E. H. L., Vaughan, M. H., Rich, A. 1963. The isolation and characterization of DNA associated with chloroplast preparations. *J. Mol. Biol.* 7: 130–41
6. Sager, R., Ishida, M. R. 1963. Chloroplast DNA in *Chlamydomonas. Proc. Nat. Acad. Sci. USA* 50:725–30
7. Luck, D. J. L., Reich, E. 1964. DNA in mitochondria of *Neurospora crassa. Proc. Nat. Acad. Sci. USA* 52:931–38
8. Kirk, J. T. O., Tilney-Bassett, R. A. E. 1967. *The Plastids.* San Francisco: Freeman. 608 pp.
9. Sager, R. 1972. *Cytoplasmic Genes and Organelles.* New York: Academic. 405 pp.
10. Wilkie, D. 1964. *The Cytoplasm in Heredity.* London: Methuen. 115 pp.
11. Roodyn, D. B., Wilkie, D. 1968. *The Biogenesis of Mitochondria.* London: Methuen. 123 pp.
12. Slater, E. C., Tager, J. M., Papa, S., Quagliariello, E., Eds. 1968. *Biochemical Aspects of the Biogenesis of Mitochondria.* Bari: Adriatica Editrice. 494 pp.
13. Boardman, N. K., Linnane, A. W., Smillie, R. M., Eds. 1971. *Autonomy and Biogenesis of Mitochondria and Chloroplasts.* Amsterdam: North-Holland. 511 pp.
14. Miller, P. L., Ed. 1970. *Control of Organelle Development.* Soc. Exp. Biol. Vol. 24. Cambridge: Cambridge Univ. Press. 524 pp.
15. Harris, P. J., Ed. 1969. *Biological Ultrastructure: The Origin of Cell Organelles.* Biology Colloquium Proceedings. Corvallis: Oregon State Univ. Press

16. Van den Bergh, G. S., Borst, P., Slater, E. C., Eds. 1973. *Mitochondria, Biogenesis and Bioenergetics.* Amsterdam: North-Holland
17. Borst, P. 1972. Mitochondrial nucleic acids. *Ann. Rev. Biochem.* 41:334–76
18. Borst, P., Kroon, A. M. 1969. Mitochondrial DNA: Physicochemical properties, replication and genetic function. *Int. Rev. Cytol.* 26:107–89
19. Ashwell, M., Work, T. S. 1970. The biogenesis of mitochondria. *Ann. Rev. Biochem.* 39:251–90
20. Borst, P., Grivell, L. A. 1971. Mitochondrial ribosomes. *FEBS Lett.* 13:73–88
21. Baxter, R. 1971. Origin and continuity of mitochondria. In *Origin and Continuity of Cell Organelles,* ed. W. Beerman, J. Reinert, H. Ursprung, pp. 46–64. New York: Springer-Verlag
22. Beattie, D. 1971. The synthesis of mitochondrial proteins. *Subcell. Biochem.* 1:1–23
23. Getz, G. S. 1972. Organelle biogenesis. In *Membrane Molecular Biology,* ed. C. F. Fox, A. Keith, p. 386. Stamford, Conn.: Sinauer
24. Kroon, A. M., Agsteribbe, E., de Vries, H. 1972. Protein synthesis in mitochondria and chloroplasts. In *The Mechanism of Protein Synthesis and Its Regulation,* ed. L. Bosch, pp. 539–82. Amsterdam: North-Holland
25. Linnane, A. W., Haslam, J. M. 1970. The biogenesis of yeast mitochondria. *Curr. Top. Cell. Regul.* 2:102–72
26. Linnane, A. W., Haslam, J. M., Lukins, H. B., Nagley, P. 1972. The biogenesis of mitochondria in microorganisms. *Ann. Rev. Microbiol.* 26:163–98
27. Mahler, H. 1973. Biogenetic autonomy of mitochondria. *Crit. Rev. Biochem.* 1:381–460
28. Nass, M. M. K. 1969. Mitochondrial DNA: Advances, problems, and goals. *Science* 165:25–35
29. Nass, S. 1969. The significance of the structural and functional similarity of bacteria and mitochondria. *Int. Rev. Cytol.* 25:55–129
30. Rabinowitz, M., Swift, H. 1970. Mitochondrial nucleic acids and their relation to the biogenesis of mitochondria. *Physiol. Rev.* 3:376–427
31. Tzagoloff, A., Rubin, M. S., Sierra, M. F. 1973. Biosynthesis of mitochondrial enzymes. *Biochim. Biophys. Acta* 301:71–104

32. Schatz, G. 1970. Biogenesis of mitochondria. In *Membranes of Mitochondria and Chloroplasts,* ed. E. Racker, pp. 251–314. New York: Van Nostrand-Reinhold

33. Wagner, R. P. 1969. Genetics and phenogenetics of mitochondria. *Science* 163:1026–31

34. Boulter, D., Ellis, R. J., Yarwood, A. 1972. Biochemistry of protein synthesis in plants. *Biol. Rev.* 47:113–75

35. Kirk, J. T. O. 1972. The genetic control of plastid formation. Recent advances and strategies for the future. *Subcell. Biochem.* 1:333–61

36. Kirk, J. T. O. 1971. Chloroplast structure and biogenesis. *Ann. Rev. Biochem.* 40:161–96

37. Kirk, J. T. O. 1970. Biochemical aspects of chloroplast development. *Ann. Rev. Plant Physiol.* 21:11–42

38. Levine, R. P. 1969. Analysis of photosynthesis using mutant strains of algae and higher plants. *Ann. Rev. Plant Physiol.* 20:523–40

39. Levine, R. P., Goodenough, U. W. 1970. Genetics of photosynthesis and of the chloroplast of *Chlamydomonas reinhardi. Ann. Rev. Genet.* 4:397–408

40. Tewari, K. K. 1971. Genetic autonomy of extranuclear organelles. *Ann. Rev. Plant Physiol.* 22:141–68

41. Preer, J. 1971. Extrachromosomal inheritance: Hereditary symbionts, mitochondria, chloroplasts. *Ann. Rev. Genet.* 5:361–406

42. Adoutte, A., Beisson, J. 1970. Cytoplasmic inheritance of erythromycin resistant mutations in *Paramecium aurelia. Mol. Gen. Genet.* 108:70–77

43. Adoutte, A., Beisson, J. 1972. Evolution of mixed populations of genetically different mitochondria in *Paramecium aurelia. Nature* 235:393–95

44. Adoutte, A., Balmfrezol, M., Beisson, J., Andre, J. 1972. The effects of erythromycin and chloramphenicol on the ultrastructure of mitochondria in sensitive and resistant strains of *Paramecium. J. Cell Biol.* 54:8–19

45. Beale, G. 1969. A note on the inheritance of erythromycin resistance in *Paramecium. Genet. Res.* 14:341–42

46. Beale, G., Knowles, J., Tait, A. 1972. Mitochondrial genetics in *Paramecium. Nature* 235:396–97

47. Tait, A. 1972. Altered mitochondrial ribosomes in an erythromycin resistant mutant of *Paramecium. FEBS Lett.* 24:117–19

48. Ephrussi, B. 1953. *Nucleo-cytoplasmic Relations in Microorganisms.* Oxford: Clarendon. 127 pp.

49. Chen, S.-Y., Ephrussi, B., Hottinguer, H. 1950. Nature genetique des mutants a deficience respiratoire de la souche B-II de la levure de boulangerie. *Heredity* 4:337–51

50. Sherman, F. 1963. Respiration deficient mutants of yeast. I. Genetics. *Genetics* 48:375–85

51. Sherman, F., Slonimski, P. P. 1964. Respiration deficient mutants of yeast. II. Biochemistry. *Biochim. Biophys. Acta* 90:1–15

52. Beck, J. C., Parker, J. H., Balcavage, W. X., Mattoon, J. R. 1971. Mendelian genes affecting development and function of yeast mitochondria. See Ref. 13, pp. 194–204

53. Sanders, H. K., Mied, P. A., Briquet, M., Hernandez-Rodriquez, J., Gottal, R. F., Mattoon, J. R. 1973. Regulation of mitochondrial biogenesis: yeast mutants deficient in synthesis of δ-aminolevulinic acid. *J. Mol. Biol.* 80:17–39

54. Gunge, N., Sugimura, T., Iwasaki, M. 1967. Genetic analysis of a respiration-deficient mutant of *Saccharomyces cerevisiae* lacking all cytochromes and accumulating coproporphyrin. *Genetics* 57:213–26

55. Sugimura, T., Okabe, K., Nagao, M., Gunge, N. 1966. A respiration-deficient mutant of *Saccharomyces cerevisiae* which accumulates porphyrins and lacks cytochromes. *Biochim. Biophys. Acta* 115:267–75

56. Ebner, E., Mennucci, L., Schatz, G. 1973. Mitochondrial assembly in respiration-deficient mutants of *Saccharomyces cerevisiae.* I. Effect of nuclear mutations on mitochondrial protein synthesis. *J. Biol. Chem.* 248:5360–68

57. Ebner, E., Mason, T. L., Schatz, G. 1973. Mitochondrial assembly in respiration-deficient mutants of *Saccharomyces cerevisiae.* II. Effect of nuclear and extrachromosomal mutations on the formation of cytochrome *c* oxidase. *J. Biol. Chem.* 248:5369–78

58. Ebner, E., Schatz, G. 1973. Mitochondrial assembly in respiration-deficient mutants of *Saccharomyces cerevisiae.* III. A nuclear mutant lacking mitochondrial adenosine triphosphatase. *J. Biol. Chem.* 248:5379–84

59. Ephrussi, B., de Margerie-Hottinguer, H., Roman, H. 1955. Suppressiveness: a

new factor in the genetic determinism of the synthesis of respiratory enzymes in yeast. *Proc. Nat. Acad. Sci. USA* 41: 1065–71

60. Ephrussi, B., Grandchamp, S. 1965. Études sur la suppressivité des mutants a deficience respiratoire de la levure. I. Existence au niveau cellulaire de divers "degrés de suppressivité." *Heredity* 20:1–7

61. Rank, G. H., Person, C. 1969. Reversion of cytoplasmically-inherited respiratory deficiency in *Saccharomyces cerevisiae. Can. J. Genet. Cytol.* 11: 716–28

62. Ephrussi, B., Jakob, H., Grandchamp, S. 1966. Études sur la suppressivité des mutants à déficience respiratoire de la levure. II. Étapes de la mutation grande en petite provoqueé par le facteur suppressif. *Genetics* 54:1–29

63. Wright, R. E., Lederberg, J. 1957. Extranuclear transmission in yeast heterokaryons. *Proc. Nat. Acad. Sci. USA* 43: 919–23

64. Ephrussi, B., L'Héritier, P., Hottinguer, M. 1949. Action de l'acriflavine sur les levures. VI. Analyse quantitative de la transformation des populations. *Ann. Inst. Pasteur Paris* 77:1–20

65. Slonimski, P. P., Perrodin, G., Croft, J. H. 1968. Ethidium bromide induced mutation of yeast mitochondria: complete transformation of cells into respiratory deficient non-chromosomal "petites." *Biochem. Biophys. Res. Commun.* 30:232–39

66. Goldring, E. S., Grossman, L. I., Krupnick, D., Cryer, D. R., Marmur, J. 1970. The petite mutation of yeast: loss of mitochondrial deoxyribonucleic acid during induction of petites with ethidium bromide. *J. Mol. Biol.* 52: 323–35

67. Nagley, P., Linnane, A. W. 1970. Mitochondrial DNA deficient petite mutants of yeast. *Biochem. Biophys. Res. Commun.* 39:989–96

68. Bernardi, G., Carnevali, F., Nicolaieff, A., Piperno, G., Tecce, G. 1968. Separation and characterization of a satellite DNA from a yeast cytoplasmic petite mutant. *J. Mol. Biol.* 37:493–505

69. Bernardi, G., Faures, M., Piperno, G., Slonimski, P. P. 1970. Mitochondrial DNA's from respiratory-sufficient and cytoplasmic respiratory-deficient mutant yeast. *J. Mol. Biol.* 48:23–42

70. Carnevali, F., Morpurgo, G., Tecce, G. 1969. Cytoplasmic DNA from petite colonies of *Saccharomyces cerevisiae:* a hypothesis on the nature of the mutation. *Science* 163:1331–33

71. Mounolou, J. C., Jakob, H., Slonimski, P. P. 1966. Mitochondrial DNA from yeast "petite" mutants: Specific changes of buoyant density corresponding to different cytoplasmic mutations. *Biochim. Biophys. Res. Commun.* 24: 218–24

72. Michaelis, G., Douglass, S., Tsai, M.-J., Criddle, R. S. 1971. Mitochondrial DNA and suppressiveness of petite mutants in *Saccharomyces cerevisiae. Biochem. Genet.* 5:487–95

73. Hollenberg, C. P. et al 1972. The unusual properties of mtDNA from a "low density" petite mutant of yeast. *Biochim. Biophys. Acta* 277:44–58

74. Mehrotra, B. D., Mahler, H. R. 1968. Characterization of some unusual DNAs from the mitochondria from certain "petite" strains of *Saccharomyces cerevisiae. Arch. Biochem. Biophys.* 128:685–703

75. Nagley, P., Linnane, A. W. 1972. Biogenesis of mitochondria. XXI. Studies on the nature of the mitochondrial genome in yeast: the degenerative effects of ethidium bromide on mitochondrial genetic information in a respiratory competent strain. *J. Mol. Biol.* 66:181–93

76. Wilkie, D., Saunders, G., Linnane, A. W. 1967. Inhibition of respiratory enzyme synthesis in yeast by chloramphenicol: relationship between chloramphenicol tolerance and resistance to other antibacterial antibiotics. *Genet. Res. Cambridge* 10:199–203

77. Thomas, D. Y., Wilkie, D. 1968. Inhibition of mitochondrial synthesis in yeast by erythromycin: cytoplasmic and nuclear factors controlling resistance. *Genet. Res. Cambridge* 11:33–41

78. Linnane, A. W., Saunders, G. W., Gingold, E. B., Lukins, H. B. 1968. The biogenesis of mitochondria. V. Cytoplasmic inheritance of erythromycin resistance in *Saccharomyces cerevisiae. Proc. Nat. Acad. Sci. USA* 59:903–10

79. Coen, D., Deutsch, J., Netter, P., Petrochilo, E., Slonimski, P. P. 1970. Mitochondrial genetics. I. Methodology and phenomenology. See Ref. 14, pp. 449–96

80. Gingold, E. B., Saunders, G. W., Lukins, H. B., Linnane, A. W. 1969. Biogenesis of mitochondria. X. Reassortment of the cytoplasmic determinants for respiratory competence and erythro-

mycin resistance in *Saccharomyces cerevisiae. Genetics* 62:735–44
81. Deutsch, J. et al 1974. Mitochondrial genetics. VI. The petite mutation in *Saccharomyces cerevisiae:* interrelations between the loss of the ρ^+ factor and the loss of the drug resistance mitochondrial genetic markers. *Genetics* 76:195–219
82. Sherman, F. 1959. The effect of elevated temperatures on yeast. II. Induction of respiratory deficient mutants. *J. Cell Compar. Physiol.* 54:37–52
83. Sugimura, T., Okabe, K., Imamura, A. 1966. Number of cytoplasmic factors in yeast cells. *Nature* 212:304
84. Maroudas, N. G., Wilkie, D. 1968. Ultraviolet irradiation studies on the cytoplasmic determinant of the yeast mitochondrion. *Biochim. Biophys. Acta.* 166:681–88
85. Allen, N. E., MacQuillan, A. M. 1969. Target analysis of mitochondrial genetic units in yeast. *J. Bacteriol.* 97:1142–48
86. Mahler, H. R., Perlman, P. S., Mehrotra, B. D. 1971. Mitochondrial specification of the respiratory chain. See Ref. 13, pp. 492–511
87. Williamson, D. H. 1970. The effect of environmental and genetic factors on the replication of mitochondrial DNA in yeast. See Ref. 14, pp. 247–76
88. Michaelis, G., Petrochilo, E., Slonimski, P. P. 1973. Mitochondrial genetics III. Recombined molecules of mitochondrial DNA obtained from crosses between cytoplasmic *petite* mutants of *Saccharomyces cerevisiae:* physical and genetic characterization. *Mol. Gen. Genet.* 123:51–65
89. Saunders, G. W., Gingold, E. B., Trembath, M. K., Lukins, H. B., Linnane, A. W. 1971. Mitochondrial genetics in yeast: segregation of a cytoplasmic determinant in crosses and its loss or retention of the petite. See Ref. 13, pp. 185–93
90. Van Kreijl, C. F., Borst, P., Flavell, R. A., Hollenberg, C. P. 1972. Pyrimidine tract analysis of mtDNA from a "low-density" petite mutant of yeast. *Biochim. Biophys. Acta* 277:61–70
91. Hollenberg, C. P., Borst, P., van Bruggen, E. F. J. 1972. Mitochondrial DNA from cytoplasmic petite mutants of yeast. *Biochim. Biophys. Acta* 277:35–43
92. Hollenberg, C. P. et al 1972. The unusual properties of mtDNA from a

"low-density" petite mutant of yeast. *Biochim. Biophys. Acta* 277:44–58
93. Reijnders, L., Kleisen, C. M., Grivell, L. A., Borst, P. 1972. Hybridization studies with yeast mitochondrial DNAs. *Biochim. Biophys. Acta* 272:396–407
94. Halbreich, A., Rabinowitz, M. 1971. Isolation of *Saccharomyces cerevisiae* mitochondrial formyltetrahydrofolic acid: methionyl-tRNA transformylase and the hybridization of mitochondrial fmet-tRNA with mitochondrial DNA. *Proc. Nat. Acad. Sci. USA* 68:294–98
95. Casey, J., Cohen, M., Rabinowitz, M., Fukuhara, H., Getz, G. S. 1972. Hybridization of mitochondrial transfer RNA's with mitochondrial and nuclear DNA of grande (wildtype) yeast. *J. Mol. Biol.* 63:431–40
96. Robberson, D., Aloni, Y., Attardi, G., Davidson, N. 1972. Expression of the mitochondrial genome in HeLa cells. VIII. The relative position of ribosomal RNA genes in mitochondrial DNA. *J. Mol. Biol.* 64:313–17
97. Wu, M., Davidson, N., Attardi, G., Aloni, Y. 1972. Expression of the mitochondrial genome in HeLa cells. XIV. The relative positions of the 4S RNA genes and of the ribosomal genes in mitochondrial DNA. *J. Mol. Biol.* 71:81–93
98. Cohen, M., Casey, J., Rabinowitz, M., Getz, G. S. 1972. Hybridization of mitochondrial transfer RNA and mitochondrial DNA in petite mutants of yeast. *J. Mol. Biol.* 63:441–51
99. Nagley, P., Gingold, E. B., Lukins, H. B., Linnane, A. W. 1973. Biogenesis of mitochondria. XXV. Studies on the mitochondrial genomes of petite mutants of yeast using ethidium bromide as a probe. *J. Mol. Biol.* 78:335–50
100. Faye, G. et al 1973. Mitochondrial nucleic acids in petite colonie mutants: deletions and repetitions of genes. *Biochimie* 55:779–92
101. Thomas, C. 1966. Recombination of DNA molecules. *Progr. Nucl. Acid Res.* 5:315–37
102. Slonimski, P. P. 1968. Discussion on biochemical studies of mitochondria in cytoplasmic mutants. See Ref. 12, p. 475
103. Rank, G. H., Person, C. 1969. Reversion of spontaneously arising respiratory deficiency in *Saccharomyces cerevisiae. Can. J. Genet. Cytol.* 11:716–28
104. Rank, G. H. 1970. Genetic evidence for "Darwinian" selection at the molecular

level. I. The effect of the suppressive factor on cytoplasmically-inherited erythromycin-resistance in *Saccharomyces cerevisiae. Can. J. Genet. Cytol.* 12:129–36

105. Rank, G. H. 1970. Genetic evidence for "Darwinian" selection at the molecular level. II. Genetic analysis of cytoplasmically-inherited high and low suppressivity in *Saccharomyces cerevisiae. Can. J. Genet. Cytol.* 12:340–46

106. Rank, G. H., Bech-Hansen, N. T. 1972. Genetic evidence for "Darwinian" selection at the molecular level. III. The effect of the suppressive factor on nuclearly and cytoplasmically inherited chloramphenicol resistance in *S. cerevisiae. Can. J. Microbiol.* 18:1–7

107. Mills, D. R., Peterson, R. L., Spiegelman, S. 1967. An extracellular Darwinian experiment with a self-duplicating nucleic acid molecule. *Proc. Nat. Acad. Sci. USA* 58:217–24

108. Bolotin, M. et al 1971. La recombinaison des mitochondries chez *Saccharomyces cerevisiae. Bull. Inst. Pasteur* 69:215–39

109. Avner, P. R., Coen, D., Dujon, B., Slonimski, P. P. 1973. Mitochondrial genetics. IV. Allelism and mapping studies of oligomycin resistant mutants in *S. cerevisiae. Mol. Gen. Genet.* 125:9–52

110. Wolf, K., Dujon, B., Slonimski, P. P. 1973. Mitochondrial genetics. V. Multifactorial mitochondrial crosses involving a mutation conferring paromomycin resistance in *Saccharomyces cerevisiae. Mol. Gen. Genet.* 175:50–90

111. Dujon, B., Slonimski, P. P., Weill, L. 1974. Mitochondrial genetics. IX: A model for recombination and segregation of mitochondrial genomes in *Saccharomyces cerevisiae. Int. Congr. Genet. Symp., XIII,* Vol. 1: In press

112. Thomas, D. Y., Wilkie, D. 1968. Recombination of mitochondrial drug-resistance factors in *Saccharomyces cerevisiae. Biochem. Biophys. Res. Commun.* 30:368–72

113. Lukins, H. B., Tate, J. R., Saunders, G. W., Linnane, A. W. 1973. The biogenesis, of mitochondria. 26. Mitochondrial recombination: the segregation of parental and recombinant mitochondrial genotypes during vegetative division in yeast. *Mol. Gen. Genet.* 123:17–25

114. Mitchell, C. H., Bunn, C. L., Lukins, H. B., Linnane, A. W. 1973. Biogenesis of mitochondria. 23. The biochemical and genetic characteristics of two oligomycin resistant mutants of *Saccharomyces*

cerevisiae under the influence of cytoplasmic genetic modification. *Bioenergetics* 4:161–77

115. Trembath, M. K., Bunn, C. L., Lukins, H. B., Linnane, A. W. 1973. Biogenesis of mitochondria. 27. Genetic and biochemical characterisation of cytoplasmic and nuclear mutations to spiramycin resistance in *Saccharomyces cerevisiae. Mol. Gen. Genet.* 121:35–48

116. Howell, N., Trembath, M. K., Linnane, A. W., Lukins, H. B. 1973. Biogenesis of mitochondria. 30. An analysis of polarity of mitochondrial gene recombination and transmission. *Mol. Gen. Genet.* 122:37–51

117. Kleese, R. A., Grotbeck, R. C., Snyder, J. R. 1972. Two cytoplasmically inherited chloramphenicol resistance loci in yeast (*Saccharomyces cerevisiae*). *Can. J. Genet. Cytol.* 14:713–15

118. Kleese, R. A., Grotbeck, R. C., Snyder, J. R. 1972. Recombination among three mitochondrial genes in yeast (*Saccharomyces cerevisiae*). *J. Bacteriol.* 112:1023–25

119. Wilkie, D., Thomas, D. Y. 1973. Mitochondrial genetic analysis by zygote cell lineages in *Saccharomyces cerevisiae. Genetics* 73:367–77

120. Rank, G. H., Bech-Hansen, N. T. 1972. Somatic segregation, recombination, asymmetrical distribution and complementation tests of cytoplasmically-inherited antibiotic-resistance mitochondrial markers in *S. cerevisiae. Genetics* 72:1–15

121. Avner, P. R., Griffiths, D. E. 1973. Studies on energy-linked reactions: Genetics of oligomycin-resistant mutants of *Saccharomyces cerevisiae. Eur. J. Biochem.* 32:312–21

122. Rank, G. H. 1973. Recombination in 3-factor crosses of cytoplasmically inherited antibiotic-resistance mitochondrial markers in *S. cerevisiae. Heredity* 30:265–71

123. Suda, K., Uchida, A. 1972. Segregation and recombination of cytoplasmic drug-resistance factors in *Saccharomyces cerevisiae. Japan J. Genet.* 47:441–44

124. Wakabayashi, K., Kamei, S. 1973. Oligomycin resistance in yeast. Linkage of the mitochondrial drug resistance. *FEBS Lett.* 33:263–65

125. Smith, D. G., Wilkie, D., Srivastava, K. C. 1972. Ultrastructural changes in mitochondria of zygotes in *Saccharomyces cerevisiae. Microbios* 6:231–38

126. Grivell, L. A., Netter, P., Borst, P., Slonimski, P. P. 1973. Mitochondrial antibiotic resistance in yeast: ribosomal mutants resistant to chloramphenicol, erythromycin and spiramycin. *Biochim. Biophys. Acta* 312:358–67

127. Bunn, C. L., Mitchell, C. H., Lukins, H. B., Linnane, A. W. 1970. Biogenesis of mitochondria. XVIII. A new class of cytoplasmically determined antibiotic resistant mutants in *Saccharomyces cerevisiae. Proc. Nat. Acad. Sci. USA* 67:1233–40

128. Avner, P., Griffiths, D. E. 1973. Studies on energy-linked reactions isolation and characterization of oligomycin-resistant mutants of *Saccharomyces cerevisiae. Eur. J. Biochem.* 32: 301–11

129. Somlo, M., Avner, P. R., Cosson, J., Dujon, B., Krupa, M. 1974. Oligomycin sensitivity of ATPase studied as a function of mitochondrial biogenesis, using mitochondrially determined oligomycin-resistant mutants of *Saccharomyces cerevisiae. Eur. J. Biochem.* 42:439–45

130. Kutzleb, R., Schweyen, R. J., Kaudewitz, F. 1973. Extrachromosomal inheritance of paromomycin resistance in *Saccharomyces cerevisiae. Mol. Gen. Genet.* 125:91–98

131. Visconti, N., Delbrück, M. 1953. The mechanism of genetic recombination in phage. *Genetics* 38:5–33

132. Sager, R. 1954. Mendelian and non-Mendelian inheritance of streptomycin resistance in *Chlamydomonas. Proc. Nat. Acad. Sci. USA* 40:356–62

133. Gillham, N. W. 1969. Uniparental inheritance in *Chlamydomonas reinhardi. Am. Naturalist* 103:355–88

134. Surzycki, S. J., Gillham, N. W. 1971. Organelle mutations and their expression in *Chlamydomonas reinhardi. Proc. Nat. Acad. Sci. USA* 68:1301–6

135. Mets, L. J., Bogorad, L. 1971. Mendelian and uniparental alterations in erythromycin binding by plastid ribosomes. *Science* 174:707–9

136. Bastia, D., Chiang, K.-S., Swift, H. 1969. Chloroplast de-differentiation and redifferentiation during zygote maturation and germination in *Chlamydomonas reinhardi. J. Cell Biol.* 43:11a

137. Cavalier-Smith, T. 1970. Electron microscopic evidence for chloroplast fusion in zygotes of *Chlamydomonas reinhardi. Nature* 228:333–35

138. Chiang, K.-S., Sueoka, N. 1967. Replication of chromosomal and cytoplasmic DNA during mitosis and meiosis in the eucaryote *Chlamydomonas reinhardi. J. Cell Physiol.* 70: Suppl. 1, 89–112

139. Wells, R., Sager, R. 1971. Denaturation and the renaturation kinetics of chloroplast DNA from *Chlamydomonas reinhardi. J. Mol. Biol.* 58:611–22

140. Sager, R., Ishida, M. R. 1963. Chloroplast DNA in *Chlamydomonas. Proc. Nat. Acad. Sci. USA* 50:725–30

141. Bastia, D., Chiang, K.-S., Swift, H., Siersma, P. 1971. Heterogeneity, complexity and repetition of chloroplast DNA of *Chlamydomonas reinhardi. Proc. Nat. Acad. Sci. USA* 68:1157–61

142. Bastia, D., Chiang, K.-S., Swift, H. 1971. Studies on the ribosomal RNA cistrons of chloroplast and nucleus in *Chlamydomonas reinhardtii. Abstr. Ann. Meet. Am. Soc. Cell Biol., 11th,* p. 25

143. Ryan, R. S., Grant, D., Chiang, K.-S., Swift, H. 1973. Isolation of mitochondria and characterization of the mitochondrial DNA of *Chlamydomonas reinhardtii. J. Cell Biol.* 59:297a

144. Ryan, R. S., Grant, D., Chiang, K.-S., Swift, H. Personal communication

145. Kolodner, R., Tewari, K. K. 1972. Physicochemical characterization of mitochondrial DNA from pea leaves. *Proc. Nat. Acad. Sci. USA* 69:1830–34

146. Sager, R., Lane, D. 1972. Molecular basis of maternal inheritance. *Proc. Nat. Acad. Sci. USA* 69:2410–13

147. Chiang, K.-S. 1968. Physical conservation of parental cytoplasmic DNA through meiosis in *Chlamydomonas reinhardi. Proc. Nat. Acad. Sci. USA* 60:194–200

148. Chiang, K. S. 1971. Replication, transmission and recombination of cytoplasmic DNAs in *Chlamydomonas reinhardi.* See Ref. 13, pp. 235–49

149. Siersma, P. W., Chiang, K.-S. 1971. Conservation and degradation of cytoplasmic and chloroplast ribosomes in *Chlamydomonas reinhardtii. J. Mol. Biol.* 58:167–85

150. Reich, E., Luck, D. J. L. 1966. Replication and inheritance of mitochondrial DNA. *Proc. Nat. Acad. Sci. USA* 55:1600–8

151. Swinton, D. C., Hanawalt, P. C. 1972. In vivo specific labeling of *Chlamydomonas* chloroplast DNA. *J. Cell Biol.* 54:592–97

152. Lee, R. W., Jones, R. F. 1973. Induction of Mendelian and non-Mendelian streptomycin resistant mutants during the synchronous cell cycle of

Chlamydomonas reinhardtii. Mol. Gen. Genet. 121:99–108

153. Gillham, N. W. 1965. Induction of chromosomal and nonchromosomal mutations in *Chlamydomonas reinhardi* with N - methyl - N' - nitro - N - nitroso-guanidine. *Genetics* 52:529–37

154. Cerda-Olmedo, E., Hanawalt, P. C., Guerola, N. 1968. Mutagenesis of the replication point by nitrosoguanidine: Map and pattern of replication of the *Escherichia coli* chromosome. *J. Mol. Biol.* 33:705–19

155. Alexander, N. J., Gillham, N. W., Boynton, J. E. 1974. The mitochondrial genome of *Chlamydomonas:* Induction of minute colony mutations by acriflavin and their inheritance. *Mol. Gen. Genet.* In press

156. Mason, T. L., Schatz, G. 1973. Cytochrome *c* oxidase from Baker's yeast II. Site of translation of the protein components. *J. Biol. Chem.* 248:1355–60

157. Weiss, H., Sebald, W., Schwab, A. J., Kleinow, W., Lorenz, B. 1973. Contribution of mitochondrial and cytoplasmic protein synthesis to the formation of cytochrome *b* and cytochrome aa_3. *Biochimie* 55:815–21

158. Stegeman, W. J., Hoober, J. K. 1973. Stimulation of the synthesis of a mitochondrial gene product after exposure of dark-grown *Chlamydomonas reinhardi* Y-1 to light. *J. Cell Biol.* 59:334a

159. Flechtner, V. R., Sager, R. 1973. Ethidium bromide induced selective and reversible loss of chloroplast DNA. *Nature New Biol.* 241:277–79

160. Nass, M. M. K., Ben-Shaul, Y. 1973. Effects of ethidium bromide on growth, chlorophyll synthesis, ultrastructure, and mitochondrial DNA in green and bleached mutant *Euglena gracilis. J. Cell Sci.* 13:567–90

161. Boynton, J. E., Burton, W. G., Gillham, N. W., Harris, E. H. 1973. Can a non-Mendelian mutation affect both chloroplast and mitochondrial ribosomes? *Proc. Nat. Acad. Sci. USA* 70:3463–67

162. Burton, W. G. 1972. Dihydrospectinomycin binding to chloroplast ribosomes from antibiotic-sensitive and -resistant strains of *Chlamydomonas reinhardtii. Biochim. Biophys. Acta* 272:305–11

163. Mets, L., Bogorad, L. 1972. Altered chloroplast ribosomal proteins associated with erythromycin-resistant mutants in two genetic systems of *Chlamydomonas reinhardi. Proc. Nat. Acad. Sci. USA* 69:3779–83

164. Ohta, N., Inouye, M., Sager, R. 1974. Identification of a chloroplast ribosomal protein coded by a chloroplast gene in *Chlamydomonas. J. Biol. Chem.* Submitted

165. Schlanger, G., Sager, R., Ramanis, Z. 1972. Mutation of a cytoplasmic gene in *Chlamydomonas* alters chloroplast ribosome function. *Proc. Nat. Acad. Sci. USA* 69:3551–55

166. Schlanger, G., Sager, R. 1974. Localization of five antibiotic resistances at the subunit level in chloroplast ribosomes of *Chlamydomonas. Proc. Nat. Acad. Sci. USA* 71:1715–19

167. Togasaki, R. K., Levine, R. P. 1970. Chloroplast structure and function in *ac-20,* a mutant strain of *Chlamydomonas reinhardi.* I. CO_2 fixation and ribu-lose-1,5-diphosphate carboxylase synthesis. *J. Cell Biol.* 44:531–38

168. Levine, R. P., Paszewski, A. 1970. Chloroplast structure and function in *ac-20,* a mutant strain of *Chlamydomonas reinhardi.* II. Photosynthetic electron transport. *J. Cell Biol.* 44:540–46

169. Goodenough, U. W., Levine, R. P. 1970. Chloroplast structure and function in *ac-20,* a mutant strain of *Chlamydomonas reinhardi.* III. Chloroplast ribosomes and membrane organization. *J. Cell Biol.* 44:547–62

170. Boynton, J. E., Gillham, N. W., Burkholder, B. 1970. Mutations altering chloroplast ribosome phenotype in *Chlamydomonas.* II. A new Mendelian mutation. *Proc. Nat. Acad. Sci. USA* 67:1505–12

171. Bourque, D. P., Boynton, J. E., Gillham, N. W. 1971. Studies on the structure and cellular location of various ribosome and ribosomal RNA species in the green alga *Chlamydomonas reinhardi. J. Cell Sci.* 8:153–83

172. Boynton, J. E., Gillham, N. W., Chabot, J. F. 1972. Chloroplast ribosome deficient mutants in the green alga *Chlamydomonas reinhardi* and the question of chloroplast ribosome function. *J. Cell Sci.* 10:267–305

173. Harris, E. H., Boynton, J. E., Gillham, N. W., Togasaki, R. K. 1973. Positive selection for mutants in *Chlamydomonas reinhardtii* with defects in chloroplast protein synthesis. *Genetics* 74: S109–10

174. Surzycki, S. J. 1969. Genetic functions of the chloroplast of *Chlamydomonas reinhardi:* effect of rifampicin on chloroplast DNA-dependent RNA polyme-

rase. *Proc. Nat. Acad. Sci. USA* 63: 1327–34

175. Surzycki, S. J., Goodenough, U. W., Levine, R. P., Armstrong, J. J. 1970. Nuclear and chloroplast control of chloroplast structure and function in *Chlamydomonas reinhardi.* See Ref. 14, pp. 13–37

176. Surzycki, S. J., Rochaix, J. D. 1971. Transcriptional mapping of ribosomal RNA genes of chloroplast and nucleus of *Chlamydomonas reinhardi. J. Mol. Biol.* 62:89–109

177. Goodenough, U. W., Togasaki, R. K., Paszewski, A., Levine, R. P. 1971. Inhibition of chloroplast ribosome formation by gene mutation in *Chlamydomonas reinhardi.* See Ref. 13, pp. 224–34

178. Harris, E. H., Boynton, J. E., Gillham, N. W. 1974. Chloroplast ribosome biogenesis in *Chlamydomonas:* Selection and characterization of mutants blocked in ribosome formation. *J. Cell Biol.* In press

179. Davies, J., Nomura, M. 1972. The genetics of bacterial ribosomes. *Ann Rev. Genet.* 6:203–34

180. Chu-Der, O. M. Y., Chiang, K.-S. 1974. The interaction between Mendelian and non-Mendelian genes in *Chlamydomonas reinhardtii* I. The regulation of the transmission of non-Mendelian genes by a Mendelian gene. *Proc. Nat. Acad. Sci. USA* 71:153–57

181. Sager, R., Ramanis, Z. 1973. The mechanism of maternal inheritance in *Chlamydomonas:* biochemical and genetic studies. *Theor. Appl. Genet.* 43:101–8

182. Arber, W., Linn, S. 1969. DNA modification and restriction. *Ann. Rev. Biochem.* 38:467–500

183. Sager, R., Ramanis, Z. 1967. Biparental inheritance of nonchromosomal genes induced by ultraviolet irradiation. *Proc. Nat. Acad. Sci. USA* 58:931–37

184. Sager, R. 1970. Genetic studies of chloroplast DNA in *Chlamydomonas.* See Ref. 14, pp. 401–17

185. Sager, R., Ramanis, Z. 1971. Methods of genetic analysis of chloroplast DNA in *Chlamydomonas.* See Ref. 13, pp. 250–59

186. Sager, R., Ramanis, Z. 1971. Formal genetic analysis of organelle genetic systems. *Stadler Symposia* Vols. 1 & 2, pp. 65–73

187. Gillham, N. W., Boynton, J. E., Lee, R. W. 1974. Segregation and recombination of non-Mendelian genes in *Chlam-*

ydomonas. Int. Congr. Genet. Symp., XIII, Vol. 1. In press

188. Sager, R., Ramanis, Z. 1970. A genetic map of non-Mendelian genes in *Chlamydomonas. Proc. Nat. Acad. Sci. USA* 65:593–600

189. Sager, R., Ramanis, Z. 1963. The particulate nature of nonchromosomal genes in *Chlamydomonas. Proc. Nat. Acad. Sci. USA* 50:260–68

190. Sager, R., Ramanis, Z. 1965. Recombination of nonchromosomal genes in *Chlamydomonas. Proc. Nat. Acad. Sci. USA* 53:1053–61

191. Sager, R., Ramanis, Z. 1968. The pattern of segregation of cytoplasmic genes in *Chlamydomonas. Proc. Nat. Acad. Sci. USA* 61:324–31

192. Gillham, N. W. 1965. Linkage and recombination between nonchromosomal mutations in *Chlamydomonas reinhardi. Proc. Nat. Acad. Sci. USA* 54: 1560–67

193. Gillham, N. W., Fifer, W. 1968. Recombination of non-chromosomal mutations: a three-point cross in the green alga *Chlamydomonas reinhardi. Science* 162:683–84

194. Gillham, N. W. 1963. The nature of exceptions to the pattern of uniparental inheritance for high level streptomycin-resistance in *Chlamydomonas reinhardi. Genetics* 48:431–39

195. Schimmer, O., Arnold, C. G. 1969. Untersuchungen zur lokalisation eines ausserkaryotischen gens bei *Chlamydomonas reinhardi. Arch. Microbiol.* 66:199–202

196. Schimmer, O., Arnold, C. G. 1970. Untersuchungen über reversions-und segregations verhalten eines ausserkaryotischen gens von *Chlamydomonas reinhardii* zur bestimmung des erbträgers. *Mol. Gen. Genet.* 107:281–90

197. Schimmer, O., Arnold, C. G. 1970. Über die zahl der kopien eines ausserkaryotischen gens bei *Chlamydomonas reinhardii. Mol. Gen. Genet.* 107: 366–71

198. Schimmer, O., Arnold, C. G. 1970. Hin-und rücksegregation eines ausserkaryotischen gens bei *Chlamydomonas reinhardii. Mol. Gen. Genet.* 108:33–40

199. Clowes, R. C. 1972. The molecular structure of bacterial plasmids. *Bacteriol. Rev.* 36:361–405

200. Uzzo, A., Lyman, H. 1971. The nature of the chloroplast genome of *Euglena gracilis. Proc. Int. Congr. Photosynth., 2nd,* ed. G. Forti, 3:2585–2601. The Hague:Junk

THE ORIGIN OF LIFE   ❖3079

N. H. Horowitz and Jerry S. Hubbard[1]

Biology Division, California Institute of Technology, Pasadena, California 91109

INTRODUCTION

The origin of life is in a sense a genetic problem, for, as H. J. Muller pointed out many years ago, the essential attribute that identifies living matter is its capacity to replicate itself and its variants (1). Because this uniquely biological property has its physical basis in proteins and nucleic acids, the goal of modern work on the origin of life is to discover the manner of origin of these polymers and of the interactions between them that constitute the genetic mechanism. In attempting to review this subject in a limited space, we cannot undertake an exhaustive treatment. Rather, we summarize work published principally since 1970 in the following areas, with emphasis on those aspects that are of greatest current interest: 1. precambrian paleontology, 2. chemical evolution of genetically important monomers, 3. prebiotic dehydration-condensation reactions, 4. organic compounds in meteorites and interstellar space, and 5. biological exploration of the planets.

A large number of review articles (2–5), critical and theoretical discussions (6–8), books (9–16), and conference proceedings (17–21) dealing with the origin of life have appeared in recent years. In addition, a new serial, the *Journal of Molecular Evolution,* publishing papers on this and related subjects, appeared in 1971; the journal *Space Life Sciences* has been renamed *Origins of Life;* and a society, the International Society for the Study of the Origin of Life, was recently founded.

PRECAMBRIAN PALEONTOLOGY

Until about 20 years ago it was thought that no fossils remained of the life that populated the earth before the opening of the Cambrian era 600 million years ago. For the first 85% of Earth's history there were thus no data on which estimates of the rate of evolution and the epoch of the origin of life might be based. The situation has changed with the discovery of a dozen or more Precambrian sediments, some of them exceeding 3×10^9 years in age, containing a variety of fossilized microorganisms. These findings have been reviewed by Schopf (22).

---

[1]Present address: School of Biology, Georgia Institute of Technology, Atlanta, Georgia 30332.

The oldest objects generally accepted as fossils at the present time are unicellular organisms resembling bacteria and coccoid blue-green algae, found in cherts of the Fig Tree Series of South Africa (23, 24). These cherts have been dated at 3.1 X 10^9 years. A still older series of rocks of the same general sequence, the Onverwacht Group, with an age of approximately 3.3 X 10^9 years (25), contains microstructures that have been interpreted with varying degrees of confidence as fossils (26–29). The Onverwacht structures are more variable in size and, if fossils, are also less well preserved than those of the Fig Tree sediments.

Fossil identifications in these ancient rocks are based on morphology (including its continuity with that of better-preserved and more abundant fossils in younger sediments) and on the organic chemistry of the host rock. Precambrian cherts typically contain 0.5–1.0% organic matter, the bulk of it in the form of an insoluble, unextractable polymer ("kerogen"). Most of the chemical work has naturally been done on the small extractable fraction. This has been found to contain a characteristically biological assemblage of hydrocarbons, including pristane and phytane (known breakdown products of chlorophyll), and amino acids. Recent work has cast doubt on the significance of these findings, because it appears that the soluble fraction may not be syngenetic with the rocks. Various types of evidence indicate that the cherts are sufficiently permeable to have been infiltrated by soluble material over geologic time and that the soluble fraction is largely or entirely of recent origin (30–33).

Similar doubts have not arisen with regard to the insoluble fraction which, being less mobile, is more likely to be indigenous to the rocks. Pyrolytic analysis shows that this fraction consists mainly of polymerized aliphatic and aromatic hydrocarbons (34, 35). Carbon isotope ratios of kerogens from the Fig Tree and upper Onverwacht sediments show the deficiency of ^{13}C that is typical of photosynthetically derived organic matter (32, 36) and thus support the morphological evidence which suggests that blue-green alga-like forms were extant in Fig Tree times. The carbon from lower Onverwacht kerogens is anomalously heavy, however, a finding that led Oehler et al to speculate that an unusual geological or biological event, such as the onset of photosynthesis or even the origin of life, occurred in Onverwacht times (36). The heavy kerogen, it was suggested, may be left over from a prephotosynthetic or prebiotic era. This is an extraordinarily interesting result, but, to quote the authors, "the data are so few and their potential implication so far-reaching that no firm conclusion should be drawn at the present time."

Whatever the outcome of this issue, it seems established that life existed on the earth 3.1 X 10^9 years ago, and it appears that at least one species was carrying out photosynthesis at that time. Because the capacity for photosynthesis represents a long evolutionary development from the presumed heterotrophy of the primordial organism, the implication is that life originated at a much earlier time, possibly during the first billion years of the earth's 4.5 billion year history.

Evidence is beginning to accumulate on the time of origin of the eukaryotic cell. Eukaryotic cells seem to be well established in the Bitter Springs Formation of Australia, about 900 million years old (22), but the remains of possibly nuclear structures have been discovered in the much older Beck Spring dolomite of California, estimated to be 1.3×10^9 years of age (37). Recent findings of relatively complex multicellular microfossils in the McArthur Group formation of Australia, thought to be 1.6×10^9 years of age, suggest that eukaryotic organisms may have existed at this even earlier time (38).

CHEMICAL EVOLUTION OF GENETICALLY IMPORTANT MONOMERS

The synthesis of amino acids, purines, pyrimidines, and sugars from cosmically abundant gases has been demonstrated in numerous investigations since Stanley Miller (39) performed the first experiments in this field (see 3, 10, 11, 15 for reviews of these investigations). It is clear that there are many pathways from primitive gas mixtures to the genetically important monomers, and it is unlikely that any one pathway was exclusively involved in their synthesis on the primitive earth. Recent work in this area is concerned with the investigation of various alternative prebiotic conditions, energy sources, and starting materials (reviewed below) and with the study of some particularly difficult syntheses, e.g. methionine (40), thymine (41), and aromatic amino acids (42).

Long-Wavelength uv as Energy Source

Although the original Miller experiment used a spark discharge as energy source, solar radiation is now and probably was in primitive times the major source of energy reaching the earth (6, 43). However, only a small fraction of sunlight is in wavelengths below 200 nm where CH_4, CO, NH_3, N_2, H_2O, and H_2 have almost their entire absorption. The large amount of energetic radiation between 200 and 300 nm is photochemically useless in these gases without a photosensitizing agent that can absorb and transfer energy in this spectral range. Hg vapor is such an agent (44), but it is not a plausible constituent of the primitive atmosphere.

Sagan & Khare (45) have found that H_2S, which absorbs in the 200–300 nm region, photosensitizes the synthesis of glycine, alanine, serine, glutamic acid, aspartic acid, and cystine, together with an orange-brown polymer, from mixtures of CH_4, C_2H_6, NH_3, and H_2O irradiated with the 254 nm line of a Hg source. Ethane was essential, for reasons that are not understood. Ethane would occur only in traces in a reducing atmosphere at equilibrium. It can be generated from methane by electric discharges or short uv, but these are the energy sources for which the experiment is seeking an alternative. Similarly, the partial pressure of H_2S in the primitive atmosphere would have been limited by the formation of metal sulfides on the earth's surface and by photolysis in the atmosphere (46).

In similar experiments with H_2S, Becker and co-workers have confirmed the production of amino acids, but without formation of the orange polymer (47). Most important, these workers did not find a requirement for C_2H_6. CH_4 was sufficient as a carbon source to produce amino acids. The quantum yields calculated by the two groups of experimenters are comparable: 10^{-4}–10^{-5}. No explanation has been established for the discrepancy in results. If CH_4 in fact suffices for amino acid production, this reaction may have been an important primary source of biomonomers on the primitive earth.

Acoustic Energy

It is known from the work of Elpiner (48) that ultrasonic vibrations can produce organic compounds, including HCN and HCHO, in water saturated with N_2, CO (or CH_4), and H_2. The reactions occur in cavitation bubbles which, on collapsing, generate intense hydraulic shocks accompanied by brief but very high pressure and temperature pulses. Anbar (49) showed that cavitation occurs in systems simulating ocean waves and falling water; amino acids were produced when the water contained NH_3 and CH_4. Recently Bar-Nun et al found that amino acids are synthesized in gas mixtures by shock waves simulating the entry of meteorites into the atmosphere (50). The efficiency was extraordinarily high: 5×10^{10} molecules of amino acid per erg of shock energy. The authors estimate that 30 kg of organic matter per cm^2 of the earth's surface could have been produced in the first 10^9 years of the earth's history by meteorite- and thunder-induced shock waves. Unfortunately, the gas mixture used in the experiments of Bar-Nun et al contained ethane, so that it is difficult to evaluate the significance of the high efficiency they observed.

CO as a Carbon Source

The ocean and atmosphere were produced by outgassing from the molten interior of the earth. CO is stable at magmatic temperatures and would have been the principal form of outgassed carbon. Methane is the stable form of carbon in a reducing atmosphere at low temperatures, however, and would have formed slowly from CO on cooling. A continuing output of CO from volcanoes would have maintained a steady-state pressure of CO in the primitive atmosphere [see Abelson (51) and Van Trump & Miller (52) for further discussion]. CO is also a constituent of the Martian atmosphere, and it is a major component of interstellar clouds (see section on organic compounds in meteorites and interstellar space, p. 402).

The reduction of CO by H_2 is accelerated by various metallic catalysts, this being the basis of the Fischer-Tropsch process for making hydrocarbons. Anders and co-workers (53) have proposed that a Fischer-Tropsch–like reaction in which NH_3 participates along with CO and H_2 can account more satisfactorily than can the Miller-Urey reaction for some prebiotic syntheses. (Fischer-Tropsch–like reactions are discussed further on pp. 402–3.)

A different type of surface-catalyzed reaction involving CO has been discovered by Hubbard et al in experiments designed initially to simulate the surface environment of Mars (54, 55). It was found that uv irradiation of siliceous materials or alumina in atmospheres containing CO and water vapor (diluted in a large volume of CO_2 or N_2) results in the synthesis of simple organic compounds on the solid surfaces. The major product is formic acid, with smaller amounts of formaldehyde, acetaldehyde, and glycolic acid. The most remarkable feature of this reaction is that it is effected by wavelengths as long as 300 nm, although the gaseous reactants do not absorb wavelengths longer than 200 nm. Photosensitization is provided by the solid surfaces, which also protect the photosensitive products from destruction by uv. The available evidence suggests that the reaction is initiated by absorption of the excitation energy by the solid substratum with cleavage of silanol (Si–OH) bonds in the case of siliceous substrata to form surface OH radicals (56). Reaction of the OH radical with CO could yield surface COOH radicals, the precursor of formic acid.

On highly effective substrata such as volcanic ash shale or clay minerals, this photocatalysis can produce tens of nanomoles of product per cm^3 per day (55). Photodestruction of the organic products occurs, as well as synthesis, and eventually a steady state is attained in which the rates of formation and decomposition are equal.

If NH_3 is added to the above system, urea, formamide, and formaldehyde are produced photocatalytically from CO and NH_3 (57). This reaction apparently depends on the formation of a complex between NH_3 and the surface material which absorbs at longer wavelengths than does free NH_3. Because the reaction between CO and NH_3 is inhibited by water vapor, its relevance for terrestrial chemistry is questionable, but it may offer a mechanism for the synthesis of interstellar molecules on siliceous grains (see p. 402).

Stability of Ammonia in the Primitive Atmosphere

NH_3 is the thermodynamically favored form of nitrogen in a reducing atmosphere at ordinary temperatures (43). The lack of an ozone screen in such an atmosphere would, however, permit rapid photolysis of NH_3 (to $N_2 + H_2$) by uv wavelengths shorter than 230 nm (51). The question of the actual lifetime of NH_3 in the primitive atmosphere is thus of interest. Some years ago, Bada & Miller (58) made an ingenious estimate of the lower limit of NH_4^+ in the primitive ocean based on the fact that NH_4^+ is necessary to prevent the deamination of aspartate in solution. This estimate, about $10^{-3}M$, leads to a minimum pressure of about 10^{-6} atm NH_3 in the atmosphere. In recent theoretical studies, Ferris & Nicodem (46, 59) find that photolysis of NH_3 is so rapid that its lifetime would not have exceeded a few millions of years under conditions usually considered realistic for the primitive earth. The presence of H_2 at high pressure (> 50 torr) could have maintained NH_3 by regenerating it, as could the slow release of NH_3 by the hydrolysis of urea formed from NH_3 and CO. In the absence of these mechanisms, it appears that either NH_3 had no role in the origin of life, or its role was completed within the first few million years of the earth's history.

PREBIOTIC DEHYDRATION-CONDENSATION REACTIONS

The condensation of amino acids into polypeptides and of bases, sugars, and phosphate into mono- and polynucleotides involves dehydrations that are endergonic by 2–6 kcal per mole. Attempts to find plausible prebiotic mechanisms for these energy-consuming reactions have been carried out under three principal conditions: homogeneous aqueous, heterogeneous aqueous, and nonaqueous thermal.

Homogeneous Aqueous Dehydrations

A number of simple derivatives of HCN have been found to promote dehydration-condensations in aqueous solution, e.g. cyanamide, cyanogen, cyanamide dimer, and others (3, 10, 11). These compounds are energy-rich with respect to their hydration products. They add water in reactions similar to those entered into by carbodiimides in synthetic organic chemistry:

$$RN = C = NR' + H_2O \longrightarrow RHN\text{-}\overset{\overset{\displaystyle O}{\|}}{C}\text{-}NHR'.$$

Although a variety of HCN derivatives may well have existed on the primitive earth, their importance for prebiological dehydration reactions is questionable because of the fact that that they react readily with solvent water. (In synthetic chemistry, carbodiimides are used in anhydrous solvents.) Because of this lack of specificity, very high concentrations of HCN derivatives are needed to bring about significant peptide- and ester-bond formation in aqueous solution. A critical review of these findings, with a bibliography of important papers, is given by Miller & Orgel (15).

Alkaline solutions of HCN form, on standing, an insoluble black material referred to as "polymer." Acid hydrolysis of the colored supernatant from these preparations yields a complex mixture of ninhydrin-positive substances, including amino acids (60). Matthews and co-workers (61 and earlier papers cited there) have proposed that HCN polymer is a polypeptide formed directly from HCN and water without the intervention of free amino acids. On this view, the manner of formation and polymerization of amino acids is irrelevant to the question of the origin of proteins. Ferris et al (62) find, however, that amino acids are not released from HCN polymer by proteolytic enzymes, nor can a dinitrophenyl derivative or a positive biuret reaction be elicited. It is difficult to reconcile these results with a peptidic structure. It appears that the numerous products derivable from HCN polymer are formed by hydrolysis and/or oxidation-reduction of an equilibrium mixture of HCN and its dimer, trimer, and tetramer that forms spontaneously in alkaline HCN solutions (63).

Heterogeneous Aqueous Dehydrations

One of J. D. Bernal's principal contributions to the study of chemical evolution was the idea that dilute solutions of biomonomers in the primitive ocean could have been concentrated by adsorption on the surface of clay minerals, or at air-water interfaces, where further reactions among them might take place (9, 64). Until now, the best example of such a process has been the polymerization of alanyl adenylate to polyalanine on the surface of montmorillonite, discovered by Paecht-Horowitz et al (65). Aminoacyl adenylates react in free solution to yield short peptide chains and adenylic acid, but hydrolysis to monomeric amino acid predominates. In the presence of montmorillonite, a common aluminosilicate clay mineral having a layer-lattice structure and a large absorptive capacity, the predominant reaction is polymerization. Polymers containing up to 56 alanyl residues are formed in 24 hr under mild conditions.

In a recent paper, the same authors have taken the process a step further by showing that the synthesis of alanyl adenylate from alanine and ATP is catalyzed by a synthetic zeolite (66). In a system containing alanine, ATP, zeolite, and montmorillonite, the amino acid is polymerized quantitatively. It is doubtful whether ATP in large amounts was available in prebiological times, although small quantities were probably produced by reactions described below. It would be interesting to know whether polyphosphates, energetically equivalent to ATP and possibly more abundant on the primitive earth, can serve as the energy source for this remarkable polymerization.

It has recently been claimed that kaolinite has the property of adsorbing amino acids from dilute solutions at 90°C and polymerizing them to polypeptides, with a strong preference for the L-optical isomers (67, 68). Bonner & Flores (69, 70) have tried unsuccessfully to repeat these observations; they could find evidence neither for asymmetric adsorption of amino acids nor for polymerization. Along similar lines, Amariglio & Amariglio (71) have summarized their careful attempts to duplicate experiments claiming to demonstrate optical asymmetry in reactions catalyzed by optically active quartz. The results were uniformly negative. The most reasonable explanation for the optical asymmetry of living matter is that it is a biologically determined property which has selective value, but the choice of optical isomers was originally a matter of chance. Wald advanced this view many years ago (72).

Nonaqueous Thermal Dehydrations

The energy requirement for dehydration condensations can be reduced by carrying the reactions out under anhydrous conditions. S. W. Fox discovered some years ago that dry amino acid mixtures heated to 150°C or higher condense to form polypeptides if the mixture contains sufficiently large amounts of glutamic or aspartic acid, or lysine. The properties of these thermal polypeptides have been the subject of a large number of investigations by Fox and his associates, who have based a theory of the origin of life on them (14). That the thermal polymerization of amino acids played an important role in chemical evolution has been questioned on two grounds: First, amino acids and polypeptides are unstable at 150–200°C, except for brief exposures. It is difficult to see how a correctly timed exposure to such a temperature —long enough to effect the reaction, but not so long as to destroy the products— could have been attained on a large scale under natural conditions. Second, it is doubtful whether mixtures of pure amino acids like those used in these experiments ever existed on the primitive earth. Yet it is questionable whether polymerization would occur in a complex and more realistic mixture of organic chemicals, where the growth of peptide chains could be stopped by reactions with a variety of plausible molecules (9, 15, 73). It should be added that almost all so-called prebiotic syntheses are subject to one or the other of these objections in some degree, but the thermal synthesis of peptides seems particularly vulnerable to them.

Nonaqueous dehydrations that proceed at temperatures below 100°C have been described in a number of recent papers. Neuman et al (74, 75) find that adenosine is phosphorylated to a mixture of 2', 3', and 5'-AMPs when heated to 95°C with inorganic pyrophosphate and apatite (the generic name for common calcium phosphate minerals). When 5'-AMP was treated in the same way, 5'-ADP and 5'-ATP were identified among the products. Both apatite and pyrophosphate were required for these reactions.

Orgel and co-workers have found that many nonaqueous dehydrations are catalyzed by simple organic or inorganic substances, of which urea, imidazole, and Mg^{2+} are particularly effective. Thus, purine and pyrimidine nucleosides are phosphorylated at temperatures of 65–100°C in dry mixtures with urea, ammonium chloride, and inorganic phosphate (76, 77). Hydroxylapatite can serve as the source of phosphate. Cyclic nucleoside 2', 3'-phosphates and nucleoside 5'-phosphates are formed in good yields in this reaction. If ammonium dihydrogen phosphate is included in the system, oligonucleotides are produced, largely with the biologically important 3'-5' linkage (78). Adenosine cyclic 2', 3'-phosphate polymerizes to oligonucleotides when dried in the presence of diamines (79). Urea and ammonium dihydrogen phosphate heated together form long-chain polyphosphates (80). A large excess of urea is used in these reactions.

The mechanism of the urea-mediated condensations is not clear. Osterberg & Orgel (80) favor the view that urea acts as an acid-base catalyst. In an earlier study Lundstrom & Whittaker (81) proposed that urea functions as a dehydrating agent by undergoing hydrolysis to CO_2 and NH_3. Urea is readily produced from cosmically abundant gases and is a frequently observed product of simulated prebiotic experiments (see 57 and 82 for recent examples).

The addition of Mg^{2+} to the ammonium phosphate-urea system causes the precipitation of $MgNH_4PO_4 \cdot 6H_2O$, or struvite, a mineral that may have precipitated from the primeval ocean (83–85). When nucleotides are heated with struvite and urea at 65–100°C, nucleoside-5'-diphosphates are formed in good yield, with small amounts of nucleoside triphosphates (86). Mg salts are also involved in dry-phase syntheses of purine nucleosides from ribose and base discovered by Fuller et al (87); one of these is catalyzed by sea salts (88).

In a recent note (89), Lohrmann & Orgel report that Mg^+ catalyzes the reaction, in the dry state, of ATP with imidazole to form adenosine-5'-phosphorimidazolide (ImpA). The latter had been shown previously to condense to short oligonucleotides on a poly-U template (90). It is now found that ImpA heated with glycine generates oligopeptides. The synthesis of short peptides is accomplished in one operation by heating ATP, glycine, imidazole, and Mg^{2+} together. The authors propose a general scheme for the prebiotic synthesis of oligonucleotides and oligopeptides based on Mg^{2+}-, urea-, and imidazole-catalyzed reactions. Several syntheses of imidazole from simple precursors are known (see 89 for references).

ORGANIC COMPOUNDS IN METEORITES AND INTERSTELLAR SPACE

Amino Acids in Meteorites

Before 1970, little significance was attached to findings of amino acids and other compounds of biological interest in meteorites, because it was difficult to distinguish between indigenous compounds and contaminants acquired during entry of the meteorite into the earth's atmosphere, impact with the ground, and subsequent handling, storage, and processing for analysis. The early findings have been reviewed elsewhere (91). The situation has changed in recent years as a result of the development of ultrasensitive analytical methods based on combined gas chromatography and mass spectrometry (GCMS), which can not only identify individual amino acids present in meteorites, but in addition determine their optical purity. Also important has been recognition of the need for rapid recovery of observed meteorite falls and their handling and storage under chemically clean conditions.

Only a minor fraction of recovered meteorites (the carbonaceous chondrites) contains organic carbon, and not all of these have sufficient nitrogen to yield detectable amino acids. Of particular interest in recent years has been the Murchison meteorite, a carbonaceous chondrite that fell near Murchison, Australia in September 1969. This meteorite contains 2% carbon and 0.16% nitrogen (92). Much of the organic carbon is present as a complex mixture of aliphatic and aromatic hydrocarbons (92, 94). Carbon isotope analysis indicates that this material has a nonbiological source (92).

Water extraction of the meteorite followed by HCl hydrolysis yields a fraction containing about 20 μg of amino acids per gram of meteorite. Analysis by GCMS of the N-trifluoroacetyl-D-2-butyl esters (a derivative that volatilizes amino acids and converts enantiomers into diastereoisomers with chemically distinguishable properties) revealed 18 amino acids, of which 6 were identified as "normal" protein amino acids (glycine, alanine, valine, proline, glutamic and aspartic acids) and 12 were nonprotein amino acids (92, 93). All of the amino acids with asymmetric carbon atoms whose diastereoisomers could be resolved (9 out of 18) were found in nearly racemic mixtures. These findings by Kvenvolden et al have been confirmed by a different approach in which amino acid enantiomers were separated on optically active columns in a GCMS procedure (94). The results clearly indicate a nonbiological and extraterrestrial origin of meteoritic amino acids, because terrestrial contamination would introduce predominantly optically active amino acids of the kinds found in proteins.

Almost identical results have been obtained with the Murray chondrite which fell in the US in 1950 (95, 96). It has also been possible by the GCMS method to demonstrate the presence of nonprotein amino acids, at least one of them racemic, in the Orgueil meteorite (97). This carbonaceous chondrite, one of the most famous in the world, fell in France in 1864 and has been extensively studied. Pasteur is said to have examined it for bacteria, with negative results (98). Because it is heavily contaminated, it has not been possible until now to establish that some of its amino acids are indigenous to the meteorite.

It is the usual practice to hydrolyze the aqueous extracts to maximize the yield of amino acids. It has been shown, however, that free amino acids are present in unhydrolyzed extracts of Murchison and Murray, although in substantially lower amounts than in hydrolyzed samples (96). The nature of the hydrolyzable precursor is uncertain.

Purines and Pyrimidines in Meteorites

The heterocyclic bases of meteorites have received less attention than the amino acids. Hayatsu found evidence for adenine and guanine in the Orgueil (99), but analyses on the Murchison meteorite, known not to be seriously contaminated, are more convincing. Folsome et al (100) found 4-hydroxypyrimidine and two of its methylated derivatives in Murchison, but no purines. However, in a preliminary report Anders et al have detected both adenine and guanine in Murchison (53). These authors suggest that the discrepancy between their results and those of Folsome et al may be attributable to different extraction procedures. This important result needs confirmation.

Interstellar Molecules

In recent years, radioastronomers employing microwave spectroscopy have discovered a large number of molecules of biological importance in interstellar space (reviewed in 101, 102). These molecules are found in association with dust clouds that occur in different regions of the sky, notably in the direction of the great nebula in Orion. It is generally supposed that the dust grains have a catalytic role in the formation of the molecules from the rarefied interstellar gas, and it is certain that they shield them from photolysis by stellar ultraviolet radiation. Since there is reason to think that dust clouds are regions where stars and planetary systems form, the detection in them of molecules that have been identified in laboratory experiments as likely evolutionary precursors of amino acids and nucleotides is a matter of unusual interest.

Of approximately 27 molecules that have been identified in interstellar clouds, the following are of particular biological significance: H_2, H_2O, NH_3, H_2S, CO, HCN, cyanoacetylene, methanol, formaldehyde, formic acid, formamide, acetonitrile, and acetaldehyde. All of these molecules were identified by microwave spectroscopy except H_2, which was found by optical astronomy. As Oró has pointed out (103), this list includes what are generally considered to be the most important prebiotic precursors of amino acids, purines, pyrimidines, and sugars.

General Implications

The remarkable discoveries summarized above constitute the most important advance of recent years in the study of the origin of life. They show that carbon chemistry very similar to that found in laboratory simulations of the prebiotic earth occurs on a large scale in our galaxy. HCN and cyanoacetylene, for example, are the most abundant primary N-containing products of the Miller spark-discharge experiment (104, 105). Aldehydes are also formed in the spark and react with HCN and NH_3 to yield amino acids in a Strecker synthesis (104). HCN and cyanoacetylene are the starting points for syntheses of adenine (106, 107) and cytosine (105, 108), respectively. Miller and his associates have shown that all of the amino acids found (and some not found) in the Murchison and Murray meteorites are synthesized in a spark-discharge experiment that differs from the classical Miller experiment in that most of the NH_3 is replaced by N_2 (109, 110). Aminonitriles are produced, and these must be hydrolyzed at the end of the run to yield amino acids; in the original Miller reaction, conditions were sufficiently alkaline to hydrolyze the aminonitriles directly.

An alternative route for the synthesis of organic compounds of meteorites, interstellar clouds, and the primitive earth has been proposed by Anders et al, e.g.

Fischer-Tropsch–like reactions of CO, H_2, and NH_3 (53). These reactions take place spontaneously on the surface of catalysts at temperatures below 600°K. Magnetite, hydrated silicates, and other minerals expected to be cosmically abundant can serve as catalysts. Fischer-Tropsch–like reactions are not as efficient as the Miller synthesis for amino acids, but they appear to be superior in other respects, for example, in purine, pyrimidine, and hydrocarbon syntheses. Critical discussions of the Fischer-Tropsch–like process have been given by Miller & Orgel (15) and Oró (103).

After H_2, CO is the most abundant gas in interstellar clouds. It is also plentiful in volcanic gases. It seems clear that both CO and CH_4 can yield biologically important molecules under various plausible conditions, and it is likely that both did so on the primitive earth. A recent discussion of the fate of CO in the primitive atmosphere is given by van Trump & Miller (52).

Chemical evidence strongly suggests that carbonaceous chondrites are relatively unaltered condensates from the solar nebula (111). If so, the organic compounds in these objects date back to the time of formation of the solar system and are a record of the carbon chemistry of that era. An apparently similar chemistry, but in a different place and time, is seen in the interstellar dust clouds. Taken in conjunction with the known chemical makeup of terrestrial organisms, these observations suggest that wherever life occurs in the galaxy it will be built out of similar (although not necessarily identical) molecules.

The question of how large a contribution was made by meteorites and interstellar organic matter to the pool of prebiological organic compounds on the earth has been discussed (112, 113). It is unlikely to have been more than a minor fraction of the total, because most incoming organic matter would have been pyrolyzed during entry into the earth's atmosphere or on impact with the ground. The bulk of prebiological organic compounds was probably synthesized from methane or CO outgassed from the interior of the earth.

BIOLOGICAL ASPECTS OF THE PLANETS

In considering the possibilities for life on other planets, it is assumed that any such life must be based on carbon chemistry (2). The properties of the carbon atom that peculiarly fit it for the construction of living matter have often been remarked (see for example 114). No credible alternative to carbon-based life has yet been proposed. The cosmic abundance of the light elements, including carbon, and of their compounds (reviewed above) adds further weight to this assumption.

If the foregoing postulate is correct, then most of the bodies in the solar system can be eliminated as possible habitats of life on the basis of two corollaries: First, because low molecular weight compounds of carbon are usually volatile, they would be lost from objects that cannot retain an atmosphere. Yet it seems inescapable that such compounds would be produced by or from living systems, would enter the atmosphere, and would in fact have an essential role in the cycling of matter that must occur on any life-bearing planet. It follows that bodies without an atmosphere are unsuitable for life. The second corollary is that planets whose temperatures are higher than about 150°C are also unsuitable for life, since carbon compounds become unstable (on a geological time scale) at this temperature.

We can say immediately that life is not possible on the moon or most of the other satellites of the solar system, nor on Mercury or the asteroids, because none of these objects has an appreciable atmosphere. In the case of the moon, this conclusion has been verified by direct observation of the Apollo samples. Venus is also excluded as an abode of life because of its high surface temperature—in the neighborhood of 400–500°C.

Jupiter

Jupiter has a deep atmosphere and temperatures that range from 150°K or less at the visible cloud tops to an estimated 7000°K or more at the central core of the planet. The temperature structure of the Jovian atmosphere results from an internal heat source, whose presence is revealed by the fact, confirmed by the Pioneer 10 flyby (115), that Jupiter radiates some 2.5 times more energy than it receives from the sun. The high temperatures required by the currently accepted physical model of the planet exclude the possibility that life, or even organic matter, exists on the surface, if in fact there is a surface.

It has been generally assumed that the high atmosphere of Jupiter provides a favorable environment for the production of organic matter. Not only are solar radiation and reasonable temperatures available, but also the composition of the atmosphere (H_2, CH_4, and NH_3 observed; H_2O inferred at lower depths) is qualitatively identical to the Urey atmosphere which has been employed in numerous prebiological simulation experiments, starting with that of Miller (39). In experiments designed specifically to simulate Jovian conditions, Woeller & Ponnamperuma (116) and Sagan & Khare (117) demonstrated the expected production of biologically interesting compounds, together with colored organic products which they suggested could account for the red, yellow, and brown coloration of Jupiter. In reviewing these and other data, Sagan (118) concludes that "it does not seem at all out of the question that life has originated on Jupiter."

Serious doubts have been raised about the significance of these experiments in a critical review by Lewis & Prinn (119). These authors make the point that, contrary to the popular view, conditions on Jupiter are exceedingly hostile to the production of organic matter and the origin of life. As a consequence of its thermal structure, the atmosphere is highly convective. (The visible turbulence of the clouds of Jupiter is an indication of this convection.) Any organic compounds formed in the high atmosphere would be transported convectively, in a time that is short relative to chemical evolution, to deep, hot regions where they would be destroyed. Convective mixing does not extend into the topmost part of the atmosphere, above the tropopause, but little organic synthesis can occur here, because the buildup of complex organic molecules is prevented by uv photolysis and by reactions with H_2—by far the most abundant gas in the Jovian atmosphere—which tend to reduce all carbon compounds to CH_4 and other simple hydrides. Lewis & Prinn conclude that none of the reported experiments are relevant to conditions on Jupiter. The same general inferences apply to the other Jovian planets: Saturn, Uranus, and Neptune.

Titan

We are left with Mars, Pluto, and the larger satellites of the Jovian planets as the surviving candidate habitats of extraterrestrial life in the solar system. Of the objects beyond Mars, Titan, the largest satellite of Saturn, with an atmosphere more massive than that of Mars, is currently the most interesting biologically. Methane was identified years ago as a constituent of Titan's atmosphere. The mean surface temperature of a body of Titan's albedo and distance from the sun is expected to be about 80°K, or too low for solution chemistry. Recent infrared measurements, however, have suggested the possibility of an atmospheric greenhouse effect which would produce much higher temperatures and allow biologically interesting models of the surface to be considered (120). Serious studies of the physics and chemistry of Titan are still in their early stages. It seems likely that answers to some of the major questions about Titan will be forthcoming in the near future.

A very tenuous atmosphere on Io, the innermost satellite of Jupiter, is implied by recent results from Pioneer 10 (121). The estimated pressure at the surface is 10^{-8} to 10^{-10} atmospheres.

Mars

Mars fulfills the minimal conditions for a life-bearing planet, but these are necessary conditions, not sufficient ones. There is no evidence suggesting that Mars is an abode of life. On the contrary, the indications are on the negative side, but not conclusively so. A definitive answer can only be had by placing a probe on the planet. The first US landing on Mars is planned for 1976 with a pair of sophisticated unmanned spacecraft called Viking. The Viking instruments, which are biased toward experiments of biological interest, have been described in an issue of *Icarus* (122).

The highly successful Mariner 9 orbiter showed Mars to be more active geologically than had previously been supposed. A series of papers comprising a "final" report on the results of the Mariner 9 mission has been published in the *Journal of Geophysical Research* (123). Insofar as the new data are concerned with biological questions, they relate to the past history of Mars and the possibility that more favorable environmental conditions existed in earlier epochs. Of particular interest is the geological evidence, supported by theoretical computations on secular fluctuations of the obliquity of the planet (124), suggesting periodic climatic changes (125). Depending on the assumptions one makes about the thickness of the frozen volatile deposits in the polar caps, these changes can produce periods of nearly earthlike conditions in the Martian equatorial region (126) or only very slight climatic effects (127).

One of the surprising discoveries of the Mariner 9 photography was that of various meandering, braided, and dendritic features strongly resembling old water courses (128). Liquid water cannot exist on Mars, except in concentrated salt solutions. The vapor pressure of water in the Martian atmosphere is far below the triple point, and the mean surface temperature is only 210°K, compared with 288°K for the earth. Although ice deposits may well underlie the surface, ice can evaporate but cannot melt under present Martian conditions (129). The source of the water that cut the presumed channels (not all observers are agreed on this interpretation of the features) is thus a challenging problem. Climatic amelioration, mentioned above, is one possibility. Another is subsurface water in the form of carbon dioxide hydrate, $CO_2 \cdot 6H_2O$ (130). This compound, a clathrate, dissociates on the release of pressure such as might occur with deep fracturing of the crust.

Mars is probably losing water to space in the form of atomic hydrogen and oxygen (131). Classical escape theory does not explain the loss of O atoms, but a recently discovered nonthermal escape mechanism does (132). Ionized oxygen derived from H_2O or CO_2 can recombine with an electron to yield energetic O atoms: $O_2^+ + e \longrightarrow O + O$. These hot atoms can leave Mars if they are moving in the right direction. A similar mechanism operates for nitrogen (133) and may account for the fact that this element has not been detected in the Martian atmosphere (131). This mechanism is not important for the earth because of its stronger gravitational field.

The extreme dryness of Mars and the possibility that the planet has been severely depleted of nitrogen pose difficult problems for a Martian biota. The discovery of abiotic areas in the far less hostile Antarctic desert, where scarcity of water is also the life-limiting factor, has suggested that Martian life could not be built on a terrestrial model (134).

ACKNOWLEDGMENT

We wish to acknowledge support from the National Aeronautics and Space Administration, research grant NGR 05–002–308.

Literature Cited

1. Muller, H. J. 1929. The gene as the basis of life. *Proc. Int. Congr. Plant Sci.* 1:897–921
2. Horowitz, N. H., Drake, F. D., Miller, S. L., Orgel, L. E., Sagan, C. 1970. The origins of life. In *Biology and the Future of Man*, ed. P. Handler, 163–201. New York: Oxford Univ. Press. 967 pp.
3. Lemmon, R. M. 1970. Chemical evolution. *Chem. Rev.* 70:95–109
4. Ponnamperuma, C. 1971. Primordial organic chemistry and the origin of life. *Quart. Rev. Biophys.* 4:77–106
5. Stephen-Sherwood, E., Oró, J. 1973. Recent syntheses of bioorganic molecules. *Space Life Sci.* 4:5–31
6. Hulett, H. R. 1969. Limitations on prebiological synthesis. *J. Theor. Biol.* 24:56–72
7. Eigen, M. 1971. Self-organization of matter and the evolution of biological macromolecules. *Naturwissenschaften* 58:465–523
8. Black, S. 1973. A theory on the origin of life. *Advan. Enzymol.* 193–234
9. Bernal, J. D. 1967. *The Origin of Life.* Cleveland: World Publ. Co. 345 pp.
10. Calvin, M. 1969. *Chemical Evolution.* New York: Oxford Univ. Press. 278 pp.
11. Kenyon, D. H., Steinman, G. 1969. *Biochemical Predestination.* New York: McGraw-Hill. 301 pp.
12. Cairns-Smith, A. G. 1971. *The Life Puzzle.* Edinburgh: Oliver & Boyd. 165 pp.
13. Rutten, M. G. 1971. *The Origin of Life by Natural Causes.* New York: Elsevier. 420 pp.
14. Fox, S. W., Dose, K. 1972. *Molecular Evolution and the Origin of Life.* San Francisco: Freeman. 359 pp.
15. Miller, S. L., Orgel, L. E. 1973. *The Origins of Life on the Earth.* Englewood Cliffs, NJ: Prentice-Hall. 229 pp.
16. Orgel, L. E. 1973. *The Origins of Life.* New York: Wiley. 237 pp.
17. Margulis, L., Ed. 1970. *Origins of Life.* I. New York: Gordon & Breach. 376 pp.
18. Margulis, L., Ed. 1971. *Origins of Life.* II. New York: Gordon & Breach. 238 pp.
19. Buvet, R., Ponnamperuma, C., Eds. 1971. *Chemical Evolution and the Origin of Life.* Amsterdam: North-Holland. 560 pp.
20. Kimball, A. P., Oró, J., Eds. 1971.

Prebiotic and Biochemical Evolution. Amsterdam: North-Holland. 296 pp.

21. Margulis, L., Ed. 1973. *Origins of Life: Chemistry and Radioastronomy.* New York: Springer. 291 pp.

22. Schopf, J. W. 1970. Precambrian microorganisms and evolutionary events prior to the origin of vascular plants. *Biol. Rev.* 45:319–52

23. Barghoorn, E. S., Schopf, J. W. 1966. Microorganisms three billion years old from the Precambrian of South Africa. *Science* 152:758–63

24. Schopf, J. W., Barghoorn, E. S. 1967. Alga-like fossils from the early Precambrian of South Africa. *Science* 156: 508–12

25. Hurley, P. M., Pinson, W. H., Nagy, B., Teska, T. M. 1972. Ancient age of the middle marker horizon, Onverwacht Group, Swaziland Sequence, South Africa. *Earth Planet. Sci. Lett.* 14: 360–66

26. Engel, A. E. J. et al 1968. Alga-like forms in Onverwacht Series, South Africa: Oldest recognized lifelike forms on earth. *Science* 161:1005–8

27. Nagy, B., Nagy, L. A. 1969. Early pre-Cambrian Onverwacht microstructures: possibly the oldest fossils on earth? *Nature* 223:1226–29

28. Nagy, L. A. 1971. Ellipsoidal microstructures of narrow size range in the oldest known sediments on earth. *Grana* 11:91–94

29. Brooks, J., Muir, M. D., Shaw, G. 1973. Chemistry and morphology of Precambrian microorganisms. *Nature* 244: 215–17

30. Abelson, P. H., Hare, P. E. 1968. Recent amino acids in the Gunflint chert. *Carnegie Inst. Wash. Yearb.* 67:208–10

31. Nagy, B. 1970. Porosity and permeability of the Early Precambrian Onverwacht chert: origin of the hydrocarbon content. *Geochim. Cosmochim. Acta* 34:525–27

32. Smith, J. W., Schopf, J. W., Kaplan, I. R. 1970. Extractable organic matter in Precambrian cherts. *Geochim. Cosmochim. Acta* 34:659–75

33. Oró, J., Nakaparksin, S., Lichtenstein, H., Gil-Av, E. 1971. Configuration of amino acids in carbonaceous chondrites and a Pre-Cambrian chert. *Nature* 230:107–8

34. Simmonds, P. G., Shulman, G. P., Stembridge, C. H. 1969. Organic analysis by pyrolysis-gas chromatography-mass spectrometry, a candidate experiment for the biological exploration of Mars. *J. Chromatograph. Sci.* 7:36–41

35. Scott, W. M., Modzeleski, V. E., Nagy, B. 1970. Pyrolysis of early pre-Cambrian Onverwacht organic matter (> 3 $\times 10^9$ yr old). *Nature* 225:1129–30

36. Oehler, D. Z., Schopf, J. W., Kvenvolden, K. A. 1972. Carbon isotopic studies of organic matter in Precambrian rocks. *Science* 175:1246–48

37. Cloud, P. E. Jr., Licari, G. R., Wright, L. A., Troxel, B. W. 1969. Proterozoic eucaryotes from eastern California. *Proc. Nat. Acad. Sci. USA* 62:623–30

38. Muir, M. D. 1973. *Microfossils from the Middle Precambrian McArthur Group, Northern Territory.* Presented at 4th Int. Conf. Origin of Life, Barcelona

39. Miller, S. L. 1953. A production of amino acids under possible primitive earth conditions. *Science* 117:528–29

40. van Trump, J. E., Miller, S. L. 1972. Prebiotic synthesis of methionine. *Science* 178:859–60

41. Stephen-Sherwood, E., Oró, J., Kimball, A. P. 1971. Thymine: a possible prebiotic synthesis. *Science* 173:446–47

42. Friedmann, N., Haverland, W. J., Miller, S. L. 1971. Prebiotic synthesis of the aromatic and other amino acids. See Ref. 19, pp. 123–35

43. Miller, S. L., Urey, H. C. 1959. Organic compound synthesis on the primitive earth. *Science* 130:245–51

44. Groth, W. 1957. Photochemische Bildung von Aminosäuren und anderen organischen Verbindungen aus Mischungen von H_2O, NH_3, und den einfachsten Kohlenwasserstoffen. *Angew. Chem.* 69:681

45. Sagan, C., Khare, B. N. 1971. Long-wavelength ultraviolet photoproduction of amino acids on the primitive earth. *Science* 173:417–20

46. Ferris, J. P., Nicodem, D. E. 1974. NH_3: Did it have a central role in chemical evolution? In *The Origin of Life and Evolutionary Biochemistry,* ed. G. Deborin, K. Dose, S. Fox, M. Kritsky. New York: Plenum. In press

47. Hong, K., Hong, J., Becker, R. S. 1974. Chemical reactions of interest in chemical evolution and interstellar chemistry initiated by hot hydrogen atoms. *Science* 184:984–87

48. Elpiner, I. E. 1964. *Ultrasound: Physical, Chemical, and Biological Effects.* Transl. F. L. Sinclair. New York:Consultants Bur. 371 pp.

49. Anbar, M. 1968. Cavitation during impact of liquid water on water: geochemical implications. *Science* 161:1343–44

50. Bar-Nun, A., Bar-Nun, N., Bauer, S. H., Sagan, C. 1970. Shock synthesis of amino acids in simulated primitive environments. *Science* 168:470–73
51. Abelson, P. H. 1966. Chemical events on the primitive earth. *Proc. Nat. Acad. Sci. USA* 55:1365–72
52. van Trump, J. E., Miller, S. L. 1973. Carbon monoxide on the primitive earth. *Earth Planet. Sci. Lett.* 20:145–50
53. Anders, E., Hayatsu, R., Studier, M. H. 1973. Organic compounds in meteorites. *Science* 182:781–90
54. Hubbard, J. S., Hardy, J. P., Horowitz, N. H. 1971. Photocatalytic production of organic compounds from CO and H_2O in a simulated Martian atmosphere. *Proc. Nat. Acad. Sci. USA* 68:574–48
55. Hubbard, J. S., Hardy, J. P., Voecks, G. E., Golub, E. E. 1973. Photocatalytic synthesis of organic compounds from CO and water: Involvement of surfaces in the formation and stabilization of products. *J. Mol. Evol.* 2:149–66
56. Tseng, S. S., Chang, S. 1974. Photoinduced free radicals on simulated Martian surface. *Nature* 248:575–77
57. Ferris, J. P., Williams, E. A., Nicodem, D. E., Hubbard, J. E., Voecks, G. E. 1974. Photolysis of CO-NH_3 mixtures and the Martian atmosphere. *Nature* 249:437–39
58. Bada, J. L., Miller, S. L. 1968. Ammonium ion concentration in the primitive ocean. *Science* 159:423–25
59. Ferris, J. P., Nicodem, D. E. 1972. Ammonia photolysis and the role of ammonia in chemical evolution. *Nature* 238:268–69
60. Lowe, C. U., Rees, M. W., Markham, R. 1963. Synthesis of complex organic compounds from simple precursors: formation of amino acids, amino-acid polymers, fatty acids and purines from ammonium cyanide. *Nature* 199:219–22
61. Matthews, C. N. 1971. The origin of proteins: heteropolypeptides from hydrogen cyanide and water. See Ref. 19, pp. 231–35
62. Ferris, J. P., Donner, D. B., Lobo, A. P. 1973. Possible role of hydrogen cyanide in chemical evolution: investigation of the proposed direct synthesis of peptides from hydrogen cyanide. *J. Mol. Biol.* 74:499–510
63. Ferris, J. P., Donner, D. B., Lobo, A. P. 1973. Possible role of hydrogen cyanide in chemical evolution: the oligomeriza-

tion and condensation of hydrogen cyanide. *J. Mol. Biol.* 74:511–18
64. Bernal, J. D. 1960. Reply to D. E. Hull. *Nature* 186:694–95
65. Paecht-Horowitz, M., Berger, J., Katchalsky, A. 1970. Prebiotic synthesis of polypeptides by heterogeneous polycondensation of amino-acid adenylates. *Nature* 228:636–39
66. Paecht-Horowitz, M., Katchalsky, A. 1973. Synthesis of amino acyladenylates under prebiotic conditions. *J. Mol. Evol.* 2:91–98
67. Degens, E. G., Matheja, J., Jackson, T. A. 1970. Template catalysis: asymmetric polymerization of amino-acids on clay minerals. *Nature* 227:492–93
68. Jackson, T. A. 1971. Preferential polymerization and adsorption of L-optical isomers of amino acids relative to D-optical isomers on kaolinite templates. *Chem. Geol.* 7:295–306
69. Bonner, W. A., Flores, J. 1973. On the asymmetric adsorption of phenylalanine enantiomers by kaolin. *Curr. Mod. Biol.* 5:103–13
70. Bonner, W. A., Flores, J. J. 1973. *Experiments on the Origins of Optical Activity.* Presented at 4th Int. Conf. Origin of Life, Barcelona
71. Amariglio, A., Amariglio, H. 1971. Unsuccessful attempts of asymmetric synthesis under the influence of optically active quartz. Some comments about the possible origin of the dissymmetry of life. See Ref. 19, pp. 63–69
72. Wald, G. 1957. The origin of optical activity. *Ann. NY Acad. Sci.* 69:352–68
73. Horowitz, N. H., Miller, S. L. 1962. Current theories on the origin of life. *Fortschr. Chem. Org. Naturst.* 20:423–59
74. Neuman, M. W., Neuman, W. F., Lane, K. 1970. On the possible role of crystals in the origins of life. III. The phosphorylation of adenosine to AMP by apatite. *Curr. Mod. Biol.* 3:253–59
75. Neuman, M. W., Neuman, W. F., Lane, K. 1970. On the possible role of crystals in the origins of life. IV. The phosphorylation of nucleotides. *Curr. Mod. Biol.* 3:277–83
76. Lohrmann, R., Orgel, L. E. 1971. Urea-inorganic phosphate mixtures as prebiotic phosphorylating agents. *Science* 171:490–94
77. Bishop, M. J., Lohrmann, R., Orgel, L. E. 1972. Prebiotic phosphorylation of thymidine at 65°C in simulated desert conditions. *Nature* 237:162–64

78. Österberg, R., Orgel, L. E., Lohrmann, R. 1973. Further studies of urea-catalyzed phosphorylation reactions. *J. Mol. Evol.* 2:231–34
79. Verlander, M. S., Lohrmann, R., Orgel, L. E. 1973. Catalysts for the self-polymerization of adenosine cyclic 2',3'-phosphate. *J. Mol. Evol.* 2:303–16
80. Österberg, R., Orgel, L. E. 1972. Polyphosphate and trimetaphosphate formation under potentially prebiotic conditions. *J. Mol. Evol.* 1:241–48
81. Lundstrom, F. O., Whittaker, C. W. 1937. Chemical reactions in fertilizer mixtures. *Ind. Eng. Chem.* 29:61–68
82. Lohrmann, R. 1972. Formation of urea and guanidine by irradiation of ammonium cyanide. *J. Mol. Evol.* 1:263–69
83. Handschuh, G. J., Orgel, L. E. 1973. Struvite and prebiotic phosphorylation. *Science* 179:483–84
84. McConnell, D. 1973. Precipitation of phosphates in a primeval sea. *Science* 181:582
85. Handschuh, G. J., Orgel, L. E. 1973. Reply to McConnell. *Science* 181:582
86. Handschuh, G. J., Lohrmann, R., Orgel, L. E. 1973. The effect of Mg^{2+} and Ca^{2+} on urea-catalyzed phosphorylation reactions. *J. Mol. Evol.* 2:251–62
87. Fuller, W. D., Sanchez, R. A., Orgel, L. E. 1972. Studies in prebiotic synthesis. VI. Synthesis of purine nucleosides. *J. Mol. Biol.* 67:25–33
88. Fuller, W. D., Sanchez, R. A., Orgel, L. E. 1972. Studies in prebiotic synthesis. VII. Solid-state synthesis of purine nucleosides. *J. Mol. Evol.* 1:249–57
89. Lohrmann, R., Orgel, L. E. 1973. Prebiotic activation processes. *Nature* 244:418–20
90. Weimann, B. J., Lohrmann, R., Orgel, L. E., Schneider-Bernloehr, H., Sulston, J. E. 1968. Template-directed synthesis with adenosine-5'-phosphorimidazolide. *Science* 161:387
91. Hayes, J. M. 1967. Organic constituents of meteorites—a review. *Geochim. Cosmochim. Acta* 31:1395–1440
92. Kvenvolden, K. et al 1970. Evidence for extraterrestrial amino acids and hydrocarbons in the Murchison meteorite. *Nature* 228:923–26
93. Kvenvolden, K. A., Lawless, J. G., Ponnamperuma, C. 1971. Nonprotein amino acids in the Murchison meteorite. *Proc. Nat. Acad. Sci. USA* 68:486–90
94. Oró, J., Gibert, J., Lichtenstein, H., Wikstrom, S., Flory, D. A. 1971.

Amino acids, aliphatic and aromatic hydrocarbons in the Murchison meteorite. *Nature* 230:105–6
95. Lawless, J. G., Kvenvolden, K. A., Peterson, E., Ponnamperuma, C., Moore, C. 1971. Amino acids indigenous to the Murray meteorite. *Science* 173:626–27
96. Cronin, J. R., Moore, C. B. 1971. Amino acid analyses of the Murchison, Murray, and Allende carbonaceous chondrites. *Science* 172:1327–29
97. Lawless, J. G., Kvenvolden, K. A., Peterson, E., Ponnamperuma, C., Jarosewich, E. 1972. Evidence for amino acids of extraterrestrial origin in the Orgueil meteorite. *Nature* 236:66–67
98. Becquerel, P. 1924. La vie terrestre provient-elle d'un autre monde? *L'Astronomie* 38:393–417
99. Hayatsu, R. 1964. Orgueil meteorite: organic nitrogen contents. *Science* 146:1291–92
100. Folsome, C. E., Lawless, J., Romiez, M., Ponnamperuma, C. 1971. Heterocyclic compounds indigenous to the Murchison meteorite. *Nature* 232:108–9
101. Rank, D. M., Townes, C. H., Welch, W. J. 1971. Interstellar molecules and dense clouds. *Science* 174:1083–1101
102. Buhl, D. 1971. Chemical constituents of interstellar clouds. *Nature* 234:332–34
103. Oró, J. 1972. Extraterrestrial organic analysis. *Space Life Sci.* 3:507–50
104. Miller, S. L. 1957. The formation of organic compounds on the primitive earth. *Ann. NY Acad. Sci.* 69:260–75
105. Sanchez, R. A., Ferris, J. P., Orgel, L. E. 1966. Cyanoacetylene in prebiotic synthesis. *Science* 154:784–85
106. Oró, J., Kimball, A. P. 1961. Synthesis of purines under possible primitive earth conditions. *Arch. Biochem. Biophys.* 94:217–27
107. Sanchez, R., Ferris, J., Orgel, L. E. 1966. Conditions for purine synthesis: Did prebiotic synthesis occur at low temperatures? *Science* 153:72–73
108. Ferris, J. P., Sanchez, R. A., Orgel, L. E. 1968. Studies in prebiotic synthesis. 3. Synthesis of pyrimidines from cyanoacetylene and cyanate. *J. Mol. Biol.* 33:693–704
109. Ring, D., Wolman, Y., Friedmann, N., Miller, S. L. 1972. Prebiotic synthesis of hydrophobic and protein amino acids. *Proc. Nat. Acad. Sci. USA* 69:765–68
110. Wolman, Y., Haverland, W. J., Miller, S. L. 1972. Nonprotein amino acids from spark discharges and their comparison with the Murchison meteorite

amino acids. *Proc. Nat. Acad. Sci. USA* 69:809–11

111. Anders, E. 1971. Meteorites and the early solar system. *Ann. Rev. Astron. Astrophys.* 9:1–34

112. Sagan, C. Interstellar organic chemistry. 1972. *Nature* 238:77–80

113. Breger, I. A., Zubovic, P., Chandler, J. C., Clarke, R. S. 1972. Occurrence and significance of formaldehyde in the Allende carbonaceous chondrite. *Nature* 236:155–58

114. Edsall, J. T., Wyman, J. 1958. *Biophysical Chemistry,* 23–25. New York: Academic. 699 pp.

115. Opp, A. G. 1974. Pioneer 10 mission: summary of scientific results from the encounter with Jupiter. *Science* 183: 302–3

116. Woeller, F., Ponnamperuma, C. 1969. Organic synthesis in a simulated Jovian atmosphere. *Icarus* 10:386–92

117. Sagan, C., Khare, B. N. 1971. Experimental Jovian photochemistry:Initial results. *Astrophys. J.* 168:563–69

118. Sagan, C. 1971. The solar system beyond Mars: An exobiological survey. *Space Sci. Rev.* 11:827–66

119. Lewis, J. S., Prinn, R. G. 1971. Chemistry and photochemistry of the atmosphere of Jupiter. *Theory and Experiment in Exobiology* 1:125–42

120. Hunten, D. M., Ed. 1974. *The Atmosphere of Titan.* Washington DC: Nat. Aeronaut. Space Admin. 177 pp.

121. Kliore, A., Cain, D. L., Fjeldbo, G., Seidel, B. L., Rasool, S. I. 1974. Preliminary results on the atmospheres of Io and Jupiter from the Pioneer 10 S-band occultation experiment. *Science* 183: 323–24

122. Various authors. 1972. The Viking missions to Mars. *Icarus* 16:1–227

123. Various authors. 1973. Scientific results of the Mariner 9 mission to Mars. *J. Geophys. Res.* 78:4007–4440

124. Ward, W. R. 1973. Large-scale variations in the obliquity of Mars. *Science* 181:260–62

125. Cutts, J. A. 1973. Nature and origin of layered deposits of the Martian polar regions. *J. Geophys. Res.* 78:4231–49

126. Sagan, C., Toon, O. B., Gierasch, P. J. 1973. Climatic change on Mars. *Science* 181:1045–49

127. Murray, B. C., Malin, M. C. 1973. Polar volatiles on Mars—theory versus observation. *Science* 182:437–43

128. Milton, D. J. 1973. Water and processes of degradation in the Martian landscape. *J. Geophys. Res.* 78:4037–47

129. Ingersoll, A. P. 1970. Mars: occurrence of liquid water. *Science* 168:972–73

130. Milton, D. J. 1974. Carbon dioxide hydrate and floods on Mars. *Science* 183:654–55

131. Barth, C. A. 1974. The atmosphere of Mars. *Ann. Rev. Earth Planet. Sci.* 2:333–67

132. McElroy, M. B. 1972. Mars: an evolving atmosphere. *Science* 175:443–45

133. Brinkmann, R. T. 1971. Mars: has nitrogen escaped? *Science* 174:944–45

134. Horowitz, N. H., Cameron, R. E., Hubbard, J. S. 1972. Microbiology of the dry valleys of Antarctica. *Science* 176: 242–45

GENE CONTROL OF ♦3080
MAMMALIAN DIFFERENTIATION

Beatrice Mintz
Institute for Cancer Research, Fox Chase, Philadelphia, Pennsylvania 19111

1 INTRODUCTION

The generally accepted concept that differentiation rests on temporal changes in specific gene activity subsumes occurrence of some changes concurrently in all cells and of others differentially in certain cells as a basis for cellular diversification. Examples of the former, in mammalian development, have already been reviewed elsewhere (24, 217). It is with the latter, for which models are lacking in unicellular organisms or in mammalian cell cultures, that we are chiefly concerned here. A differentiated multicellular individual may be regarded as an organized natural "mosaic" of genetically similar cells with diverse gene expressions and orderly cell interactions. Animal models with unambiguous *genotypic* cellular mosaicism throughout development can now be produced experimentally and have been providing new ways of probing the origins and consequences of differential gene action in vivo (123, 124).

The present review summarizes and evaluates progress made with this approach and indicates some possibilities for future work.

Genetic *mosaicism* will be used here as a general term designating constitutional cellular differences in an individual. Although some investigators have described as "mosaics" only those individuals with genotypic changes within a single zygote lineage, while those with cells from different zygotes have been called "chimeras" —after the lion-goat-serpent monster of Greek mythology—there has been considerable inconsistency and interchangeability in this usage, and it has often proved impossible to diagnose the etiology of a case and to choose between the two terms (12, 41). Continued insistence on the distinction thus serves little purpose, and invokes a metaphor inapt for harmoniously integrated, nonfictional, genotypic mosaics. According to the ancient Greeks, Bellerophon overcame and slew the chimera; may it rest in peace.

Cases of spontaneous genotypic mosaicism are relatively rare in mammals (12, 41) and have been variously attributed to somatic cell mutation, somatic segrega-

411

tion, chromosomal or cytokinetic accidents, fertilization of both egg and polar body, adhesion between cells from separate embryos, and transplacental exchange of cells between fetuses or between mother and fetus (43, 102, 162, 9, 29, 220, 27, 176, 167, 13, 4). The few cases proposed as candidates for somatic crossing over in the mouse (64) may have arisen instead by other means, such as mutation. There is little expectation that mammalian geneticists will reap a rich harvest of genetic mosaics comparable to that found in *Drosophila,* where phenomena such as somatic crossing over or postzygotic chromosome loss are relatively frequent or inducible (193, 186, 46, 75). In mammalian work, it was Owen's discovery of spontaneous erythrocyte mosaicism in fraternal cattle twins with shared placental blood vessels (167) that first stimulated interest in the possibility of experimental intergenotypic transfer of cells by injecting or grafting donor cells into fetal or postnatal hosts (15, 16, 42). Such methods have since been widely and effectively employed (16), especially in immunology and hematology (the literature will not be reviewed here). Mosaicism in these cases is, however, necessarily restricted to certain tissues and to relatively advanced stages. Other methods were required to enable cells of different genotypes to be confronted before differentiation, so that they could participate jointly in development as coevals rather than as graft and host.

The first successful technical resolutions of the problem were independently devised by Tarkowski (196) and Mintz (107) and consisted of two different means of achieving the same end: formation of genotypic composites in cleavage stages by denuding separate embryos and aggregating the blastomeres. (Egg "fusion" does not take place, frequent references to "fusion" notwithstanding.) Another practical route to mosaicism was subsequently contributed by Gardner (47) and involved injection of young embryo cells into the cavity of the blastocyst. The methods summarized in Figure 1 (106–113, reviewed in 122) have been the most widely used because of their simplicity and high yield. In all, over 3000 viable genotypically composite mice, as well as a few rabbits (51a), rats (92), and sheep (204), have thus far been produced.

The original survivors were dubbed "quadriparental" mice (111). Some workers chose to change this Greek-derived name to a Latin-Greek hybrid, "tetraparental" (211). In either case, such terms are awkward when referring to the animals' cells, and there is also a possibility that more than two cellular genotypes, and therefore more than four parents, might be involved. The original term was therefore replaced by "allophenic" (113), which not only distinguishes these from other mosaic mammals, but focuses on the different phenotypic subpopulations of cells coexisting within any tissue and serving as the analytical starting point. [Allophenic was also chosen because of its close relationship to, yet distinguishability from, "allogenic" or "allogeneic," terms used to describe graft-host cellular associations involving antigenic differences (183).] The double-arrow symbol↔was introduced as the special connecting element between component genotypes (110).

There are no immunogenetic barriers to a free experimental choice of contributing cell genotypes in allophenic animals (140). Even lethal genotypes may sometimes be "rescued" (119, 36). Thus, a wide range of questions previously inaccessible to investigation in vivo has come under experimental study in mammals.

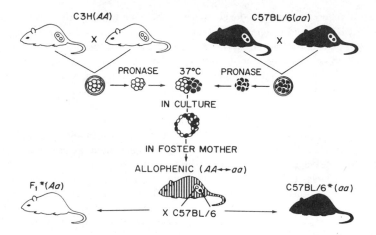

Figure 1 Experimental production of allophenic mice by the methods of Mintz (reviewed in 122). The example shows two cleavage-stage embryos derived, respectively, from gametes of a pair of C3H and a pair of C57BL/6 inbred-strain parents. The enveloping zona pellucida of each explanted embryo is lysed in pronase and the embryos are aggregated by incubation at 37°C (or by exposure to phytohemagglutinin, as described in 134) and cultured for a day. The resultant composite double-size blastocyst is then surgically transferred to the uterus of a pseudopregnant recipient previously mated with a sterile vasectomized male. Embryo size regulation occurs soon after implantation, and development continues normally to birth. If both cell strains are adequately represented in the coat of the C3H↔C57BL/6 allophenic animal, a pattern of fine transverse bands, representing the component *agouti (A/A)* and *nonagouti (a/a)* hair follicle clones, is seen. Other tissues, including the germ line (diagrammatically shown), may also comprise both cell strains. Each cell, except for skeletal myoblasts, retains its individuality. This is especially striking in the germ line: breeding tests with ordinary animals of the recessive color strain yield C57BL/6 and F₁ progeny differing(*) from ordinary animals only in their strange history and in the possibility of effects of the original foster mother or of the allophenic environment.

2   INITIATION OF DIFFERENTIAL GENE ACTIVITY IN EARLY DEVELOPMENT

The mammalian embryo shows its first signs of differential change in the late morula, when a fluid-filled cavity, the blastocoel, appears and ushers in the blastocyst stage. The blastocoel is created by formation of vacuoles in some of the morula cells and the discharge of their contents (112). An internal coherent group of cells, the inner cell mass, remains on one side of the blastocoel and, after implantation, forms the embryo and extraembryonic membranes, including those that contribute to the placenta. The blastocyst is covered by a layer of trophoblast cells which differ from the inner mass cells in various cytological features (178), cell-surface properties (49), and other characteristics, and which will contribute giant cells and ectoplacental cone to the placenta (184). We may roughly define three early periods of differential change: the cleavage period, characterized by complete developmental lability and cellular totipotency, and terminating with formation of the two major parts of the blastocyst; the mature blastocyst period, when inner cell mass and trophoblast have irreversibly diverged, but each still retains some internal developmental lability; and a period shortly after implantation, when determination of the embryo itself takes place.

Cleavage Period

There has been general agreement that local (biochemical or biophysical) differences may be responsible for triggering initial divergences in cellular gene activity at some time in the cleavage period. But according to an earlier hypothesis, championed especially by Dalcq (28), regional cytoplasmic differences were preformed in the egg and compartmentalized by cleavage virtually from the outset. A later view, presented by Mintz (110), held that the differences arise epigenetically in the mammal, following a substantial period of complete developmental lability throughout cleavage; after this, regional differences inevitably arise within the relatively thick mass, in which the cells have become "packaged into smaller units which must then have disparate microenvironments" (112).

The earlier hypothesis stated that polarity and symmetry were intrinsic in the cytoplasm of the mammalian egg, and predicted that disturbance of the egg architecture as cleavage progressed would lead to anomalies. Observations cited in support of the hypothesis were: inhomogeneous distribution of certain cytochemically visualizable cytoplasmic components (28); abnormal development, in some cases, when single rat, rabbit, or mouse blastomeres were isolated during early cleavage, or when some blastomeres were destroyed in situ (157, 179, 195, 150); and failure of aggregated pairs of mouse embryos to form unitary blastocysts if paired in the late eight-cell period or beyond (196, 197). The cytochemical observations have yet to be confirmed in sectioned material (as against thick whole-mounts) and would, in any event, not establish a causal basis for differentiation, because cytoplasmic inhomogeneity is a general property of cells. And the fact that the blastomere isolation or damage experiments had their share of normal survivors furnished unambiguous evidence of embryo lability; the failures could have been due to injury or technical inadequacies.

The problem was reinvestigated experimentally with more effective methods of recombining blastomeres at 37°C; conjoined mouse embryos were then found to yield normal, though giant, unitary blastocysts even when the contributing members were late morulae or had small blastocoel cavities, or when as many as 16 embryos were aggregated in one group (109, 110, 112). The contributors were also spatially rearranged in other ways certain to perturb any "intrinsic" architecture, and some cells were prelabeled in ^{3}H-thymidine so that their final positions could be seen in autoradiographs and compared with their initial ones. Despite these extensive modifications, normal development was maintained in a large majority of the artificial composites. Even when normal blastomeres were combined with abnormal ones of the homozygous t^{12}/t^{12} genotype, which is lethal in the morula stage, blastocysts were formed through the efforts of the normal cells, despite the impediment of the intermingled moribund ones (110).

Early lability has since been repeatedly documented in further experiments involving mouse blastomere isolations (203) and rearrangements (62, 74, 83, 151, 189, 200, 216). While isolation of single blastomeres (203) is the more direct test of cellular totipotency, the innate limitation of this method is soon reached when the donor embryo has only eight, or even fewer, cells, because the biosynthetic changes that characterize all the cells during cleavage (109, 217) follow an approximate "developmental clock" (109, 110), even in an isolated blastomere. Therefore, when a cell is taken from an eight-cell embryo, its mitotic progeny physiologically enter the blastocyst stage and some begin to form a cavity when there are insufficient cells to constitute the normal geometry of a blastocyst. The rearrangement experiments have again utilized temperature-induced blastomere aggregations and have focused on placing cells in interior or exterior starting positions in the artificial composites. "Inside-outside" differential locations were marked by prelabeling the central or peripheral group with ^{3}H-thymidine (62, 74), by choosing contributing cell strains with allelic enzymatic differences used to analyze the resultant parts (74), or by injecting silicone droplets as cell markers (189, 216). While the position of cells in the morula is fortuitously arrived at, only cells located in the interior form inner cell mass while those on the outside become trophoblast. The critical differences in these microenvironments are still unknown. Even in the early blastocyst, there may still be some measure of lability for trophoblast and inner cell mass formation (not necessarily by interconversion), as complete cell disaggregation and reaggregation is compatible with formation of a normal blastocyst (187).

Blastocyst Period

In the mature blastocyst period, the cells have apparently all become committed as trophoblast or inner cell mass: when the mouse trophoblast layer is isolated, it forms only trophoblastic fluid-containing vesicles in vitro; the isolated inner mass does not cavitate and also does not reconstitute a complete blastocyst (49). Similarly, trophoblast cells no longer have the option of contributing to embryo formation when they are taken from mature blastocysts and injected into the blastocoel of recipients with genetic markers; only transferred inner mass cells contribute to the resultant embryo (48). Nevertheless, neither the inner cell mass nor the trophoblast has yet undergone further internal differentiation, for when part of the inner mass is removed (82) or the entire blastocyst is sectioned, leaving only half of the inner mass and trophoblast (48), normal postimplantation embryos are obtained. The inner mass alone cannot induce uterine implantation changes, hence it cannot develop autonomously; the trophoblast alone can induce some uterine reactions but soon fails to proliferate (49). When genetically distinct inner cell mass and trophoblast were recombined (52), they developed normally in vivo and, despite some puzzling results, the genotypes of embryo plus membranes on the one hand, and of trophoblast on the other, largely confirmed the classical picture of their origins, as previously deduced from histological sections (184).

Determination and Size Regulation of the Embryo

Determination of the embryo as a derivative of part of the inner cell mass has thus not yet taken place in the blastocyst and must occur some time after implantation, i.e. on day 4 or later in the mouse (counting the date of copulation-plug detection as day 0). Since the general structure of the embryo begins to be visible on day 5 (184), embryo determination has presumably already occurred at some time on day 4–5.

It is of particular interest that this time period coincides with regulation of giant-size composites to normal size, and it may be that embryo-determination and size-regulation mechanisms are interrelated. Direct inspection of double-size embryos in utero shows that they have become normal size shortly after implantation (113), and actual measurements (19) indicate normal size on day 5, after the egg cylinder stage. While the mechanisms governing size regulation in giant composites, or in half-size blastocysts, are unknown, they are likely to be the same mechanisms as ensure a species-specific range of size normalcy under ordinary conditions (113). (In fact, the known normal size of identical twins led originally to the expectation that paired whole embryos would form normal-size allophenic mice.) A proposed mechanism for size regulation (119) is that "embryo-determining" loci may become activated in some constant small number of cells, irrespective of whether the inner cell mass was initially miniature, normal, or giant; physiological and extrinsic factors, such as uterine crypt size, might then regulate rates of proliferation of extraembryonic components. Such a "counting out" of clonal initiator cells would not be unique to embryo determination: as later sections will show, each specialized kind of cell population seems to arise from a small fixed pool of primordial cells. The fact that mosaicism was found in only 75% of a large series of allophenic mice derived from the same strain pair led to the suggestion that there might be as few as three primordial embryo cells; the model is of course idealized in that it assumes there was complete admixture of the two cell strains at the time of embryo determination, and no selection in the totality of tissues genetically analyzed. The presence of two cellular genotypes in all tissues of an allophenic mouse clearly demonstrates that the embryo proper must originate from at least two cells; as yet there have been no reports of more than two cell strains incorporated and identified in single individuals. The observation (48) that injection of a single inner mass cell into a recipient blastocyst can yield a mouse with extensive tissue contributions of the donor strain is also suggestive that the embryo may come from very few cells. The speculation that the embryo derives specifically from 10–12 cells (19) is based on previous models of dependence of inner cell mass determination on an "inside" location at the end of cleavage (62). However, there is insufficient knowledge to consider this possibility, as embryo determination occurs at an unknown time and from a population of unknown cell numbers, among the "inside" cells, from which various other, extraembryonic, tissues also form. Ultimately, the number of cell strains recoverable in allophenic animals should resolve the question.

From other allophenic-mouse studies (to be discussed later), it has been inferred that many specific tissues originate after day 4 and before day 7 of embryonic life (116, 119, 124–127). Independent evidence for loss of totipotency by day 7 comes from the interesting experiments of Stevens on experimental induction of mouse teratocarcinomas—tumors with multipotential stem cells analogous to blastomeres and capable of giving rise to numerous differentiated tissues. Embryos grafted to ectopic sites, e.g. under the testis capsule, at 6 days of age or younger were able to form teratocarcinomas; similar grafts of older embryos yielded only differentiated tissues. Stevens concluded that after 6 days "most if not all the cells have already become determined" (190). Moustafa & Brinster (146, 147), on the other hand, have claimed that 8-day embryo cells (randomly picked after dissociation) are still multipotential. After injecting genotypically black cells from 8-day embryos into 4-day albino blastocysts, 4 fetuses (out of a total of 221 fetuses and neonates in this or reversed color-marker transfers) had ocular pigmentary mosaicism at autopsy on days 15–17. Three had died earlier and, as no developmental stages were given for any of the four, all may still have been at a stage of incomplete ocular pigmentation when the eyes appear mottled in pigmented controls. In the absence of any other evidence of mosaicism, it would be difficult to sustain the authors' conclusion, as uninjected blastocysts of unspecified genotype accompanied some of the experimentals, and embryos at 8 days' gestation often include developmentally younger ones. The results from injections of "5½"-day cells are also inconclusive, as the cell donors were retarded blastocysts, hence developmentally less than their chronological age. Therefore, 4½-day donor cells are still the oldest shown to be multipotential by the blastocyst-injection test, as previously demonstrated by Gardner (48). This does not rule out the possibility that slightly older embryo cells may still possess some developmental versatility: if donor cells could not be incorporated into the inner cell mass of a substantially younger blastocyst, the test would simply fail for technical reasons.

Interspecific mammalian cell transfers have proved functional in some cases, e.g. rat marrow cells have been grafted to, and have rescued, irradiated mouse hosts (160). Allophenic combinations of rat and mouse blastomeres may offer a means of histochemically visualizing early cell deployments by means of antigenic differences. Unitary blastocysts have been formed from blastomere aggregates of the two species (188, 219); success was more likely if both donors were at the same cleavage stage rather than the same age, as development is slower in the rat than in the mouse. Some postimplantation development has been obtained by injecting rat inner mass cells into younger mouse blastocysts of equivalent stage and transferring them to the uterus of a mouse (50); these embryos clearly had both species of cells, by chromosomal and immunofluorescence criteria. Thus far, they have been composed chiefly of mouse cells and the rat cells have formed large patches. This points up a possible caveat in interpreting "maps" of these embryos, as cells of the more slowly developing species may tend to be selected against or to lag behind in their movements and collect in atypical groupings.

3 SOMITES AND THEIR DERIVATIVES

Somite Differentiation

Somites are of special interest because they are among the first morphological features to appear in vertebrate embryos and because they give rise to a number of major parts of the body, including muscle, from the myotome portion of the somite;

cartilage and bone, from the sclerotome; and dermal parts of skin and hair, from the dermatome.

In the mouse, somites begin to appear on late day 7 of embryonic life as tiny paired blocks separating out of longitudinal bands of mesoderm on either side of the dorsal midline. Within a few days, approximately 30 pairs form anteroposteriorly in the body region and some 35 pairs appear in the tail. Although somites are not visible in the head of mammalian embryos, except occasionally in the posterior occipital region, anatomists have long held that "invisible" head somites, comparable to structures actually observed in lower vertebrates, probably form but immediately disperse.

Somite-cell origins have been investigated by means of genotypic analyses of single somites in allophenic embryos (56). This was possible, despite the minute amounts of tissue involved, thanks to a sensitive method for detection of strain-specific electrophoretic variants of a ubiquitous enzyme, glucosephosphate isomerase (GPI), coded for by alleles at a single locus. In embryos with different homozygous cell strains, both pure-strain variants were usually found in individual somites. A somite must therefore arise from at least two cells and is not a single clone.

Further studies (summarized immediately following) indicate that each of the major somite derivatives seems to have originated separately from a relatively small number of precursor cells—smaller than the numbers of cells visible in the earliest stages of somites or their subdivisions. This suggests that the cells comprising the somite are themselves the mitotic progeny, rather than the original cells in which myotome, sclerotome, and dermatome were genetically determined. Thus, somite specializations may have had their genetic functional origins at some earlier, *presomite,* stage, before 7 days of embryonic life.

Myogenesis and Cell Fusion

Individual muscles generally originate from many somites, but each of the small muscles that move the eyeball is unusual in that each arises from the myotome of a single somite. These muscles in allophenic mice thus furnish a unique opportunity to trace retrospectively the developmental lineage of myotome: If each eye muscle in an allophenic mouse always contains only one cellular genotype, the myogenic component in the somite of origin must have developed as a clone from a single initiator cell; if two genotypes are sometimes included in one muscle, its myotome source must have differentiated from two or more precursor cells. When eye muscles were tested for strain-specific GPI variants, both genotypes were in fact represented in single muscles (56). The myotome portion of a somite must therefore be multiclonal in origin.

These and all other skeletal muscles, unlike most other tissues, are multinucleated, although they are known to arise from uninucleated myoblasts. Whether the definitive state is achieved in vivo (as it is in vitro) by cell fusion, or by repeated nuclear division without cytoplasmic division, was long a matter of controversy. The question was finally resolved by a new approach: a search for hybrid enzyme molecules in skeletal muscle of allophenic mice derived from two inbred strains with electrophoretic variants of a marker enzyme. Enzymes such as GPI, isocitrate dehydrogenase, and malate dehydrogenase are each polymers with subunit variants coded for by alleles at a single locus; in vivo, heteropolymers between different subunits can occur only in an F_1 hybrid cell or a heterokaryon from fused cells of the two pure types. Allophenic muscle was in fact found to contain hybrid forms of each of these enzymes, along with the pure-strain types, in widely varying proportions (130, 2, 56). This result, in a multinucleated structure with diploid nuclei, could be accounted for only if myoblasts of different genotypes had fused to form myotubes which developed into mature muscle heterokaryons. Occurrence of heteropolymeric enzyme in an eye muscle means that at least two lineages of myoblasts coexisted in that somite, *prior* to fusion. While the actual number of myoblast clones may be larger than two per somite, the high frequency (33%) of double-genotype eye muscles lacking hybrid enzyme suggests rather small numbers of precursor cells, as this result in such small muscles is most easily understood if the muscles consisted of only a few coherent clonal patches, hence had relatively few interfaces for fusions between myoblast cells of dissimilar genotypes.

Allophenic muscle has interesting prospects for future study of mitochondrial protein synthesis in heterokaryons with some mitochondrial and some nuclear-encoded proteins. It also has particularly attractive possibilities for discovering the primary lesion in genetic dystrophies and other myopathies: if allophenic cases are found in which a phenotypically abnormal muscle is genetically normal (or vice versa), the primary defect would not be in the muscle and might then be identified elsewhere. Combinations of normal cells and cells with homozygous (autosomal or X-linked) muscle defects could also provide models for X-linked heterozygotes, as in human females heterozygous for the X-linked Duchenne type of muscular dystrophy (37). The models involving all-homozygous myoblasts could account for variability, and for phenotypically normal human carriers, in terms of fusion of widely varying proportions of the two functional classes of myoblasts, and/or selection favoring the functionally normal variant, in X-linked heterozygotes with single-allele activity per nucleus.

Skeletal muscle is thus far unique among allophenic mouse tissues in evidence for cell fusion in vivo. Virtually all soft tissues have been examined in allophenics from pure strains with the sensitive GPI marker, with other protein markers, or in metaphase preparations from mosaics with translocation ($T6/T6 \leftrightarrow +/+$) or sex chromosome (X/X $\leftrightarrow$ XTY) markers (118, 136). The results demonstrate that allophenic animals are truly cellular mosaics rather than hybrids: cells do not detectably fuse in vivo, except during myogenesis. Absence of heteropolymers in allophenic cardiac muscle confirms that heart muscle resemblance to a syncytium is only superficial, and reveals that enzyme monomers, and perhaps other macromolecules, do not traverse the intercalated discs (130). Absence of hybrid molecules in liver indicates that the binucleation and polyploidy common in that organ must arise by endoreduplication rather than by cell fusion (130). An absence of heteropolymers in the placenta establishes that giant trophoblast cells also originate by endoreduplication (57, 23). With other kinds of constitutional mosaics, there have been two reports of possible fusion of hematopoietic or lymphoid cells in vivo:

one involving lethally irradiated mice rescued with an inoçulum of chromosomally marked bone marrow cells (80), the other dealing with fraternal cattle twins whose originally separate antigenic types of erythrocytes were interchanged prenatally through placental anastomoses (192). However, the cells presumed to be hybrids may in both cases have arisen by a number of means other than cell fusion.

Skeletal Development

Of all the parts of the skeleton, the vertebral column lends itself particularly well to a retrospective study of development in constitutional mosaics as it is regularly segmented and each bone in the series bears a constant relation to the original somite segmentation. Through the use of a special stereomicroscope enabling the images of two bones to be superimposed and compared in any plane or orientation, many differences in vertebral morphology became distinguishable and could be reliably discriminated between the C57BL/6 and C3H inbred strains, so that the vertebrae of C57BL/6↔C3H experimentals could then be evaluated in relation to the controls (145). One would anticipate that, for any vertebra, some morphologically definable "parts" would display variability of strain-specific phenotype in allophenic specimens, independently of the strain-type of the rest of that vertebra or of neighboring bones. Such a part by definition possesses some developmental autonomy, for which the most logical explanation is development as a clonal lineage from a single precursor cell of one or the other contributing strain. Only in rare experimental individuals—those in which every clone differed fortuitously in strain composition from its nearest clonal neighbors—would all clones be simultaneously detectable.

The analyses showed that left and right halves of any vertebra could vary in strain-type independently of the other half or of the anterior and posterior neighbors; a lateral half could also comprise mixed characteristics of both strains. A vertebra is known to form by coalescence of the caudal sclerotome parts of a pair of somites and the cranial sclerotome parts of the next succeeding pair. Thus, the developmental *archetype* that best fits the observations is that at least two cell lineages form each lateral half-vertebra; each of this minimum of four clones in a vertebra may correspond to a cranial-or-caudal sclerotome contribution on the left or right side (145). This estimated minimum of four may in fact be close to the actual number of precursor cells for each vertebra (see the following paragraph), in which case it would appear that far fewer cells initiate development of the vertebral column than are seen in the sclerotome parts of somites. Those sclerotome cells may therefore be the progeny of primordial cells first distinguished by specific gene function for vertebral determination at some *presomite* stage, hence before day 7 of embryonic life in the mouse.

This clonal hypothesis of vertebral development has the advantage that it is testable (in single-genotype mice) because it makes two predictions for the effects of teratogenic agents or mutant genes: 1. The first prediction is that the period when clones are initiated, but morphogenesis is not yet evident, should be an especially "sensitive period" to teratogens, as cell destruction or damage would be maximally "magnified" during later clonal expansion. This has in fact been borne out in recent teratogenic studies based on administration of a DNA antimetabolite (5-fluoro-2'-deoxycytidine, or FCdR) to mice at various times during pregnancy. One sensitive period was identified at a day prior to somite morphogenesis, another a day before sclerotome appearance (79, 194). 2. The second prediction is that the presumed individual clonal units (cranial and caudal sclerotomes on each side) should be independently susceptible to deletion or damage. Mouse embryos of the homozygous *undulated* genotype have already been cited as a possible case in point (145), as their vertebral malformations result from deficient cell condensations specific for the cranial sclerotomes. A more extensive test of the prediction has recently been undertaken by an interesting use of the teratological data from FCdR experiments on mice of a normal inbred strain (79). It was reasoned that since the body and arch of a given vertebra are contributed by separate cranial and caudal sclerotome elements, it should follow—if the latter truly constitute separate developmental cell lineages—that a given class of induced anomalies, such as fusions, would occur independently in the vertebral arch and body of the same segment. A statistical analysis of the incidence and location of such fusions revealed that the observed results were an excellent fit to the calculated expectations based on the model. These investigators have concluded: "The hypothesis concerning the clonal origins of somites and sclerotomes is in line with the teratological findings. The clonal theory can better explain the data than could be done before" (78).

Individual skull bones from C57BL/6↔C3H allophenics were also found to vary independently on left and right sides and to include lateral phenotypic intermediates indicative of multiclonal derivation of each component from at least two cells whose mitotic progeny were present in varying proportions. In the occipital region of the experimental skulls, the component parts sometimes failed to fuse properly; this constitutes further evidence for genotypically different head-somite sclerotome contributions with autonomous growth rates (145).

Intermediate skeletal phenotypes have also been observed in allophenic mice with combinations of wild-type cells and cells homozygous for certain autosomal mutant genes (68). One of the mutations, *vestigial-tail*, ordinarily produces marked abnormalities and reduction in numbers of vertebrae in the posterior body region and tail; the allophenics included cases with milder defects. The other mutation, *short-ear*, ordinarily includes among its effects reduction in rib number and bifurcation of the sternal process (a remnant of its bilateral origin); some of the allophenics showed intermediate degrees of these defects. Neither mutation is strictly recessive, as each shows partial effects in an F_1 heterozygote. Thus, the allophenic group had members resembling heterozygotes, but with more variance, as would be expected from occurrence of two cellular genotypes to varying extents in the former. No clonal analyses were undertaken in these studies with mutants.

An interesting possible future use of the clonal models in genetic mosaics would be in identification of sclerotome clones in which skeletal defects are expressed, in the case of strictly dominant or recessive mutations. If the mutant-phenotype cells were at a selective disadvantage in competition with normal cells, as would be expected, those clones in which the mutant expression was focused should be relatively more consistently and selectively replaced by normal ones.

Hair Follicle Clones

Hairs are complex structures whose external part is ectodermal (epidermal) in origin; the dermal papilla at the base of the follicle is mesodermal and is one of the derivatives of the dermatome portion of somites. Hairs are especially favorable for developmental studies in the mouse: they are affected by many known genes; there is experimental evidence of an inductive effect of the mesoderm on the ectoderm in hair morphogenesis (165); and there are known indirect effects of the hair-follicle environment on pigment production by the melanoblasts that invade the follicle (181). Allophenic mice with appropriate cell markers might be expected to shed further light on cell lineages and interactions in hair follicle development.

When homozygous *fuzzy* (*fz/fz*) and normal (+/+) cells were combined in allophenic mice, a striking coat pattern of narrow transverse bands was seen (115, 119). Each band was primarily of the abnormal or normal phenotype and they were separate (i.e. sometimes of different phenotypes) on each side of the dorsal midline. Band numbers and widths varied, but narrow, apparently *unit-width,* bands were seen in each axial position, if one considered all the animals together. Numbers of these unit bands on each side of the body corresponded to the somite number, in those body regions where somites (approximately 30 per side) were visible in the embryo (115, 119, 124). Each of the narrowest bands on the left or right side therefore seemed to represent a clone derived mitotically from a single cell of somite origin; ostensibly wider bands could be readily accounted for as comprising two (or more) adjacent clones which were, fortuitously, of identical strain composition. The fact that approximately 18 pairs of bands (at first transverse, then oblique, as the face grew forward) were also present in the head region was taken as concrete evidence for correctness of the long-standing theory that mammalian embryos might have anatomically "invisible" head somites, corresponding to those actually visible in lower vertebrates, but characterized by unusually rapid dispersal. In view of the known origin of dermatome from somites, and of the histologically documented mediolateral spread of somite cells, the logical conclusion is that the observed pattern is in all probability based on its mesodermal, rather than its ectodermal, history. Progressive dorsoventral blurring of band definition is also consistent with the known migratory direction of somite dermatome cells, and it reinforces the inference that a band is a coherent clone rather than a secondary assemblage of cells from separate cell lineages. The entire clonal history of hair follicle mesoderm, not merely its terminal phases, seems to be exposed to view.

The developmental archetype for the coat mesoderm appears to consist of derivation from as few as approximately 166 cells (after extrapolating, from somite numbers, to about 35 per side in the tail); this small number would be more consistent with determination at a presomite stage rather than in the somites (119). While even earlier determination in a still smaller number of cells cannot be ruled out, there has been no compelling evidence for such a model. (Criticisms of the present clonal interpretation are discussed in the next section, together with those of the melanoblast clonal model.) The fact that the frequency of bands conforms to their mesodermal history while their phenotypes are seen in the ectodermal component of the hairs strongly suggests that development of the ectoderm is dependent on inductive influences from the underlying mesoderm, though it need not follow that the epidermis is passive in this interaction.

The same archetypal pattern of fine transverse bands was obtained in allophenic mice with *agouti* (*A/A*) and nonagouti (*a/a*) cell strains (111, 115, 119). Genes at the *agouti* locus are known from skin graft exchanges between genotypes to express themselves in hair follicle cells and to influence coat color only secondarily by causing melanocytes near the base of the follicle to produce some yellow phaeomelanin, as in the subterminal yellow zone of some hairs in *A/A*, or eumelanin, as in black *a/a* (181). The somite-like frequency of bands in *A/A*↔*a/a* animals further localizes expression of these genes primarily to the mesodermal or dermal papilla, rather than the epidermal part of the follicle (115, 119). This has been confirmed by dermal-epidermal genotypic recombinations in skin grafts (94), although such experiments have also suggested that the epidermis may play some role (170).

While individual bands are predominantly of one or the other component phenotype in the allophenic coats, each includes not only hairs of that pure type but also a wide range of individual hairs intermediate between the morphologically aberrant *fuzzy* and normal, in *fz/fz*↔+/+, and with less than the usual *agouti* amount of yellow in *A/A*↔*a/a* cases (135,141). Such intermediates were shown, by double-marker experiments, to be due to degrees of admixture of autonomously behaving cells of the two strains. For example, *agouti*- and *H-2*-locus differences were incorporated simultaneously in allophenics whose skin was then grafted to the parental strains; the surviving cells in each host, after selective rejection of histoincompatible ones, gave rise to characteristically *agouti* or *nonagouti* hairs in the respective hosts (141, 142). The same experiment also demonstrates that each predominantly *agouti* or *nonagouti* band was indeed genotype-specific and was not merely an ambiguous developmental modulation. Spontaneously occurring fine-banded coat patterns resembling those in experimental allophenic mice and thought to involve two genotypic cell strains have also been observed in a mouse possibly originating by double fertilization or embryo aggregation in vivo (176) and in sheep with areas of presumed somatic cell mutation (43).

Since the double-marker allophenic experiments show intermediate individual-hair phenotypes to be due to two cell strains, the dermal component of a single hair follicle must be derived from at least two cells (119). This has interesting consequences for differentiation in ordinary single-genotype animals. If in fact cells from phenotypically different clones of hair-follicle dermal cells coexisted, despite uniformity of cell genotype, intermediate hair phenotypes could result. The semi-isolation of each hair follicle would maximize effects of localized cell selection and drift, analogous to evolutionary changes in small isolated natural populations, and would augment the prospects of phenotypic variability. Other kinds of small morphological units in multicellular organisms could similarly have enhanced the potential for phenotypic variability, and therefore for developmental and physiological adaptability, without a corresponding increase in genome size.

Is there in fact any evidence of phenotypically different clonal variants, termed *phenoclones* (119, 123) in single-genotype animals? Two cell lineages, each with single-allele function, have been proposed in X-linked heterozygotes (86). In human hairs, the epidermal part of a single hair follicle may have one or both types of X-linked glucose-6-phosphate dehydrogenase (the enzyme is not present in the dermal cells); therefore, this component also arises from at least two cells (53) and may vary phenotypically according to their proportions. In mice, X-linked heterozygotes with hair-follicle markers offer the opportunity to compare coat patterns with the allophenic model in order to learn whether they correspond to the latter and are therefore likely to indicate phenoclonal lineages in single-genotype animals. The marked resemblance of patterns in $A/A \leftrightarrow a/a$ allophenic mice to the finely banded pattern of females heterozygous for the X-linked *tabby* gene ($Ta/+$) has been presented as evidence for this view (119, 124); the great range of genotypic proportions in the allophenic coats means, of course, that only those in a relatively narrow part of the range closely resemble $Ta/+$. When $Ta/Ta \leftrightarrow +/+$ allophenics were produced and in fact showed the same pattern, with two known homozygous cell populations, as seen in the $Ta/+$ heterozygote (22), the multiclonal explanation for the $Ta/+$ heterozygote (86) was strongly reinforced. The experiment thus failed to sustain the "complemental-X" alternative presented by Grüneberg (66), who has argued that phenotypically different areas in X-linked heterozygotes are due to partial and variable action of *both* alleles within each cell and to threshold phenomena.

Parallelism of the allophenic model with the heterozygotes has been disputed in another study, on effects of the *tabby* gene on other tissues (e.g. tail rings). Greater phenotypic uniformity was seen in heterozygotes than in the allophenics and this was taken to mean that the heterozygotes did not include two cell lineages in these tissues or had basic differences in clonal history, e.g. smaller clonal patches, within normalizing range of a hypothetical diffusible gene product (100). These conclusions are open to question, inasmuch as heterozygotes are being compared with allophenics in which the component cell strains had quite different genetic backgrounds and many other genes with tissue-specific expression that may have affected cell growth rates and selection differentially in the various tissues.

Activity of another X-linked gene, *greasy* (*Gs*), which has various effects on hair structure, has also been examined in $Gs/Gs \leftrightarrow +/+$ allophenics in comparison with $Gs/+$ heterozygotes, and the results have once more supported the view that the single-genotype $Gs/+$ animals have phenoclonal cellular differences (33).

While single-allele activity per cell is a likely explanation for phenoclone patterns in X-linked heterozygotes, coat patterns in some other single-genotype mice also resemble the clonal model but are caused by autosomal genes in heterozygotes (e.g. *viable yellow* A^{vy}/a) or homozygotes (e.g. *mottled agouti* a^m/a^m) and, for at least the latter, some other mechanisms must be sought. The hypothesis has been presented that such intragenotypic variability among clones of a tissue may in fact be widespread in all tissues and in many genotypes, and may be a particularly advantageous innovation in the evolution of higher animals (119, 123, 124).

The allophenic model of hair-follicle mesoderm origin from some 166 somite cells implies that at the time these cells were set aside, the embryo proper must have had at least that number of cells. The period before day 5 would probably have to be ruled out on these grounds. The sharply delineated autonomy of clones on left and right sides of the dorsal midline further suggests physical separation of the two sides at the time of clonal initiation. Day 7 or earlier, before midline fusion begins, has therefore been estimated to be the latest possible time of determination. The similarity of X-linked heterozygous patterns would also mean that differentiation of two functional classes of cells with respect to X-linked genes may have occurred in this same time period (119). It thus seems that a relatively long period may elapse between inception of genetic specialization for hair follicle formation and actual morphological appearance of hair follicles in the fetus. The delay might be due to the necessity for many biochemical pathways to mature, and to a need to accumulate sufficient cellular mitotic progeny for morphogenesis and effective cell interactions. The developmental picture that emerges for this mammalian system, as well as for melanoblast, retinal, and other tissue histories to be discussed later, is one of early postimplantation determination in small cell numbers, followed by a long proliferative period of clonal stability. Inasmuch as tissue-specific genes are not necessarily functioning at all times in the cell cycle during cloning, there may be special mechanisms which "lock in" capacity for potential function at specific loci.

Genes in mice affect not only the structure and melanizing influences of hairs but also loss or lack of hair, and this presents another new area: analysis of systemic vs local effects in allophenic animals. An allophenic mouse with cells of homozygous *nude* (*nu/nu*) and of normal genotype has proved to have some transverse "nude" areas as well as bands with normal hairs (91). Therefore, this particular defect is due to a local rather than a systemic physiological lesion.

4   CLONAL ORIGINS OF MELANOBLASTS

The first clonal history described for a mammalian tissue was that of the melanoblasts in the mouse coat (113, 115). Their development appears to exemplify many principles broadly applicable to development of other tissues.

The Melanoblast Clonal Pattern

Melanoblast lineages have been visualized in allophenic mice by means of cellular homozygous allelic color combinations, as in *brown* (*b/b*)↔*black* (*B/B*), *dilute* (*d/d*)↔(nondilute (*D/D*), *leaden* (*ln/ln*)↔*nonleaden* (*Ln/Ln*), *albino* (*c/c*)↔*colored* (*C/C*). Two-color animals from each paired combination exhibited a wide spectrum of ostensibly different patterns. Yet the variability could readily be accounted for by such factors as individual differences in proportions of the two genotypic populations; selective advantage of one cell strain over the other; invasion of territory "vacated" by occasional cell death; phenotypic identity vs nonidentity of adjacently situated lineages; frequent interpenetration at boundaries between clones; and ordinary developmental "noise" due to variability in cell movements, proliferation, and survival. When all patterns were arranged in an orderly series, it was evident that all were in fact permutations and derivations of a single *archetypal,* thematic, or standard pattern consisting of a series of wide transverse bands on left and right sides of the body.

Autonomy of the two sides was seen where colors were mismatched or out of register; the latter phenomenon also demonstrated that the bands did not occupy a fixed position with reference to any anteroposteriorly distributed structures. Each independently visualizable band was interpreted as a clone mitotically descended from a single primordial melanoblast (113, 115). On each side, there are approximately 3 clones in the head, 6 in the body, and 8 in the tail, 34 clones in all. Only in those animals with *chance* alternation of band phenotypes would all clones be visible simultaneously; each was independently visualizable in at least some animals of the total group. The original impression, which I stated in the first report (113), was that the animals with about half of each color tended to have alternation of band phenotypes, which therefore appeared to be nonrandomly arranged. It should be emphasized that this was *not* borne out in a larger sample. The conclusion, as later amended (119), was that "adjacent melanoblast clones may sometimes be of the same, rather than different, cellular phenotypes or colors; the width of the visible band is then approximately two (or more) times that of one clone. In other words, the full number of clones can be detected only in an animal that happens to have the maximal possible clonal phenotypic differentiation." Wolpert & Gingell (218) were apparently unaware of this retraction when they attempted to produce a model to explain the earlier published impression.

Questions have also arisen as to whether *any* developmental archetype could be deduced from the observed variability in melanoblast distributions, and similar questions have been raised regarding the underlying standards or "signals," despite the "noise," claimed for development of hair follicles (115, 119), sensory retina (123, 139), or pigment retina (123, 139). For example, McLaren & Bowman (98) have since examined some allophenic mice with pigmentary markers and have noted "some irregularities" in color distribution, denying any reality of the standard pattern "postulated by Mintz." These authors seem to have misinterpreted the "standard" to mean a universally visible, rather than an underlying or ideal pattern, as intended by its original juxtaposition to "derived" patterns (113). Nevertheless, the photographs of McLaren & Bowman's allophenics (98) clearly show the banded components of the pattern in various multicolored areas of different animals. Similarly, Mystkowska & Tarkowski (153) have also claimed that their observations "do not indicate that there is any regularity in distribution of pigment cells," and have asserted that the coloration "does not exhibit any definite and constant pattern." But a striking approximation of the archetype may be seen in their material (Figure 9 in 153). (Flat skin preparations, on which most of their observations were based, tend to be unfavorable for pattern analysis, especially if taken from adults, as coat patterns become blurred in older animals.) Admittedly, the patterns underlying the melanoblast (113) or other cell lineages are not immediately self-evident in allophenic mice. Yet assertions of lack of order, especially after examination of only small amounts of material, are surprising in view of the obvious repeatability of developmental norms. As Stent has stated (185), in discussing recognition of patterns (in the broad sense) and the search for common denominators, "every ensemble of real events contains some noise. And so the basic problem of scientific investigation is to recognize a significant structure of an ensemble of events above its inevitable background noise."

The phenotypic differences between the bands in color-labeled allophenic mice are not developmental modulations but accurate and autonomous expressions of the specific marker locus, as seen in the coisogenic $B/B \leftrightarrow b/b$ combination on a standard strain background (113) and in double-label genetic experiments in which accompanying histocompatibility differences of melanoblasts led to specific rejection only of the incompatible color after allophenic skin was grafted to parental hosts (141).

The predominantly dorsoventral direction of melanoblast migration was apparent from progressive diminution of color intensity; some cell movements in other directions would also be required to explain band width. The allophenic patterns thus conform to the results of classical graft experiments on mouse embryos by Rawles (172), from which she inferred that melanoblasts must be derived from neural crest cells on each side of the dorsal midline, whence they migrate laterally between days 8–12 of embryonic life. Coat pigmentation is not actually produced until a few days after birth; the allophenic patterns suggest that melanoblast determination has in fact occurred long before definitive differentiation and has remained stable over a long proliferative period. Some estimate has been made of the time of clonal initiation (119). In order for left and right sides to behave autonomously, as in allophenic clones out of register, the neural folds would still have to be open so that the two sides would be physically separated. Therefore, day 7 would presumably be the latest possible time of melanoblast determination. If there are 34 primordial melanoblasts, there would have to be at least that many embryo-forming cells, and preferably some additional ones for other tissues, so that day 5, i.e. the period immediately following implantation, is likely to be the earliest time of clonal initiation.

Schaible (177) has analyzed, and selectively bred for, certain coat patterns in mice with piebald (white-spotted) genotypes or somatic cell mutations and has concluded that they also support the idea of melanoblast origin from a small number of determined cells. He has suggested, however, that there are 14 cells, each of which proliferates in a separate "center," including 3 paired and 1 medial center in the head, 3 pairs in the body, and 1 medial one in the tail. A basic difference between this model and the 34-clone transverse-banding model of Mintz (113) should be pointed out, quite apart from the differences in numbers and distribution of the clones: Schaible's center model requires that all but the medial primordial melanoblasts first migrate laterally for some distance beyond the original neural crest position before beginning clonal proliferation, following which each forms a radially expanding clone. Mintz's banding model implies that melanoblasts originate close to the dorsal midline, and that clonal proliferation and migration then occur in a predominantly mediolateral direction. Mintz's model is more consistent with Rawles' embryological observations (172), with the banding and mediolateral direction of color gradients in allophenic mice with two positive colors, and with pigmented-and-white patterns in allophenics whose white areas are due to albinism (pigment-free melanoblasts) or to piebald spotting (absence of melanoblasts).

The presence of intermediate-color areas and of variable patch size has led Lyon to the opinion that a band (or patch) in an allophenic coat, an X-linked heterozygote, or other patterned genotypes may come from two or more precursor cells rather than a single cell, and that the total number of precursor cells is therefore not known (86–88). She regards the possibility of continued random cell movements without clonal contiguity and of extensive cell intermingling as very real. This view would apply not only to the melanoblast clonal interpretation but also to the finely banded transverse hair follicle clones (115, 119) and radiating retinal clones (123, 139).

There are various lines of evidence that dispute Lyon's proposal and favor the view that single bands come from single precursor cells. When hairs were sampled from $B/B \leftrightarrow b/b$ allophenics, the proportions of black, brown, and mixed hairs were distributed in an edge-to-center gradient consistent only with the interpretation that intermediate phenotypes were due to mixtures of adjacent clonal lineages of single-cell origins, and with absence of diffusible products (124, 135). Results of mutation studies also provide evidence that each "unit-width" band is an entire coherent clone. An example is the p^{un}/p^{un} genotype (102) (formerly p^m/p^m) characterized by black somatic revertant areas on a beige ground color; the distribution and frequency of revertant areas (see Figure 6 in 116 and Figure 4 in 119) are consistent with the probability that a black band or "spot" is a clone resulting from a single mutational event in a primordial cell, rather than from a much larger number of mutations in cells assembled from separate lineages. Somatic revertant patches in other genotypes similarly support the clonal interpretation (177). Lyon has also stated that, as "it is natural to think of somites" as a cause of transverse banding effects in mammals, the transversely banded melanoblast pattern is somehow related to somites, whose number (up to the base of the tail) is taken as 35 (86, 87); she has accordingly derived various simulated patterns purporting to show that more than one foundation cell per somite can give a pattern roughly like the hair-follicle or the melanoblast pattern, neither of them markedly transverse and both highly variable in patch size. These models, however, simply do not resemble the patterns seen in allophenic mice or in patterned genotypes. This point is fundamental, especially as Lyon has expressed doubt that Mintz's "archetypal" pattern of transverse bands is observable. It is therefore noteworthy that in a more recent paper (22), of which Lyon is a co-author, her view appears to have changed, as she states full acceptance of the archetypal pattern (see below).

Returning to the simulation models, it should be further pointed out that they involve a number of incorrect assumptions. The neural crest-derived melanoblasts have no known relationship to somite (or intersomite) frequency or location, and their clones do not occupy an anatomically fixed position. In addition, the total number of somite pairs in a model must include those in the head and tail, not just those in the body, if the total number of bands (34 claimed by Mintz for melanoblasts in head, body, and tail, approximately 166 for hair follicles) is to be tested in the model. The *tabby* (*Ta*) gene (in X-linked heterozygotes), for example, generates a somite-derived pattern like that of allophenics with mesodermal hair-follicle markers but it is not derived from only 35 pairs of somites. And the number of melanoblast bands located only in the body region is 12, or only 6 per side, rather than 34 per side, as Lyon implies.

The resemblance of the transversely banded melanoblast pattern in allophenic mice to patterns of certain X-linked heterozygotes has been pointed out by Mintz (119, 124), who has taken this as supporting evidence for the hypothesis by Lyon (86) and Russell (175) that single-allele function has created two cell phenotypes in the heterozygotes and that threshold phenomena postulated by Grüneberg (66) are not required. The resemblance has been further substantiated in a comparison of $C/C \leftrightarrow c/c$ allophenics with females heterozygous for the *flecked* X-autosome translocation (22). These allophenics are described by Lyon and co-workers as "clearly showing banding" and as "being in full accord with Mintz's archetypal melanoblast pattern." Moreover, the patterns of the allophenics are reported as "indistinguishable" from those of the translocation heterozygotes, in which "the archetypal melanoblast pattern is also found," "hence there is no reason to suspect that differing mechanisms may be responsible" and "there is no need to seek postulates beyond the developmental consequences [described by Mintz] that can modify the archetypal patterns."

On the basis of such similarities, and on the assumptions that each band is initiated by only *one* cell and that the initiator cells represent the stage of determination, Mintz had previously proposed (119, 124) that the same line of reasoning that led to the conclusion that the melanoblast pattern of allophenics was determined in 34 cells after implantation at some time in the day 5–7 period, should also apply to establishment of the pattern in X-linked heterozygotes, despite previous suggestions (86) that inactivation occurred earlier. The results of injecting blastocysts with single cells from embryo donors with X-linked heterozygous color markers are thus of special interest, as the resultant coat colors show that X-chromosome inactivation had not yet occurred on day 4½ (51; unpublished data cited in 88). Objections have been raised by others (81, 97) that the stage and degree of fixity of development to which the allophenic-derived numbers apply cannot be ascertained and is theoretically subject to other interpretations, including relatively late-stage cessation of cell movements in lineages actually established earlier (an idea of Lyon's already discussed above). One such interpretation goes so far as to propose that the cellular genetic markers in allophenics may merely reflect the situation at the time the composite was assembled, that they are no different in principle from the dabs of vital dyes applied by classical embryologists, and that they "provide rather the same type of information," the advantage in the allophenic being "simply that its cells are marked by a label which does not fade, leak, or get diluted" (81). Given the still-obscure nature of differentiation in multicellular organisms, one can only say that these various theoretical considerations and alternatives may in fact not exhaust the possibilities but do not appear to lead to a better reconciliation of the observed facts or to new and persuasive insights into the problem.

The full spectrum of archetypal and modified allophenic melanoblast patterns has been found to comprise not only the X-linked patterns but also those of various natural autosomal heterozygous and homozygous genotypes, of which at least the latter must be accounted for by some mechanism other than single-allele activity (116, 119, 123, 124). Thus, multicolor coats in single-genotype mice have *clonal* differences in pigmentation. As in the analogous allophenic situation, selection may then occur between these "phenoclones." Phenoclonal heterogeneity and clonal selection—previously considered the sole prerogatives of the immune system (20) —have therefore been proposed as possible universal attributes of differentiation and aging of all mammalian tissues (119).

The apparent simplicity and versatility of the melanoblast clonal scheme has suggested broad evolutionary adaptiveness; possible examples of the same pattern have been pointed out in other species (119, 124), including some unusual and "revealing" human color mosaics (27, 29). An allophenic rabbit (51a) and rat (92) have also been produced and show evidence of transverse banding.

Preprogrammed Clonal Death

White-spotted ("piebald") genotypes (unlike the amelanotic albino) lack pigment cells in the white areas, and this lack has been variously attributed to restricted proliferation of pigment cells (177), neural crest abnormalities, or, in other geno- types, failure of melanoblasts to differentiate or survive in the skin environment of certain specific areas (93). The results of allophenic experiments (115, 116, 119, 124) have led to still another hypothesis for the genotypes tested (with *Mi-* and *W*-locus markers). In the latter, the allophenics were comprised of all-homozygous wild-type (pigmented) and mutant cells (single-genotype mutant homozygotes are all-white). The allophenic patterns were exact (though more variable) copies of the white- spotted heterozygous ones. White areas resembled transverse clones or parts of clones, and were sometimes bordered by a thin pigment fringe. The conclusion was that the "null" areas were ghosts of deceased clones which had proliferated exten- sively and possibly normally before dying at an advanced stage when neighboring viable clones had only limited migratory capacities remaining. In order for all cells of a clone to die at once, they may have shared a cell heredity (in the heterozygote, of only certain phenoclones) handed down from the clonal initiator cell for a specific life span, i.e. a "preprogrammed cell death," a model possibly applicable to some aspects of aging in other situations. Other markers in the skin unequivocally showed both strains in white as well as colored areas, so that white could not be attributed to a genotype-specific regional adverse skin environment. This model has been challenged (30) in favor of regional tissue differences, because of dissimilarities in pigmentation among iris, retina, and other areas of natural genotypes. However, such dissimilarities, which have long been known (90), relate to different kinds of pigmentary tissues with entirely separate recent clonal histories from those in the coat. There is no reason, therefore, to expect that all would be governed by the same tissue-specific gene expressions at these marker loci or at other loci capable of affecting clonal longevity in some tissues.

Melanoblast Subclones

Genes at the *agouti* locus, expressed in hair follicles, indirectly affect melanoblasts and cause eumelanin vs phaeomelanin production (181). Thus, the physical overlap of each melanoblast clone with several hair follicle clones causes the former to show differentiation into subclones (119, 124). Allophenics in which each tissue is genetically marked show independence of these lineages: the strain of a melanoblast clone may be different from that of the local hair clones or the skin. These results are inconsistent with the suggestion (163) that skin cells in patterned X-autosome translocation heterozygotes have the same X inactivated as do the unrelated pigment cells in that area.

The simultaneous visualization of the highly variable overlapping but independent hair and melanoblast patterns in allophenic mice has called attention to similar overlapping and hitherto undetected "double patterns" in some natural multicolored genotypes, for both X-linked and autosomal genes (119, 124). This has indicated that such loci are expressed in *both* the hair follicles and the melanoblasts, independently and with obviously different phenotypes, and has raised the question of whether they may not be complex loci.

5   NERVOUS SYSTEM AND SENSORY ORGANS

Brain

According to the clonal view of differentiation (124), each specialization in a multicellular organism is initiated in a small number of precursor cells, presumably by genetic functional commitment, and that capacity for specific activity is retained (unless further specialization occurs) in the mitotic lineage comprising each clone; a clone thus has some autonomy as a developmental unit of perpetuated specific gene expression.

This view has especially intriguing implications for a new kind of study of the brain. A knowledge of the clonal history of the brain, i.e. of its underlying developmental organization, as distinct from its apparent anatomical organization, would disclose cellular communities with some measure of gene-controlled autonomy and with the potentiality for functional subspecializations, unit interactions, or preferred circuitry, in relation to behavior.

A useful clonal analysis of the brain in allophenic animals requires that the tissue architecture be preserved; therefore, histochemical or related means must be used. Markers for such an analysis of the cerebrum have not yet been identified; preliminary steps have been taken to investigate markers in the cerebellum. In an exploratory study, we have utilized strain differences in histocompatibility antigens (not restricted to the *H-2* locus), as these have been visualizable in cryostat sections of control C3H and C57BL/6 mouse cerebellum by an indirect fluorescent antibody technique (58). After treatment of controls with appropriate anti-C3H or anti-C57BL/6 antisera, green fluorescence was detectable in the cytoplasm of the giant Purkinje cells and, less intensely, in the cells of the inner granular layer. A limitation in applicability of this technique to allophenic material is that not all cells of these types show staining in the controls, although a large majority do. When cerebellum sections from C3H↔C57BL/6 allophenics were examined, rough complementarity was found with the two reagents in adjacent sections and there was considerably more variability between experimental animals than between controls (59).

Small sequences of Purkinje cells and granular cells were stained with one or the other reagent and may therefore represent clonal patches or parts of clones.

Quantitative allelic variants, analogous to the β-glucuronidase differences expressed in other tissues, may provide a more precise means of visualizing cell lineages in the brain; promising candidates are the lysosomal enzymes in the *beige* mutant, which has anomalous lysosomes in various tissues, including parts of the brain (164). Neurological mutations with autonomous, cytologically distinctive manifestations in the brain should also prove useful in future studies. It is unlikely, however, that all genetic behavioral disorders are due to primary brain lesions; some may result from sensory, muscle, or other still unrecognized defects. One of the most interesting future uses of allophenic mice would be in identifying the anatomical focus of primary expression of these disorders by testing genotypic composition of brain and other tissues in relation to the behavioral phenotype, in an approach similar in principle to that being used in Drosophila by Benzer and his colleagues (75).

Sensory Retina

The retina is the only sensory organ thus far examined in any detail in constitutional mosaics, although other sensory structures will doubtless be amenable to detailed study with appropriate markers. Eye morphogenesis begins with the appearance of a pair of optic vesicles in the anterior part of the neural plate on about the eighth day of gestation in the mouse. These grow out as a pair of bulbs from the forebrain and each invaginates to make a double-layered optic cup whose inner layer forms the sensory or "neural" retina, retaining its connection to the brain via the optic nerve, while the outer layer becomes the pigment retina. The sensory retina differentiates further into photoreceptor cells (chiefly rods, in the case of the mouse) and neurons.

Secondary degenerative defects (whether of sensory or other tissues) are more favorable for mapping cell lineages in mosaics than are defects involving partial or total primary agenesis, because degeneration leaves a discernible null or "ghost" area in a mosaic tissue, whereas agenesis of one of the prospective contributors results in the other member's preempting the available space and essentially generating nonmosaicism. The hereditary defect known as *retinal degeneration (rd/rd)* in the mouse is an example of the former, and has been successfully used to trace photoreceptor cell lineages in allophenic mice. In *rd/rd* controls (e.g. the C3H inbred strain), the retina continues to develop normally until 10–12 days after birth, by which time all the 8–12 rows of photoreceptor cell nuclei are formed but the rod outer segments are still incomplete. Degeneration then begins in the rod outer segments, and by approximately 3 weeks of age the entire visual cell layer of rods and their nuclei has disappeared, except for a single row of nuclei (although the neural network is preserved). Histological examination of retinas from *rd/rd* ↔ +/+ allophenic mice has revealed localized areas in which the cells followed the degenerative timetable seen in pure-strain *rd/rd* controls (139, 213). This was considered consistent with autonomous behavior of the *rd/rd* cells (139), despite the occurrence of intermediate numbers of outer nuclear rows at the margins of the intact patches, a phenomenon which might have been due to displacement.

In one study of 6 retinas from allophenics (213), no particular geometric pattern of the two cell strains was discerned, although this is not surprising, as 2 of the 6 retinas were entirely of the normal genotype, 2 others were almost entirely normal, and only a small part of another retina was actually reconstructed from serial sections. In another study of 20 retinas (139, 123, 132)—all of them genotypically mosaic, and all completely reconstructed (in two dimensions) from drawings of serial sections—a pattern suggestive of cell lineages emerged as an underlying developmental archetype or "theme," with numerous "variations" or developmental perturbations. The theme seemed to consist of 10 sectors of visual cells radiating from a center in each retina (see Figure 5 in 123); each sector was interpreted as a separate cell lineage or clone in which specificity of differentiation was stable throughout a long proliferative history. Variations on the theme were attributable not only to individual differences in ratios of the two cell strains and in the spatial permutations (i.e. different lineages would be detectable only where adjacent ones were, fortuitously, of separate genotypes), but also to selection, to developmental "noise" resulting from nonproliferation of some cells or irregularities in morphogenetic cell movements, and to distortions due to the reconstruction method itself. Selection generally favored +/+ over rd/rd cells on the particular genetic backgrounds studied. However, a relative widening of rd/rd sectors toward the periphery (leaving the central areas of the mosaic retinas less adversely affected than the peripheral ones) implied genotype-specific growth curve inflections during radial outgrowth, with more selective advantage accruing to +/+ cells during early, and to rd/rd cells during late, stages of clonal proliferation. The interpretation that the radiating sectors represent radially expanding clones is harmonious with histological descriptions, from classical embryology, in which mitoses were found to progress from center to periphery during retinal development. Individual differences in left and right eyes of an allophenic are consistent with independent origin of each side, although divergence from an unpaired earlier primordium is not ruled out.

A retrospective tentative "fate map" that would represent the determinative stage of visual cell development has been proposed (139), based on the adult patterns, and consists of a small circle of 10 visual precursor cells on each side. Since the earliest distinguishable primordium of the sensory layer of the optic cups has substantially more than 10 cells in each, this would mean that determination of the primordial visual cells may have occurred at a still earlier time, perhaps even before optic vesicle formation, hence before day 8 of embryonic life in the mouse. This possibility does not seem remote when one considers the classical fate maps of embryos of birds and lower vertebrates, constructed by testing transplanted pieces of increasing age for autonomous development in new surroundings. In those species, the basic determinations, including that of the optic primordia, were in fact largely complete by the neurula stage, when the embryo had become a "mosaic" of preprimordia incapable of accommodation to a new environment. The classical maps based on such graft tests (rather than on movements of color-marked areas) were largely made before the modern emphasis on diversified gene function in differentiation, but it may well be that the "determinative" stages of all primordia as defined in those tests actually represent inception and stabilization of specific gene activity in the clonal initiator cells of a particular tissue.

In mice of *rd/rd* genotype, the degenerative disease becomes manifest at a time when the visual cells have just differentiated to the extent which enables them to show the first excitatory response to light, as indicated in the electroretinogram (ERG) experimentally elicited by a xenon flash (158). In normal (+/+) mice, there are rapid changes in the pattern and threshold stimulus of the ERG as the rods mature, while the ERG of *rd/rd* mice becomes increasingly abnormal and is finally extinguished with the death of the cells. Analyses of visual function were undertaken in retinas of *rd/rd*↔+/+ animals, as the results could be useful for investigating retinal physiology and might ultimately provide models for partial visual defects of a hereditary nature in man. A study of the electrical responses in each eye of 59 allophenic adults showed a range of ERG patterns from normal to extinct (159). The rank order of the physiological measurement was the same as the rank order for extent of degeneration, in a sample of 19 of the eyes removed for serial sectioning and reconstruction of normal and defective areas. However, the ERG was always more reduced than would be expected from a linear relationship between area of functional retina and ERG amplitude. This result raises a number of interesting possibilities for future investigation. One is that the unique spatial distribution of degenerated areas (including their prevalence at the periphery) in the allophenic retinas accounts for the disproportionately great drop in their ERG. Alternatively, some intact +/+ visual cells may be functionally adversely affected. Or the pigment epithelium, which is known to contribute to the electrical response, may have a lowered resistance near impaired visual areas, thereby shunting away current.

Pigment Retina

The pigment epithelium of the retina not only arises in close anatomical relationship to the sensory epithelium but also becomes functionally interrelated with it in a number of ways. It is therefore interesting to know whether or not there is evidence for a shared origin, perhaps from a more generalized group of "retinal" precursor cells.

If we consider first the lineages of the pigment layer itself, it is apparent that these might be "revealed" by the use of pigmentary cellular markers in allophenic mice. The initial report of mosaic retinas, with cells of black-eyed and pink-eyed strains, described the pigmented and relatively unpigmented cells as showing a "high degree of intermingling" (198), but no particular order or pattern of distribution was noted, probably because the retinas were examined only in sections and reconstructions were not made from the sections. In a subsequent study (139), mosaic retinas from allophenics with pigmented and albino cell strains were prepared as flattened whole mounts and an archetypal pattern then became evident. As in the case of the visual retina, the developmental "signal" had many manifestations of "noise": patches of each genotype varied greatly in size and distribution as would be expected from irregularities of morphogenetic cell movements, failure of division or occasional death of cells, competition to fill gaps, other forms of selection, and strain-specific growth differences, during the proliferative period. Nevertheless, a distinct underlying pattern of radiating cell streams was apparent (see Figure 6 in 123). The number of radiating sectors (seen in the maximally patterned retinas, where neighboring lineages are more likely to show fortuitous differences in composition) is relatively small but may be greater than the number in the visual layer of the retina.

When double-label allophenic experiments were performed (139), with the rd/rd and $+/+$ distinction among cells in the visual retina and the colored-albino differential in the pigment retina cells, the areas of albinism and degeneration differed in proportions and locations in the overlying layers. Therefore, although each of the two retinal layers seems to arise by clonal proliferation from a circlet of precursor cells determined prior to the anatomically discernible optic primordium stage (before day 8), the pigment and visual layers of the retina are probably determined in separately compartmented groups of clonal lineages.

If the radial sectors of the pigment retina are indeed individually proliferating clones, as proposed (139), rather than secondary aggregations of like-type cells from separate lineages, the presumed clonal units should also be units of perpetuated mutation. That is, if mutant-color small patches or completely mutant radial sectors were occasionally found in nonallophenic mice, each such patch or sector could be accounted for by only a single mutational event, occurring respectively in a single cell during the period of clonal proliferation or in the initiator cell, whereas multiple events, hence a very much higher mutation rate, would be required if the mutated cells in that sector had come together from independent mutations in separate lineages. We would also expect that the greater number of cells in an older than in a younger clonal stage would result in more of the smaller patches than of the completely mutated sectors. With this in mind, we have examined the retinas of mice of the recessive *pink-eyed dilution unstable* (p^{un}/p^{un}) genotype (formerly designated p^m/p^m). These animals have pink eyes and pale fur in which occasional spontaneous somatic-cell reversion to the dominant wild-type (black) is detectable (102, 116); the mutational nature of the somatic change is inferred from the fact that the reversion sometimes occurs in the germ line and then gives rise to completely wild-type progeny. The revertant areas in the pigment epithelium of the eye were found to vary in size, and were more often small—as would be expected from variable occurrence of the mutation at different times in clonal proliferation—but sometimes involved entire radiating sectors corresponding to those seen in the allophenic retinas (128).

In another study of pigment retinal development in allophenic mice, by Deol & Whitten (31), pigmented and unpigmented "discrete patches" (of any size) were counted in retinas of allophenics from a pigmented-albino combination of two unrelated inbred strains (C57BL/10 ↔ SJL). Numbers were compared with those of nonallophenic mice heterozygous for Cattanach's translocation, in which an autosomal segment with the wild-type allele at the albino locus has been inserted into the X chromosome and has resulted in a mottled retina with both pigmented and albino cells, presumably due to random activity of the translocation (or the entire X bearing it) in some retinal melanocytes and inactivity in others (86). Inasmuch as the great variability among the allophenics is attributable to retinal formation from a small number of previously coexisting cell strains, the authors assumed (31) that if both types of functionally differentiated X chromosomes in the translocation heterozygotes had also predated retinal determination, there should be great variability among them as well. From the fact that there was actually much less variability in the latter group, they concluded that single-X activation must have occurred fairly late, on the 7th day or later, i.e. almost concurrent with optic vesicle formation, and was therefore much later than had been estimated by others (51).

However, this conclusion (31) is open to challenge on several grounds: First, while the residual genotype was identical in the two phenotypic color strains of retinal melanocytes in the case of the translocation heterozygotes, it was certainly different in the two unrelated color strains of the allophenics; one would therefore expect selective pressures and strain-specific growth differences, mediated by a number of loci, to cause much greater developmental variability in the allophenics, irrespective of time of origin of the retina or of single-X activation. In addition, it is questionable whether every "discrete patch," no matter how objectively defined and tabulated, is a *biologically* meaningful unit of measurement when one is attempting to deduce developmental histories and cell lineages. For example, if an ideal single clone of specifically determined retinal melanocytes consisted definitively of 1000 cells, any or all of which might have become spatially separated from each other (by cell death, nonproliferation, selection, etc) as that clone proliferated onto an extensive hemispherical surface, counts of "discrete patches" might well vary enormously in number in different retinal examples. To refer to any proliferating population of cells, during the complex morphogenetic movements of organogenesis, as "nonmigratory," as these authors have done, is an oversimplification. The same arguments apply to similar speculations concerning admittedly migratory melanoblasts of neural crest origin in the iris and other parts of the eyes and inner ears of the same animals (32).

6   HEMATOPOIETIC SYSTEM

Occurrence of complex stages in hematopoiesis has raised questions of possible shared origins of diverse blood cells and of the developmental basis for hereditary blood diseases. Cell markers in mosaics are helping to investigate these questions.

Normal Hematopoiesis

The normal succession of hematopoiesis in embryonic yolk-sac, fetal liver and spleen, and postnatal bone marrow has been reviewed (166, 174). Multipotential hematopoietic stem cells seem, from cloning experiments in vivo, to give rise to myeloid blood cells, including erythrocytes, various granulocytes, and megakaryocytes, and also to lymphoid cells. Examination of human females heterozygous for glucose-6-phosphate dehydrogenase has shown that the mosaic composition of erythrocytes, granulocytes, and lymphocytes is the same (45); this supports the earlier view that they originated from a common pool of multipotential cells. Further comparisons of the composition of these tissues with skin and skeletal muscle in the same subjects again showed high positive correlations which were taken to imply a common pool for all, calculated at approximately 16 cells (38). However, it is quite unlikely that such diverse tissues had a recently shared developmental origin, and this raises the problem of whether the correlation per se among the kinds of blood cells proved their divergence from a hematopoietic stem cell pool. In another study (155), cultured cells from various minced tissues (lung, spleen, fascia, thymus) of mice heterozygous for a translocation to the X chromosome were scored for inactivation of the translocated or normal X; again these correlations (possibly chiefly reflecting only connective tissue components in this case) of such disparate tissues are unlikely to mean a recent shared precursor pool. Another interpretation proposed for the data is that correlations by these methods reflect ultimate derivation from embryoblast and may indicate that single-X inactivation occurred when small numbers of embryo-forming cells were determined, with subsequent specific divergence of cell types in somewhat larger numbers of cells in each pathway (38).

While random X-inactivation generates approximately equal frequencies of each cell class, there is the possibility of much more variability in input of cellular genotypic proportions among allophenic mice. Though the latter tend to have widespread rough similarities among tissues of an individual, i.e. either large or small representation of each genotype, there are enough exceptions so that good correlation is not necessarily seen among tissues being compared. Therefore, the unusually high correlations between erythrocytes and lymphocytes (or their products) (137, 211) and between circulating red and white blood cells (123) support the view that shared hematopoietic precursor cells gave rise to erythrocytes, granulocytes, and lymphocytes.

Genetic Anemias

There are many genetically caused anemias in mice (174). In allophenic combination with normal cells, selection would be expected to replace defective with normal cells after the locus in question was expressed, but not before. This method should reveal the cell type, developmental stage, and time (yolk sac, liver, or bone marrow period) in hematopoiesis in which the locus first becomes active. Genotypic analyses could also disclose whether the defect is autonomous in the blood cells or influenced by their tissue environment. In preliminary experiments along these lines, the lethal macrocytic hypoplastic anemia W/W has been combined with normal-strain cells. In the few viable allophenic adults thus far obtained, not only red but also white blood cells were of the normal strain, which suggests a possible initial action of the gene in a hematopoietic stem cell stage (123, 136). The microcytic anemia mk/mk is occasionally lethal but may be successfully compensated postnatally by increased formation of erythrocytes. A number of viable $mk/mk \leftrightarrow +/+$ allophenics have been produced and all have had entirely or largely $+/+$ red blood cells, along with both strains of white blood cells (131). This result is consistent with expression of the gene in only the erythroid cell line. However, it paradoxically implies relatively inferior proliferative capacity in mk/mk cells, despite their known postnatal compensatory ability to proliferate excessively. Further allophenic experiments may hold additional surprises and should help to clarify the etiology of such genetic diseases.

7   SEX DIFFERENTIATION

Primary Sex Development

Sex modification is one of the chief means of analyzing the basis for normal male and female sex differentiation. The cattle freemartin, a sterile female cotwin to a male, was the first mammalian example of an aberration in which constitutional sex chromosome mosaicism might play a part; however, transport of the male's humoral factors through anastomoses in the placenta, rather than admixture of the two cellular sex types in gonads or other tissues, is the probable cause of the limited female development (see reviews 12, 34). Occasional examples of more severely modified individuals, i.e. true or pseudo-hermaphrodites, in which coexisting X/X and X/Y cells are evidently at the root of the disorder, have been reported in various mammalian species, including man (12). Allophenic mice present the first opportunity to produce such mammals experimentally and also to combine cell strains with other genetic differences which might influence sex development or function. The following observations on primary and secondary sex differentiation and on germ cell proliferation and selection in allophenic mice are, however, not necessarily generalizable for all mammals, as other examples of mammalian species differences in gene control of sex development are already known.

Primary sex differentiation might have been expected to deviate from normal in approximately 50% of allophenic mice derived from two sets of blastomeres, inasmuch as this is the frequency with which the contributors would be fortuitously of the X/X⟷X/Y sex chromosome combination. From the sex distribution and unexpectedly low incidence of hermaphrodites among a total of 40 animals (20% ♀ : 10% ♀̄ : 70% ♂), it was at first suggested that X/X⟷X/Y mosaics may become either hermaphrodites or normal males, but not females (196, 197, 199, 153). But a much larger series of 746 allophenic mice yielded a sex ratio much closer to normal, with 46% ♀ : 1% ♀̄ : 53% ♂ (114, 116). This same ratio describes a more recent tally covering our first 1000 allophenic mice and comprising 47 different genotypic combinations (124). These results have since been confirmed in another study of 308 allophenic mice (44% ♀ : 1% ♀̄ : 55% ♂) (149). The sex distribution strongly suggests that cellular sex chromosome mosaicism is compatible with all morphological sex phenotypes. This has been confirmed by karyological observations (114, 153). Intersexual X/X⟷X/Y animals have all been sterile and have had either bilateral ovotestes, a modified ovary and a normal testis, a testis and an ovotestis, or an ovary and an ovotestis, with various combinations and modifications of male and female sex ducts (199, 114). X/X⟷X/Y phenotypic males and females have only rarely been sterile (114). Among the fertile ones, the sex ratios and strains of the progeny show that only the germ cell strain corresponding to the phenotypic sex actually functioned reproductively (114, 116, 153, 201, 202). It therefore seems certain that germ cells of X/X or X/Y chromosomal sex can no longer be functionally reversed in animals this high in the evolutionary scale, despite the fact that *complete* reversal can occur in lower vertebrates such as amphibians, where gonia of either genetic sex depend solely on influences from the gonadal soma for their phenotype. Some human sex chromosome mosaics have been described (e.g. 27, 220) and have also varied in sex phenotype. Among the possible explanations for these cases is that some may have arisen in a spontaneous manner in vivo, similar to the in vitro experimental formation of allophenic mice. A mechanism has been described in the mouse and consists of premature lysis of the zona pellucida of two embryos by a normal uterine proteolytic factor, followed by adhesion between the denuded embryos and subsequent unitary development (121).

The entire germ line in the mouse is clonally derived from approximately 2–9 primordial germ cells (114) first seen in the yolk sac and adjacent extraembryonic locations, from which they migrate into the developing germinal ridges (138). Gonial proliferation permanently ceases and meiosis begins in the female on about day 13 of embryonic life, though the male gonia continue to divide indefinitely. This would suggest that X/Y gonia might have a marked selective advantage over X/X gonia when the two coexist, thereby contributing to a lessened incidence of hermaphroditism and tending to favor male over female development (116). One would also expect that other genes would cause selection and influence the outcome. Indirect evidence consistent with both predictions was seen when the sex ratio was tabulated *only* in multicolored animals, as they all have strain mosaicism in at least some tissues: In one inbred strain combination, this group comprised 41% females; in another combination, 36% females (116). In another study of various genotypic combinations by Mullen & Whitten (149), the multicolored allophenics in which one genotype predominated (so-called "unbalanced" pairs) had normal 1:1 sex ratios, suggesting that the phenotypic sex was primarily influenced by the predominating component. In the multicolored ones in which neither predominated ("balanced"), the sex ratios approached 3:1 males:females, suggesting that X/X⟷X/Y cases usually developed as males.

Nonreversal of germ cell phenotype in X/X⟷X/Y animals implies a block in germ cell development, and degeneration seems to occur in the nonfunctional strain at the end of meiotic prophase (153, 154, 116, 96, 99). An autosomal mutation (*Sxr*) in mice causes X/X genetic females to develop as phenotypic males (21). Here also the germ cells do not survive, but they do undergo a limited spermatogonial development attributed to possession of Y-chromosome-like functions.

In X/X⟷X/Y allophenics, the cellular genetic sex varies in tissues outside the reproductive tract and proportions are unrelated to the sex phenotype; this is not surprising and signifies that sex-determining genes are not active in those tissues (116). While some authors (148) have based their interpretations of mosaics on an expectation that all tissues should have an equal distribution of X/X and X/Y cells, this expectation is not borne out by the facts.

The relationship of germ cells to the somatic cells in the gonad has been investigated by comparing the reproductive cell genotype(s) (from progeny tests) in males with the total genotypic composition of their gonads (from strain-specific isozymes); some males were X/Y⟷X/Y, including some germinal mosaics, and others were X/X⟷X/Y fertile or sterile cases (116). In all categories, the genotypic strain and/or sex chromosome composition of the whole gonad often differed greatly from that of the gametes. This demonstrates that major aspects of gametogenesis depend on genes autonomously expressed within the germ cells themselves. It also suggests selection in opposite directions, due to some different batteries of expressed genes, in germ cells and in soma. In some fertile males, X/Y spermatogonia were able to undergo complete spermatogenesis in a testis apparently containing very large numbers of X/X somatic cells, though sterility resulted when the latter became an overwhelming majority. The partially X/X gonadal soma was histologically typically male in some cases and was thus developmentally labile, though the germ cells were not. It was concluded that primary sex development in the mouse involves gene expressions in both germ cells and soma, and cell interactions. Initial determination of sex cell type, and early stages of gametogenesis may both rely in good part on genes expressed in the germ line; other genes expressed chiefly in the gonadal soma may then indirectly influence aspects of further gametogenesis. Other workers (104) have hypothesized that both germ cells and gonadal soma produce "diffusible factors," with critical threshold levels mutually required for normal development; abnormal X/X←→X/Y individuals might have aberrant levels of the factors.

Accessory and Secondary Sex Characters

Considerable lability exists in the development and responses of these structures, according to genetic (89) and embryological studies, and from observations in allophenic mice. For example, seminal vesicles and epididymis with X/X cells were found in sex chromosomal mosaic males (116). And seminal vesicles composed of X/X and X/Y cells of different inbred strains actually secreted the female as well as the male strain variants of the tissue-specific normal male seminal vesicle protein (*Svp* locus) (133). A sterile female was also found to contain male and female cells in mammary glands in which the characteristically female conversion to neoplasia was occurring (143). Behavioral studies of sex chromosome mosaics would be of interest in future studies.

Germ Cell Selection

Striking examples of selection have been observed in the reproductive cells of allophenic mice. In germinal mosaic C3H←→C57BL/6 mice, shown by matings to the recessive-color C57BL/6 strain to have both germ lines of the same genetic sex, the majority of females had over half their gametes of the C57BL/6 strain, while the majority of males had most of their gametes of the C3H strain (114, 128). The ratios remained stable in individual females but shifted further toward C3H with age in the males. Haploid-stage germ cell selection at fertilization is known in mice (35), but the age-related change in males—which alone maintain proliferative diploid gonia—established that diploid-stage selection was operating. Some selection, though less extreme, may even occur in congenic combinations. The question has therefore been raised (114) whether "hidden" selection of this sort may not have occurred against some spontaneously mutated cells in ordinary mice progeny-tested for spontaneous mutations, thus presenting a spuriously low estimate of mutation rate. The kinetics of the shift in the allophenic males suggested autonomous strain differences in germ cell proliferative patterns, traceable to embryonic life, and existence of these differences has been verified in the primordial germ cells of the control strains (128).

8   IMMUNE SYSTEM

The fact that antigenically disparate cell strains and lymphoid strains with different immunological responsiveness can develop as coevals, rather than as host and donor, in allophenic mice, has opened new possibilities for in vivo experimental analysis of ontogeny, function, and disease of the immune system. Studies relating to self-tolerance, immunodifferentiation, the immune response, and genetically caused immune diseases have been carried out in these animals.

Self-Tolerance

The first indication that cells with different antigens could coexist with impunity if first associated before immunological maturity was in cattle cotwins whose placental blood vessels had anastomosed in utero, resulting in hematopoietic tissue exchange and consequent erythrocyte mosaicism for certain blood group antigens (167). Owen's appreciation of the significance of this "natural experiment" ushered in a major new era of immunological investigation, in which mosaicism was experimentally established by postnatal transplantation of tissue. Observations of occasional spontaneous blood group mosaicism, probably also due in some cases to placental fusions, have since been made in a number of other mammalian species. Recently, erythrocyte mosaicism for blood group antigens has been experimentally produced prenatally in allophenic sheep (204).

Cellular mosaicism for histocompatibility antigens in mice has been the chief tool in the experimental study of the basis for discrimination between "self" and "nonself" and for self-tolerance. Until recently, mosaicism was imposed after development (even if not full competence) of the immune system (15, 16). Allophenic mice have offered the first possibility of investigating a possible model of natural self-tolerance, in which cells with potential immunogens coexist even before differentiation.

When allophenic mice comprise genotypic cell lineages with different histocompatibility alloantigenic determinants, including those governed by the major *H-2* locus, these are in fact expressed just as they are in the strains of origin. Skin cells, hair follicle cells, melanoblasts, mammary gland epithelium, and erythrocytes are among the tissues shown, by graft or serological tests, to contain two allelic types of histocompatibility antigens in allophenic animals. When skin from such donors was test-grafted to parental-strain recipients, each host selectively rejected cells with antigens foreign to it and spared cells with familiar antigens (140); the specificity of choice was verified by genetic double-label experiments in which strain-associated allelic differences, e. g. alternative colors, in the case of melanoblasts, were included along with allelic *H-2* differences. The doubly labeled transplants to parental strains also revealed that the graft rejection process is strikingly localized to target cells, as seen in strain-specific cell destruction when antigenically diverse hair-follicle or melanoblast cell strains resided within the narrow confines of single donor hair bulbs (141). Similarly, when halves of tiny single hyperplastic (premalignant) nodules from mosaic mammary glands of mammary-tumor-susceptible and nonsusceptible strains were transplanted to parental hosts, mixed acceptance-and-rejection patterns were found, and the ultimate malignancy or normalcy of the surviving portion confirmed the autonomous cell-strain specificity of antigen production (143). Two populations of erythrocytes with different histocompatibility antigens have also been identified in allophenic mice by means of anti-*H-2* antisera in hemagglutination and absorption tests (137).

In view of the normal expression of cellular antigens which would ordinarily lead to histoincompatibility between strains—expressed as graft rejection or, in the case of immunocompetent lymphoid-cell grafts, as destructive graft-vs-host (GVH) reactions—are allophenic mice in fact truly tolerant of their alloantigenic differences? As tolerance is a purely in vivo phenomenon, the answer must be ascertained operationally by appropriate tests in vivo, such as skin graft tests and examination for GVH disease. Moreover, the enormous variability in genotypic composition of allophenic mice (137) requires that they be tested individually. We have test-grafted over 100 allophenic mice with parental skin from alloantigenic strain pairs (140, 129) and three categories of results were obtained: (a) Some animals summarily rejected skin from one of the two parental strains, hence were nontolerant, and showed no evidence of that cell strain in any of the 6–10 tissues genotypically analyzed with other markers. In the development of these individuals, one of the two strains may have failed initially to be incorporated into the small group of embryo-forming cells of the inner cell mass (119) or may have been lost at some later time. (b) Some other (albeit very few) individuals also rejected grafts of one of the parental strains, but were found to possess a very minor amount of that strain in only one of their genotypically tested tissues (137). Such nontolerant allophenic animals are reminiscent of cases of skin graft rejection between dizygotic cattle twins with erythrocyte mosaicism (191) or skin allograft rejection by irradiated mice with donor-type γ-globulin-producing cells (209) or erythrocytes and lymphocytes (17). Skin graft rejection in these instances could be due either to presence of a tissue-specific transplantation alloantigen expressed on skin cells but not on lymphoid or bone-marrow cells (in a host lacking the rejected strain in its own skin), or to "sequestering" of a small cache of cells, or to minor representation of cells with antigen levels inadequate for immune recognition by the animal's lymphoid cells. (c) Finally, most of the graft-tested allophenics accepted skin grafts of both alloantigenic parental strains; various strain phenotypes revealed both cell strains in one or more of their tissues and tissues from some were also directly tested serologically or by grafts to parental strains and showed positivity for both H-2 types (140, 137). That tolerance in these animals was not due to indiscriminate immunological failure was seen from their ability to manifest a normal rejection of "third-party" skin grafts. [It may be added that they did not reject F_1 hybrid skin grafts, thereby demonstrating absence of a "hybrid antigen" in the F_1 (117).] Thus, retention of cellular mosaicism is a necessary, but not a sufficient, condition for operational immunological tolerance of both parental strains in allophenic mice, although it should be emphasized that nontolerant mosaics appear to be rare and to have a very minor representation of the nontolerated strain. Other (nonallophenic) mice made tolerant by neonatal inoculation of allogeneic cells have been found to abrogate tolerance if the donor antigen-producing cells are lost (85) and it may therefore be possible for a tolerant allophenic mouse to lose its tolerance if one strain of antigen-producing cells were secondarily lost. However, we have not yet observed a case in which a *demonstrably* tolerant allophenic animal, as defined by actual acceptance of skin grafts from both strains, has ever later rejected one of the grafts.

Barnes et al (8), after graft-testing only one animal (of a total of 4 NZB↔CFW allophenics studied by them) and observing its rejection of an NZB skin graft, have repeatedly claimed that self-tolerance is lacking in allophenic mice. Because this potentially important statement has been frequently quoted as a generalization, and seems to have gained currency in the re-telling, it should be pointed out that the one mouse in question may in fact be irrelevant to a consideration of tolerance as it may have totally lacked any living cells of the rejected strain. This animal (their case C3) had only four of its tissues tested for mosaicism: of these, the coat appeared all white (CFW strain) and was described as having previously lost a very minor amount of the other color; the germ line also showed only CFW; the erythrocytes were described as "originally" of both types and "later" of only the CFW type, but no ages, amounts, or other data were given. The serum allotype tests showed CFW and questionable presence of NZB, but even if NZB strain allotype were present, mosaicism in the γ-globulin-producing cells is not adequate for tolerance, as already pointed out by others (209). Nor might this animal be said to exemplify a "break-down" of tolerance, as prior tolerance was never actually demonstrated to have existed in it. In the same study (8, 5), Barnes et al described 3 other allophenic animals with genotype mosaicism (albeit they were not graft-tested for tolerance), all of which ultimately developed the spontaneous autoimmune disease (of unknown etiology) known to characterize the NZB control strain. The clinical picture of the NZB-type autoimmune disease is widely regarded as being in certain respects indistinguishable from the GVH disease seen in some acquired-tolerance experiments. In spite of this resemblance, which the authors readily grant, they repeatedly express an unsupported preference for the interpretation that the evidence "slightly favours" presence of GVH disease. Their preferred interpretation is also not supported by the observations of others. For example, in our own studies of a much larger number of allophenics with both NZB cells and normal cells, there has been permanent retention of skin grafts from both strains, even when the NZB type of autoimmunity has become well established, and no special evidence for GVH disease (144). In addition, any pathological changes which may overlap with GVH symptoms have not been more frequent in allophenics of *any* strain combination than in pure-strain controls (120, 132). It is therefore questionable whether there is any bona fide evidence from allophenic mice to justify Barnes' extrapolation (8) "that both theories of tolerance and self-recognition should be re-examined."

It is important at this point to take up the question of whether the shifts in genotypic composition commonly seen with age in some allophenic mice reflect an allogeneic effect and abrogation of tolerance. The evidence is clearly against this interpretation for a number of reasons. A particularly striking one is the fact that shifts may go in *opposite* directions in different tissues of animals of a given inbred strain combination. For example, in C3H↔C57BL/6 allophenics, the C3H strain tends to be favored over C57BL/6 in germ cells of males (but not females) (114) and in the liver of both sexes (120), whereas C57BL/6 cells are favored in the blood (137) and increasingly predominate with age in the coats of the same animals. Such tissue-specific selection in allophenic mice may mean that various unrecognized loci, unrelated to histocompatibility, differentially express themselves in, and mediate, cell growth and proliferation in specific tissues, and have allelic strain differences capable of causing selective shifts when the cell strains coexist (120). Another kind of evidence negating abrogation of tolerance is that changes may even occur between coisogenic or congenic strain pairs with coat color markers but without histocompatibility discriminants (113).

The immunological tolerance that Silvers and I have documented in allophenic mice (140) differs from the well-known "acquired" tolerance induced by allogeneic cell inoculations in neonatal or irradiated adult hosts (16) in that the advent of new proteins of a potentially antigenic nature developmentally antedates appearance of the immune system. In this important respect, tolerance in the allophenics is like normal self-tolerance, except that two "selves"—in the frame of reference of the histocompatibility-antigen student, if not of the embryo—are mutually involved. We have therefore designated this as "intrinsic" immunological tolerance and have presented it as a model for natural self-tolerance (140). On the allophenic model, "self" cannot be recognized as "foreign" and anti-self immune reactions simply do not ordinarily develop. Thus, discrimination between "self" and "nonself" arises as a strictly epigenetic phenomenon, "learned" during development rather than being obligatory in the genotype. The model therefore supports the previous views of self-tolerance as nonreactivity, first advanced by Burnet (20), whose clonal selection theory holds that if a maverick clone of specifically self-reactive lymphoid cells arises, e.g. through mutation, it is normally somatically eliminated, hence "forbidden." Medawar and colleagues also concluded on experimental grounds, from acquired-tolerance studies, that absence of self-reactive cells constitutes the basis for tolerance (16).

This concept of tolerance as nondevelopment of anti-self clones has recently been challenged and an alternative view introduced by Voisin (206) and expanded by the Hellströms (71, 72). They propose that specifically self-reactive lymphoid cells (Burnet's "forbidden clones") are in fact present, but are prevented from attacking prospective target cells by the interposition of "blocking" factors in the serum. (Various speculations have been advanced as to whether the presumed blocking factors might be antigens, antibodies, or antigen-antibody complexes.) The important question therefore arises whether allophenic mice, even though operationally tolerant (as defined by their permanent acceptance of parental-strain grafts and lack of GVH disease), do actually harbor self-reactive cells. In addition to their experiments involving animals with acquired tolerance, the Hellströms, with Wegmann and collaborators, have examined this hypothesis in intrinsically tolerant allophenic mice, whose lymphoid cells and serum were tested in vitro. We will consider this work in some detail, along with that of others who have since obtained contrary results in experiments with allophenic mice.

Wegmann et al (212) reported that lymph node cells from allophenic mice appeared to be capable, in cytotoxicity assays, of destroying fibroblast target cells of the respective allogeneic pure strains, and that this destruction was prevented or blocked by allophenic serum but not by pure-strain serum. Mixed lymphocyte cultures of the two pure parental strains of spleen cells, which showed an allogeneic proliferative response, were also ostensibly inhibited by allophenic serum from showing that response (168). Allophenic spleen cells were themselves not stimulated by parental cells in mixed lymphocyte cultures and seemed to prevent parental cells from stimulating each other, i. e. they caused reduced ^{3}H-thymidine uptake in the parental cell mixtures; therefore, allophenic spleen cells were believed to be conducting active and continuous suppression in the in vivo situation of intrinsic tolerance (169).

The evidence for "suppressor cells" (169) is perhaps the most questionable of these proposals, as it rests on rather small differences in ^{3}H-thymidine counts in a complex, unanalyzed, and not readily quantifiable situation involving background counts taken up by the allophenic cells, counts due to stimulation of parental cells to which one of the allophenic strains is foreign, and other unknown factors. The remaining experiments, relating to self-reactive cells and blocking factors in allophenics, raise a number of other questions. It should first be pointed out that none of the allophenics in these studies (168, 169, 212) was actually tested for tolerance, although all showed coat color mosaicism and, in some cases, erythrocyte mosaicism. More important is the fact that the lymphoid cells were assumed, without tests, to include both genotypes. However, judging from other data (137, 120, 211, 103), it is highly probable that the animals differed widely in proportions of their component strains and in the genotypic composition of specific tissues.

While this variability greatly enhances the potential interest of the experiments, it means that the genotypic makeup of any tissue in any individual must be determined empirically and not assumed; this would appear to be especially important in studies of the immune system. A tendency for extensive mosaicism, or for minor representation of one strain, generally characterizes tissues within allophenic individuals, but important exceptions have been repeatedly noted, especially in developmentally unrelated tissues. For example, melanoblasts and hair follicle cells, in which coat color phenotypes are expressed, develop quite independently of cells in the immune system and, not surprisingly, may have a different genotypic composition from the latter; some animals with two-color coats may actually lack one genotype in their immune system (137). Thus, if lymphoid cells of each strain, after removal from allophenic mice, were in fact capable, in the absence of their own serum, of overtly exhibiting "self-reactivity" against the allogeneic cell strain in vitro, we would expect that cells from individuals with different ratios of the component strains should behave differently. We would also expect that complementarity should be demonstrable in tests against the two parental target strains. For example, lymphoid cells from two C3H↔C57BL/10 mice, one with a 90:10 and the other with a 30:70 ratio of the two strains, should show, respectively, a very high anti-C57 and very low anti-C3H, and an appreciable but relatively low anti-C57 and moderately high anti-C3H specificity of destruction of target cells in the cytotoxicity tests. Appropriate controls, apart from parental cell strains and F$_1$ hybrids, hence should include a series of artificial lymphoid cell mixtures of the two parental types tailored to match the actual genotypic ratios in each of the allophenics, so as to learn allogeneic control results under actual test conditions. Lymphoid genotypic ratios in the allophenics themselves could be determined with *H-2* and other lymphoid markers or, in certain combinations, with isozymes such as glucosephosphate isomerase.

These kinds of experiments and controls were not done in the studies under discussion. In addition, very small numbers of allophenics were included and, in most of the experiments, cells or sera were pooled (168) or individual test data were largely omitted or pooled (168, 212). In the first study (212), for example, only 4 allophenics were involved. Each was tested separately against the two parental strains, but detailed data are given for only one of the eight tests. Another of the eight was discarded for technical reasons, another was discounted because it gave exceptional results, and some of the remainder appear to be possibly of borderline significance. Some of the controls in that study also gave puzzling results, e.g. pure-strain C3H lymph node cells (rows 5 and 6 in Table 2 of 212) were not cytotoxic against allogeneic C57 target cells (without allophenic serum). The "blocking" effect of allophenic serum was therefore documented in detail in only one relevant case.

In another study, the authors state that individual sera from 7 allophenics "showed varying amounts of blocking activity" (168), but individual data are not given. In the tests of suppressive effects of allophenic spleen cells on the parental mixed lymphocyte culture response (169), a few individual cases are presented in which one parental strain was much more suppressed than the other, but, in the absence of genotypic analyses of the allophenic material, these results cannot be interpreted as reflecting genotype-specific complementarity rather than experimental variability. Finally, one of the studies (168) involved the SJL strain, which is known to have proliferative immunological abnormalities and a high incidence of reticulum cell sarcomas (207).

The ambiguities in the reports discussed above have been largely avoided in two subsequent studies, by Matsunaga and collaborators (91, 103) and by Harris et al (69), on the basis for tolerance in allophenic mice. Much larger numbers of allophenic animals have been individually examined in both studies. Both have led to the conclusion that neither self-reactive cells nor serum "blocking factors" are present in tolerant allophenic mice. In one set of experiments (91, 103) allophenics of the C3H↔C57BL/6 combination were checked for presence of both allotypes of the 7Sγ2a immunoglobulin class, as evidence of both genotypic strains of antibody-producing plasma cells, and some were tested with monospecific anti-*H-2* antisera. In unidirectional and bidirectional mixed lymphocyte cultures with cells from lymph nodes and spleens, cells from allophenics were tested as possible responders or as stimulators (in the latter case, they were X-irradiated). The allophenic lymphoid cells, without allophenic serum, remained unresponsive to cells of the parental strains or the F_1 hybrid; the immunological specificity of the unresponsiveness was evident from the fact that they invariably responded to cells of an unrelated "third party." Thus, immunocompetent allophenic cells, even when removed from the organism, continued to regard the parental strains as "self" rather than "nonself," just as they seem to do in vivo. Allophenic cells did not stimulate F_1 hybrid lymphoid cells but showed full capacity to consistently stimulate "third-party" cells; one or both pure parental types responded, to very varying extents, and reflected specific recognition of the "foreign" or allogeneic component of the stimulators.

Allophenic serum with both allotypes was tested for ability to block the mixed lymphocyte reaction between allogeneic cells of pure strains, and its effect was compared with that of commercially obtained pooled mouse serum. While some cultures showed a reduced proliferative response, the reduction was clearly *nonspecific*. Matsunaga et al have also discounted the possibility of in vitro production of blocking humoral factors or the existence of suppressor cells, inasmuch as full responsiveness was observed in allophenic-parental bidirectional tests (91).

Harris et al (69) have also conducted a large study in which they have used spleen or lymph node cells from 17 individually tested allophenics of the C57BL/6↔ BALB/c combination in cytotoxicity assays against parental-type target tumor cell lines; target cells were labeled with radioactive chromium whose release was a measure of lysis. While the method gave high cytotoxic activity between pure-strain allogeneic cells, cells from allophenic mice gave no activity against one of the target lines (C57); the variable low activity against the other was also found with control BALB/c or F_1 hybrid spleen cells. Whether new reactive cells are produced in vitro that would either not have been produced or not have survived in vivo is not known. In any event, the allophenic cells clearly did not differ significantly from the controls. Allophenic serum had "no ability to specifically block appropriate cytotoxic activity or to block its in vitro generation." They also performed in vivo tests of the ability of allophenic spleen cells to induce GVH reactions, as measured by the Simonsen spleen weight gain assay, in F_1 hybrid hosts. Although control parental-strain spleen cells caused significant reactions, allophenic cells from 9 separate donors were without effect, and were unable to suppress the GVH effects of pure-strain parental cells.

These failures of more extensive and better controlled tests to confirm existence of self-reactive "forbidden" clones, "blocking factors," or "suppressor" cells in allophenic mice appear to place the burden of proof on the exponents of those views and, at least for the present, to sustain the earlier proposal that allophenic mice constitute a valid model for normal self-tolerance via simple absence (i.e. non-development or elimination) of anti-self lymphoid clones. In any definitive resolution of the problem, the question should perhaps be raised whether in vitro "killer" lymphocytes in the blocking experiments might have resulted from an artifactual change induced by culture conditions. It should be added that new in vitro (14, 63, 215) and in vivo (18) investigations of acquired tolerance (in neonatal animals or irradiated adults inoculated with allogeneic cells) have also failed to confirm self-reactive cells and blocking factors. These recent experiments have included direct in vivo tests of tolerance (e.g. skin allograft acceptance, lack of GVH reaction) and tests of degrees of cellular mosaicism in each animal, information which, as Medawar has pointed out (101), was lacking in earlier evidence purporting to show immunologically activated lymphocytes and serum blocking factors in tolerant hosts. In authentically fully tolerant and mosaic animals, these phenomena could not be reproduced. In animals given low dosages of allogeneic lymphoid cells, inadequate to confer full tolerance or full reactivity, cytotoxic cells were not evident, but a minority had serum that interfered in vitro with the cytotoxic effect of experimentally presensitized lymphoid cells (14). If future results are consistent with these observations, the possibility arises that although blocking factors may play no part in normal self-tolerance, they may exist as a manifestation peculiar to precarious experimental states of acquired tolerance.

The proposal that blocking factors might explain tolerance in allophenic mice was, in a sense, a logical extension by the Hellströms of their earlier hypothesis of "allogeneic inhibition" (70) advanced originally as a possible surveillance mechanism for elimination of neoplastic or other variant cells. The hypothesis held that allogeneic cells or cells with other surface-structure incompatibilities can mutually destroy or inhibit each other upon direct contact, without immune-system intervention. The lifelong coexistence of cells with diverse histocompatibility alloantigens in healthy and fully tolerant allophenic mice is obviously inconsistent with this hypothesis, as the authors themselves have noted (73). They have suggested that the paradox is resolved if tolerance is mediated by blocking antibodies, "since allogeneic inhibition can be blocked by antisera directed against cell surface antigens." A different view, which we have already expressed (140), is that the phenomenon of allogeneic inhibition, which has thus far been observed only in very unusual in vitro and transplant circumstances, may be strictly *sui generis* and not applicable to the biological situation more characteristic of a normal intact organism. A similar possibility may be considered for blocking factors.

Ontogeny of Immunocompetent Cells

The specificity of immune responses in mammals appears to be due largely to two main classes of cells, T and B lymphocytes, and their subclasses (see reviews 166 and 171). The development of these lymphoid populations is not entirely clear but it is believed that both are ultimately derived from multipotential hematopoietic stem cells which also give rise to the myeloid types of blood cells (erythrocytes, granulocytes, and megakaryocytes), in addition to renewing the stem cell population. The stem cells first appear in the yolk sac, next in fetal liver and spleen, and finally in bone marrow; it is possible, though not certain, that the earlier sites progressively seed later ones. In the definitive stages, precursor cells from the bone marrow enter the thymus; T (thymus-derived) lymphocytes develop from these and emigrate to, and subsequently recirculate between, blood and peripheral lymphoid tissues (lymph nodes, spleen, and Peyer's patches on gastrointestinal surfaces). Other precursor cells in bone marrow are thought to give rise to B (possibly equivalent to chick bursa-derived) lymphocytes; these also recirculate, though to a more limited extent, through the lymphoid tissues. According to the now widely accepted clonal selection hypothesis of Burnet (20), lymphocytes have (intrinsic or mutational) individual commitments, perpetuated in clones, to respond to only a single (or small number) of antigens, for which they have antigen-specific receptors on their surfaces. Those lymphocytes with appropriate receptors are selected for response to a given foreign antigen, and the antigen-receptor interaction activates the specific lymphocytes to divide and differentiate further. B lymphocytes seem to have as receptors immunoglobulin molecules (i.e. antibodies which resemble those that the cell is capable of secreting); the activated cells divide and mature into plasma cells which secrete the specific antibodies into the circulation. Antigen-activated T lymphocytes also divide and differentiate but their various products do not include antibodies. These cells are responsible for the relatively slow, cell-mediated, kinds of immunity (e.g. allograft rejections, GVH reactions), in which they are assisted by macrophages. The T cells also have another important function: they somehow help B cells to become active in humoral antibody production. This may come about by T-cell binding to the immunogenic "carrier" part of the antigen, followed by presentation of the "hapten" part to the immunoglobulin receptors on the B cells. The nature of the antigen-recognition receptors that confer specificity on T cells is unknown (though it has not been ruled out that they may include immunoglobulins); we return to this question in the section on gene control of the immune response.

The high correlation in genotypic composition between erythrocytes and lymphocytes (or their antibody products) in individual allophenic mice, despite lower correlation of the latter with some other tissues, is consistent with possible common origin of erythroid and lymphoid cells from a pool of more generalized precursors. This has been shown by comparing strain-specific $H-2$ (137) or hemoglobin types (211) of erythrocytes with serum allotypes of $7S\gamma2a$ immunoglobulins (due to alleles at the $Ig-1$ locus). A high correlation has also been found between genotypes of circulating erythrocytes and white blood cells (with allelic electrophoretic variants of glucosephosphate isomerase); this also supports the possibility of a shared lymphoid and myeloid embryological origin (123). These studies agree on relative independence of coat color and γ-globulin quantitative ratios, as would be expected from developmental unrelatedness of those tissues.

Genotypic composition has been found to be highly correlated in thymus, bone marrow, and spleen of individual allophenic mice with the T6 translocation marker, visible in metaphases, in one component strain (61, 136). However, this similarity is to be expected, in view of the great mobility of lymphoid cells in and out of these organs, and has no special bearing on the question of their origin. Although Nesbitt, using an X-chromosome inactivation marker (155), noted a general correlation among many tissues, including spleen, the correlations of these tissues with thymus were low. Mukherjee & Milet (148) also found among a small number of allophenic mice some animals whose lymphoid organs differed significantly in genotypic composition. It is possible that the lower correlations among lymphoid tissues in both these studies can be ascribed to the use of cell cultures rather than direct preparations from these organs. This is an important difference, inasmuch as it is chiefly the stromal cells that grow in culture, whereas the more direct studies based on samples from the organism reveal chiefly the more relevant specialized (parenchymal) cells. In addition, cells growing in vitro may undergo selective proliferation during the culture period, so that the end results do not even necessarily reflect the in vivo composition of the stromal component in the tissue of origin. These difficulties are not restricted to lymphoid organs but also pertain to some of the other tissues for which attempts have been made (54, 155, 156) to extrapolate, from two in vitro connective-tissue-derived populations, the in vivo ratios of other specialized cells, by application of mathematical models (76), and to ascertain the size of the precursor cell pool from which they were derived.

Gene Control of the Immune Response

The ability of an animal to generate a strong immune response to a specific antigen, both by cellular immunity and antibody production, has been found to be controlled by the so-called immune response genes (11). Synthetic polypeptides of different L-amino acids, and their hapten conjugates, have afforded a way of presenting the immune system with a highly specific challenge of restricted heterogeneity and have contributed substantially to defining immune response genes, especially in guinea pigs and mice. The immunogens most extensively used in mice are branched multi-chain copolymers, of which an example is (T, G)-A--L, a compound with a back-bone of L-lysine with D, L-alanine side chains, L-glutamic acid, and L-tyrosine. Inbred mouse strains differ quantitatively in their capacity to make antibodies in response to these antigens and the differences are due to dominant genes at autosomal loci. Responsiveness to certain (not all) polymers, including (T, G)-A--L, is controlled by the *Ir-1A* locus, situated within the *H-2* region controlling major histocompatibility antigens.

The mechanism of action of immune response genes is still not known. The close association of *Ir-1A* and *H-2* genes has suggested one hypothesis: that the ability to make an immune response might be determined by the degree of immunologic cross-reactivity between the animal's own histocompatibility antigens and the specific administered antigen (11). For example, if a mouse strain of a certain *H-2* antigenic type were a *non*responder to a specific synthetic antigen, this might be due to cross-reactivity of part of that *H-2* antigenic complex with the administered polymer, in other words, to tolerance of that antigenic determinant as "self." An argument against this "cross-tolerance" as a mechanism of nonresponse is that the F_1 hybrid between specific responder and nonresponder strains is a responder, despite the fact that it possesses the *H-2* antigens of both parental strains. Allophenic mice have provided further evidence against the hypothesis, in experiments by McDevitt and collaborators (95, 44). The input strains had different *H-2* antigenic types and were high and low responders, respectively, to (T, G)-A--L; at least some of the allophenics were high responders, while presumably tolerant to both *H-2* antigenic types.

On the other hand, Warner et al (208), in studies involving two other synthetic polymers administered to allophenic mice of a variety of strain combinations, have concluded that cross-tolerance between histocompatibility and administered antigens may in fact be a cause of unresponsiveness in their cases. While it is possible that different mechanisms may explain unresponsiveness to different antigens in different strains, there is another important difference between these two studies that may account for the disparity in results. In the work by McDevitt et al (95, 44), genotypic mosaicism in the immune system was directly tested for in each animal by markers for strain-specific allotypes of total immunoglobulins of the subclass 7Sγ2a (due to alleles at the *Ig-1* locus), and by allotypy tests of specific anti-(T, G)-A--L antibody. In the experiments by Warner et al (208), tests were not made of immune-system genotypes. The authors assumed that the admixture of coat colors quantitatively reflected mosaicism in the immune systems of their two-color allophenics. This assumption is basic to their conclusions. While there is a tendency for many tissues within allophenic individuals to have either extensive representation of both cell strains or a similarly unbalanced representation of both strains, many specific examples of discordance in genotypic composition of tissues—especially of developmentally unrelated ones—within individuals have by now been published, and we have observed many others (129). A reported example is that 20% of 20 two-color allophenics in one study had no detectable 7S γ-globulin allotype of one of the strains (137). The capacity for genotypic variation within animals is in fact one of the experimentally valuable features of allophenic mice (120). The requirement for conducting specific-tissue analyses with unambiguous markers thus cannot be overemphasized and is very much at the heart of the problem in all studies of the immune system in allophenic animals.

Another hypothesis on the mechanism of action of immune response genes is based on observations that responsiveness is a cellular function (i.e. it is passively transferred with immunocompetent cells) and is lost after thymectomy (11). It has therefore been proposed that this is a T-cell function. Possibly the gene products may themselves be the postulated antigen receptors on T cells, or may be otherwise concerned with T-cell receptor structure or function, so that the genetic lesion in nonresponders or low responders would lead to deficient cognitive capacity of T cells and a deficiency in their ability to stimulate B cells to produce antibody against that specific antigen. The possibility that some T-cell helper function is at issue has been investigated in allophenic mice by McDevitt et al in the interesting experiments already alluded to (44). They administered (T,G)-A--L to mice of an $Ir\text{-}1A^{hi}$ (high responder)$\leftrightarrow Ir\text{-}1A^{lo}$ (low responder) strain combination, with the expectation that if the relevant $Ir\text{-}1A$ gene were expressed solely in T cells, some of the allophenics should show a specific anti-(T,G)-A--L antibody response from *low*-responder-strain B cells (evident from allotype of the specific antibody produced) and such individuals should be detectable by their greater production of this specific antibody than is manifest in the low-responder controls. Some allophenic individuals did in fact yield results consistent with this expectation and these animals were presumed to have high-responder-strain T cells, though the genotypes of their helper T cells have not yet been tested, and there may be other unidentified factors influencing the outcome, inasmuch as the mechanisms of collaboration between T and B cells are still largely unknown.

Another possible complication has been raised by Katz et al (77), who have reported that histoincompatible T and B cells cannot cooperate to produce antibody in response to hapten-protein conjugates in irradiated pure-strain hosts with host-type T cells and allogeneic donor-type B cells of specifically primed response capacities. But they found that parental T and B cells could cooperate if transferred to F_1 irradiated hosts. They hypothesized that there are genetic restrictions for cooperation, possibly through a need for T and B cells to have common gene products of the *H-2* region. Since McDevitt's allophenics comprised cell strains with different *H-2* alleles, Katz et al have suggested that the antibody of low-responder allotype in these allophenics may have arisen through an "allogeneic effect" rather than through true physiologic cooperation of T and B cells. However, in more recent allophenic experiments (10), congenic strains were involved, one with the $H\text{-}2^k$ allele and the other an F_1 hybrid between $H\text{-}2^k$ and $H\text{-}2^b$; results were comparable to the earlier ones with more diverse inbred strain pairs. The continuing debate underscores the need to resolve definitively the tolerance status of allophenic mice, irrespective of the situation in irradiated graft-hosts. As discussed in the section on self-tolerance, the weight of the evidence supports the view that allophenics are truly tolerant and that their coexisting histoincompatible lymphocytes are not self-reactive but are competent against foreign stimuli. If this is in fact the case, these cells are not obligatorily antigenic and would not cause an allogeneic effect in the allophenic experiments dealing with high vs low immune responsiveness.

Genetic Diseases of the Immune System

Allophenic mice are being used to probe the etiology of several genetically influenced diseases of the immune system. These studies are surveyed only briefly, as they are still in preliminary stages.

The NZB inbred mouse strain is of special interest because virtually all its members succumb to spontaneous age-related autoimmune disease characterized by Coombs positivity, antinuclear antibodies, splenomegaly and generalized lymphoproliferation, hemolytic anemia, and severe immune complex glomerulonephritis; some also develop malignant lymphomas. The disease partly resembles human systemic lupus erythematosus. NZB mice carry the Gross leukemia virus and gradually lose tolerance to the viral antigens, although it is not clear what role this loss plays. Barnes et al (8, 5) have studied 3 demonstrably mosaic allophenics of the NZB↔CFW (random-bred) strain combination; Mintz and collaborators (144) have observed 15 allophenics of the NZB↔C57BL/6 inbred strain combination. In both studies, there is a lessened incidence and a later onset of autoimmunity in the allophenics than in NZB controls. It has not yet been determined whether the amelioration involves a dilution effect or an active inhibition of the disease. Barnes has interpreted the lesions in his allophenics as indicating both NZB autoimmune disease and allogeneic or GVH disease, the latter presumed to result from loss of tolerance to allogeneic cells. However, we have found that the lesions in the allophenics are indistinguishable from those in autoimmune NZB controls and that there is no special evidence for GVH disease. In addition, all of our animals have kept their skin grafts and other cells of both strains, and have thus remained tolerant of both, even after the NZB autoimmune disease had developed. Barnes did not test-graft his three allophenics but has presented a complex model, involving GVH activity, of active suppression of the disease. Our studies are still in progress and will rely on a combination of genotypic and clinical tests to rule tested components in or out of the chain of events leading to the disease.

Diseases of the immune system include malignancies. We discuss, in the section on tumorigenesis, work being done on viral thymic-derived lymphomas and leukemia in allophenic mice with some cells of the hereditarily susceptible AKR strain.

The study of plasma cell tumors (myelomas) has also been undertaken in BALB/c↔C57BL/6 allophenic mice (132), partly to attempt to obtain C57BL/6-type myeloma proteins. These tumors rarely occur spontaneously; they are readily inducible by intraperitoneal injections of certain mineral oils in the BALB/c strain (but not in C57BL/6 or most other strains) and often secrete strain-type immunoglobulin-like myeloma proteins. Plasmacytomas have been successfully induced in the BALB/c↔C57BL/6 mice, though with a lower frequency than in BALB/c controls, and the tumors and their myeloma proteins have thus far resembled those in BALB/c controls. Some histological indication of formation of a possible plasmacytoma occurred in tissue of apparent C57BL/6 genotype, judging from its ability to grow in C57BL/6 and not in BALB/c hosts, but the transplants subsequently gave rise to other tumors, including histiocytomas, rather than to plasmacytomas. It will be interesting to learn whether this is an actual transition, perhaps made possible by common derivation from primitive reticuloendothelial cells.

Allophenic mice also offer interesting future possibilities for study of the causes and nature of immune-deficiency states due to genetic lesions. An example is the *nude* (*nu/nu*) or athymic mouse. This abnormality need not necessarily be due to a total absence or defect of all thymus-forming components; perhaps only one component is deficient embryologically, in a system requiring cell interactions for development as well as for later function. We would expect that if stem cells capable of becoming T cells were present, an allophenic combination of *nu/nu* with +/+ (normal) cells might lead to thymic morphogenesis and to T cell formation to which the *nu/nu* cells could contribute. One allophenic *nu/nu*↔+/+ individual has recently been produced (91) and offers preliminary support for this view.

9   GENE ACTION IN MOLAR DEVELOPMENT

Teeth permanently record their morphogenesis. As the molars are morphologically well differentiated in the mouse, genetic mosaicism offers ways of tracing the developmental history of these structures and of learning how genes affect them. They have complex ectodermal and mesodermal origins and suspected interactions, and first appear as thickenings of the oral ectoderm in the presumptive molar region around day 11 of embryonic life, although tooth-forming properties seem present in tissue taken from that region a day earlier. Differentiation results in a characteristic pattern of cusps, separated by fissures and ridges, for the crown of each molar. There are numerous morphological differences associated with inbred strains and with specific genes.

Effects of the sex-linked gene *tabby* (*Ta*) are of special interest in relation to the hypothesis of single-allele activity in heterozygotes (86). *Ta/Ta* females and *Ta/–* males have reduced first and second molars, and third molars often absent, in each quadrant; *Ta/+* females have unusually variable morphological characteristics, including some normal molars, some frankly mutant ones, some mixed, and an occasional supernumerary one. Grüneberg has based much of his claim of nonvalidity of single-gene activity of X-linked genes on his analysis of this dental syndrome (65); his proposed "complemental-X" alternative to two cellular phenotypes is intracellular partial expression of both alleles and determination of final phenotype by "threshold mechanisms." In a subsequent collaborative study (67) of the molars of *Ta/Ta* (or *Ta/–*)   +/+ allophenic mice, this position is somewhat modified. With the mixtures of mutant and normal cells, molar phenotypes similar to those in *Ta/+* females were obtained. The greater overall preponderance of *tabby*-like attributes, albeit in only a small number of allophenics, as compared with *Ta/+*, is presented as possible evidence either that the heterozygote does not involve two clonal classes or that the clonal histories in allophenics and heterozygotes are somehow different, i.e. there may be a smaller patch size and greater cell interaction in the heterozygotes. This suggestion does not take into account the fact that the genetic backgrounds differed greatly in the two allophenic cellular components under consideration but not in the heterozygous cells. It is therefore unlikely that selective forces mediated by other loci would be comparable in the two situations. The same objections apply to interpretations of pattern differences between other tissues of $Ta/Ta \longleftrightarrow +/+$ and $Ta/+$ by McLaren et al (100) or of eye pigmentary patterns in other allophenics and X-linked heterozygotes by Deol & Whitten (31, 32).

In X-linked heterozygotes, mutant-phenotype clones would be expected to be initially as frequent as wild-type ones; while they might then show "dominance," "codominance," or "recessiveness" and modify the proportions of the two, the residual cellular genotypes are the same, and a narrow spectrum of phenotypes should and does result. In allophenics, on the other hand, the initial cellular proportions may differ widely, irrespective of whether different or coisogenic strain pairs are involved. Selection at the marker locus and/or other loci would then augment the phenotypic variability to different extents in different genetic situations, but in any case, greater variability would be expected than in any single-genotype situation. Only one or another narrow part of the total allophenic spectrum can justifiably be compared with any specific single-genotype case (119, 124). Indeed, there is as yet no unequivocal basis for suspecting that allophenics and X-linked heterozygotes have different archetypal or underlying clonal histories, though they may indeed have different and consistent developmental modifications and degrees of "developmental noise," according to the genes involved.

In the preceding study of allophenic teeth (67), old animals were used and the normal erosion of the crowns with age may have limited the level of resolution of the analysis. In another study, aimed at investigating in greater detail the teeth of a large number of young allophenics, without having to sacrifice the animals, we have developed a miniaturized technique, emulating human dental practice, for making dental impressions from living mice (105). Positive models were also cast from some of the impressions. In a test series, morphological details of the molars corresponded closely to those of the teeth themselves. Individual cusps (and other features) were scored in C57BL/6$\longleftrightarrow$C3Hf and C57BL/6$\longleftrightarrow$BALB/c allophenics and were often found to be indistinguishable from one or the other pure strain, while some cusps were intermediate to varying degrees between the donor strains, some differed from either strain, and some cusps were abnormally large or small. Each cusp, in at least some individuals, varied in strain phenotype independently of its immediate neighbors. The simplest interpretations of these results are that each molar was derived from a small number of specific precursor cells, i.e. prior to the stage of relatively large numbers of cells first seen on day 11, and that an individual cusp includes at least two clones or terminal parts of clones. Unusual cusp appearance or size could reflect inherent asynchrony in time of appearance or rate of development of two clonal strains in interacting components. The range and nature of the variations in these cases, where *tabby* or other abnormality-causing genes are not involved at all, demonstrates that the intracusp clonal units of tooth development are sufficiently small as to account for pronounced morphological deviations and abnormalities in simple clonal terms. This model not only sustains the clonal interpretation of tooth phenotypes in Ta/+ but might be applicable to soft tissues and to many other genotypes.

10   LIVER MORPHOGENESIS

Genotypic mosaicism has been identified biochemically in liver homogenates of allophenic mice with strain-specific allelic electrophoretic variants of isocitrate dehydrogenase (130) and malate dehydrogenase (2, 120), and with high- vs low-activity variants of β-glucuronidase (210). Entire lobes of a given liver sometimes differ in their genotypic composition, and variations found in still smaller samples have suggested that single-genotype clonal units may exist within lobes.

The focal point of interest in liver morphogenesis is the specialized parenchymal population of hepatocytes, derived from endoderm. However, the liver is an especially complex tissue comprising a number of kinds of cells in addition to hepatocytes. All of the preceding methods for biochemical analyses of mosaicism in homogenates thus are complicated to unknown degrees by the admixture of cell types of unidentified specific contributions and also by possible regional variations in activity of the hepatocytes themselves. Examples have been observed histochemically (see below) in the case of β-glucuronidase: hepatocytes of strains with high activity of this enzyme have relatively lower specific activity in central than in periportal areas of lobules; in additon, the reticuloendothelial cell population within the liver includes cells with paradoxically low, or high, activity in the high and low activity strains, respectively. It is therefore questionable whether biochemical analyses of overall enzyme activity in homogenates furnish an accurate quantitative basis for calculating the size of the primordial precursor pool of liver parenchyma cells, although this has been attempted (210).

These difficulties do not exist in histochemical methods, which permit direct in situ visualization of strain differences in particular kinds of specialized cells. A histochemical method for staining of β-glucuronidase activity has been applied to tissue sections of allophenic liver (25). Inbred strains (e.g. C3H) homozygous for the g allele at the controlling locus have higher sensitivity of the enzyme to denaturing agents, so that all hepatocytes appear essentially unstained while, under the same conditions, all hepatocytes of wild-type control strains (e.g. C57BL/6 or BALB/c) have a diffuse orange-red stain and/or intense red granules. Thus, the spatial distribution of the respective cell strains can be seen in genotypically mosaic livers. [This method is of course applicable only in allophenic studies involving those specialized cell types, of which hepatocytes and mammary gland epithelium (126) are among the few examples, with sufficiently high specific activity of the enzyme in high-activity strains so as to be unequivocally visualizable in *all* cells of that type in the high-activity controls.] Observations of genotypically mosaic adult tissues of course reveal only those morphogenetic changes that are conserved in the terminal-stage record. The embryological stages of liver formation are not amenable to study with this technique because full β-glucuronidase activity matures only postnatally, after morphogenesis is complete (26).

The histochemical preparations of allophenic livers showed great variability and complexity of distribution of the two cell strains (25, 26), as one would expect from the complex histogenesis of this organ, and three-dimensional reconstructions will be required for a more complete picture. Nevertheless, some indications of cell lineages have already emerged. In areas with maximal representation of both geno-types (i.e. with maximum chance that adjacent clones may be of different strains and will therefore be detectable in one or another location), the two hepatocyte strains were not uniformly or finely interspersed but were arranged in relatively small patches of irregular shape. A small patch of a given genotype was smaller than a lobule and was sometimes clearly aligned along the radius of a lobule, i.e. as a hepatic cord. The hepatic lobules, which are the architectural units of the liver, therefore cannot be considered to be the developmental units or cell lineages. The cells of a given hepatic cord, on the other hand, may in fact comprise a developmen-tal unit and these small groupings may represent clones. Alternatively, the cords may be subclones, or "fanned-out" termini of a much smaller number of initial cell lineages (at least two, because of mosaicism) that gave rise to the liver at the time the hepatic diverticulum budded out from the endodermal lining of the gut; isolation and selection may have caused subsequent disparities among lobes. These results are interesting to consider in light of Bakemeir's suggestion (1), based on multiplicity and nonuniform distribution of steroid reductases in rat liver, that liver hepatocytes might be multiclonal in origin, with the various clones possessing some functional specialization. In the more general hypothesis of *phenoclones* (116, 119, 123, 124), I have suggested that the clones comprising *any* specialized cell type may perhaps be differentiated from each other in minor but significant ways, and that such phenotypic subdifferentiations at the clonal level may account for some of the heterogeneity and functional subspecializations in complex tissues such as liver and brain.

The foregoing biochemical and histochemical procedures for analysis of mosa-icism are all restricted to biopsy or autopsy specimens of liver. Another useful approach has been introduced (25, 3) for the purpose of determining genotypic ratios in the livers of *living* allophenic mice, without any surgical intervention. This is based on the fact that a major protein fraction normally excreted in mouse urine is manufactured in the liver and occurs in electrophoretically distinguishable allelic forms (due to the *Mup-1* locus). Our results with the *Mup-1* marker in urine samples taken just before autopsy are in good agreement with direct genotypic analyses of the same liver at autopsy. Therefore, *Mup-1* presumably reflects paren-chyma composition and may be used as a simple and indirect means of diagnosing liver cell genotypic ratios throughout life in allophenic mice. We have also used it before and after partial hepatectomy to learn whether one genotype tends to out-grow the other during liver regeneration and during formation of liver tumors (25, 3).

11 TUMORIGENESIS

Tumorigenesis, no less than the genesis of normal tissue diversity, must entail many functional genetic changes, irrespective of whether structural genetic change may also be involved. Numerous genetic and other biological questions relating to differential divergence of gene function from "normal" to "abnormal" in certain cells will be answerable only from in vivo observations. Constitutional mosaics provide many possibilities, some of them unique, for pursuit of these questions in vivo. Examples are presented and discussed below.

Single-Cell vs Multiple-Cell Tumor Origin

Does a tumor arise as a clone from a single aberrant cell or is it multicellular in origin? An ancillary question is: Are metastases or recurrences clonal? Evidence from mosaics has disclosed that some kinds of tumors seem in fact to be of single-cell and others of multiple-cell origin. The best known example of a human malignancy of probable clonal origin is chronic myelogenous leukemia. This conclusion was based on chromosomal mosaicism, typified by the prevalence of cells with the minute "Philadelphia" chromosome in bone marrow and peripheral blood of persons with the disease (161). However, unique marker chromosomes have not yet proven to be consistently characteristic of any other human or animal tumors, despite extensive searches.

Many kinds of tumors in mice occur spontaneously, or are inducible, in specific inbred strains; some strains have actually been developed through selective breeding for high incidence of a particular kind of tumor. Allophenic mice derived from two different strains, each independently capable of forming a certain tumor, offer a simple test of tumor cell lineage. If only one or the other strain is invariably found in the tumor, while normal parts of that organ may have both strains present in relatively small-size sample areas, such a tumor is likely to have originated from only one cell. On the other hand, if both genotypic strains are commonly present in single tumors, there is a presumption of multicellular origin, provided that "contamination" of the tumor with normal cells can be ruled out. The genotypic composition of tumors in allophenic mice has been determined in the following ways: 1. biochemically, by means of strain differences in proteins produced in many tissues (e.g. allelic electrophoretic forms of various isoenzymes) or in a specific tissue (e.g. myeloma immunoglobulins) (120, 132); 2. karyologically, with the T6/T6$\leftrightarrow$+/+ translocation chromosome marker or with fortuitous X/X$\leftrightarrow$X/Y sex chromosome mosaicism (143); 3. by grafting pieces of the tumor to parental-strain recipients for acceptance-or-rejection tests of antigenic strain-specific histocompatibility differences in the grafts (143); or 4. histochemically, by in situ visualization of high vs low activity of the enzyme β-glucuronidase in strains of the +/+$\leftrightarrow$g/g genotypic combination. The last method is of course limited to those tissues having quite high levels of enzyme activity in the wild-type strain; included are liver (25) and mammary gland (126). Histochemical localization of cell strains in allophenics is an invaluable tool for revealing the progress of normal histogenesis as well as of tumorigenesis. In the latter, pathology, architecture, and genotypes of the tissue are simultaneously apparent, thus making it possible to rule out spurious diagnoses of multicellular tumor origin when normal and tumor cells are admixed, and to discern whether the genotype of normal tissue adjacent to the tumor has any influence on tumor growth.

Results of experiments in which many tumors from allophenic mice have been genetically analyzed have usually shown single-strain tumor composition and are consistent with the interpretation that a given mammary tumor or lung tumor arises as a clone (120, 132). In the case of liver tumors, the large majority were found to have only one genotype, but a few with two genotypes were found and both their cell strains were histochemically diagnosed as tumorous (25). It should be added that one of the two input control strains is characterized by lower tumor incidence than the other in the experiments under discussion. Therefore, the mosaic liver tumors seem especially significant and suggest that even a liver tumor of uniform strain composition, whether in an allophenic animal or a pure-strain control, may be a genetically complex entity containing different clones of transformed cells.

The mammary tumor studies are of special interest because they have revealed that coexistence of normal and transformed cells (in the allophenic model, of different genotypes) during early stages of tumorigenesis may be critical for sustaining tumor progression (143). Mammary tumors originate as premalignant hyperplastic nodules or as outgrowths resembling normal mammary gland of late pregnancy but differing from the latter in their irreversible commitment to malignancy. In allophenic females from a combination of the high mammary tumor C3H and low mammary tumor C57BL/6 strains, premalignant nodules morphologically indistinguishable from those in C3H were found. However, when single nodules were split and the halves grafted to pure-strain hosts, whose alleles differ at the strong *H-2* locus, the selective acceptance of some cells and rejection of others in grafts in each host strain demonstrated that both high- and low-tumor cell strains had coexisted in the nodules. After recovery from the semirejection, the genotypically "purified" grafts in hosts of the corresponding strain then grew out in autonomous, strain-specific growth patterns: C3H cells grew as hyperplastic and eventually formed tumors; C57BL/6 cells now formed a completely normal mammary tree. Both strains survived in F_1 hybrid hosts until tumor progression became advanced, when the normal (C57) cells disappeared. The overtly hyperplastic morphology of the original genetically mosaic nodules had therefore obscured the presence of appreciable numbers of normal cells. The hypothesis was proposed that premalignant nodules in *single*-genotype animals might also be a phenotypic mosaic of hyperplastic and normal cells, and that this association might in fact be necessary for survival of some premalignant components until the stage of frank malignancy is attained. This hypothesis has since been confirmed in experiments involving experimental reassociations of premalignant and normal mammary transplant lines (182).

Functional, rather than constitutional, genetic mosaicism has been for some time extensively used to resolve the origins of tumors by utilizing as subjects women heterozygous for electrophoretic variants of X-linked glucose-6-phosphate dehydrogenase (G-6-PD). The stability of single-allele enzyme activity per cell permits an experimental approach analogous to the one described in allophenic mice. These studies have been especially fruitful in the hands of Gartler and of Fialkow and their collaborators, who have examined many kinds of tumors. Those of apparent single-cell origin have included leiomyomas (84). "Hereditary" trichoepitheliomas (55) and neurofibromas (40) have been among tumors of multicellular origin and may indicate universality of cell susceptibility in such genetic circumstances. Analyses of multiple tumors in individuals with Burkitt's lymphoma led to the hypothesis that this disease arose at one focus and metastasized to other parts of the body (39). Comparison of initial tumors and recurrences after therapy indicated that most early recurrences were regrowths of original clones while some late ones were due to new malignant clones.

Suitable X-linked markers are not available for comparable analyses of tumor cellular origin in mice. For study of this particular question, the human G-6-PD heterozygotes have the advantage over genotypically mosaic mice of different strain pairs that the two phenotypically distinguishable human cell populations have the same genomes and equal susceptibility. The human X-linked marker of course has the restriction that it does not permit tumors of males to be studied. Nor does it lend itself to study of most of the other questions enumerated below.

Cell Fusion in Tumorigenesis?

Does cell fusion play any appreciable role in tumor origin or progression? Recent experiments with transplantable tumors have yielded examples of fusion between some tumor cells and host cells. Isozymic, antigenic, or chromosomal differences between donor and host cells were used to document cell hybridization after a human lymphoma was grafted to the immunologically privileged cheek pouch of the hamster (60) and after mouse L-cell and other tumor lines were grafted to irradiated newborn mice (214). Selective in vitro growth conditions were used to increase the frequency of the cell hybrids over their very low spontaneous in vivo occurrence. These results have resurrected a speculation that cell fusion might ordinarily be occurring in tumorigenesis in vivo and might establish a basis for increased genetic variation, upon which selection could act, in a growing tumor.

Thus far, examination of a large number and variety of spontaneous and induced tumors formed in situ in allophenic mice has failed to disclose any evidence for cell fusion in vivo (120, 132). If any cell fusion had occurred, it must have been rare and was not selectively perpetuated under these circumstances. The questions whether the cell fusion found in certain transplantable tumors is a biological curiosity of the transplantation situation, and whether it is significant at the level found in vivo, therefore remain open for further investigation.

Genetic Susceptibility to Tumors

High incidence of specific tumors characterizes diverse inbred strains of mice and offers convincing proof of a genetic basis for susceptibility to malignancy in this species. For certain tumors, attribution of susceptibility control to particular loci is becoming possible through genetic studies (173), but specific genes have not yet been recognized in the etiology of most murine tumors.

Allophenic mice composed of cells from susceptible and relatively nonsusceptible strains have provided a means of determining the focal tissues in which "tumor susceptibility" genes are differentially expressed, whether or not the relevant loci have been identified (120, 132, 25, 143). The experiments exploit the fortuitous existence of variability in genotypic composition of tissues among allophenic individuals of a given strain-pair combination. The animals themselves thus conduct the experiment: they present us with permutations and combinations of tissue genotypes, from which we may become apprised of which tissue(s) *must* be of the susceptible genotype for a given kind of tumor to develop. They also may inform us whether susceptibility is adoptively conferred on cells of the nontumor strain in the prospectively tumorous organ, or conversely, whether tumor-strain cells in that organ are substantially diverted from malignancy by close association with cells of the low-tumor strain.

Populations of allophenic mice have been constructed with tumor-susceptible cells from strains with high incidence of virally caused mammary tumors in females (C3H or C3Hf inbred strains), of hepatomas in males (C3H or C3Hf), lung tumors (BALB/c), mineral-oil-induced plasma cell tumors (BALB/c), or leukemia (AKR). Various low-tumor-strain companion cells were used. The animals were kept alive until tumors or other diseases occurred. The tumors themselves, and many other tissues from each animal, were genetically analyzed with the sorts of markers summarized in the first part of this section; in addition, serum allotypes, histocompatibility-antigen markers, allelic hemoglobin variants, and other differentials were used. Genetic susceptibility to formation of these tumors has proved to be due chiefly to gene expression in the potentially tumorous tissue; i.e. tumors formed if susceptible-strain cells were present in that tissue, regardless of variations in genotypic composition of the other tissues analyzed (120, 132, 25).

A matter for further investigation in viral tumors is the mode of action of genes differentially expressed in cells of the target tissue, perhaps by influencing adsorption or replication of oncogenic viruses in those cells. Allophenic mice may contribute to these studies through experiments on horizontal transmission of such viruses between cells of different genotypes.

The overwhelming majority of all tumors were composed only of cells of the more susceptible genotype. Some tumors of the low-susceptibility strain were found; they also occur to some extent in controls of the latter strains. The notable observation of two genotypically mosaic hepatomas has been discussed in the first part of this section. For liver, mammary, and lung tumors, the age of onset paralleled that of controls (132). Thus, the two cell strains displayed remarkably autonomous behavior despite their long and intimate association in the allophenics and their exposure to common systemic influences. Direct evidence of cell-localized genotypic susceptibility was also seen histochemically in the formation of susceptible-strain hepatomas amidst liver tissue of both genotypes (25) and with histochemical (126) and antigenic (143) demonstrations of two genotypes in premalignant mammary tissue from which susceptible-strain tumors were usually formed (132, 143).

It has been pointed out in the preceding studies that the coexistence of cells of different histocompatibility types was also without influence on tumorigenesis. The in vivo data from allophenic mice are therefore contrary to the expectation from the hypothesis of "allogeneic inhibition" (70), which was proposed as a surveillance mechanism against neoplasia and which predicted that neoplastic cells with variant *H-2* antigens would be destroyed upon contact with other cells, outside of the immune system.

Allophenic mice may shed new light on tumors of the immune system, which may feature cell interactions even if there were cell-localized expression of susceptibility genes. In preliminary observations on a few mice partly derived from the AKR strain, known for its viral leukemia of thymic origin, it was noted that AKR lymphocytes predominated and might enjoy a proliferative advantage, perhaps even before leukemia set in (132), a possibility strengthened by subsequent data from a larger series (205). The animals nevertheless showed a decreased incidence and later onset of leukemia than did AKR controls (6, 7, 205). Whether this is due to dilution of AKR cells below a critical mass or to active inhibition of AKR leukemia by the normal "tumor recognition and control processes" of the other strain, as some authors (7) believe, remains to be clarified by further experiments, especially in light of the fact that simple reduction in thymic mass by subtotal thymectomy is known to delay and reduce incidence of the disease in AKR mice (180).

Tumors Associated with Mosaicism

On the whole, tumors found in allophenic mice have resembled those of the parental strains and have been comparably or less frequent. A notable exception is gonadal tumors, including teratomas. Spontaneous ovarian or testicular teratomas, which are exceedingly rare in almost all strains of mice, are significantly more frequent in allophenic mice, and a number of the affected animals have been identified as sex chromosome mosaics (131). It is therefore possible that the coexistence of gonadal cells of both sexes may be a contributing cause of in situ parthenogenesis leading to teratoma formation from germ cells. The fact that both male and female cells have been identified in some individual human teratomas (152) is consistent with this view.

12   TISSUE-SPECIFIC GROWTH CONTROL

Known differences in cell-type-specific histories of cell growth and proliferation suggest that they are controlled by locally expressed loci. Such "growth" genes are largely unrecognized, as they do not have easily recognizable products or endpoints, though they may well play a major role in cell developmental diversification and in aging. New clues regarding these genes in vivo are obtainable in allophenic mice by seeking evidence of cell selection.

Development and Cell Selection

When two inbred strains, e.g. C3H and C57BL/6 (without specific lethal genes), were combined and the tissues genetically analyzed in a large number of allophenics, a tissue-specific profile for the entire population was obtained, despite considerable individual variation (120). Thus, the liver of this "statistical allophenic mouse" was primarily of the C3H strain, the erythrocytes and γ-globulin-producing cells were weighted toward C57, the germ cells favored C57 slightly in females and C3H heavily in males, etc. This has been taken to mean that different batteries of loci controlling growth and proliferation are active in each tissue, and allelic differences at the relevant loci in these particular strains militate selection differentially in favor of one or the other in each tissue. If each strain is separately combined with another, the selective "dominance" trends may change (129), as would be expected.

A change in selective advantage from one to the other strain in the course of development in vivo appears to indicate some functional genetic change, e.g. expression of some new controlling loci or inactivation of others. An example has been cited in the case of the vertebral column, where the C57BL/6 type is more frequent in anterior, and the C3H type in posterior, regions (145). This distribution probably reflects some changes at growth-controlling loci expressed in sclerotome cells in the early forming vs the later-forming (more posterior) somites.

Aging

Apart from malignancies and other age-related diseases, "normal" aging is clearly influenced by the genotype. It is likely to implicate some tissues more directly than others, though the sites of primary decline in a particular genotype are not always evident because physiological changes in them may soon entrain deleterious consequences in other tissues. Thus, an understanding of aging in terms of differential or tissue-specific primary genetic change has obvious parallels with an analysis of development in the same terms, irrespective of whether structural, in addition to functional, genetic change is involved in the former. Relatively short-lived and long-lived strains may be combined in allophenic mice. In tissues still capable of proliferation, the pivotal "prematurely" aging tissues may be identified in each of the former strains by their selective and precocious replacement by cells of the latter type.

Local differences in selection, indicative of tissue-specific aging differences within a genotype, are seen even in relatively long-lived strains, e.g. in the C3H *agouti* *(A/A)* and C57BL/6 *nonagouti (a/a)* strains. When these are combined in allophenic mice, there is a progressive shift in the hair follicle dermal component toward *nonagouti* (135), while the reproductive cells (in males) shift toward C3H (114). Even single-locus differences may mediate selection with age within a tissue, as in a trend toward *black (B/B)* and away from *brown (b/b)* in the coat melanoblast population of a coisogenic (C57BL/6 background) strain pair (128).

Lethals

Some lethal genes may affect basic processes shared by all cells. Lethals of this kind could only be "rescued" in allophenic combination if a diffusible product were involved and could be supplied by accompanying normal cells. The t^{12}/t^{12} genotype, fatal in the morula stage of mouse embryos, apparently affects all the cells but does not involve such a product: In combination with $+/+$ cells, composite blastocysts were formed, but the $+/+$ cells seemed to account for the limited developmental progress, and the lethal cells continued their decline (110).

Other lethal genes may be primarily expressed in only one or a few tissues with important physiological roles and such tissues may be "replaced" in allophenics. Here, it is necessary to determine directly the strain(s) comprising the tissue in question before the mode of rescue (cell autonomy vs intercellular effects) can be ascertained. The *W/W* genotype, with a lethal macrocytic anemia, has been "rescued" in allophenics; they contained only normal-genotype blood cells but had both cell strains in other tissues (119). Similarly, the sometimes inviable *mk/mk* strain, with a microcytic anemia, has been consistently viable in allophenics (131). The X-linked lethal neurological mutation *jimpy (jp)*, which is lethal in hemizygous males, persists in tissues such as germ cells of allophenics, and matings involving *jp* sperm from such males have enabled testing for allelism with another X-linked gene also causing myelin deficiency (36). Another question presently being investigated in a number of laboratories is whether an inviable haploid (parthenogenetic) mouse embryo can be rescued in an allophenic combination with a normal diploid one.

ACKNOWLEDGMENT

The author's studies were supported by USPHS grants HD-01646, CA-06927, and RR-05539, and by an appropriation from the Commonwealth of Pennsylvania.

Literature Cited

1. Bakemeir, R. F. 1961. A possible cellular explanation of the multiplicity of steroid reductases. *Cold Spring Harbor Symp. Quant. Biol.* 26:379–87
2. Baker, W. W., Mintz, B. 1969. Subunit structure and gene control of mouse NADP-malate dehydrogenase. *Biochem. Genet.* 2:351–60
3. Baker, W. W., Mintz, B. In preparation
4. Barnes, R. D., Tuffrey, M. 1971. Maternal cells in the newborn. In *Schering Symp. Intrinsic Extrinsic Factors Early Mammalian Develop. Advan. Biosci.,* ed. G. Raspé. Oxford: Pergamon. 6:457–73
5. Barnes, R. D., Tuffrey, M. 1973. Allotype-specific anti-autoantibody activity in tetraparental NZB mice. *Eur. J. Immunol.* 3:60–61
6. Barnes, R. D., Tuffrey, M., Ford, C. E. 1973. Suppression of lymphoma development in tetraparental AKR mouse chimaeras derived from ovum fusion. *Nature New Biol.* 244:282–84
7. Barnes, R. D., Tuffrey, M., Kingman, J. 1972. The delay of leukaemia in tetraparental ovum fusion-derived AKR chimaeras. *Clin. Exp. Immunol.* 12: 541–45
8. Barnes, R. D., Tuffrey, M., Kingman, J., Thornton, C., Turner, M. W. 1972. The disease of the NZB mouse. I: Examination of ovum fusion derived tetraparental NZB:CFW chimaeras. *Clin. Exp. Immunol.* 11:605–28
9. Beatty, R. A. 1957. *Parthenogenesis and Polyploidy in Mammalian Develop-*ment. Cambridge: Univ. Press 132 pp.
10. Bechtol, K. B., Herzenberg, L. A., McDevitt, H. O. 1974. Cellular expression of *Ir-1A* in tetraparental mice with a hemizygous *H-2* difference. *Fed. Proc.* 33:773
11. Benacerraf, B., McDevitt, H. O. 1972. Histocompatibility-linked immune response genes. *Science* 175:273–79
12. Benirschke, K. 1970. Spontaneous chimerism in mammals: A critical review. *Curr. Top. Pathol.* 51:1–51
13. Benirschke, K., Brownhill, L. E. 1962. Further observations on marrow chimerism in marmosets. *Cytogenetics* 1: 245–57
14. Beverley, P. C. L., Brent, L., Brooks, C., Medawar, P. B., Simpson, E. 1973. *In vitro* reactivity of lymphoid cells from tolerant mice. *Transplant. Proc.* 5:679–84
15. Billingham, R. E., Brent, L., Medawar, P. B. 1953. 'Actively acquired tolerance' of foreign cells. *Nature* 172:603–6
16. Billingham, R. E., Brent, L., Medawar, P. B. 1956. Quantitative studies on tissue transplantation immunity. III. Actively acquired tolerance. *Phil. Trans. Roy. Soc. London, Ser. B* 239:357–414
17. Boyse, E. A., Lance, E. M., Carswell, E. A., Cooper, S., Old, L. J. 1970. Rejection of skin allografts by radiation chimaeras: Selective gene action in the specification of cell surface structure. *Nature* 227:901–3
18. Brent, L., Brooks, C., Lubling, N., Thomas, A. V. 1972. Attempts to dem-

onstrate an *in vivo* role for serum blocking factors in tolerant mice. *Transplantation* 14:382–87

19. Buehr, M., McLaren, A. 1974. Size regulation in chimaeric mouse embryos. *J. Embryol. Exp. Morphol.* 31:229–34

20. Burnet, F. M. 1959. *The Clonal Selection Theory of Acquired Immunity.* Nashville, Tenn.: Vanderbilt Univ. Press. 209 pp.

21. Cattanach, B. M., Pollard, C. E., Hawkes, S. G. 1971. Sex-reversed mice: XX and XO males. *Cytogenetics* 10: 318–37

22. Cattanach, B. M., Wolfe, H. G., Lyon, M. F. 1972. A comparative study of the coats of chimaeric mice and those of heterozygotes for X-linked genes. *Genet. Res.* 19:213–28

23. Chapman, V. M., Ansell, J. D., McLaren, A. 1972. Trophoblast giant cell differentiation in the mouse: Expression of glucose phosphate isomerase (GPI-1) electrophoretic variants in transferred and chimeric embryos. *Develop. Biol.* 29:48–54

24. Church, R. B. 1973. Differential gene activity in the pre- and post-implantation mammalian embryo. In *Curr. Top. Develop. Biol.,* ed. A. Moscona, A. Monroy, 8:179–202. New York: Academic

25. Condamine, H., Custer, R. P., Mintz, B. 1971. Pure-strain and genetically mosaic liver tumors histochemically identified with the β-glucuronidase marker in allophenic mice. *Proc. Nat. Acad. Sci. USA* 68:2032–36

26. Condamine, H., Mintz, B. In preparation

27. Corey, M. J., Miller, J. R., MacLean, J. R., Chown, B. 1967. A case of XX/XY mosaicism. *Am. J. Hum. Genet.* 19:378–87

28. Dalcq, A. M. 1957. *Introduction to General Embryology.* London: Oxford Univ. Press. 177 pp.

29. Davis, D. G., Shaw, M. W. 1964. An unusual human mosaic for skin pigmentation. *N. Engl. J. Med.* 270:1384–89

30. Deol, M. S. 1973. The role of the tissue environment in the expression of spotting genes in the mouse. *J. Embryol. Exp. Morphol.* 30:483–89

31. Deol, M. S., Whitten, W. K. 1972. Time of X-chromosome inactivation in retinal melanocytes of the mouse. *Nature New Biol.* 238:159–60

32. Deol, M. S., Whitten, W. K. 1972. X-chromosome inactivation: Does it occur at the same time in all cells of the embryo? *Nature New Biol.* 240:277–79

33. Dunn, G. R. 1972. Expression of a sex-linked gene in standard and fusion-chimeric mice. *J. Exp. Zool.* 181:1–16

34. Dunn, H. O., Kenney, R. M., Lein, D. H. 1968. XX/XY chimerism in a bovine true hermaphrodite: An insight into the understanding of freemartinism. *Cytogenetics* 7:390–402

35. Edwards, R. G. 1955. Selective fertilization following the use of sperm mixtures in the mouse. *Nature* 175:215–16

36. Eicher, E. M., Hoppe, P. C. 1973. Use of chimeras to transmit lethal genes in the mouse and to demonstrate allelism of the two X-linked male lethal genes *jp* and *msd. J. Exp. Zool.* 183:181–84

37. Emery, A. E. H. 1964. Lyonisation of the X chromosome. *Lancet* i:884

38. Fialkow, P. J. 1973. Primordial cell pool size and lineage relationships of five human cell types. *Ann. Hum. Genet.* 37:39–48

39. Fialkow, P. J., Klein, E., Klein, G., Clifford, P., Singh, S. 1973. Immunoglobulin and glucose-6-phosphate dehydrogenase as markers of cellular origin in Burkitt lymphoma. *J. Exp. Med.* 138:89–102

40. Fialkow, P. J., Sagebiel, R. W., Gartler, S. M., Rimoin, D. L. 1971. Multiple cell origin of hereditary neurofibromas. *N. Engl. J. Med.* 284:298–300

41. Ford, C. E. 1969. Mosaics and chimaeras. *Brit. Med. Bull.* 25:104–9

42. Ford, C. E., Hamerton, J. L., Barnes, D. W. H., Loutit, J. F. 1956. Cytological identification of radiation-chimaeras. *Nature* 177:452–54

43. Fraser, A. S., Short, B. F. 1958. Studies of sheep mosaic for fleece type. I. Patterns and origins of mosaicism. *Aust. J. Biol. Sci.* 11:200–8

44. Freed, J. H., Bechtol, B., Herzenberg, L. A., Herzenberg, L. A., McDevitt, H. O. 1973. Analysis of anti-(T,G)-A--L antibody in tetraparental mice. *Transplant. Proc.* 5:167–71

45. Gandini, E., Gartler, S. M. 1969. Glucose-6-phosphate dehydrogenase mosaicism for studying the development of blood cell precursors. *Nature* 224:599–600

46. Garcia-Bellido, A., Merriam, J. R. 1969. Cell lineage of the imaginal discs in *Drosophila* gynandromorphs. *J. Exp. Zool.* 170:61–76

47. Gardner, R. L. 1968. Mouse chimaeras obtained by the injection of cells into the blastocyst. *Nature* 220:596–97

48. Gardner, R. L. 1971. Manipulations on the blastocyst. See Ref. 4, pp. 279–301
49. Gardner, R. L., Johnson, M. H. 1972. An investigation of inner cell mass and trophoblast tissues following their isolation from the mouse blastocyst. *J. Embryol. Exp. Morphol.* 28:279–312
50. Gardner, R. L., Johnson, M. H. 1973. Investigation of early mammalian development using interspecific chimaeras between rat and mouse. *Nature New Biol.* 246:86–89
51. Gardner, R. L., Lyon, M. F. 1971. X chromosome inactivation studied by injection of a single cell into the mouse blastocyst. *Nature* 231:385–86
51a. Gardner, R. L., Munro, A. J. 1974. Successful construction of chimaeric rabbit. *Nature* 250:146–47
52. Gardner, R. L., Papaioannou, V. E., Barton, S. C. 1973. Origin of the ectoplacental cone and secondary giant cells in mouse blastocysts reconstituted from isolated trophoblast and inner cell mass. *J. Embryol. Exp. Morphol.* 30:561–72
53. Gartler, S. M., Gandini, E., Hutchison, H. T., Campbell, B., Zechhi, G. 1971. Glucose-6-phosphate dehydrogenase mosaicism: Utilization in the study of hair follicle variegation. *Ann. Hum. Genet., London* 35:1–7
54. Gartler, S. M., Nesbitt, M. N. 1971. Sex chromosome markers as indicators in embryonic development. See Ref. 4, pp. 225–54
55. Gartler, S. M. et al 1966. Glucose-6-phosphate dehydrogenase mosaicism as a tracer in the study of hereditary multiple trichoepithelioma. *Am. J. Hum. Genet.* 18:282–87
56. Gearhart, J. D., Mintz, B. 1972. Clonal origins of somites and their muscle derivatives: Evidence from allophenic mice. *Develop. Biol.* 29:27–37
57. Gearhart, J. D., Mintz, B. 1972. Glucosephosphate isomerase subunit-reassociation tests for maternal-fetal and fetal-fetal cell fusion in the mouse placenta. *Develop. Biol.* 29:55–64
58. Gervais, A. G. 1970. Transplantation antigens in the central nervous system. *Nature* 225:647
59. Gervais, A. G., Mintz, B. In preparation
60. Goldenberg, D. M., Bhan, R. D., Pavia, R. A. 1971. *In vivo* human-hamster somatic cell fusion indicated by glucose 6-phosphate dehydrogenase and lactate dehydrogenase profiles. *Cancer Res.* 31:1148–52
61. Gornish, M., Webster, M. P., Wegmann, T. G. 1972. Chimaerism in the immune system of tetraparental mice. *Nature New Biol.* 237:249–51
62. Graham, C. F. 1971. The design of the mouse blastocyst. In *Control Mechanisms of Growth and Differentiation. 25th Symp. Soc. Exp. Biol.,* ed. D. D. Davies, M. Balls. 25:371–78. Cambridge: Univ. Press
63. Grant, C. K., Leuchars, E., Alexander, P. 1972. Failure to detect cytotoxic lymphoid cells or humoral blocking factors in mouse radiation chimaeras. *Transplantation* 14:722–27
64. Grüneberg, H. 1966. The case for somatic crossing over in the mouse. *Genet. Res.* 7:58–75
65. Grüneberg, H. 1966. The molars of the tabby mouse, and a test of the 'single-active X-chromosome' hypothesis. *J. Embryol. Exp. Morphol.* 15:223–44
66. Grüneberg, H. 1969. Threshold phenomena versus cell heredity in the manifestation of sex-linked genes in mammals. *J. Embryol. Exp. Morphol.* 22:145–79
67. Grüneberg, H., Cattanach, B. M., McLaren, A., Wolfe, H. G., Bowman, P. 1972. The molars of tabby chimaeras in the mouse. *Proc. Roy. Soc. London, Ser. B* 182:183–92
68. Grüneberg, H., McLaren, A. 1972. The skeletal phenotype of some mouse chimaeras. *Proc. Roy. Soc. London, Ser. B* 182:9–23
69. Harris, A. W., Röllinghoff, M., Holmes, M. C. Personal communication
70. Hellström, I., Hellström, K. E. 1966. Recent studies on the mechanisms of the allogeneic inhibition phenomenon. *Ann. NY Acad. Sci.* 129:724–34
71. Hellström, I., Hellström, K. E., Allison, A. C. 1971. Neonatally induced allograft tolerance may be mediated by serum-borne factors. *Nature New Biol.* 230:49–50
72. Hellström, I., Hellström, K. E., Trentin, J. J. 1973. Cellular immunity and blocking serum activity in chimeric mice. *Cell. Immunol.* 7:73–84
73. Hellström, K. E., Hellström, I. 1970. Immunological enhancement as studied by cell culture techniques. *Ann. Rev. Microbiol.* 24:373–98
74. Hillman, N., Sherman, M. I., Graham, C. 1972. The effect of spatial arrangement on cell determination during mouse development. *J. Embryol. Exp. Morphol.* 28:263–78

75. Hotta, Y., Benzer, S. 1972. Mapping of behaviour in *Drosophila* mosaics. *Nature* 240:527–35

76. Hutchison, H. T. 1973. A model for estimating the extent of variegation in mosaic tissues. *J. Theor. Biol.* 38:61–79

77. Katz, D. H., Hamaoka, T., Benacerraf, B. 1973. Cell interactions between histoincompatible T and B lymphocytes. II. Failure of physiologic cooperative interactions between T and B lymphocytes from allogeneic donor strains in humoral response to hapten-protein conjugates. *J. Exp. Med.* 137:1405–18

78. Kleinebrecht, J. Personal communication

79. Kleinebrecht, J., Degenhardt, K.-H., Fränz, J., Nishimura, H., Svejcar, J. 1973. Realization of malformations induced by FCdR. In *4th Int. Conf. Birth Defects*, ed. A. G. Motulsky, Int. Congr. Ser. 297, p. 21. Amsterdam: Excerpta Medica

80. Lengerová, A. 1967. Radiation chimaras and genetics of somatic cells. *Science* 155:529–35

81. Lewis, J. H., Summerbell, D., Wolpert, L. 1972. Chimaeras and cell lineage in development. *Nature* 239:276–79

82. Lin, T. P. 1969. Microsurgery of inner cell mass of mouse blastocysts. *Nature* 222:480–81

83. Lin, T. P., Florence, J. 1970. Aggregation of dissociated mouse blastomeres. *Exp. Cell Res.* 63:220–24

84. Linder, D., Gartler, S. M. 1965. Glucose-6-phosphate dehydrogenase mosaicism: Utilization as a cell marker in the study of leiomyomas. *Science* 150:67–69

85. Lubaroff, D. M., Silvers, W. K. 1973. The importance of chimerism in maintaining tolerance of skin allografts in mice. *J. Immunol.* 111:65–71

86. Lyon, M. F. 1968. Chromosomal and subchromosomal inactivation. *Ann. Rev. Genet.* 2:31–52

87. Lyon, M. F. 1970. Genetic activity of sex chromosomes in somatic cells of mammals. *Phil. Trans. Roy. Soc. London, Ser. B* 259:41–52

88. Lyon, M. F. 1972. X-chromosome inactivation and developmental patterns in mammals. *Biol. Rev.* 47:1–35

89. Lyon, M. F., Hawkes, S. G. 1970. X-linked gene for testicular feminization in the mouse. *Nature* 227:1217–19

90. Markert, C. L., Silvers, W. K. 1956. The effects of genotype and hair environment on melanoblast differentiation in the house mouse. *Genetics* 41:429–50

91. Matsunaga, T. 1973. Self tolerance in allophenic mice studied by the mixed lymphocyte culture reaction. PhD thesis. Basel Institute for Immunology, Basel, Switzerland

92. Mayer, J. F. Jr., Fritz, H. I. 1974. The culture of preimplantation rat embryos and the production of allophenic rats. *J. Reprod. Fert.* 39:1–9

93. Mayer, T. C. 1970. A comparison of pigment cell development in albino, steel, and dominant-spotting mutant mouse embryos. *Develop. Biol.* 23:297–309

94. Mayer, T. C., Fishbane, J. L. 1972. Mesoderm-ectoderm interaction in the production of the agouti pigmentation pattern in mice. *Genetics* 71:297–303

95. McDevitt, H. O., Bechtol, K. B., Grumet, F. C., Mitchell, G. F., Wegmann, T. G. 1971. Genetic control of the immune response to branched synthetic polypeptide antigens in inbred mice. In *Progress in Immunology*, ed. B. Amos, pp. 495–508. New York: Academic

96. McLaren, A. 1971. Germ cell differentiation in artificial chimaeras of mice. In *Proc. Int. Symp. Genet. Spermatozoon*, ed. R. A. Beatty, S. Gluecksohn-Waelsch, pp. 313–23

97. McLaren, A. 1972. Numerology of development. *Nature* 239:274–76

98. McLaren, A., Bowman, P. 1969. Mouse chimaeras derived from fusion of embryos differing by nine genetic factors. *Nature* 224:236–40

99. McLaren, A., Chandley, A. C., Kofman-Alfaro, S. 1972. A study of meiotic germ cells in the gonads of foetal mouse chimaeras. *J. Embryol. Exp. Morphol.* 27:515–24

100. McLaren, A., Gauld, I. K., Bowman, P. 1973. Comparison between mice chimaeric and heterozygous for the X-linked gene *tabby*. *Nature* 241:180–83

101. Medawar, P. B. 1973. Tolerance reconsidered—a critical survey. *Transplant. Proc.* 5:7–9

102. Melvold, R. W. 1971. Spontaneous somatic reversion in mice. Effects of parental genotype on stability at the p-locus. *Mutat. Res.* 12:171–74

103. Meo, T., Matsunaga, T., Rijnbeek, A. M. 1974. On the mechanism of self-tolerance in embryo-fusion chimeras. *Transplant. Proc.* 5:1607–10

104. Milet, R. G., Mukherjee, B. B., Whitten, W. K. 1972. Cellular distribution and possible mechanism of sex-differentiation in XX/XY chimeric mice. *Can. J. Genet. Cytol.* 14:933–41

105. Miller, W. A., Mintz, B. In preparation
106. Mintz, B. 1962. Experimental study of the developing mammalian egg: Removal of the zona pellucida. *Science* 138:594–95
107. Mintz, B. 1962. Formation of genotypically mosaic mouse embryos. *Am. Zool.* 2:432
108. Mintz, B. 1962. Experimental recombination of cells in the developing mouse egg: Normal and lethal mutant genotypes. *Am. Zool.* 2:541–42
109. Mintz, B. 1964. Synthetic processes and early development in the mammalian egg. *J. Exp. Zool.* 157:85–100
110. Mintz, B. 1964. Formation of genetically mosaic mouse embryos, and early development of "lethal (t^{12}/t^{12})-normal" mosaics. *J. Exp. Zool.* 157:273–92
111. Mintz, B. 1965. Genetic mosaicism in adult mice of quadriparental lineage. *Science* 148:1232–33
112. Mintz, B. 1965. Experimental genetic mosaicism in the mouse. In *Ciba Found. Symp. Preimplantation Stages Pregnancy*, ed. G. E. W. Wolstenholme, M. O'Connor, pp. 194–207. London: Churchill
113. Mintz, B. 1967. Gene control of mammalian pigmentary differentiation. I. Clonal origin of melanocytes. *Proc. Nat. Acad. Sci. USA* 58:344–51
114. Mintz, B. 1968. Hermaphroditism, sex chromosomal mosaicism and germ cell selection in allophenic mice. *8th Bien. Symp. Animal Reprod. J. Animal Sci.* 27:51–60
115. Mintz, B. 1969. Gene control of the mouse pigmentary system. *Genetics* 61:41
116. Mintz, B. 1969. Developmental mechanisms found in allophenic mice with sex chromosomal and pigmentary mosaicism. In *Birth Defects: Orig. Art. Ser. 5, First Conf. Clin. Delineation Birth Defects*, ed. D. Bergsma, V. McKusick, pp. 11–22. New York: Nat. Found.
117. Mintz, B. 1970. Allophenic mice as test animals to detect tissue-specific histocompatibility alloantigens or F_1 hybrid antigens. *Transplantation* 9:523–25
118. Mintz, B. 1970. Do cells fuse *in vivo?* In *Advan. Tissue Cult. In Vitro* 5:40–47
119. Mintz, B. 1970. Gene expression in allophenic mice. In *Control Mech. Expression Cellular Phenotypes, Symp. Int. Soc. Cell Biol.* ed. H. Padykula, 9: 15–42. New York: Academic
120. Mintz, B. 1970. Neoplasia and gene activity in allophenic mice. In *Genet.*

Conc. Neoplasia, 23rd Ann. Symp. Fundam. Cancer Res. M. D. Anderson Hosp. Tumor Inst. pp. 477–517. Baltimore: Williams & Wilkins
121. Mintz, B. 1971. Control of embryo implantation and survival. See Ref. 4, pp. 317–42
122. Mintz, B. 1971. Allophenic mice of multi-embryo origin. In *Methods in Mammalian Embryology*, ed. J. Daniel Jr., pp. 186–214. San Francisco: Freeman
123. Mintz, B. 1971. Genetic mosaicism *in vivo:* Development and disease in allophenic mice. In *Symp. "Mammalian Cell Hybridization." Fed. Proc.* 30: 935–43
124. Mintz, B. 1971. Clonal basis of mammalian differentiation. See Ref. 62, pp. 345–69
125. Mintz, B. 1972. Clonal differentiation in early mammalian development. In *Mol. Genet. Develop. Biol., Symp. Soc. Gen. Physiol.*, ed. M. Sussman, pp. 455–74. Englewood Cliffs, New Jersey: Prentice-Hall
126. Mintz, B. 1972. Cellular expression of genes controlling susceptibility to neoplasia in mice. In *Cell Differentiation*, ed. R. Harris, P. Allin, D. Viza, pp. 176–81. Copenhagen: Munksgaard
127. Mintz, B. 1972. Clonal units of gene control in mammalian differentiation. See Ref. 126, pp. 267–71
128. Mintz, B. In preparation
129. Mintz, B. Unpublished data
130. Mintz, B., Baker, W. W. 1967. Normal mammalian muscle differentiation and gene control of isocitrate dehydrogenase synthesis. *Proc. Nat. Acad. Sci. USA* 58:592–98
131. Mintz, B., Custer, R. P. In preparation
132. Mintz, B., Custer, R. P., Donnelly, A. J. 1971. Genetic diseases and developmental defects analyzed in allophenic mice. *Int. Rev. Exp. Pathol.* 10:143–79
133. Mintz, B., Domon, M., Hungerford, D. A., Morrow, J. 1972. Seminal vesicle formation and specific male protein secretion by female cells in allophenic mice. *Science* 175:657–59
134. Mintz, B., Gearhart, J. D., Guymont, A. G. Phytohemagglutinin-mediated blastomere aggregation and development of allophenic mice. *Develop. Biol.* 31:195–99
135. Mintz, B., Kindred, B. In preparation
136. Mintz, B., Morrow, J., Hungerford, D. A., Custer, R. P. In preparation
137. Mintz, B., Palm, J. 1969. Gene control of hematopoiesis. I. Erythrocyte mosa-

icism and permanent immunological tolerance in allophenic mice. *J. Exp. Med.* 129:1013–27

138. Mintz, B., Russell, E. S. 1957. Gene-induced embryological modifications of primordial germ cells in the mouse. *J. Exp. Zool.* 134:207–38

139. Mintz, B., Sanyal, S. 1970. Clonal origin of the mouse visual retina mapped from genetically mosaic eyes. *Genetics* 64:43–44

140. Mintz, B., Silvers, W. K. 1967. "Intrinsic" immunological tolerance in allophenic mice. *Science* 158:1484–87

141. Mintz, B., Silvers, W. K. 1970. Histocompatibility antigens on melanoblasts and hair-follicle cells: Cell-localized homograft rejection in allophenic skin grafts. *Transplantation* 9:497–505

142. Mintz, B., Silvers, W. K., Lappé, M. In preparation

143. Mintz, B., Slemmer, G. 1969. Gene control of neoplasia. I. Genotypic mosaicism in normal and preneoplastic mammary glands of allophenic mice. *J. Nat. Cancer Inst.* 43:87–95

144. Mintz, B., Talal, N., Dauphinée, M., Asofsky, R., Greespun, R. In preparation

145. Moore, W. J., Mintz, B. 1972. Clonal model of vertebral column and skull development derived from genetically mosaic skeletons in allophenic mice. *Develop. Biol.* 27:55–70

146. Moustafa, L. A., Brinster, R. L. 1972. The fate of transplanted cells in mouse blastocysts *in vitro. J. Exp. Zool.* 181:181–92

147. Moustafa, L. A., Brinster, R. L. 1972. Induced chimaerism by transplanting embryonic cells into mouse blastocysts. *J. Exp. Zool.* 181:193–202

148. Mukherjee, B. B., Milet, R. G. 1972. Nonrandom X-chromosome inactivation—an artifact of cell selection. *Proc. Nat. Acad. Sci. USA* 69:37–39

149. Mullen. R. J., Whitten, W. K. 1971. Relationship of genotype and degree of chimerism in coat color to sex ratios and gametogenesis in chimeric mice. *J. Exp. Zool.* 178:165–76

150. Mulnard, J. G. 1965. Studies of regulation of mouse ova *in vitro.* In *Ciba Found. Symp. Preimplantation Stages Pregnancy,* ed. G. E. W. Wolstenholme, M. O'Connor, pp. 123–44. London: Churchill

151. Mulnard, J. G. 1971. Manipulation of cleaving mammalian embryo with special reference to a time-lapse cinematographic analysis of centrifuged and fused mouse eggs. See Ref. 4, pp. 255–77

152. Myers, L. M. 1959. Sex chromatin in teratomas. *J. Pathol. Bacteriol.* 78: 43–55

153. Mystkowska, E. T., Tarkowski, A. K. 1968. Observations on CBA-p/CBA-T6T6 mouse chimeras. *J. Embryol. Exp. Morphol.* 20:33–52

154. Mystkowska, E. T., Tarkowski, A. K. 1970. Behaviour of germ cells and sexual differentiation in late embryonic and early postnatal mouse chimeras. *J. Embryol. Exp. Morphol.* 23:395–405

155. Nesbitt, M. N. 1971. X-chromosome inactivation mosaicism in the mouse. *Develop. Biol.* 26:252–63

156. Nesbitt, M. N., Gartler, S. M. 1971. The applications of genetic mosaicism to developmental problems. *Ann. Rev. Genet.* 5:143–62

157. Nicholas, J. S., Hall, B. V. 1942. Experiments on developing rats. II. The development of isolated blastomeres and fused eggs. *J. Exp. Zool.* 90:441–59

158. Noell, W. K. 1965. Aspects of experimental and hereditary retinal degeneration. In *Biochemistry of the Retina,* ed. C. N. Graymore. pp. 51–72. New York: Academic

159. Noell, W. K., Mintz, B. In preparation

160. Nowell, P. C., Cole, L. J., Habermeyer, J. G., Roan, P. L. 1956. Growth and continued function of rat marrow cells in X-radiated mice. *Cancer Res.* 16:258–61

161. Nowell, P. C., Hungerford, D. A. 1960. A minute chromosome in human chronic granulocytic leukemia. *Science* 132:1497

162. Ohno, S. 1966. Cytologic and genetic evidence of somatic segregation in mammals, birds, and fishes. In *Advan. Tissue Culture. In Vitro* 2:46–60

163. Ohno, S., Cattanach, B. M. 1962. Cytological study of an X-autosome translocation in *Mus musculus. Cytogenetics* 1:129–40

164. Oliver, C., Essner, E. 1973. Distribution of anomalous lysosomes in the beige mouse: A homologue of Chediak-Higashi syndrome. *J. Histochem. Cytochem.* 21:218–28

165. Oliver, R. F. 1967. The experimental induction of whisker growth in the hooded rat by implantation of dermal papillae. *J. Embryol. Exp. Morphol.* 18:43–51

166. Owen, J. J. T. 1972. The origins and development of lymphocyte populations. In *Ontogeny of Acquired Immu-*

nity, Ciba Found. Symp. ed. R. Porter, J. Knight, pp. 35–54. Amsterdam: North-Holland

167. Owen, R. D. 1945. Immunogenetic consequences of vascular anastomoses between bovine twins. *Science* 102:400–1

168. Phillips, S. M., Martin, W. J., Shaw, A. R., Wegmann, T. G. 1971. Serum-mediated immunological non-reactivity between histoincompatible cells in tetraparental mice. *Nature* 235:146–48

169. Phillips, S. M., Wegmann, T. G. 1973. Active suppression as a possible mechanism of tolerance in tetraparental mice. *J. Exp. Med.* 137:291–300

170. Poole, T. W. 1974. Dermal-epidermal interactions and the site of action of the yellow (A^y) and nonagouti (a) coat color genes in the mouse. *Develop. Biol.* 36:208–11

171. Raff, M. C. 1973. T and B lymphocytes and immune responses. *Nature* 242: 19–23

172. Rawles, M. E. 1947. Origin of pigment cells from the neural crest in the mouse embryo. *Physiol. Zool.* 20:248–66

173. Rowe, W. P. 1973. Genetic factors in the natural history of murine leukemia virus infection. *Cancer Res.* 33:3061–68

174. Russell, E. S., Bernstein, S. E. 1966. Blood and blood formation. In *Biology of the Laboratory Mouse,* ed. E. L. Green, pp. 351–72. New York: McGraw-Hill. 2nd ed.

175. Russell, L. B. 1964. Genetic and functional mosaicism in the mouse. In *The Role of Chromosomes in Development, 23rd Symp. Soc. Study Develop. Growth,* ed. M. Locke, pp. 153–81. New York: Academic

176. Russell, L. B., Woodiel, F. N. 1966. A spontaneous mouse chimera formed from separate fertilization of two meiotic products of oogenesis. *Cytogenetics* 5:106–19

177. Schaible, R. H. 1969. Clonal distribution of melanocytes in piebald-spotted and variegated mice. *J. Exp. Zool.* 172:181–99

178. Schlafke, S., Enders, A. C. 1963. Observations on the fine structure of the rat blastocyst. *J. Anat.* 97:353–60

179. Seidel, F. 1960. Die Entwicklungsfähigkeiten Isolierter Furchungszellen aus dem Ei des Kaninchens *Oryctolagus cuniculus. Roux' Arch. f. Entw. Mech.* 152:43–130

180. Siegler, R., Rich, M. A. 1966. Influence of thymic mass on murine viral leukaemogenesis. *Nature* 209:313–14

181. Silvers, W. K., Russell, E. S. 1955. An experimental approach to action of genes at the agouti locus in the mouse. *J. Exp. Zool.* 130:199–220

182. Slemmer, G. 1972. Host response to premalignant mammary tissues. *Nat. Cancer Inst. Monogr.* 35:57–71

183. Snell, G. D. 1964. The terminology of tissue transplantation. *Transplantation* 2:655–57

184. Snell, G. D., Stevens, L. C. 1966. Early embryology. See Ref. 174, pp. 205–45

185. Stent, G. S. 1969. *The Coming of the Golden Age.* Garden City, New York: Natural History Press. 146 pp.

186. Stern, C. 1936. Somatic crossing over and segregation in *Drosophila melanogaster. Genetics* 21:625–730

187. Stern, M. S. 1972. Experimental studies on the organization of the preimplantation mouse embryo. II. Reaggregation of disaggregated embryos. *J. Embryol. Exp. Morphol.* 28:255–61

188. Stern, M. S. 1973. Chimaeras obtained by aggregation of mouse eggs with rat eggs. *Nature* 243:472–73

189. Stern, M. S., Wilson, I. B. 1972. Experimental studies on the organization of the preimplantation mouse embryo. I. Fusion of asynchronously cleaving eggs. *J. Embryol. Exp. Morphol.* 28: 247–54

190. Stevens, L. C. 1970. The development of transplantable teratocarcinomas from intratesticular grafts of pre- and postimplantation mouse embryos. *Develop. Biol.* 21:364–82

191. Stone, W. H. et al 1971. Long-term observations of skin grafts between chimeric cattle twins. *Transplantation* 12: 421–28

192. Stone, W. H., Friedman, J., Fregin, A. 1964. Possible somatic cell mating in twin cattle with erythrocyte mosaicism. *Proc. Nat. Acad. Sci. USA* 51:1036–44

193. Sturtevant, A. H. 1929. The claret mutant type of *Drosophila simulans,* a study of chromosome elimination and cell-lineage. *Z. Wiss. Zool.* 135:323–56

194. Svejcar, J., Kleinebrecht, J. 1974. Mucopolysaccharides in morphogenesis of vertebral malformations induced by 5-fluoro-2'-deoxycytidine in mice. *Develop. Biol.* In press

195. Tarkowski, A. K. 1959. Experimental studies on regulation in the development of isolated blastomeres of mouse eggs. *Acta Theriologica* 3:191–267

196. Tarkowski, A. K. 1961. Mouse chimaeras developed from fused eggs. *Nature* 190:857–60

197. Tarkowski, A. K. 1963. Studies on mouse chimeras developed from eggs fused *in vitro. Nat. Cancer Inst. Monogr.* 11:51–67

198. Tarkowski, A. K. 1964. Patterns of pigmentation in experimentally produced mouse chimaerae. *J. Embryol. Exp. Morphol.* 12:575–85

199. Tarkowski, A. K. 1964. True hermaphroditism in chimaeric mice. *J. Embryol. Exp. Morphol.* 12:735–57

200. Tarkowski, A. K. 1965. See Ref. 150, pp. 183–216

201. Tarkowski, A. K. 1969. Consequences of sex chromosome chimerism for sexual differentiation in mammals. *Ann. d'Embryol. Morphog.* Suppl., 1:211–22

202. Tarkowski, A. K. 1970. Germ cells in natural and experimental chimeras in mammals. *Phil. Trans. Roy. Soc. London, Ser. B* 259:107–11

203. Tarkowski, A. K., Wróblewska, J. 1967. Development of blastomeres of mouse eggs isolated at the 4- and 8-cell stage. *J. Embryol. Exp. Morphol.* 18:155–80

204. Tucker, E. M., Moor, R. M., Rowson, L. E. A. 1974. Tetraparental sheep chimaeras induced by blastomere transplantation. *Immunology* 26:613–21

205. Tuffrey, M., Barnes, R. D., Evans, E. P., Ford, C. E. 1973. Dominance of AKR lymphocytes in tetraparental AKR ↔ CBA-T6T6 chimaeras. *Nature New Biol.* 243:207–8

206. Voisin, G. A., Kinsky, R. G., Maillard, J. 1968. Démonstration d'une réactivité immunitaire spécifique chez des animaux tolérants aux homogreffes. Rôle possible dans le maintien de la tolérance. In *Advances in Transplantation,* ed J. Dausset, J. Hamburger, G. Mathé, pp. 31–40. Copenhagen: Munksgaard

207. Wanebo, H. J., Gallmeier, W. M., Boyse, E. A., Old, L. J. 1966. Paraproteinemia and reticulum cell sarcoma in an inbred mouse strain. *Science* 154:901–3

208. Warner, C. M., Fitzmaurice, M., Maurer, P. H., Merryman, C. F., Schmerr, M. F. 1973. The immune response of tetraparental mice to two synthetic amino acid polymers: "high-conjugation" 2,4 dinitrophenyl-glutamic acid[57]-lysine[38]-alanine[5] (DNP-GLA[5])

and glutamic acid[60] alanine[30] tyrosine[10] (GAT[10]). *J. Immunol.* 111:1887–93

209. Warner, N. L., Herzenberg, L. A., Cole, L. J., Davis, W. E. 1965. Dissociation of skin homograft tolerance and donor type gamma globulin synthesis in allogeneic mouse radiation chimaeras. *Nature* 205:1077–79

210. Wegmann, T. G. 1970. Enzyme patterns in tetraparental mouse liver. *Nature* 225:462–63

211. Wegmann, T. G., Gilman, J. G. 1970. Chimerism for three genetic systems in tetraparental mice. *Develop. Biol.* 21:281–91

212. Wegmann, T. G., Hellström, I., Hellström, K. E. 1971. Immunological tolerance: "Forbidden clones" allowed in tetraparental mice. *Proc. Nat. Acad. Sci. USA* 68:1644–47

213. Wegmann, T. G., LaVail, M. M., Sidman, R. L. 1971. Patchy retina degeneration in tetraparental mice. *Nature* 230:333–34

214. Wiener, F., Fenyö, E. M., Klein, G., Harris, H. 1972. Fusion of tumour cells with host cells. *Nature New Biol.* 238:155–59

215. Wilson, D. B., Nowell, P. C. 1970. Quantitative studies on the mixed lymphocyte interaction in rats. IV. Immunologic potentiality of the responding cells. *J. Exp. Med.* 131:391–407

216. Wilson, I. B., Bolton, E., Cuttler, R. H. 1972. Preimplantation differentiation in the mouse egg as revealed by microinjection of vital markers. *J. Embryol. Exp. Morphol.* 27:467–79

217. Wolf, U., Engel, W. 1972. Gene activation during early development of mammals. *Humangenetik* 15:99–118

218. Wolpert, L., Gingell, D. 1970. Striping and the pattern of melanocyte cells in chimaeric mice. *J. Theor. Biol.* 29:147–50

219. Zeilmaker, G. H. 1973. Fusion of rat and mouse morulae and formation of chimaeric blastocysts. *Nature* 242:115–16

220. Zuelzer, W. W., Beattie, K. M., Reisman, L. E. 1964. Generalized unbalanced mosaicism attributable to dispermy and probable fertilization of a polar body. *Am. J. Hum. Genet.* 16:38–51

SUBJECT INDEX

A

Accessory chromosomes, 243-61
characterization of, 244
chromatin elimination
  induced by B chromosomes
  in maize, 257-58
effects in new gene environments, 259-60
influence of
  on chiasma frequency,
    258
  on chromosome pairing,
    258
  on ecological adaptations,
    254-55
  on fertility and vigor, 254-
    55
interaction with A chromosomes, 257-60
in mammals, 245-46
mechanisms of numerical
  increase, 247-55
numerical increase of in
  animals
  by differential nondisjunc-
    tion, 252
  at embryonic mitosis, 252-
    53
  mechanisms of, 252-53
  by preferential segregation,
    253
numerical increase of in
  plants
  by endomitotic reduplica-
    tion, 252
  by meiotic preferential
    distribution, 251
  by postmeiotic nondisjunc-
    tion, 247-50
  by somatic nondisjunction,
    251
occurrence of in different
  biological groups, 245-
    46
studies of with molecular-
  genetic methods, 261-62
transfer of
  from diploid to autotetra-
    ploid rye, 259
  from rye to wheat, 259
Ac element of maize
  different states of, 34-35
Allophenic mice
  experimental production of,
    412-13
  see also Gene control of
    mammalian differentiation
Amino acid transport in

bacteria
  coupling of energy to, 126-
    28
  genetics of, 103-29
  membrane vesicle systems
    comparison with intact
      cells, 124-26
    description of, 124-26
    specificity of transport sys-
      tems, 104-23
    transport system for
      alanine, 104-6
      arginine, 116-18
      asparagine, 123
      aspartic acid, 118-23
      cystine, 112-13
      diaminopimelic acid, 112-
        13
      glutamic acid, 118-23
      glutamine, 123
      glycine, 104-6
      histidine, 113-16
      hydroxyproline, 111-12
      isoleucine, 106-9
      leucine, 106-9
      lysine, 116-18
      methionine, 112
      ornithine, 116-18
      phenylalanine, 109-11
      proline, 111-12
      serine, 104-6
      threonine, 104-6
      tryptophan, 109-11
      tyrosine, 109-11
      valine, 106-9
Amino acid transport in isolated bacterial membrane
  vesicles, 124-29
Antibiotic modification of
  protein synthesis, 135-
    48
Antibiotic-resistant mutants
  of Chlamydomonas, 369-
    71
L-arabinose gene-enzyme
  complex and positive
  control
  see Positive control sys-
    tems of bacteria

B

Bacteria
  biochemical genetics of,
    79-95
  genetics of amino acid
    transport in, 103-29
  regulation by positive con-
    trol systems in, 220-40
  see also Biochemical

genetics; Genetic regula-
  tory units of bacteria
B chromosomes
  see Accessory chromosomes
Binding proteins of membrane
  transport systems, 128-
    29
Biochemical genetics of
  bacteria, 79-95
  biosynthesis
    of arginine, 86-88
    of cysteine, 88-90
    of histidine, 83-84
    of isoleucine, 82-83
    of leucine, 82-83
    of methionine, 88-90
    of purine nucleotides, 90-
      92
    of pyrimidines, 92
    of tryptophan, 84-86
    of valine, 82-83
  histidine utilization, 94-
    95
  nucleoside catabolism, 93-
    94
Biological aspects of the
  planets, 403-6
Biosynthesis
  see Biochemical genetics of
    bacteria

C

cAMP
  activation of catabolite-
    sensitive operons by,
    237-38
Chlamydomonas reinhardtii
  chloroplast genetics of, 347,
    365-83
Chloroplast genetics of
  Chlamydomonas
  antibiotic resistance, 369-
    71
  chloroplast genome, 365
  inheritance of
    chloroplast DNA, 365-67
    chloroplast mutations,
      365
    mitochondrial mutations,
      368-69
    non-Mendelian inheritance
      in, 365
    organelle ribosomes, 369-
      71
  recombination
    mapping of chloroplast
      genes, 379-83
    methodology and terminol-
      ogy, 374-75

471

CUMULATIVE INDEXES

CONTRIBUTING AUTHORS VOLUMES 4-8

CHAPTER TITLES VOLUMES 4-8